THE LITHOSPHERE: GEOCHEMISTRY, GEOLOGY AND GEOPHYSICS

THE LITHOSPHERE: GEOCHEMISTRY, GEOLOGY AND GEOPHYSICS

JAROD E. ANDERSON
AND
ROBERT W. COATES
EDITORS

Nova Science Publishers, Inc.
New York

LIBRARY OF CONGRESS CATALOGING-IN-PUBLICATION DATA
Available upon request.

ISBN 978-1-60456-903-2

Published by Nova Science Publishers, Inc. ✢ New York

CONTENTS

PREFACE

The lithosphere is the outer solid part of the earth, including the crust and uppermost mantle. The lithosphere is about 100 km thick, although its thickness is age dependent (older lithosphere is thicker). The lithosphere below the crust is brittle enough at some locations to produce earthquakes by faulting, such as within a subducted oceanic plate. This new book presents leading research in the field from around the globe.

Some physical problems related to modeling the conditions of the formation and evolution of the lithosphere are discussed. It is shown that if we take into account both the effects of thermal expansion and compressibility we could receive results with no change or even an increase of density under the *P-T* conditions within the lithosphere. During planetary accretion and differentiation of Earth, the planet could have been entirely molten and at some point of its evolution was entirely covered by a magma ocean. The formation and composition of the early lithosphere were mostly related to processes of the differentiation of matter and the rate of cooling of the magma ocean. The process of the differentiation of the magma ocean would begin during its formation and continue until its solidification. This stratification of the composition of Earth caused an initial state of separation of rocks and minerals into slabs within the upper mantle and crust, which was strictly regulated by their density, whether solid or melted. The difference in density between felsic, intermediate, mafic, and ultramafic magmatic slabs within the magma ocean was enough to prevent the exchange of matter between them, and therefore mantle-wide convection could not have taken place. The solidification of the magma ocean, upon which the process of the formation of the lithosphere is dependent, most likely began with the formation of a forsterite (or forsterite-rich peridotite) slab at a depth of about 100 km followed by the solidification of Earth's surface, cooled by heat radiation from the surface and the cooling effect of the early atmosphere. It is shown in Chapter 1 that under the thermal conditions of the magma ocean, carbonate rocks were unstable and decomposed, releasing carbon dioxide into the atmosphere. Water could also not exist in its liquid state at the time of the magma ocean, and together with the carbon dioxide would form a thick and dense early atmosphere. Formation of the water ocean was under the constraints of the boiling point of water at the pressure of the early atmosphere and the critical point of water. Cooling rates of the magma ocean and the early lithosphere are discussed in comparison with the cooling rates of the mantle and numerous magmatic and/or metamorphic complexes. The main periods of the appearance of komatiites and the formation of the first large igneous provinces (LIPs) indicate temperature maximums in the mantle at about 2.8-2.7 Ga and a maximum volume of mafic magmatism related to the formation of LIPs at about 2.5 Ga. Analysis of the periods of the formation of granulites suggests an increase in temperature at a depth of about 30 km from ~3.0 Ga to ~2.7 Ga, with its continuing increase to ~2.5 Ga. It is shown that at the end of the Archean, the thickness of the lithosphere was ≤100 km, including a solid forsterite slab at ~100 km depth with possible pockets of magma above it.

As explained in Chapter 2, during the last decades of the XX century several long-range seismic profiles with large chemical and Peaceful Nuclear Explosions (PNE) were carried out by Russian institutions in the Atlantic Ocean and in the Eurasia continent. They are the Angola-Brazil geotraverse with the investigation depth of 100 km and a system of PNE profiles in Russia with the wave penetrating depth of 700 km.

2-D lithosphere velocity models were constructed for all these profiles using a common methodology for the wave field interpretation. They show that along the geotraverse the oceanic basin lithosphere is of 60-70 km thickness and it is underlined by a low velocity layer (the asthenosphere). Beneath the mid-oceanic ridge instead of the asthenosphere uplift, several local low velocity zones (asthenolites) are revealed at depth of 20-50 km. The seismic velocities between these zones are too high (up to 8.5 km/s) for such high heat flow area, they may be explained by the anisotropy effects.

In Eurasia structural peculiarities of the upper mantle are difficult to describe in the classical lithosphere-asthenosphere system. The asthenosphere can not be traced as a low velocity layer, on the contrary the lithosphere is rheologically stratified. It follows from the data of 25 PNEs which were used to compile a 3-D upper mantle velocity model for the central part of the continent. Five basic boundaries were traced over the study area: N1 and N2 boundaries at a depth around 100 km, L boundary at a depth of 180-240 km and H boundary at 300-330 km. The depth maps for each boundary and the velocity distribution map in the uppermost mantle were compiled.

In general, the old and cold cratons have higher velocities in the lithosphere than the young platforms with higher heat flows. Mostly horizontal inhomogeneity is observed in the uppermost mantle: the velocities change from the average 8.0-8.1 km/s to 8.3-8.4 km/s in some blocks of the Siberian Craton and the Urals. At the depth of 100-120 km the local high velocity blocks disappear and low velocity layers are often observed. The velocity inversions are characterised by higher electrical conductivity and it means that many of them may be a result of fluids concentration. These structural features propose that the depth of 100-120 km is a bottom of a brittle part of the lithosphere.

The visible changes of the matter plasticity are observed also at the depths of around 250 km (beneath the L boundary) where the mantle structural pattern is changed too. The H boundary has a mirror form relative to the upper boundaries (its depths are greater beneath L boundary uplifts). At these depths the Q-factor is decreased. Both these features indicate an increasing of the matter plasticity, which makes an isostatic equilibrium of the upper mantle. Thus, the L boundary may be considered as a bottom of the continental lithosphere which agrees with the heat flow data.

The rheological stratification of the lithosphere follows also from the regional boundary structure. All the boundaries are not simple discontinuities, they are heterogeneous (thin layering) zones which generate multiphase reflections. Many of them may be a result of fluids concentration at some critical PT levels. The fluids change mechanical properties of the matter, they initiate partly melting and metasomatism of the mantle material. The matter flow along weak layers results in origin of the seismic anisotropy and variation of low and high velocities.

Standard geodynamic models envisage subducted oceanic crust to be part isolated in thermal boundary layers and part remixed into the depleted mantle. The two fates for such material result from combining remixing models based on observation from orogenic lherzolites, with theoretical models for the generation of intraplate volcanism by mantle plumes. The concepts were combined because high $^{3}He/^{4}He$ ratios in intraplate basalts were interpreted to require a primitive source component and the convecting mantle was considered unable to retain ^{3}He on melting. However, high $^{3}He/^{4}He$ ratios may reflect low U+Th sources, and differences in $^{3}He/^{4}He$ between MORB and OIB can be explained by sampling of the convecting mantle. Interpretations of high $^{186}Os/^{188}Os$ in intraplate lavas as evidence for interaction with the core are likewise tenuous, as the signatures can be explained by pyroxenites or mantle sulphides. Instead, the remixing models should have been combined with models for the tapping of shallow mantle sources by plate tectonic processes, to give an explanation for the origin of intralate volcanism from the convecting mantle without plumes. Pyroxenitic sources for intraplate volcanism may be generated at convergent margins if subducted oceanic crust undergoes melting in the back-arc region, or along the flanks of convective upwellings beneath ocean ridge systems as melts from altered eclogite or sediment components of recycled crust react with peridotites of the depleted mantle. Generation of intraplate melts occurs in off-axis regions as a result of fluxing of the pyroxenite-veined mantle with fluids derived from dehydration or decarbonation of later generations of subducted slabs in the shallow mantle, as explained in Chapter 3.

As discussed in Chapter 4, a linear zone with high strain rates along the Sea of Japan coast, the Nigata-Kobe Tectonic Zone (NKTZ), is considered to be associated with rheological heterogeneity in the lower crust and/or upper mantle, which may be caused by the upwelling of aqueous fluid and/or melt related to subduction of the Philippine Sea and Pacific Plates. In order to elucidate the geographic distribution of $^{3}He/^{4}He$ ratios along the NKTZ, new helium isotope data from hot spring gases and water samples were determined. In the southern NKTZ, $^{3}He/^{4}He$ ratios lower than the atmospheric value indicate that radiogenic helium dominates over any mantle helium input from aqueous fluids generated during the dehydration of the subducting Philippine Sea slab because a mantle wedge, the potential source of mantle helium, appears to be absent. Higher $^{3}He/^{4}He$ ratios are observed in the central NKTZ where active volcanoes are concentrated, suggesting the existence of magmatic fluids in the lower crust and upper mantle. The $^{3}He/^{4}He$ ratios of most hot springs in the northern NKTZ, a non-volcanic region, can be interpreted as a three-component mixture of mantle helium associated with magmatism of Middle Miocene age, radiogenic crustal helium and atmospheric helium. However, the $^{3}He/^{4}He$ ratios of gases close to active faults in the northern NKTZ are similar to those near active volcanoes in the central NKTZ, suggesting that active faults may facilitate transfer of mantle helium carried by aqueous fluids derived from the subducting Pacific Plate slab from the lower crust to the Earth's surface.

Chapter 5 discusses volatiles, which play important roles in chemical differentiation of the Earth, in concentrating economic metals, and in regulating earth's surface environments by means of magmatism, metasomatism, degassing and recycling. Mantle rocks and rocks derived from the mantle such as basalts, mantle xenoliths and ophiolitic peridotites are materials available to investigate the ways in which volatiles may store in the mantle, their compositions, and probable histories. The laser Raman spectroscopy, Infrared spectrometry and ion microprobe in combination with micro-thermometry are non-destructive methods to analyze volatile compositions trapped in fluid inclusions. On the other hand, vacuum crushing and stepwise heating are methods employed to extract the volatiles and measure their chemical and isotopic compositions using mass spectrometry. An improved vacuum stepwise heating technique can effectively separate volatiles in different occurrence modes in mantle materials, which in combination with mass spectrometry can yield excellent and highly reproducible analytical data.

Volatiles in the mantle occur in various forms such as free element or molecular species along grain boundaries, carbonate, sulfide or hydrous minerals, fluid inclusions or charged species dispersed in mineral structures (e.g., OH^-), structural defects or vacancies. Volatiles trapped in structural defects and vacancies are volumetrically significant. Large amount of hydrogen occurs as free H_2 species, not OH^- as previously thought.

Volatiles in the mantle are mixtures of primordial volatiles and recycled volatiles with characteristic chemical compositions. Volatiles in the sub-continental lithospheric mantle (SCLM) vary with depth and mantle reservoirs. Deep portions of mantle lithosphere in the diamond stability field have higher contents of reduced volatile species such as H_2 and CO etc., whereas at shallow levels, the mantle lithosphere as reflected in mantle xenoliths displays varying volatile compositions; initial volatiles trapped during primary crystallization stage are dominated by reduced species like CO, H_2. In contrast, metasomatic volatiles are more oxidized such as CO_2 and SO_2 etc.

Volatiles in mantle source regions of oceanic basalts are all dominated by H_2O and CO_2 with minor CO, CH_4, N_2, and H_2; the abundances vary with tectonic settings. MORB are depleted in volatiles as a result of source depletion in its history, whereas abundant volatiles in IAB and BABB are probably originated from subduction devolatilization. OIB may have abundant volatiles inherited form the undegassed mantle with a recycled component. Volatiles in ancient oceanic lithosphere as recorded in ophiolites are all dominated by CO_2 with minor amounts of other volatile species.

Lithospheric plate assumed originally to be "rigid" in the plate tectonics theory is in fact deformable. To explain intraplate deformation geoscientists have proposed a variety of mechanical models, in which the "slip-line field" and "viscous flow" models, as two influential ones, emphasize the deformation to be controlled mainly by shear localization and viscous flow, respectively. The "netlike plastic-flow (NPF)" model for continental dynamics argues that shear localization can occur in viscous flow field, forming NPF, which includes a plastic-flow network in the flow field, and it is the NPF in the lower lithosphere that plays a major role for the long-range transmission of driving forces and therefore controls the intraplate deformation, resulting in the netlike distributions of various geological-geophysical phenomena. Plastic-flow network is similar to but different from the slip-line network which is regarded, in the classical plasticity mechanics, only as a critical yield state. Based on the geometrical and mechanical features of plastic-flow network the "conjugate-angle-bisector" and "conjugate-angle-increment" methods have been suggested for estimating the stress directions and strains in the lower lithosphere, respectively. The results of the studies dealing with NPF are stated briefly in Chapter 6, involving the mechanism of NPF, its physical and numerical simulations, and the NPF-controlled network patterns, multi-layered tectonic deformation and stress/strain fields in the central-eastern Asian continent. It is mentioned that NPF and its controlling action are also exist in other regions of the Eurasian and other plates which subject to indentation at their boundaries. The NPF model provides an effective approach to understanding the deformable lithospheric plates.

The highest deformation rate along the Africa/Eurasia convergence zone is well documented in the Aegean area, being >4 cm/yr. However, it is still under question whether continental deformation is distributed along major faults which extend through the whole lithosphere or over large areas. Furthermore, our knowledge concerning the implication of lithosphere-asthenosphere coupling in lithospheric plates driving forces is poor. These questions can not easily be answered as most of the available information is mainly located at or close to the surface (geodesy, tectonics, seismicity). The high rates and type of surface continental deformation within the Aegean constitute this region particularly interesting in this perspective.

Chapter 7 is towards contributing to the better knowledge of the physical properties of the Aegean lithosphere by introducing experimental elastic and anelastic parameters inferred from long period Rayleigh wave. For this scope path-average phase velocities and attenuation coefficients of fundamental Rayleigh wave crossing the Aegean were extracted over the period range 10-100 s. The wavetrains were recorded at the temporary broadband stations installed some years ago in the Aegean region within the framework of a large scale experiment (SEISFAULTGREECE project).

The stochastic inversion algorithm has been used to derive 36 path-average models of shear velocity and 17 path-average models of inverse shear Q down to 200 km depth. Furthermore, the elastic and anelastic 1-D path-average models were combined in a continuous regionalization tomographic scheme to obtain a 3-D model of shear velocity variation and a 3-D model of Q_b^{-1} variation down to 120 km.

The most prominent features in the tomograms are: a) A low shear velocity zone in the back-arc region, particularly in the central and north Aegean. This region is located south of the North Aegean Trough (the western edge of the North Anatolian Fault) and correlates well with the derived anelastic tomograms which present high attenuation in this area. b) A high velocity/low attenuation zone in South Aegean indicating the subducted African lithosphere beneath the Aegean.

The low velocities/high attenuation zone in central and north Aegean is compatible with a region of high extensional strain rates, recent volcanism and high heat flow. These observations suggest a hot or perhaps partially molten upper lithospheric-asthenospheric mantle and/or distributed deformation beneath the study region, probably related with the slab roll-back that has accompanied back-arc extension.

Rock fracturing, fluid migration and pore pressure build-up is the sequence of a phenomenon which is believed to occur within the brittle outer shell of the earth during the preparatory processes of magmatic intrusion and earthquake generation. This phenomenon produces transient variations of the elastic parameters within wide rock volumes. Since the lithosphere is largely pervaded by fluids, we expect that high frequency seismic waves interact with these liquids and store them as "earth breath" in locations yet to be discovered.

Time resolved seismic tomography (4D) progressively demonstrates that transient Vp or Vs anomaly variations can be identifiable by seismic data. On a local scale, oil industries pilot research projects for the 4D seismic tomography in order to retrieve the depletion of fossil fuels in reservoirs during extraction. The proposed approach allows us to recognize eventual rock fracturing or pore pressure variations, which are useful for magmatic intrusion or earthquake forecasting.

In Chapter 8 we present a series of new observations which testify that both the elastic and anelastic properties of the crust vary in time during the rapidly changing tectonic process. We address clear 4D variations of velocity and, for the first time, the attenuation of body waves during the development of normal faulting seismic sequence, seismic swarms and magmatic supply at Quaternary volcanoes. Our results support the theory that physical properties of the lithosphere are not (dramatically) static but follow predictable patterns generated by the tectonic process.

At the Earth's surface the lithospheric magnetic field is some two-three orders of magnitude smaller than that of the core's contribution. It is produced by magnetic material located within a thin layer (10 km–70 km thick) of the crust and upper mantle. This source of magnetic field is thus carried by the magnetic lithosphere, whose definition is slightly different than that used in the fields of geochemistry, seismology or geology, although in the end these definitions all converge. The Earth's lithosphere is generated by both induced and remanent magnetizations that depend on the chemical composition and crystal conformation of subsurface rocks as well as on the Earth's core field history. Addressing the question of which type of magnetization takes over at each depth would, in particular, partially unveil the nature of the upper mantle. Such an investigation would require consistent magnetic information provided by ground stations, aeromagnetic and marine data, and satellite measurements. A comprehensive understanding of the magnetic lithosphere depends on the ability to combine all these kinds of information. Here, we discuss the available magnetic data for characterizing the lithospheric field, the methods used in describing the lithospheric field at global and regional scales, and the remarkable achievements made in obtaining the firstWorld Digital Magnetic Anomaly Map (WDMAM). Chapter 9 also addresses the difference between magnetic anomalies over the continents and the oceans, as well as comparing induced versus remanent magnetizations. Finally, we illustrate a few of the major and recent achievements that have been made using both the near-surface and satellite platforms, and products like WDMAM.

In: The Lithosphere: Geochemistry, Geology and Geophysics
Editors: J.E. Anderson et al, pp. 1-68
ISBN: 978-1-60456-903-2

Chapter 1

THE EARLY EARTH AND FORMATION OF THE LITHOSPHERE

Arkady Pilchin[1]
Universal Geosciences & Environmental Consulting Company, Toronto, Ontario, Canada,
Lev Eppelbaum[2]
Dept. of Geophysics & Planetary Sciences, Raymond and Beverly Sackler Faculty of Exact Sciences, Tel Aviv University, Tel Aviv, Israel,

Abstract

Some physical problems related to modeling the conditions of the formation and evolution of the lithosphere are discussed. It is shown that if we take into account both the effects of thermal expansion and compressibility we could receive results with no change or even an increase of density under the *P-T* conditions within the lithosphere. During planetary accretion and differentiation of Earth, the planet could have been entirely molten and at some point of its evolution was entirely covered by a magma ocean. The formation and composition of the early lithosphere were mostly related to processes of the differentiation of matter and the rate of cooling of the magma ocean. The process of the differentiation of the magma ocean would begin during its formation and continue until its solidification. This stratification of the composition of Earth caused an initial state of separation of rocks and minerals into slabs within the upper mantle and crust, which was strictly regulated by their density, whether solid or melted. The difference in density between felsic, intermediate, mafic, and ultramafic magmatic slabs within the magma ocean was enough to prevent the exchange of matter between them, and therefore mantle-wide convection could not have taken place. The solidification of the magma ocean, upon which the process of the formation of the lithosphere is dependent, most likely began with the formation of a forsterite (or forsterite-rich peridotite) slab at a depth of about 100 km followed by the solidification of Earth's surface, cooled by heat radiation from the surface and the cooling effect of the early atmosphere. It is shown that under the thermal conditions of the magma ocean, carbonate rocks were unstable and decomposed, releasing carbon dioxide into the atmosphere. Water could also not exist in its liquid state at the time of the magma ocean, and together with the carbon dioxide would form

[1] E-mail address: arkadypilchin@yahoo.com
[2] E-mail address: levap@post.tau.ac.il

a thick and dense early atmosphere. Formation of the water ocean was under the constraints of the boiling point of water at the pressure of the early atmosphere and the critical point of water. Cooling rates of the magma ocean and the early lithosphere are discussed in comparison with the cooling rates of the mantle and numerous magmatic and/or metamorphic complexes. The main periods of the appearance of komatiites and the formation of the first large igneous provinces (LIPs) indicate temperature maximums in the mantle at about 2.8-2.7 Ga and a maximum volume of mafic magmatism related to the formation of LIPs at about 2.5 Ga. Analysis of the periods of the formation of granulites suggests an increase in temperature at a depth of about 30 km from ~3.0 Ga to ~2.7 Ga, with its continuing increase to ~2.5 Ga. It is shown that at the end of the Archean, the thickness of the lithosphere was ≤100 km, including a solid forsterite slab at ~100 km depth with possible pockets of magma above it.

1. Introduction

The lithosphere is the outermost super-slab of Earth comprising of Earth's crust and the rigid outer part of the upper mantle. Underlying the lithosphere is the asthenosphere, which is believed to be either in a plastic condition or partially molten. The lithosphere-asthenosphere model was originally introduced by Barrell in 1914-1915 [15,158] based on analysis of gravity anomalies over continental crust; it was later further developed by Daly [105]. The Mohorovičić discontinuity (Moho) represents the border between the crust and mantle parts of the lithosphere. It is clear from the definition that both the lithosphere and the asthenosphere are characterized by different various rheology and physical, particularly mechanical, properties. The position and primary physical characteristics of both the lithosphere and asthenosphere call for such chief methods of their investigation as those pertaining to seismic, thermal and gravitational modes of inquiry, as well as analysis of the mineral composition of xenoliths and the conditions of their stability. Unfortunately, none of these methods is free of ambiguity and potential for error. Some geologists believe that mantle xenoliths from diamond pipes (kimberlites) can now provide more precise data [415] than geophysical methods. Depending on a method used for investigation of the lithosphere there are different names by which it may be called in geological literature. The most commonly used terms are thermal lithosphere [23,100,250,278,308,404,415], the chemical lithosphere [99,468], the seismic lithosphere [10,23,155,249,370,413], and the elastic or mechanical lithosphere [61,67,68,118,199,277,460,475]. It is evident that the thermal conditions of the lithosphere must have been below the liquidus temperatures and in many cases even below the solidus temperatures of composing it rocks and minerals. It led to the acceptance of a temperature maximum for the lithosphere in the range of 1473-1573 K [23,250,278,308,326,404,415]. It is also clear that in order for the lithosphere to behave as a single slab and be able to transfer stress over significant distances it must be elastic, which would require very specific mechanical properties of its rock and mineral composition. The majority of scientists agree that rocks and minerals could have such elastic properties only at specific thermal conditions, which for the oceanic lithosphere are constrained by a temperature maximum of 873 K [67,68,118,460]. However, some scientists believe that this temperature limit could be higher and reach 973 K [460], 1073 K [199] or even 1123 K [460,475]. This temperature marks the greatest depth of the mechanical lithosphere [67,68,199]. The depth of the corresponding isotherm is usually called the effective elastic thickness [61,118,277], the equivalent elastic thickness [68] or simply the elastic thickness [199]. Taking into account that rocks and minerals are very brittle at low temperatures, some researchers argue the existence of a core block in the elastic lithosphere with depth limited by geotherms 573 K and 873 K [61], or 673 K and 873 K [68]. At the same time, in continents the elastic thickness has no relation with any particular depth [67,68,277], and estimates of the elastic thickness bear little relation to specific geological or physical boundaries within the continental lithosphere [67]. This makes Geothermics essential for the investigation of the entire lithosphere or its individual parts, the crust and mantle, separately.

Seismic methods arc among the most powerful and precise methods of determining the thickness of the lithosphere, because they allow to pinpoint the position of the Low Velocity Zone (LVZ), which usually associated with the asthenosphere [10,15,158,247].

A gravity method is also of great importance for investigation of the lithosphere [218, 447,458], because this method is directly related to the study of the distribution of the masses and densities of rocks composing the lithosphere. The distribution of mass is essential to analysis of the isostatic equilibrium, which is supposed to explain the existence of different topographic heights and depressions on the Earth's surface [7]. The first models of analysis of isostatic equilibrium were offered by Airy [6] and Pratt [369] in the 19th century. In the Airy model, crustal blocks have the same density but different thickness; while in the Pratt model, crustal blocks have different densities allowing the depth at which crustal material achieves equilibrium to be the same. Both are considered to be local models. Regional models, on other hand, must take in consideration such factor as flexural rigidity. One such a model is that of Vening-Meinesz [454,458].

During the past few decades, greater importance placed on investigation of the lithosphere gained results from studies of xenoliths [22,23,71,16,124,127,148,149,185,201, 233,245,295,313, 326,329,338,388,401,415]. There is no doubt that the analysis of xenolyths and their mineral and chemical composition gives a great deal of information about the composition and structure of the lower crust and upper mantle. At the same time, methods of experimental petrology, and specifically investigations related to the determination of *P-T* conditions of the stability of rocks and minerals (thermobarometric methods), allow the opportunity to estimate the *P-T* conditions in the environments in which these xenolyths originated. Some geologists believe that xenolyths and diamonds were formed during the Archean [57,338], and delivered to the surface within kimberlite pipes at a later time. Based on this hypothesis, they argue that the deep parts of the mantle were cold enough in the Archean for the formation of diamonds.

All of the methods mentioned are useful for investigation of the present lithosphere. However, the thickness of the lithosphere of the Archean cratons and thickness of the lithosphere in the Archean or Hadean are two completely different understandings. One of the most important questions related to the lithosphere is: when its formation began? Another important question is: which point in the time of Earth's evolution should be accepted as that of the formation of early lithosphere? Geophysical methods mainly provide us with information about the present lithosphere and its conditions; analysis of xenoliths could give us information about both the present and past of the lithosphere. It is clear that from conception, the formation and evolution of the lithosphere would not stop even for a short period of time. We have lithosphere forming today in oceanic ridges, and the evolution of the lithosphere of the Archean cratons which began its formation at least in the Archean will not stop for a long period of geological time yet, because the cooling of Earth's mantle will lead to farther increase in its thickness. This process will not stop at least until the asthenosphere becomes a rigid lump.

It is widely accepted that at the end of planetary accretion, Earth was covered with magma ocean [2,3,251,279,366,378,415,422,428,436]. It is obvious that there could not be any lithosphere or crust at that point in time. However, this would also means that the formation of the lithosphere began with the solidification of the magma ocean.

The goal of the present research is to analyze the thermodynamic and most importantly thermal conditions which took place during the formation and evolution of early Earth; the formation and cooling of the magma ocean; analysis of *P-T*-conditions of some geologic processes related to the formation and evolution of the lithosphere in the Hadean and Archean; the conditions and composition of the early atmosphere and possible conditions of the formation of the water ocean.

2. Some Remarks on Physical Problems Related to Modeling Conditions of Formation and Evolution of the Lithosphere

It is obvious that any modeling of the formation or evolution of the lithosphere or the investigation of lithosphere conditions requires the use of numerous physical parameters and characteristics. A short list of these parameters would include temperature, melting point, solidus temperature, liquidus temperature, pressure, thermal constants, elastic constants, seismic velocity, density, mass, and volume. It is also canon that no model should violate any law of physics, and that laws of physics should only be used in an appropriate way in order to obtain correct and meaningful results.

The first models of isostasy were related by their authors and future ones to the distribution of masses forming topographical units and slabs in the crust and lithosphere. However, these models were actually about the equilibrium between the lithostatic pressure generated by different blocks of either the crust or the entire lithosphere. The value of lithostatic pressure P_h at any depth could be calculated using formula

$$P_h = \sum_i g_i \rho_i \Delta h_i \, . \quad (1)$$

Here i is the arbitrary layer number in a layered model of the crust or lithosphere, g_i, ρ_i and Δh_i are the average acceleration due to gravity, average density and thickness of the layer i, respectively.

It should be taken into account that lithostatic forces are related to gravity, they are strictly vertical. Analysis of the isostatic equilibrium requires a good understanding of the structure of lithospheric slabs, their petrologic composition and distribution of density with depth within each of these slabs. In many cases, for determination of density of rocks within different slabs of the lithosphere the Birch's law has been used [9,43,44,421]. Birch's law describes the variation of the compressional wave velocity V_P of rocks and minerals of a constant average atomic weight with density ρ as:

$$V_P = a\rho + b \, . \quad (2)$$

Here a and b are constants usually determined from the correlation between compressional wave velocity V_P and density ρ received experimentally for different rocks and minerals. Even though the $V_P - \rho$ correlation is high, it shows two clear trends: one for iron-poor ultramafic rocks and another for all the other rocks considered [421]. Some recearches show that the velocity into density conversion using Eq. (2) introduces an additional uncertainty [83,228]. Really, experimental data show that the increase of iron content in olivines [72,146,266,426] and dunites [86] leads to a decrease in value of both V_P and V_S while the value of their density increases with the increase of iron content. This means that iron-rich ultramafic rocks would have a negative correlation between their seismic wave velocity (V_P or V_S) and density (ρ), in contrast to the main crustal magmatic rocks, which have an increase of their seismic wave velocity values with the increase of density. This fact could lead to significant errors in the determination of density using Eq. (2), espetially for the mantle part of the lithosphere. Karato and Karki [205] demonstrate that when the cause for lateral variation in seismic wave velocities is the lateral variation in temperature, low velocity regions represent higher than average temperatures and should have lower density; while high velocity regions represent lower than average temperatures and should have higher density. However, when velocity heterogeneity is due to the heterogeneity in Fe content, the relationship between velocity heterogeneity and density heterogeneity will be opposite: slow velocity regions would imply high Fe content and hence denser regions; and high velocity regions would imply low Fe content, and hence lighter regions [205]. Another research [112] also points that positive V_S-anomalies are associated with negative temperature variations and iron depletion. Lee [244] reports of a significant increase in V_S value with increase of Mg# for olivine. Speziale et al. [427] show that at ambient conditions the calculated relative variation of compressional and shear velocities with increase of iron content due to Fe-Mg substitution, range between -0.05 and -0.46, and between -0.08 and -0.74 respectively, in the main mantle minerals (olivines, garnets, orthopyroxenes, clinopyroxenes, and periclase – wüstite mixture). All these facts demonstrate that Birch's law described in Eq. (2) works differently for crustal rocks than for rocks of the upper mantle with significant uncertainty in results derived for the uppermost mantle which contains mostly forsterite-rich olivine.

Some results of investigations indicate that for the analysis of isostatic equilibria, an important role lies in the determination of density inhomogeneties within the crust and upper mantle [19,20,200]. However, even though, the effect of solid crust density inhomogeneities can also be estimated from data on the average velocities of seismic waves, the information is less reliable than that of other datasets that include constraints on the Moho boundary position [200].

On the other hand, in a number of cases the pressure generated by different processes [355,357] has been shown to have the potential to overcome the lithostatic pressure. It was shown earlier [355], that pressure generated by high temperature during granulite facies metamorphism could have been greater than the lithostatic pressure. The pressure generated by thermal conditions during the formation of mud volcanoes and salt diapirs could also have been greater than the lithostatic pressure [354]. Even the overpressure generated in sedimentary strata by trapped heated water could in some cases have been greater than that of the lithostatic pressure [216,281]. At the same time, numerous cases point to the formation of horizontal pressure and overpressure in slabs of the crust and upper mantle [346,354-357,390]. Neither of these pressures are related to the lithostatic pressure generated by any block of the crust or lithosphere, however, they would still break the isostatic equilibrium, forcing the block to uplift in many cases without any change in the distribution of mass or values of density. This means that for the analysis of isostatic equilibrium, the distribution of pressure and stress both within and below the lithosphere is more essential than the distribution of the masses and densities within its blocks. At the same time, lithostatic pressure should be viewed as the normal pressure for the corresponding depth, and in most cases (excluding processes of immersion) it is the minimal pressure at that depth [354-357].

In the most general form, the relationships between such thermodynamic parameters as volume (V), pressure (P), and temperature (T) could be presented in the following form [357]:

$$\frac{\Delta V}{V_o} = \alpha \cdot \Delta T - \beta \cdot \Delta P . \tag{3}$$

Here V_o is the value of the initial volume, α is the volume thermal expansion coefficient, β is the compressibility coefficient, $\Delta V = V - V_o$, $\Delta P = P - P_o$, and $\Delta T = T - T_o$. Eq. (3) could be easily transferred [125,281,354-357] to

$$P = P_o + \frac{\alpha}{\beta}(T - T_o) - \frac{1}{\beta}\left(\frac{\Delta V}{V_o}\right). \tag{4}$$

As it was shown earlier [281,354,357], the initial pressure P_o is the lithostatic pressure. The second term on the right side of Eq. (4), $(\alpha/\beta)(T - T_o)$, is the additional part of the real pressure caused by temperature regime. The third element on the right side of Eq. (4), $(1/\beta)(\Delta V/V_o)$, is the part of the real pressure which is usually specifying its unloading, but in cases of tectonic compression it could lead to the reduction of volume and additional increase of pressure. From all above-mentioned and Eq. (4), it is obvious that real value of the pressure could be quite different from just its lithostatic value calculated using Eq. (1).

It should be stated that there could be different reasons of temperature (ΔT) and volume (ΔV) changes. These reasons could be physical, chemical, tectonic, or related to some geologic processes. However, in application of Eq. (4) it does not matter which process caused a change in the values of the parameters used, and only the fact of parameter and value variance is a matter.

At the same time, in Eq. (3) relative change of volume ($\Delta V/V_0$) could be replaced with the relative change of density ($\Delta\rho/\rho$) using equation:

$$\frac{\Delta V}{V_0} = -\frac{\Delta \rho}{\rho}. \tag{5}$$

In Eq. (5) density ρ is the final density and $\Delta\rho=\rho-\rho_0$, where ρ_0 is the initial density of rock or mineral.

The relationship between initial volume and initial density for a rock could be presented in form of:

$$\frac{\Delta V}{V_0} = \frac{-\Delta\rho / \rho_0}{1 + \Delta\rho / \rho_0}. \tag{6}$$

Using the condition from Eq. (5), Eq. (3) could be transferred [363] to:

$$-\frac{\Delta\rho}{\rho} = \alpha \cdot \Delta T - \beta \cdot \Delta P. \tag{7}$$

Eq. (3), Eq. (6), and Eq. (7) allow for easy analysis of the thermodynamic conditions of equilibrium ($\Delta V=0;\ \Delta\rho=0$), conditions of volume increase – density decrease ($\Delta V>0;\ \Delta\rho<0$), and condition of volume decrease – density increase ($\Delta V<0;\ \Delta\rho>0$). Eq. (7) clearly indicates that depending on *P-T*-conditions there could either be an increase or decrease of density, or its value d could remain unchanged ($\Delta\rho=0$). The case with *P-T* conditions when both volume and density remain unchanged is the most important, because it allows to determine normal *P-T* conditions (P_0,T_0) for any rock or mineral [357,358]. The conditions of equilibrium actually mean that expansion and compressibility counterbalance each other. It was noted earlier [215,357] in case of unchanged volume and density

$$\Delta V = 0;\ \Delta\rho = 0. \tag{8}$$

either of equations (3), (4), and (7) would have the following form

$$P = P_o + \frac{\alpha}{\beta}(T - T_o). \tag{9}$$

From Eq. (9) follows that if temperature T is equal to its normal value T_0 for the given depth ($T = T_0$), the real pressure p must also be equal to its normal value P_0 ($P = P_0$). The reverse is also then true, if pressure P must also be equal to its normal (hydrostatic or lithostatic) value P_0 ($P = P_0$), then temperature T must be equal to its normal value T_0 for the given depth ($T = T_0$). Moreover, Eq. (9) proves that for studied matter studied there is only one pair of values P_o, T_o for any particular depth z. Conditions of equilibrium for different rocks are presented in [357]. Pilchin [353] estimated the value of gravity anomaly caused by density change under various *P-T* conditions for the area of Kura depression (Azerbaijan).

Different researchers [8,21,202,243,268,275,276,329,343] usually use the following equation for determine the change of density with depth

$$\rho = \rho_0(1 - \alpha\Delta T). \tag{10}$$

Comparison of Eq. (10) with Eq. (7) shows that Eq. (10) completely ignores the effect of pressure (compressibility) on the value of the density of a rock or mineral. At the same time Eq. (10) cannot account for a case with an increase of density, because according to it the increase of temperature with depth in the lithosphere or any other part of Earth could only result in a decrease of density. It is also obvious that the change of density described by Eq. (10) would be different for different rocks and minerals, since the value of volume thermal expansion is different for different rocks and minerals. A graphic example of the effect of compressibility is illustrated in the statement "At 10 kilobars (equivalent to a depth of about 35 kilometers) the melt would be 4-5% denser than at the surface" (page 69 in [160]). The effect of compressibility on density, volume and seismic velocity of rocks and minerals is well known (see, for example, [10,46,76,355,356,358]), and should not be neglected. Of course, this effect would be smaller for solids, but it would be not zero. On top of that, it is unclear in Eq. (10) from which initial temperature the value of ΔT should be calculated (such temperature difference would have no any physical meaning). Some authors calculate the temperature difference from the temperature of "background" [386]. Such a "background" temperature has different value for different regions of the Earth or any other terrestrial planet. In contrast to that, the normal temperature (T_0) defined above, would be the same for a rock or mineral at the same normal (or lithostatic) pressure in any region of the Earth of any age and on any terrestrial planet. A closer comparison of Eq. (10) with Eq. (7) shows that Eq. (7) will turn to Eq. (10) only in the case of an open liquid system (water ocean, near surface ground water, etc.), in which any change of pressure would be immediately unloaded and the pressure will always be hydrostatic. This makes Eq. (10) useful only for modeling the density stratification of the liquid. This case cannot be applied to the lithosphere or even the asthenosphere, but it may be applied to the magma ocean formed at the end of the accretion of Earth.

To demonstrate, taking into consideration both the effects of thermal expansion and compressibility, the calculated variable density for granites and gabbro are presented in Tables 1 and 2, respectively:

Table 1. Variable density of granites (in kg/m^3) under different *P-T*conditions

Temperature, K	Pressure, GPa			
	0.1	**0.3**	**0.5**	**0.7**
373	-3	10	21	31
573	-26	-13	0.0	13
773	-61	-45	-32	-17
973	-103	-86	-67	-49

Table 2. Change of density of gabbro (in kg/m^3) under different *P-T* conditions

Temperature, K	Pressure, Gpa				
	0.1	**0.4**	**0.7**	**1.0**	**1.2**
373	0.0	1.5	27	40	47
673	-1.5	-5	24	43	53
973	-3.7	-9	15	38	53
1273	-6.1	-0	3	35	53

Analysis of the physical properties of olivines and peridotites indicates that forsterite and Mg-rich olivines have an extremely high thermal expansion coefficient [86,419]. Ignoring the effect of compressibility for them could yield results for their density lower up to 3% for the thermal conditions in the upper mantle. From data presented in [86] the value of the thermal expansion of forsterite at 1073 K is about 42% higher than that of the thermal expansion of fayalite, and for different temperatures the value of the thermal expansion of forsterite is about: 58.3-83.3% higher than that the value granites, 81.0-109.5% higher than that of andesites, and 134.6-171.6% higher than that of basalts, gabbros, and diabases. Our estimations show that for peridotite under *P-T* conditions of the upper mantle, if we take into account both thermal expansion and compressibility, it could actually result in an increase of its density by up to 1.0%.

Based on density changes presented in Tables 1 and 2 and real *P-T* conditions within the crust of the Kura depression (Azerbaijan), the gravity effect from "granite" and "basaltic" layers were calculated for some areas of the depression (Table 3)

Table 3. Gravity effect caused by change of density of rocks of "granite" and "basalt" layer under real P-T conditions for some areas of Kura depression, Azerbaijan

Area, tectonic structure	Average change of density in layer, kg/m^3		Gravity anomaly from layer, mGal (10^{-5} m/s^2)	
	"granitic" layer	"basaltic" layer	"granite" layer	"basaltic" layer
Dalmamedli anticline	-15	41	-1.23	15.7
Evlakh-Agdzhabedi depression				
axial zone	-15	43	-0.67	12.4
marginal zone	-13	41	-0.86	11.8
Pushkin uplift	-8	41	-1.35	12.3

It was shown earlier [352,357,365] that change of seismic velocity at different *P-T* conditions could be presented by equation:

$$dv = \left(\frac{dv}{dT}\right)_P dT + \left(\frac{dv}{dP}\right)_T dP. \tag{11}$$

Here $(dv/dT)_P$ and $(dv/dP)_T$ represent the change of the velocity with change of temperature and pressure respectively.

The most important case of Eq. (11) is related to condition of equilibrium for velocity (v):

$$dv = 0. \tag{12}$$

Physically this means that the increase of velocity (v) with increase of pressure (P) is counter-balanced by the velocity decrease with increase of temperature (T). For the condition Eq. (12), Eq. (11) would be

$$\left(\frac{dv}{dT}\right)_P dT + \left(\frac{dv}{dP}\right)_T dP = 0. \tag{13}$$

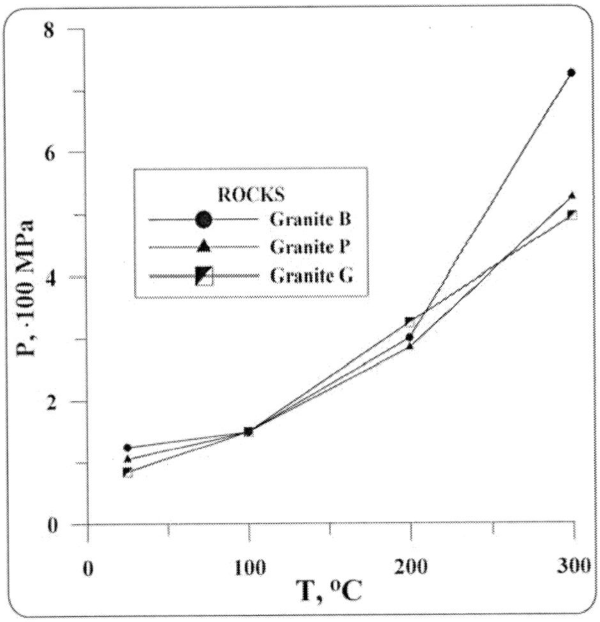

Figure 1. The condition of thermodynamic equilibrium for some granites ($dv = 0$).

Eq. (13) was used to determine the velocity equilibrium conditions for some types of granites (Figure 1). It is also obvious that for formation of Low Velocity Zone (LVZ), the change of velocity should satisfy condition:

$$dv < 0. \quad 14)$$

With this condition, Eq. (11) becomes:

$$\left(\frac{dv}{dT}\right)_P dT + \left(\frac{dv}{dp}\right)_T dp < 0. \tag{15}$$

Inequality (15) shows that velocity in different rocks and minerals could decrease under specific *P-T* conditions defined by the inequality. This means that presence of melt in the asthenosphere is not necessary for the formation of a LVZ. This is in agreement with the aforementioned above fact that both V_P and V_S velocities decrease with the increase of iron content that could explain formation of LVZ within the apper mantle without presence of any melt. This fact is supported by data showing an increase in iron content with depth within the continental part of the lithospheric mantle [295,326], and the iron-rich composition of the asthenosphere [326]. Kuskov and Kronrod [233] also report of the discovery of a solid-state zone of lower velocities on the *S* velocity profile in the depth interval of 150–260 km within the lithosphere of the Kaapvaal craton. This is in accord with cases where the asthenosphere has been found to be absent below some cratons. Lerner-Lam and Jordan [249] suggest that the tectonically stable part of northern Eurasia, including the Siberian platform, has a thick seismic lithosphere extending to a depth of about 400 km. Artemieva and Mooney [23] found the thermal lithosphere beneath the Siberian platform to be that of about 350 km thick. A high velocity layer traced down to about 300-400 km and identified as seismic lithosphere was found beneath the East-European platform [413]. For different parts of the Ukrainean schield the lithosphere thickness is 350-400 km beneath the easternmost Peri-Azovean block, about 350 km beneath the Ros-Tikich block and about 400 km beneath the Volyn-Podolie block [413]. Other examinations based on seismic data also show regions of anomalously high velocity in the mantle underlying the Archean cratons at depths of 300-400 km [155,387]. Geophysical studies of Northern Eurasia have failed to reveal a well-defined and continuous asthenosphere [331], in many cases the asthenosphere cannot be traced from the seismic data [334], and the asthenosphere-lithosphere boundary in numerous cases has not been recognized from velocity cross-sections [333].

It was also shown [352] that for the boundary between two adjacent layers, depending on the different relationships between the change of velocity with pressure and temperature of rocks composing both layers and on their *P-T*-conditions, there could be a jump in velocity at the discontinuity resulting in either decrease or increase; in some cases there could be no change to it at all. Eq. (11) can also be used for the estimation of temperatures and pressures in regions for which geothermal data are not available or geothermal measurements are not possible to obtain for some reasons [364].

It was suggested earlier [357] that if initial velocity of a rock or mineral was determined under normal *P-T* conditions (P_o,T_o), the equilibrium curve representing the unchanged values of density and velocity should coincide.

Other important processes involvied in modeling the conditions and behavior of the lithosphere are related to calculation of thermal conditions within the lithosphere and its process of cooling. For modeling of thermal conditions within the lithosphere numerous geothermal methods are used [78,79,126,204,235,259,455]. These methods are based on the solution of the heat conductivity equation using different limitations for parameters of the equation. In the most general form the heat conductivity equation could be written as [73]:

$$c\rho\frac{\partial T}{\partial t}=\frac{\partial\left(\lambda\,\partial T/\partial z\right)}{\partial z}+\frac{\partial\left(\lambda\,\partial T/\partial x\right)}{\partial x}+\frac{\partial\left(\lambda\,\partial T/\partial y\right)}{\partial y}+A(x,y,z,t)\,. \tag{16}$$

Here c is the heat capacity, ρ is the density, $T = T(x,y,z,t)$, λ is the heat conduction, and $A(x,y,z,t)$ is the heat generation in a body or layer.

Solving the heat conductivity equation is quite a complicated problem, and any solution requires some limitations to be placed on the equation in the form of specific initial and boundary conditions for the body or layer [73,78,79,126,204,235,259]. The solution of the problem of the distribution of temperature within a layer is very sensitive to values of the parameters used in Eq. (16) [78,126,204,235,259]. For example, the value of heat conductivity for any rock or mineral significantly decreases with increase of temperature [77,235,259]; as well, the sources of heat production that are the decaying of radioactive elements are only present in significant amounts within the crust, and the content of which significantly decreases with increase of depth [78,79,204,259,235, 366]. The solution of the heat conductivity equation is also sensitive to the thickness of a given layer and other sub-layers comprising it [78,126,204,235,259]. Especially in the case of solving the heat conductivity equation for the lithosphere, there are some problems which make the solution very complicated. These problems are related to the fact that the heat production by the decay of radioactive elements is very small or absent below the Moho discontinuity and that researches point to a minimum of total thermal conductivity in the upper mantle at depths of about 50-100 km [259,260]. The results of determining the temperature in the mantle part of the lithosphere could be significantly affected by uncertainty in the values of heat flow density from the upper mantle into the lower crust, as well as in determining the value of the geothermal gradient within the lowermost crust and the upper mantle. The latter is related to the fact that the geothermal gradient and heat flow density decrease with the increase of depth within any homogeneous slab [126,351,361]. However, the absence of radioactive sources for heat production within the upper mantle could create conditions where the geothermal gradient would increase with the increase of depth. This corresponds with results of some researches on paleogeotherms based on analysis of xenolith data, which show that the geotherm was found to be inflected to a higher thermal gradient at a greater depth [57,132,261,263,284]. This ambiguity in determining the geothermal gradient within the upper mantle could cause significant errors in estimating the temperature at the base of the lithosphere using geothermal data. Another problem with using geothermal methods could relate to the fact that thermal energy discharged by the decay of radioactive elements within the slabs of the crust cannot be immediately released from the Earth's surface [350,351]. This fact would contradict some researches stating that a linear relation exists between the heat flow released from Earth's surface and the heat production from radioactive elements in the form of [23,47,77,238,430]:

$$Q_0 = Q_r + DA_0 \,. \tag{17}$$

Here Q_0 is the mean surface heat flow, Q_r is the reduced heat flow, A_0 is the heat production, and D is the characteristic depth over at which most of the heat-producing elements are distributed.

The heat energy released within any layer will be partially absorbed within the same layer as well as by overlaying layers [349,350]. This would create conditions of the simultaneous heating of some layers and cooling of other [351]. This fact was confirmed by geothermal research of different sedimentary basins [125,350,351,361]. After reaching a thermal maximum or minimum, the regime of these layers would switch from that of heating to cooling or from that of cooling to heating [349,350].

Some investigators believe that cooling of the Earth's surface was primarly conducted by radiation in accordance with the Stephen-Boltzmann's law [e.g., 204,259]. The application of this law leads to a very quick cooling of Earth's surface after accretion. However, this law does not necessarily describe the process of the surface cooling correctly, because from a Physical point of view Earth is not a "black body". By definition, a "black body" absorbs all the radiation falling on it; however, Earth reflects over a third of the Sun's radiation falling upon it (~36%), and Venus reflects more than two thirds (~ 72%). Kaler [203] states that neither stars nor planets are perfect "black bodies", so by definition neither Earth nor Venus could be counted as "black bodies". It is in agreement with the statement that though ([271], p. 448): "It is perfectly true to say that $e_{BB} \propto \theta^4$ ... it is *not necessarily true* that $e \propto \theta^4$ for any body at all"; where e is the radiation per unit area per unit time, e_{BB} is the radiation per unit area per unit time for black body, and θ is the temperature are notations accepted by Marshall et al. [271]. This means that the cooling of the Earth's surface could have been much slower.

The use of xenolith data for estimating the *P-T* conditions within the parts of the lithosphere in which they were formed is based on the results of experimental petrology, and particularly methods of thermobarometry. The study of xenoliths gives us information about the composition of the lithosphere at the depths of their formation, as well as the *P-T*-conditions at those and other depths at which they were present prior to them being delivered to the surface by rising magma. The fact that the lithosphere of the Archean cratons is the thickest lithosphere, which in some places is twice as thick as the one formed during the Proterozoic, makes research of the lithosphere of the Archean cratons extremely important. Some researchers [25,58,71,148,149,313,326,338,401] attribute an exceptional role for the investigation of the lithosphere of the Archean cratons to kimberlites, which contain numerous rocks and minerals formed under ultra-high pressure conditions and could represent the composition of the deepest possible parts of the lithosphere from which xenoliths could have been delivered to the surface. Researches on xenoliths show that the absolute majority of cratonic xenoliths, especially xenoliths from kimberlites, are peridotitic xenoliths [244,283,312,338, 401]. This fact allowed some researchers to come to the conclusion that peridotitic lithologies comprise more than 95% of the mantle lithosphere [338,401]. Another estimation shows [459] shows that low-pressure phase olivine is the major phase of the upper mantle, and represents nearly 20% of the planet's volume. At the same time, even though all kimberlites within cratonic areas of southern Africa contain eclogite [157] it is still a rare type of xenolith at most localities [157,338]. Schulze [401] concluded that the amount of eclogite in the upper 200 km of the subcontinental upper mantle is perhaps less than 1% by volume. Carlson et al. [71] show that "In general, high-velocity bodies consistent with large masses of eclogite are not observed in the continental mantle" and that "... at present, eclogite appears to be a relatively minor component of the continental lithospheric mantle." (paragraph 6 in [71]). The origin of eclogite xenoliths in kimberlites is still enigmatic, because, as it was shown earlier, none of the eclogite suites can be satisfactorily explained by as having been originated from high pressure cumulates from primary melts of garnet peridotite [183]. In non-cratonic xenolith suites, eclogite is much scarcer, and garnet pyroxenites are more common [338]. Cratonic and non-cratonic eclogites have a significant difference in chemical and mineral compositions [338,348]. Some authors believe that most cratonic eclogites are of the Archean age [338]. However, even though the absolute majority of xenoliths are peridotites, it does not mean that the upper mantle is mostly composed of peridotites for the following reasons: (1) a peridotite composition of the upper mantle cannot explain the chemical composition of felsic, intermediate, or basic magmas, (2) xenoliths are delivered by ascending magmas formed at depths greater than those of their own formation, that could be hot enough to re-melt most kinds of rocks, but peridotites which have the maximum melting point among silicates could be the chief kind of rock to survive the high temperatures within the ascending magma, (3) some rocks heavier than regular peridotites within the magma could be under-represented (for example, Fe-rich dunites [21,246]), because they would sink much faster in the magma according to Stock's law.

Radiometric ages on syngenetic inclusions in diamonds show that most macrodiamonds are xenocrysts and are Archean or Proterozoic in age [338]. This raises a problem in determining the *P-T* conditions in the upper mantle necessary for the preservation of those diamonds for billions of years until their upraise to the surface within kimberlitic flows. Estimations of the temperatures at which diamonds were preserved in the lithospheric mantle gave values of 1323–1473 K [338].

The accuracy of diamond ages based on inclusions, as well as the possibility of younger diamonds encapsulating older, pre-existing inclusions, is still being debated in scientific literature [338].

The *P-T* conditions of mineral equilibria of suites of mafic to ultramafic xenoliths in alkaline basalts and kimberlites were used in numerous researches for constructing xenolith-derived mantle geotherms [56,57,62,82,132,133,189,263,264,267,284,291,313,399]. Some scientists believe that estimations of *P-T* conditions of xenoliths help to estimate the geotherm at the time of kimberlite eruption [122,313]. Other researches state that the localization of diamond-bearing kimberlites in the craton and the eruption of Archaean diamonds are evidence that these diamonds crystallized and were immobilized in a root extending to depths of 150–200 km and, based on applications of thermobarometry to silicate inclusions in the diamonds suggest that ambient temperatures at those depths 3,000 M.y. ago were 1173–1473 K [57].

However, methods of determination of *P-T* conditions based on mineral equilibria in mafic and ultramafic xenoliths are also not free of uncertainties and ambiguity. Finnerty [132] argues that different geothermometers gave good results, while in contrast numerous geobarometers produced low results with underestimates of pressure by up to 1–2 GPa, in numerous cases diamond-bearing xenoliths to fall into the stability field of graphite. Analyzing different methods of geothermometry and geobarometry applicable to garnet lherzolite, Mitchell et al. [291] show that it is not possible for an upper mantle stratigraphy to be constructed in Lesotho except in the most general terms, because of the prevailing lateral and vertical heterogeneity. Irving [186] points to considerable ambiguity in determining paleogeotherms from xenolith suites in alkaline basalts and kimberlites. He also suggests "it can be argued that xenolith suites cannot be used to define valid geotherms because the processes of magma generation and xenolith formation involve perturbations of the very geotherms we seek to measure" ([186], p. 638). Irving [186] shows that these uncertainties increase especially for low temperature conditions. At the same time, experimental studies of garnet pyroxenite series xenoliths from the Delegate basaltic pipes suggest equilibration conditions of 1.3-1.7 GPa, 1323-1423 K, which are well above the published geotherms derived from geophysical data [186]. Data of Schmidberger and Francis [399] show a significant difference between the composition of peridotites and kimberlites of the Archean Kaapvaal, Siberian and North American cratons. This would require different *P-T* conditions for the equilibration of those peridotites. Haggerty and Sautter [180] hypothesize that the peridotites originally reside deep in the mantle (at depths of perhaps 300 or 400 km) and are later transported upwards into the cratonic lithosphere (~180 km depth). Rudnick et al. [388] show that the average cratonic peridotite is not representative of the heat production of Archean mantle roots. They also show [388], that if the xenolith *P-T* data reflects equilibration to a conductive geotherm, and then the Archean lithosphere would be relatively thin (~150–200 km).

The Otrøy (Western Gneiss Region, Norway) peridotites have a similar composition to mantle xenoliths from Precambrian cratons and Phanerozoic supra-subduction-zone peridotites [122]. Moreover, some isotope data from Western Gneiss Region (WGR) peridotites suggests an age of depletion around 2.5–3.0 Ga whereas Sm–Nd dates range from the early Proterozoic to Caledonian ages [122]. These facts contribute a great deal of uncertainty in ascertaining the age of the formation these peridotites, including peridotites from Precambrian cratons. Obvious problem with age uncertainty exist in the case of eclogites from East Athabasca of the Canadian Shield; even though zircon from one eclogite sample has cores of 2.54 Ga, it was proven that the age of eclogites formation was the Early Proterozoic (~1.905 Ga) [30]. In another example from the Siberian craton, age ambiguity is related to the fact that trace element data for rare garnet and clinopyroxene inclusions trapped within diamonds occurring in eclogite xenoliths revealed that the inclusions are more depleted in incompatible trace elements and have a lower Mg# than the eclogite host minerals [185]. This means that eclogite inclusions within diamonds have possibly nothing to do with the eclogites in kimberlites themselves, and that it may be incorrect to use the age of eclogite inclusions in diamonds to estimate the age of eclogite xenoliths in kimberlites. Additional uncertainty could be furthered by the fact of the presence of high temperature (HTP) and low temperature peridotites (LTP) among xenoliths [245,338].

In some cases, the estimated equilibration temperatures exceed (>1623 K) the typical lithospheric geotherm temperatures [245]. At 5 GPa the equilibration temperatures of Koidu low-MgO eclogites are much lower (1153–1203 K) than that of Roberts Victor eclogites (1373 K) [338].

The deepest low-T xenoliths are found at the Finsch kimberlite in South Africa with equilibration pressures suggesting a depth of origin slightly in excess of 200 km [133]. This would place a temperature below 1373 K at depths greater than 200 km within this area, which is highly unlikely. Additional uncertainties in the use of kimberlite xenolith data for estimating of *P-T* conditions during the formation of those rocks and kimberlites are related to problems of the preservation of diamonds and eclogites, especially in diamond-bearing kimberlites, from the time of their formation that of the formation of the kimberlite; as well as the fact of the presence of serpentinites or serpentinized peridotites are found present in kimberlites, which can only be formed at maximum temperatures of up to 673-773 K and minimum of ~473 K [126,171,359, 360]. Low temperatures such as these cannot be reached at the great depths at which diamonds are formed and preserved. Furthermore, eclogite contains aegerine, which in turn contains ferric oxide, a compound unstable at temperatures above 723-843 K [357-360]. This fact also places strong constraints on both the conditions of formation and preservation of eclogites. The accuracy of thermobarometers for estimating the *P-T* conditions of xenoliths was discussed in some publications [74,132,417].

All mentioned numerous facts which clearly indicate that at present the use of xenolith data cannot be relied upon as a precise method for the determination of *P-T* conditions. On top of that, latest researches demonstrating the possibility of the formation of significant overpressures, reaching values of two times or higher than that of the normal lithostatic pressure [281,346,354-358,390], show that the formation of high and ultrahigh pressure rocks and minerals could take place at depths two times lower than those estimated by values of lithostatic pressure. Such overpressure could be formed by horizontal forces, which are dominant in such processes of plate tectonics as obduction and the collision of plates. This means that both eclogites and diamonds could be formed at much lower depths than those determined with the value of pressure accepted as strictly lithostatic.

Another feature, which deserves special attention, is the modeling of processes of the formation of melts and magma chambers at great depths. The existing model of "partial melting" has some abstract meaning, but little physical meaning. The main problem with this model is the formation of a magma chamber. Supposing 5% partial melting took place within a mantle slab and droplets of magma were evenly or almost evenly distributed within the slab, how would they then unite together to form a magma chamber while under conditions of a gigantic lithostatic pressure? Under such a huge pressure, there is very little chance that the droplets would be able to move in any direction, much less so in a horizontal direction. The solution to this problem is extremely important, because the formation, eruption, and solidification of magmas are ones of the most important factors in lithosphere formation.

3. Thermal Conditions during the Formation and Evolution of the Early Earth

Widely accepted planetesimal theory of terrestrial planets formation [11,39,203,211,212,241,279, and others] is the best to explain formation of Earth as a planet. However, during Earth's formation and evolution there were various sources of energy, as well as processes contributing to the loss of energy involved. Among the most important sources of energy, scientists usually consider planetary accretion, planetary differentiation, and bombardment of a planet by huge astronomic objects (planetesimals, asteroids, etc.), and the radioactive decay of short-living and long-living radioactive isotopes [39,196,211,212,241,259,279,379]. The main forms of energy loss are heat radiation [39,107,192,204,211,212,241,259,279,379] the escape of matter caused by bombardment of a planet and the escape of some gases [39,211,212,241,279,415].

Researches show that the thermal conditions at the distance of Earth from the Sun in the solar nebula were relatively high, about 650-700 K [212] and 600°K [203]. This obviously places the conditions within the temperature range of the transformation of ferrous to ferric iron (TFFI), which takes place between 473-523 K and 723-843 K [126,356,359,360] through reaction:

$$4FeO = Fe_3O_4 + Fe. \tag{18}$$

Calculation of the temperature in the solar nebula during planetary accretion at the distance of Earth from the Sun gives for different models values of 900-1400 K [111] and 1400°K [181]. It is obvious that at this stage of the accretion process, iron(III) becomes unstable, because the upper limit of its stability is within the temperature range of 473-523 K and 723-843 K, and iron(III) would be transferred into iron(II) through reactions:

$$2Fe_3O_4 \rightarrow 6FeO + O_2. \tag{19}$$

$$2Fe_2O_3 \rightarrow 4FeO + O_2. \tag{20}$$

The presence of a gigantic amount of native iron and iron-nickel alloys in Earth's core, as well as some experimental data [242], suggests that there was also another possible reaction:

$$2FeO \rightarrow 2Fe + O_2. \tag{21}$$

It can be clearly seen that during the beginning period of the Earth's accretion and at some point of the accretion process, iron oxides transformed primarily into ferrous oxide and native iron through Eqs. (19)–(21), which played a significant role in the formation of Earth's iron core and such iron containing minerals as wüstite (FeO), Fe-perovskite ($FeSiO_3$), fayalite (Fe_2SiO_4) and ferrosilite ($Fe_2Si_2O_6$) important minerals within the mantle. It was shown earlier [359,360] that these minerals could have played an important role in the preservation of oxygen for the Earth.

Analysis of energy released by the processes of accretion, formation of the core, differentiation of Earth's slabs, heavy bombardment of the planet's surface, and energy released by the decay of short-living and long-living radioactive elements shows that the amounts of energy produced was sufficient to melt the entire planet [259,366]. At the same time, different models of planetary accretion gave different time durations of the accretionary process of from ~1 to ~100 M.y. [12,13,161,190,219,259,392,450,456,463,467]. Some models adopted of even shorter timeline for the formation of the Earth and planets, of it having taken place within ~10^5-10^6 years [260]. Some authors argue that a global differentiation was completed within 30 million years of Earth's formation [60]. Even though in both the cases of rapid and relatively slow accretion, the released energy was sufficient to melt the entire planet, the time interval of the accretion is essential for determining the initial conditions of Earth's evolution as a planet. It is the case that the release of the same amount of energy over different time intervals will cause a different initial temperature of planetary matter at all depths within the planet, because during rapid accretion the portion of energy released out into space will be much smaller and the matter of the planet would be heated to much higher temperatures than during a period of relatively slow accretion. However, it should be taken into account that in both cases the melting points of different rocks and minerals originally comprising the composition of the planet at the end of the accretion period could represent merely the minimum temperatures of Earth's layers and its surface at that time. The real temperatures of the matter in Earth layers could have been much greater than their melting points, and would definitely affect the cooling process of the planet after accretion. Higher initial temperatures would require a longer cooling time for Earth's layers and would take more time to reduce temperature within the layers to and below their solidus temperature when the process of the formation of the lithosphere began.

Birch [45] has shown that for the formation of Earth's iron core, it is necessarily so that at some point in time the temperature of Earth's layers was above that of the melting point of iron.

Some researches show that Earth at some point of its early evolution was entirely covered by a magma ocean [3,196,237,251,322,366,378,379,384,378,415,422,428,436,450,456, etc.], which was up to 1000 km deep [237,322] or even deeper [384,378]. Some scientists believe that such magma ocean could exist for about 1 to 10 M.y. [428] or less [108,134,415]. Davies [108] believes that the rate of solidification of thin surface layer of a magma ocean is in the order of a few years. Other authors report that the magma ocean could have remained for 100-200 M.y. [3,366] and even longer [422].

The estimated surface temperature of the magma ocean [3] varies from 1500 to 4700 K depending on the density of the nebula, the abundance of the dust grains, and the chemical interaction between the atmosphere and materials at the surface of the proto-Earth. Brown and Mussett [64] also show that initial temperature of the Earth surface was very high and the temperature of surface of the magma ocean could have reached a few thousand K. Cameron [69] shows that after collision of the Protoearth with Mars-size body causing Moon formation, the mean temperature would be in excess of 4000 K out to about 8 Earth radii and in excess of 2000 K out to about 20 Earth radii. The maximal temperature estimation of the Earth surface right after accretion could be much higher [366]. Only the heat from short-lived radio-isotopes alone could create a surface temperature in the range of 1200-1700 K [366].

It is generally accepted that planets of the inner Solar System experienced a heavy bombardment of debris from planetary formation, which lasted from ~4.5 to 3.8 Ga [3,15,39,85, 213,259,260,378,443] and could have led to the additional heating of Earth's crust and mantle, as well as the formation of local magma oceans. This means that during the Hadean and beginning of the Archean, the surface temperature of Earth was high and was represented by entire or local magma oceans for a significant part of these time periods.

Results of our investigation [362] indicate that during formation of the magma ocean, melts formed magmatic layers ascending in order of their density and Fe-content. This differentiation of Earth's layers began with the start of planetary accretion and it continued until the solidification of the magma ocean. Other investigators [10,12,14,15] also point to the differentiation and stratification of rocks within the magma ocean. Different researches show that even a small density differences are enough to gravitationally stratify the mantle [12,106,440,441], and that at internal chemical boundaries, the density jump is expected to be only about 30 to 100 kg/m^3 [12]. It was shown that mantle-wide convection could be expected to homogenize the mantle if the various components do not differ in intrinsic density by more than 2 or 3 % [14]. The present research also indicates that the composition of the layers of the magma ocean was regulated by the *P-T* conditions of the stability of different rocks and minerals. The light felsic and intermediate magmatic rocks formed the upper layers of the magma ocean, this is in agreement with absolute majority of these rock types in crust of the Archean cratons. More dense mafic magmatic rocks formed layers below intermediate magmatic layers. Among the mafic layers, lighter Mg-rich and Fe-poor magmas would form the upper part and Fe-rich magmas would form the lower portion of mafic magmatic layers within the magma ocean. Ultramafic magmas also formed layers within the magma ocean according to their density which increases with increase of Fe-content. The difference in the density between adjacent layers would have been significant enough to prevent the exchange of magma between them.

4. Cooling of the Magma Ocean and Formation of the Lithosphere

During the evolution of Earth in the Early and Middle Precambrian, many very interesting and sometimes unique processes and features took place. Among these processes and features there are: the formation and solidification of the magma ocean; cooling of Earth and its main layers (crust and mantle), and formation of lithosphere; the formation and content of ancient atmosphere; formation of the water ocean; the origin of life; thermal evolution of the crust and upper mantle; start of plate tectonics; time span of the formation of the main kinds of metamorphic rocks; change of chemical and mineral composition of magmatic rocks during the Late Archean – Early Proterozoic, change of kinds of tectonic activity during the Late Archean – Early Proterozoic, and etc. Each of these processes itself is quite complex and could contain different processes and features in and of itself. These features and processes should be carefully analyzed to find their place and role in the evolution of Earth as a planet. Some processes could be employed as indicators of specific conditions of Earth's evolution.

There were no solid rocks or minerals at the time of the formation of the magma ocean. At that point in time, the conditions within the magma ocean were controlled by Archimedes' and Pascal's laws. According to Archimedes' law, the magma within the magma ocean must have been layered strictly by the value of its density in an ascending order of density values with depth. Estimations of the density of mantle rocks at the present time using different geophysical methods, show that their density stretches from ~3200 kg/m^3 in the uppermost mantle to ~5500-5600 kg/m^3 near the mantle – core boundary. This difference in density is big enough for the formation of numerous layers separated by density. Different researches show that even a small difference in density is enough to gravitationally stratify the mantle [12,106,440,441], and at internal chemical boundaries, the density jump is expected to be only about 30 to 100 kg/m^3 [12]. This means that during the accretion and formation of the magma ocean, the matter representing the current composition of the mantle could have been separated into dozens of separate layers with minimum interaction between them. It is important to mention that separate convection would exist in each layer, since mantle-wide convection cannot exist if the various components differ in intrinsic density by more than 2 or 3% [14].

According to the Pascal's law, the pressure within a liquid cannot exceed its hydrostatic (in our case its lithostatic) pressure at any depth ($P=P_o$), and this pressure can be calculated using Eq. (1). The change of a unit volume of matter of the magma ocean could be calculated using Eq. 3 with condition, $\Delta P = 0$, and the value of density could be calculated using Eq. (10). It is obvious that the density of the melt will be minimal at the time of a temperature maximum within the magma ocean and that the cooling of the magma ocean would cause an increase in density. This means that during the cooling of the magma ocean, the formation of overpressure (pressure greater than the lithostatic pressure) would be impossible, since any fluctuation of pressure would immediately be unloaded according to Pascal's law. Eq. (4) shows that the generation of overpressure is possible only during the heating of solid rocks, the behavior of which are not under the influence of Pascal's law; instead, the unloading of the generated pressure will be determined only by the ability of the thermal expansion of solids, if heating is taking place. In truth, it was shown earlier [357] that at great depths any volume is under huge lithostatic pressure that prevents its expansion, which is rendered only possible if the value of overpressure generated within the volume were to exceed that of the lithostatic pressure. This means that the formation of overpressure is only possible with a forming and evolving lithosphere and only under a process of heating of the corresponding layer, of in the lithosphere or presence of an external tectonic force applied to the layer by another tectonic structure. This would also mean that prior to the first heating (re-heating) process or effects of external tectonic forces, the pressure at any depth of the magma ocean or solidified layers of the magma ocean (the forming lithosphere) would be strictly lithostatic, which would also represent the minimal possible pressure.

For analysis of the processes related to the formation of the lithosphere, a few physical constraints for the matter of the magma ocean should be taken into account: (1) magma of a denser layer within the magma ocean cannot rise within molten overlaying magmatic slabs, (2) magma of the intermediate magmatic layer cannot erupt onto Earth's surface before the overlaying layers of felsic magmas solidified, (3) magma of the basic magmatic slabs cannot erupt onto Earth's surface until overlaying layers of felsic and intermediate magmas are solidified, and (4) magma of ultramafic magmatic layers cannot erupt onto Earth's surface until all overlying magmatic layers solidified. These obvious physical constraints could be used to estimate the point in time of the solidification of different layers of the magma ocean, as well as the conditions within the forming lithosphere. Really, the age of start of formation of tonalite–trondhjemite–granodiorite (TTG) rocks, later transformed to TTG gneisses means that the uppermost felsic layers were already solid, and the earliest appearance of mafic magmatic rocks means that rocks of overlaying layers of felsic and intermediate composition were already solid at that point.

It was mentioned earlier that the most important processes for the formation of the lithosphere at that period of time obviously were of course the solidification of the magma ocean and the cooling of Earth's crust and upper mantle to or below the solidus temperature of the rocks composing them. Some scientists believe that a magma ocean solidifies from the bottom but a thin chill layer may form at the surface [3,15,456]. This belief is based on results of Petrological experiments under high *P-T* conditions showing that denser minerals would solidify first and sink to the bottom. However, our research [362] shows that within the magma ocean solidification could have started with forsterite, which has an extremely high melting point. This means that the dropping temperature within magma ocean most likely crossed the point of solidification of forsterite prior to that of any other silicate rock or mineral; it is very important to note that forsterite is a relatively light mineral and would float in some Fe-rich magmas. This means that a forsterite (or forsterite-rich peridotite) layer could have started forming first at some depth within the magma ocean. There is some data supporting the presence of such a layer in the upper mantle. However, such facts as that the peridotite component is less than 1% for Archean greenstone belts [304]; that in Archean magmatic rocks of the Superior Province of Canada, the content of olivine is less than 0.4-0.9% [144]; and that among Archean rocks, olivine is either absent or extremely rare, and is not one of the main mafic minerals [355]; point to a lack of olivine/peridotite in the uppermost mantle. This is in agreement with data presented at Figure 6 of [338] showing that for most Archean cratons garnet peridotite xenoliths were delivered from depths usually greater than ~70 km and that the quantity of xenoliths significantly increases with depth, with most xenoliths originating at 100 km or deeper. Since the delivery of xenoliths to the surface is much easier from shallower depths, it also points to a lack of peridotites in the uppermost mantle.

Investigation on *P-T* conditions of the equilibration of peridotite xenoliths show that peridotites were equilibrated under pressures of 2-6 GPa [245], 2.5–5.0 GPa, and 3-7 GPa [245], which correspond to minimum depths of about 70 km, 85 km and 100 km, respectively. This means that olivine could not be present in significant amounts at shallow depths in the mantle under the forming Archean cratons, and especially at depths less than 100 km. This is also in accordance with seismic data; Pavlenkova [331] reported a regional seismic boundary in the uppermost mantle, which she named N boundary, existing at a depth of about 100 km with a P wave velocity of 8.4-8.5 km/s. Farther research indicates that this boundary represents a thin layer which is globally observed at a depth of 80 to 100 km [332,334]. This layer is located in the thermal lithosphere beneath old platforms, at the bottom of the lithosphere under active tectonic areas and inside the asthenosphere below mid-oceanic ridges [332]. Relatively high velocities (8.4 to 8.5 km/s) are typical for this boundary both in the old platforms and in the high heat flow oceanic areas [332,334]. Results of seismic examination [335] display that at a depth of about 100 km beneath the western Siberian craton, there is a velocity inversion with a strong reflecting interface at its base; however, for the entire Siberian Platform the boundary *N* at depths of 70-120 km the most persistent and often underlies low-velocity zones [333].

It is also in agreement with geothermal data showing that the total thermal conductivity of the lithosphere passes through a minimum at depths of 50-100 km [260], as well as tha a low conductivity zone (LCZ) was formed within the top thermal boundary slab in mantle convection [451]. The potential of this low conductivity zone would cause secular cooling of the mantle for a longer period of time [451].

All xenolith, geothermal, and seismic data presented above strongly support the formation of a forsterite/peridotite layer at depths of about 100 km in the Archean cratons. This is also in agreement with the abovementioned fact that forsterite has high values of *P* waves, and that the contrasts of its seismic velocity with the velocities of rocks above and below the forsterite layer it could create LVZs above and/or below the forsterite layer. Formation of the forsterite (forsterite-rich peridotite) layer could be considered as the first step or one of first steps in the formation of the lithosphere.

On the other hand, forsterite has the highest melting point among silicates of the upper mantle. This means that the forming forsterite (or forsterite-rich peridotite) layer could take a shallow position within the upper mantle, but that the temperature of the layer would be much greater than denser iron-rich underlying layers. This would create a situation where a portion of the heat of the cooling forsterite layer would be transferred to the colder underlying layers, which would gain additional heat rather than cooling. At the same time, solid forsterite has a low heat conduction coefficient and would block the transfer of heat from the upper mantle below the forsterite layer to the surface. These processes could significantly slow down the cooling of the upper mantle below the forsterite layer, causing the bottom part of the present continental lithosphere to form at a later time with significant delay. Other studies also point to the fact that a chemically stratified mantle cools more slowly than a homogeneous Earth [e.g., 15].

It is obvious that for the cooling of Earth's surface and its upper layers, a significant amount of heat energy should be lost by them. There are three main kinds of heat transfer known: those of heat conduction, heat convection, and heat radiation. However, in the case of Earth's upper slabs we are dealing with rocks, which have very small values of heat conduction coefficient, meaning that significant heat transfer could only be provided by heat radiation and convection (in atmosphere, ocean, circulating underground waters) could provide significant heat transfer. On the other hand, a planet is isolated from the matter of the Universe by space, which does not contain any substance to facilitate either the transfer or convection of heat outside Earth. This means that the only ways in which Earth can give off heat energy out into space are either heat radiation or the escape of some matter, for example hydrogen. Since hydrogen can absorb a significant amount of heat (its specific heat capacity is ~14300 J/(kg·K)), it could be a very important means of heat loss after such causes as the decomposition of water by UV rays [209,470].

Obviously, any cooling is related to the transfer of heat. Heat transfer within a body depends on the difference between the temperatures (temperature gradient *G*) of different parts of the body and is always directed from its hot side to its cold side. For example, in the case of a magma ocean, when a layer of molten magma hundreds of kilometers thick covered Earth, the geothermal gradient within it was about 0.5 K/km [259]. The corresponding value of the geothermal gradient within such a magma ocean on Moon was about 0.08 K/km [259]. Since both heat conduction and convection are strongly depend on the thermal gradient, it can be clearly seen that there was no strong convection taking place within the magma ocean. Furthermore, convection within the magma ocean would be greatly reduced by the stratification of the magma ocean with density discussed above, which would prevent the exchange of matter between magmatic layers. On the top of that, the forming (or formed) layer of forsterite within the upper mantle would create a barrier for both heat conduction and convection. This would mean that within the magma ocean, a very small quantity of thermal energy would be transferred to the surface by both heat conduction and convection until it solidified, because if there is no significant temperature gradient there would not be significant heat transfer taking place. This means that the cooling of Earth could have been a very long process. The main transfer of heat would actually only occur from the surface of the magma ocean, and only in the case of the atmosphere being colder than the surface will there be any other heat transfer in addition to radiation.

Our analysis of the conditions of cooling for the magma ocean, comparing the rates of cooling in the mantle and crust and taking into account the fact that the cooling of Earth is mainly related to the radiation of heat from the planet's surface, and that cooling of the surface was strongly dependent on the interaction with the atmosphere and water, shows that the solidification of Earth after the formation of forsterite layer would start from the surface. This is also in agreement with the fact that the melting points of mantle rocks lying below the forsterite layer would be much lower than that of forsterite and would continue to reduce with depth, since Fe content increases with depth and the increase of Fe-content of rocks is known for the reduction of their melting points (see review in [359]).

The earliest signs of formation of the lithosphere preserved until now are rocks of: the Itsaq Gneiss Complex (the 3800–3700 Ma Isua supracrustal belt; ages ≥3850 Ma on Akilia island) in the Nuuk region of Greenland [274,315,317], the Uivak Gneisses (~3800 Ma) of northeast Labrador [89,396], the Narryer Gneiss Complex (≥3550 Ma) in Western Australia [303,316], 4030–3600 Ma parts of the Acasta Gneisses, Northwest Territories, Canada [55,431], occurrences of ~3800 Ma rocks in northeastern China [254,424].

Another group indicating some of the oldest solid compounds on the Earth are the 4.4-3.6 Ga old zircons found within much younger rocks in such regions as: Mt. Narryer and the Jack Hills, Australia [137,293,462], in quartzites of China [254] and quartzites of the Beartooth Mountains, Wyoming, USA [300].

It is obvious that these rocks and zircons are possibly not the first solid compounds formed on the Earth after accretion, but they are possibly the oldest to survive the process of Earth evolution, unless older compounds are found in the future.

Table 4. Some estimations of Earth surface temperature in the Early and Middle Precambrian

Time, Ga	Surface temperature, K	Reference
4.4	< ~573	[450]
3.5–3.2	343 ± 15	[258]
after 2.9	< 333	[258]
during pre-Archaean	573 ± 100	[414]
Late Archaean - Early Proterozoic	303 less than now	[130]
3.5–3.2	328–358(ocean water)	[220,221]
Most of geologic time	382 higher than now	[232]
Archean hydrosphere	343	[239]
~0.2 after Earth formation	~373	[265]
4.3	373	[297]
3.8	343	[297]
> 4.2	373	[293,415]
Hadean	90% of surface < 0	[301]

Most scientists agree that, even though at some point Earth's surface was represented by magma ocean, it cooled very quickly by geologic time measures. Some authors suggest that Earth was cold even in the Hadean [301] and that it is possible that Earth's surface was even frozen at ~4.6 Ga [301]. Other investigators do not think that Earth's internal heat sources could have maintained Earth's surface temperature for any significant period of time. For example, Sleep et al. [415] believe that the available internal heat within the Earth could have maintained surface temperature conditions of ~373 K for at most a couple of M.y. and more likely for much less than 1 M.y. Some examples of the estimation of Earth's surface temperature in the early Precambrian are presented in the Table 4.

Unfortunately, in most of the researches such facts as the necessary time for Earth differentiation and permanent release of energy by long- and short-living radioactive elements were not taken into account. For example, geothermal researches show that under the conditions of Earth, energy released by long-living radioactive elements would create conditions of heating of the interior of Earth [259]. This is in agreement with the fact of the formation of komatiite magmas at points of Earth's evolution, which indicate a temperature maximum in the mantle. Research shows [259] that the energy released by different sources at this time would cause the melting of most of the Moon in 1 B.y. after its accretion. Some researches show [45,259] that the temperature could have increased to that of the melting point of iron during the first 0.5-1.0 B.y. after the formation of Earth was completed. All these facts do not support the theories of a fast cooling of Earth slabs.

Since TTG gneisses are among the most abundant rock type of Early Archean rocks, it means that solidification of the upper felsic layer of Earth was completed sometimes during the end of Hadean – beginning of Early Archean and intermediate magmas got a chance to rich the Earth's surface. Another important fact to the analysis of cooling of the upper layers of Earth is the magmatic activity during the Archean, including the start of the formation of Large Igneous Provinces (LIPs). According to the data posted on the official LIPs Commission website (see http://www.mantleplumes.org/TopPages/LIPTop.html) the first LIPs were formed at about 3.0 Ga, simultaneously with the first world-wide pulse of granulites formation. The total volume of magmas in LIPs significantly raised about 2.8-2.7 Ga and reached its maximum at about 2.5-2.4 Ga [362]. Condie [94] recognized the Ventersdorp (2.7 Ga; Kaapvaal craton, South Africa) and Fortescue (2.77 Ga; Pilbara craton, Australia) continental flood basalts the oldest well-documented ones. Among the most important features and processes of the evolution of Earth in the Late Archean, are the: (1) maximum temperatures in the mantle corresponding to the widespread formation of komatiites in the Middle and Late Archean, with a peak of komatiites formation at around 2.8-2.7 Ga [16,17,33,439], (2) the worldwide magmatic event during the Late Archean (2.75-2.65 Ga) [31,94,95], and (3) the second world-wide pulse of granulite metamorphism starting from about 2.7 Ga [342]. These events were followed by the superplume events of 2.51-2.45 Ga and around 2.25 Ga [31,38,168,359], and the beginning of the formation of Early Proterozoic dyke swarms started near the Archean – Early Proterozoic border [27,65,66,285].

These obviously tremendously powerful magmatic activities and thermal events could have only two possible explanations: (1) the upper mantle portion of the magma ocean had not yet solidified, or (2) the upper mantle solidified and was then reheated and melted once again.

In the first case, it is clear that the lithosphere would have been very thin and its base would likely be very hot. In the second case, the lithosphere would be thicker, but would have its base re-melted during the peak of magmatic and granulite activity at around 2.7 Ga.

Table 5. Cooling rates of Earth's mantle estimated by different authors or calculated using their data

Cooling rate, K/Ma	Reference
0.04	[195]
0.06	[229]
0.10 – 0.12	[115]
0.13	[442]
~0.16	[377]
~0.03 (for continents)	[140]
~0.06 (for no continents)	[140]
0.08	[162]
0.09	[113]
~0.04 (today's cooling rate)	[446]
0.05 (mean for last 3 Ga)	[446]
0.20 (today's cooling rate)	[93]
~ 0.12-0.15 (Archean mean cooling rate)	[366]
~0.08-0.09 (post-Archean mean cooling rate)	[366]
~0.08-0.09 (at mantle-core border)	[469]
0.04-0.10 (for South African lithosphere)	[36]
0.03 – 0.09	[340]
0.14	[377]
0.10	[451]
0.04 – 0.06	[98]
0.04-0.05	[1]

Table 6. Cooling rates of some magmatic and/or metamorphic complexes from published data

Complex, area	Cooling rate, K/Ma	Reference
Halls Creek Orogen, W.Australia	1.4-1.5	[53]
Mazury complex, NE Poland	~2.0	[117]
Las Termas belt, NW Argentina	3.0-4.0	[174]
Cherry Valley, New York	0.38	[198]
Ruhr Basin, Germany	0.1-0.2	[206]
Grenville Province, W. Québec	6.0	[272]
Central and Eastern Pyrenees	3.7	[298]
Witwatersrand Basin, South Africa	1.4	[325]
4 orogenic belts of India	0.2-25.0	[407]
Urach, SW Germany	1.0	[471]
Adirondack Highlands, New York	4.0	[197]
Grenville Province	1.0-3.0	[339]
Grenville Province	1.0-3.0	[289]
Proterozoic Adirondack terrane	0.5-1.0	[276]
Pikwitonei granulite domain, Canada	0.5	[276]
Lowlands and the Highlands, Adirondacks	1.0-2.0	[435]
metamorphism in the W Canadian Shield	~1.0	[30]
Manzano Mountains, New Mexico	1.0	[169]
Meteorite	1.0	[411]
Asteroid	1.0	[385]

All of the above points to the first significant appearance of mafic magmas sometime between 3.0 Ga and 2.7 Ga, and the appearance of komatiites (the first ultramafic magmas to appear) at about the same time with a clear maximum at about 2.8-2.7 Ga. This would indicate that by this time the uppermost layers represented by layers of felsic and intermediate composition, were solidified. Meaning that a thin young lithosphere did in fact exist at this point of Earth evolution.

Takahashi [442] has shown that that the komatiites of the latest Archean may have been produced under conditions ≥4 GPa (corresponding depth of greater or equals to about 120 km) and ≥1925 K. Some authors point out that komatiites could possibly have been formed at even greater temperatures of about 1973–2123 K [439], at depths of about 160–180 km. Since komatiites have the highest melting temperatures among all known magmatic rocks, the time of their formation can be viewed as an indicator of a maximum temperature in the upper mantle.

A comparison of the cooling rates of mantle published or calculated using available data are presented in Table 5.

Estimations of the cooling rate for Mars and Venus gave values of 0.25 K/Ma and 0.10 K/Ma respectively [452].

Some data on the cooling rates of different magmatic and/or metamorphic structures of local/regional significance are also presented (see Table 6) for comparison with the cooling rates of Earth's surface and mantle. These data (Table 6) could represent the cooling rates of the crust in different regions and at different time periods. Cases of extremely fast cooling such as the cooling of magma in underwater conditions were not included in this table, because at the time of the cooling of the magma ocean there was no water ocean present.

Some authors present results of computing showing that cooling rates are higher for high temperatures and much slower for lower temperatures [84,280,294,324,385,472,474]. For example, for the Adirondack Highlands very fast (>200.0 K/Ma) [434], slow (4.0 K/Ma) [197] and very slow (0.5-1.0 K/Ma) [276] cooling rates have been reported.

Table 7. Age of the Late Archean komatiites in different regions of Earth

Region	Age, Ga	Reference
Australia:		
Black Swan area, Yilgarn Craton	2.7	[173]
Kalgoorlie Terrane	2.707-2.666	[32]
Kurnalpi Terrane	2.715-2.698	[32]
Laverton Terrane	2.808	[32]
Yilgarn Craton, basaltic komatiites	2.716	[305]
Yilgarn Craton	2.705	[306]
Fortescue Pilbara craton	2.770	[94]
Africa:		
Rhodesia	~2.6 by	[167]
Reliance Fm., Belingwe Belt, Zimbabwe	2.7	[375]
Ngezi, Belingwe Belt, Zimbabwe	2.7	[412]
Klipriviersburg Group of the Kaapvaal Craton	2.714	[305]
Brazil:		
Crixás greenstone belt, Goiás	2.728	[18]
China:		
west Shandong, North China	~ 2.7	[81]
Canada:		
southern Abitibi belt, Ontario	~2.7	[102]
Lake Abitibe – Kirkland Lake region	2.713	[103]
Dundonald Beach, Ontario, Canada	2.7	[142]
Kidd-Munro	2.715	[142]
Alexo area	~ 2.7	[142]
Alexo Township	2.762	[142]
Whitney Township	2.7	[217]
Belts in Canada	2.7	[52]
Pyke Hill	2.7	[373]
Rae craton	2.730-2.700	[129]
Abitibi Subprovince	2.705	[306]
Abitibi greenstone belt	2.710-2.725	[464]
Baltic shield:		
Kostomuksha Greenstone Belt	~2.8	[372]
Lion Hills, Central Vetreny Belt	2.41	[371]
Kola Peninsula	2.7-2.8	[439]
Kellojärvi complex, Kuhmo greenstone belt	2.757	[328]

A comparison between the cooling rates of the mantle (see Table 5) and those of different structures, massifs, and belts within the crust (see Table 6) shows that in most cases the cooling rates of crustal structures are significantly higher than the cooling rates of the mantle. This is only to be expected, for while the crust is cooled by surface heat radiation, water (oceans and underground water), and the atmosphere; cooling of the mantle depends primarily on the heat conduction of rocks and magmatic activity. Since the temperature difference with depth was very small at the time of the magma ocean, the geothermal gradient was almost negligible. However, the temperature difference with depth drastically changed over time, resulting in the average geothermal gradient of about 30 K/km near the surface with its value reducing with depth that we have today. This change in the geothermal gradient from that time until now could only have been caused by higher cooling rates of the crust compared to the mantle.

Comparing these rates and volumes of crust structures with the volume of the hundreds of kilometers deep magma ocean, covering the entire Earth, shows that the solidification and cooling of such a magma ocean could not have been a mere of years, thousands of years or even a few million years, as some authors propose (see above), and could have been a very long process. The cooling process could have been extremely slow in the early periods of Earth's evolution, because prior to 4.4 Ga the heat produced by radioactive decay was approximately 3-6 times greater than it is today [236,259,366], and short-living radioactive elements also released significant amounts of energy during the early at that time [259]. Estimations of the energy generated by radioactive decay show [259] that the magnitude of energy released by radioactive elements during the evolution of Earth was greater than that of the energy Earth lost through heat conduction, and that a significant time interval of early Earth evolution was characterized by a period of heating of Earth's interiors. Different geothermal models show a rapid rise in the temperature of the central region in the course of the first few billion years [259,260].

One of the cases attesting to the extreme temperatures in the mantle, as it was mentioned above, is the appearance of komatiites in the Middle and Late Archean. High-magnesian komatiites, which were formed at the highest temperatures (about 1973–2123 K) were developing in most Archean cratons. Analysis of the temperatures of high-magnesian magmas formed in the Archean and the Early Proterozoic accomplished by Svetov and Smolkin [439] for the Fennoscandian Shield points to a decrease in the temperatures of magma formation from the Archean (1973–2123 K) to the Early Proterozoic (1573 K). Using results of this analysis estimated values of mantle cooling rates for the Fennoscandian Shield during the Archean – Early Proterozoic are in the order of 0.5-0.7 K/Ma. It was also shown based on analysis of komatiites that a maximum temperature (2118 K) in the Archean mantle of the Karelian craton was also reached at ~3.0 Ga [224,225]. The fact of the simultaneous appearance of komatiites at about the same time in most Archean cratons (see Table 7) indicatess that this temperature maximum within the upper mantle was world-wide. The ages of the formation of Late Archean komatiites are presented in Table 7.

Table 8. Thermodynamic conditions of some metamorphic processes

Type of metamorphism	Average T, K	Average P, GPa	Average depth of lithostatic pressure, km	Average geothermal gradient (K/km) for present surface conditions	Average P/T^*, MPa/C
Eclogite (n = 505)	858	1.90	~64	9.1	3.25
Blueschist (n = 225)	677	1.00	~35	11.5	2.48
Granulite (n = 543)	1075	0.82	~28	28.6	1.02
Amphibolite (n = 353)	925	0.77	~26-27	24.1	1.18
Greenschist (n = 188)	681	0.42	14-15	27	1.04

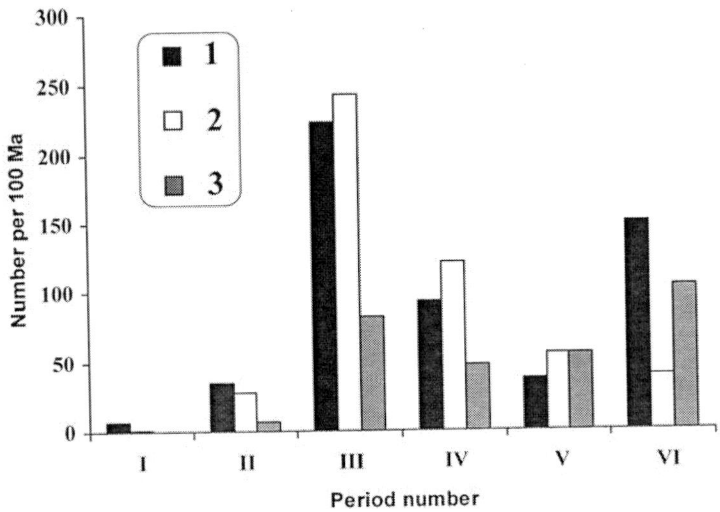

Figure 2. Number of world greenschist (1), granulite (2), and amphibolite (3) metamorphic processes per 100 Ma during different periods of the Precambrian (I) Early Archean, (II) Middle Archean, (III) Late Archean, (IV) Early Proterozoic, (V) Middle Proterozoic, (VI) Late Proterozoic.

It is possible to use data on metamorphic processes for the analysis of the thermal conditions of the lithosphere and its components the crust and the upper mantle, since the metamorphic processes of different metamorphic facies have very specific P-T conditions [355,358,359] which could characterize the conditions of the lithosphere. Analysis of the age of the appearance and P-T conditions of the most important facies of metamorphism shows [355,358,359] that the oldest metamorphic processes were related to greenschist metamorphic facies. The second and third oldest metamorphic processes to appear were the granulite metamorphic facies and the amphibolite facies of metamorphism, respectively. At the same time, metamorphic rocks of eclogite facies are not known to have existed prior to the Early Proterozoic and rocks of blueschist facies are unknown prior to the late Proterozoic

[359,360]. This fact of the absence of high-pressure metamorphic processes during the Archean points to high temperature conditions within the lithosphere at this period. The relative abundance of some metamorphic processes through time is presented in Figure 2. The results of the analysis of *P-T* conditions of different metamorphic processes are compiled in Table 8.

The data from Table 8 illustrate that the *P-T* conditions for the formation of such rocks as granulites and eclogites are completely incompatible and that these rocks cannot be formed at the same time in the same place. This fact is further proof that results of experimental petrology should be used very carefully for the interpretation of natural processes, regardless of transformations between granulites and eclogites having been achieved in different experiments [151,187]. Moreover, garnet could be formed at pressures and temperatures much lower than those in kimberlites. Depending on temperature, garnet appears in experiments at pressures between 0.96-1.52 GPa [380], and also at ~1373 K, 2.1 GPa; 1573 K, 2.4 GPa and 1773 K, 3.1 GPa [150].

Since granulites are metamorphic rocks formed at the highest temperatures and during the Early and Middle Precambrian there were three world-wide pulses of granulite formation, at ~ 3.0 Ga, ~2.7 Ga, and ~ 2.0-1.8 Ga we decided to analyze the conditions of granulite facies of metamorphism through time in greater detail (see Table 9). In addition to that, some ages of granulites formed in the Archean cratons during the second world-wide pulse of granulite metamorphism are presented in Table 10.

Table 9. Thermodynamic conditions of metamorphic processes of granulites of different origin (modified after [355,358,359])

Rock (n)	Average *T*, °K	Average *P*, GPa	Average *P/T*, MPa/K	Average depth of lithostatic pressure, km	Average geothermal gradient (K/km)
Granulites average ($n = 543$)	1073	820	1.02	~28	28.6
Granulites of:					
Early and Middle Archean	1059	880	11.42	30.3	25.3
Late Archean	1127	871	8.7	30.0	27.8
Early Palaeo-Proterozoic	1040	667	8.63	23.0	32.6
End of Palaeo-Proterozoic	1087	825	10.1	28.4	28.0

The data of Table 9 point at the maximum of thermal conditions at the Late Archean, and it can be clearly seen that the average temperature at depths of about 30 km actually significantly increased between the first and second world-wide pulses of the granulite metamorphism. If we compare the average depths of the formation of granulites with the average geothermal gradients of these periods of time, the data also shows that the temperature continued to grow and reached a maximum at the Early Paleo-Proterozoic, and continued to decline to the end of and past the Paleo-Proterozoic, known for the appearance of komatiitic and picritic magmas, and the third world-wide pulse of granulite formation at about 2.0-1.8 Ga. data on the conditions of the formation of granulites points to ultrahigh temperature conditions reaching 1273-1373 K, and in some cases granulitic rocks show signs of partial melting. Such conditions are reported for: the Eastern Ghats Granulite Belt of India [405,410], Minto block of Northern Quebec [37], the South Indian and Sri Lankan granulite terrains [230], the Napier Complex of East Antarctica [175], the Anápolis–Itauçu Complex of the Brasília Fold Belt [296], and many others. In some cases temperature reached 1223 K at depth ~26 km [37], 1273 K at 30 km [230], 1273-1383 K at 28-29 km depth [175]. These facts show that conditions with temperatures ≤1573 K allowed for maximum temperatures within the lithosphere could have reached then at depths even less than 70-100 km, or there were pockets of magma above the forsterite layer.

Table 10. Age of granulite belts of the second worldwide pulse of granulite formation

Region, tectonic unit	Age, Ga	Reference
Baltic shield:		
Pudasjärvi Granulite Belt	2.66	[302]
Jonsa block, Central Finland	2.63	[177]
Belomorian Belt	~ 2.65-2.60	[54]
Belomorian Belt	2.7	[138]
Kola peninsula	2.75±0.05	[445]
Karelia	2.65	[256]
Murmansk terrain	2.7-2.8	[29]
Belomorian Mobil Belt	2.713	[41]
Africa:		
Limpopo Belt	~ 2.69	[418]
South Marginal Zone of the Limpopo belt	2.69	[341]
Limpopo Belt	~ 2.72-2.59	[51]
Kaapvaal Craton	2.715-2.720	[400]
Mkhondo suite, Swaziland	2.75	[97]
West African Craton	~ 2.7	[367]
Uganda	2.66 (?)	[26]
Kasila group, Sierra Leone	> 2.7	[26]
Amsaga Area (Reguibat Rise)	2.73	[367]
India:		
Sausar Mobile belt, Southern granulite belt	2.672	[383]
Bhandara-Balaghat granulite	2.672±0.054	[374]
S. India	2.50-265	[309]
Dharvar metasupracrustal rocks	2.66	[145]
Highlands, S. India	2.6	[145]
China:		
Qianxi Group, Eastern Hebei	2.48-2.59	[110]
Longquanguan Group, Taihang-Wutai Mount.	~ 2.56	[156]
Taihua Complex, Henan	< ~2622	[193]
Qianan region, Hebei	2.7	[327]
Qianxi and Qian'an granulite belt east. Hebei	2.756	[327]
Qingyuan granulite area, N. Liaoning	2.65-2.61	[327]
Granulites in Bengbu, Anhui	2.65	[327]

Table 10. Continued

Region, tectonic unit	Age, Ga	Reference
Canada:		
Snowbird Tectonic Zone, N. Saskatchewan	2.60	[420]
Pikwitonei domain	2.744-2.59	[342]
Pikwitonei domain	2.641-2.648	[342]
Kapuskasing Structural zone	2.585-2.650	[342]
Kapuskasing Structural zone	2.66-2.64	[234]
Minto block	2.725	[342]
Ashuanipi Complex	2.69	[342]
Pikwitonei granulite domain	2.744-2.59	[277]
Natawahunan Lake	~2.64	[277]
Slave Province	2.64-2.58	[109]
Orma domain, Churcill Province	2.58	[194]
Crossroads domain, Churcill Province	2.60	[194]
Western English River Subprovince	2.66-2.76	[34]
Kapuskasing Structural zone	2.68-2.62	[70]
Pontiac	2.69	[70]
Wabigoon	2.71-2.695	[70]
Winnipeg River	2.7	[70]
Bird River	2.7	[70]
English River	2.68	[70]
Winnipeg River Subprovince	2.68	[104]
Wawa gneiss domain	2.66-2.637	[299]
Hopedale block, Labrador	2.74-2.78	[397]
Saglek block, Nain Province, Labrador	2.74-2.78	[398]
Minto block	2.7	[343]
Ungava Orogen	~ 2.7	[433]
Winnipeg River Subprovince	2.68	[101]
Kapuskasing uplift	2.66-2.58	[344]
N. Quebec basement, Trans-Hudson Orogen	2.7	[433]
Greenland:		
Nuuk–Maniitsoq, Saglek block	2.738	[130]
Buksefjorden	2.7-2.8	[309]
Tasiusarsuaq terrane	2.795	[90]
Tasiusarsuaq terrane	2.69-2.61	[90]
UK		
Scourian Complex	2.49-2.70	[319]
Lewisian Complex	2.6-2.8	[145]
Eastern and Northern Siberia:		
Daldyn granulite-gneiss terrane	> 2.5	[416]
Central Aldan granulite-gneiss terrane	> 2.5	[416]
Anabar Granulitic Complex	2.7	[382]
USA:		
Teton Range, WYOMING	2.675	[179]
Wind River Range, Wyoming	2.705	[80]

Table 10. Continued

Region, tectonic unit	Age, Ga	Reference
Australia:		
Yilgarn Craton	2.640-2.649	[307]
the Lake Grace Terrain, Yilgarn Craton	2.640-2.649	[120]
Wheat belt	> 2.6	[309]
North Western Gneis Terrain	2.650-2.600	[423]
southwestern Yilgarn Craton	2.64	[461]
South America:		
Jequié, Brazil	≥ 2.6	[188]

The data of Table 9 llustrate that the effective elastic thickness (*Te*) of the continental lithosphere (with temperatures between 573 and 873 K) within the Archean cratons during these periods of granulites formation was below 9-12 km, and that the total thickness of the young lithosphere was ≤100 km, including the forsterite layer with possible pockets of magma above it. This corresponds with the data of Takahashi [442] showing that the latest Archean komatiites may have been produced at depths of about 120 km. It is also in agreement with data of Doucouré and de Wit [119] showing that the lithosphere beneath the Kaapvaal craton attained its thickness of up to 120 km only after 2.0 Ga. Results of geothermal modelling indicate [260] that the smallest thickness of the solid surface layer in the period heating was at its maximum was about 10 km. Such high temperatures and so low an effective elastic thickness of the lithosphere would cause it to lose some or most of its strength and rigidity, and such lithospheric plates would not be able to be involved in processes of plate tectonics at this time. This is in agreement with data of Artyushkov et al. [24] on the loss of lithospheric rigidity in areas with high temperatures and a low effective elastic thickness. The strict stratification and differentiation of the matter of the forming crust and upper mantle during the formation and evolution of the magma ocean would also prevent plate tectonic activity in the Archean, which corresponds to data showing positive buoyancy of the upper mantle in the Archean cratons [21,23,201,329]. Another problem for the operation of plate tectonics during the Archean is related to the low Fe-content of the Archean rocks. This would prevent the formation of the highly dense rocks required for the creation of negative buoyancy. Some researches accept eclogite with density 3500 kg/m^3 as a rock capable of creating negative buoyancy of the lithosphere (see for example [343]). However, it is known that most eclogites from kimberlites, which some researches believe were formed during the Archean, are highly magnesian [149], and should then be less dense than Fe-rich eclogites. In most publications, the presented values of the density of eclogite are less than 3500 kg/m^3. For example, Hall [160] presents an average density of eclogite of 3390 kg/m^3 with a density range of 3340-3450 kg/m^3, and Kappelmeyer and Haenel [204] presented a value of density of eclogite of about 3200 kg/m^3. Data presented in [86] illustrate that only for eclogites of the California and Norway Caledonides is the density greater than 3376 kg/m^3. Meaning that only Phanerozoic eclogites have high densities. This is related to the high content of dense almandine, which is rich in iron. Taking into account that dense eclogites contain high amounts of both almandine and aegerine, while Archean rocks are known for the lowest content of iron and that during the Archean significant amounts of iron were removed from rocks and deposited as banded iron formations (BIFs), it is clear that there was not enough iron in the Archean rocks for the formation of iron-rich and dense eclogites. This means that even if there were conditions favorable to the formation of some eclogites, those would be Mg-rich and would not be very dense. On top of that massive eclogites are not known to have existed in the Archean age [360]. All these facts make the operation of plate tectonics during the Hadean and the Archean highly unlikely.

The performed examination [362] indicates that even though mafic magmas in significant amounts began to appear on continents only from the Late Archean, the FeO-content of these magmas was minimal in the Archean (FeO content <10-11%) and increased to 11.5-13.7% in the Proterozoic-Mesozoic, and to 12.25-14% in the Tertiary. The FeO-content of oceanic magmas is significantly lower at about 8.3-9.1% for mid-ocean ridge basalts, 8.5-9.7% for island arc basalts, 10.2% for ocean floor basalts, and 10.6-12.6 % for oceanic intraplate volcanism. This means that the post-Archean lithosphere would contain more iron and be denser than that of the Archean, which could play an essential role in the formation of the negatively buoyant lithosphere that is required for the start of plate tectonics. The thermal conditions of such processes as the formation of banded iron formations (BIFs), paleosols, red beds and pyrite in details discussed in [360,361]. It was shown that the formation of BIFs, paleosols and red beds requires the transformation of ferrous iron into ferric iron (TFFI), which usually takes place under temperature conditions between 723-843 K (the upper limit for TFFI) and 473-523 K (the lower limit for TFFI). These conditions would therefore be necessary at the time of the formation of these rocks.

Abe [3] remarks that Earth's surface temperature and cooling rate were both highly dependent on whether a rock vapor or steam atmosphere was present during cooling. In the absence of an atmosphere, heat radiation to space is very efficient and the deeper portion of the magma ocean would cool rapidly (10^3 years) [3,456], while the upper portion of the magma ocean (e.g., transition zone and shallow upper mantle) could have remained hot and molten for much longer. However, if an atmosphere persists, it would provide very effective thermal insulation making cooling timelines much longer (10^6–10^7 years) even for deeper portions of the magma ocean [3,456]. Other researches also show that a dense primitive atmosphere and buoyant outer layers are effective insulators and would serve to keep the crust and upper mantle from cooling [4,470] that a massive proto-atmosphere may have absorbed a significant portion of the re-radiated energy and raised the surface temperature by several hundred degrees.

Modeling of impact processes shows that the de-volatilisation and heating associated with impact would generate a hot, dense atmosphere that would serve to keep the surface temperature very high and to trap solar radiation [15] and that repeated impacts would sustain a magma ocean especially in the presence of a blanketing atmosphere [456].

All of this and the fact that the water ocean and the atmosphere are among the most important coolants of Earth's surface mean that both the composition and conditions of the early Earth's atmosphere are essential for any investigation of the cooling of the magma ocean.

5. Early Atmosphere and Possible Conditions for the Formation of the Water-Ocean

There is a great deal of controversy in models of the early Earth atmosphere composition offered by various authors. Most scientists agree that the early atmosphere was essentially anoxic [40,88, 210,223,239,292,320,336,337,368,389,394,448].

Also, many scientists believe that the concentration of carbon dioxide in the early atmosphere was much greater than it is today [2,207,208,258,297,321,436]. Some researchers believe that the early Earth atmosphere was rich in water steam [2,28,366,470], and there are researches attesting to high concentrations of both carbon dioxide and water vapour in the early atmosphere [239,314,359,415,450]. In addition to that, some scientists believe that the early atmosphere was also rich in methane [75,172,321]. Interestingly, all of these gases are among the most important greenhouse gases. Estimation of the amount of carbon dioxide and water vapour present in the early Earth atmosphere gave significant amounts for the concentrations of these gases, which are pretty small in Earth's atmosphere today.

Estimations of partial pressure generated in the early Earth's atmosphere by the different major components are presented in Table 11.

Table 11. Estimations of partial pressure generated by carbon dioxide, water and methane

Atmosphere component	Age	Partial pressure, MPa	Reference
Carbon dioxide, CO_2	Hadean	1	[297]
	after finish of accretion	a few MPa	[2]
	first several hundred Ma	1-2	[208]
	early Earth atmosphere	10-100	[436]
	early Earth atmosphere	1	[457]
Water vapor, H_2O	Hadean	all water was vaporised	[366]
	right after accretion	20-30	[2]
	early Earth atmosphere	10-30	[470]
	early Earth atmosphere	27	[450,470]
Methane, CH_4	early Earth atmosphere	Methane rich	[75,172,321]

In contrast Kasting and Ackerman [208] think that there was no significant amount of water in the atmosphere and that the early stratosphere should have been dry.

Some scientists believe that oceans on Earth were formed as early as by ~4.5 Ga [92], in Hadean [297], at ~4.4 Ga [255,340,450,462], ~4.3 Ga [257,292], by 4.2 Ga [292,415], by 3.85 Ga [318], by 3.8 Ga [239,311], by 3.6 Ga [152], ~3.5 Ga [221,314]. The oldest zircon of 4.4 Ga so far found in the Jack Hills area, Yilgarn Craton of Western Australia is now held as proof of the presence of a continental crust and oceans at the time of its formation based on its $\delta^{18}O$ values [340,462]. Analysis presented in [462] demonstrates that the zircon is zoned with respect to rare earth elements and oxygen isotope ratios ($\delta^{18}O$ values are varying from 7.4 to 5.0‰), and that some data points at its growth from a granitic melt.

However, of course, the main evidence to yield a time stamp for the formation of the water ocean lies with the features related to the formation of different kinds of rocks and rock layers that could be formed exclusively or almost exclusively in an ocean environment. Among such features, there are the formation of sedimentary layers, evaporite layers, the presence of oceanic crust fragments (ophiolites, peridotite layer), and etc. One of the most important features related to the ocean environment is the formation of carbonate rocks and carbonate layers, which could be formed by both organic and non-organic means. This means that the fact of the formation of such kinds of sedimentary rocks and fragments of oceanic crust could unambiguously indicate the point in time of the formation and existence of the ocean.

Analysis of different facts of the formation of early sediments allowed Condie et al. [96] to conclude that there is an absence of clear evidence demonstrating the existence of oceanic sediments older than of the Proterozoic age. Nutman et al. [315] show that there are no known sediments older than 3.9 Ga. Since latest researches show that plate tectonics did not operate before the Early Proterozoic [15,50,111,128,162,163,248,406], the absence of Archean sediments within Archean cratons cannot be explained by the reworking of sediments in subduction zones. Researches show that evaporites older than 1.8 Ga are also absent [96,153], excluding barite and carbonates (which are treated by some scientists as evaporites). Huston and Logan [184] demonstrate that before 3.2 Ga bedded barite deposits are relatively common, and then reappear after 1.8 Ga. Other research by Melezhik et al. [282] shows that evaporites appear as early as 2.1 Ga. It is also a known fact that peridotites, ophiolites, and serpentinites older than the Early Proterozoic age are found to be absent [114,170,355,357,465]. Some researches point at existence of the Archean carbonates. Carbonate rocks with geological attributes of marine sediments are a minor component of the Archean greenstone belts, and the greenstones containing such carbonates appear to cluster at ~ 2.8 and ~ 3.5 Ga ago [453]. Some researches [437,438] state that Late Archean carbonate platforms are rare, while large crystal pseudomorphs composed of limestone and dolomite are common; these radiate upwards to form centimeter- to meter-tall fans their depositional surfaces are coated with centimeter- to decimeter-thick calcite. However, most of these researches were conducted without analysis of the thermodynamic conditions of the formation and decomposition of carbonates.

Research shows that limestone decomposes at temperatures of 973 K [310], 1098–145 K [273], 1053 K [444]. The decomposition of calcite takes place at temperatures of 923–1173 K 923–1173 K [290], 1073–1193 K [231], 993–1093 K [409], 1158 K [48], by 773 K [402]. Decomposition of dolomite takes place at temperatures of 873–973 K [409], 773–873 K [408], and 773 K [48]. Magnesite decomposes at temperatures of 646 K [48]. Siderite decomposes at temperatures of 653–923 K [227], and 723–773 K [409].

The decomposition of K_2CO_3 takes place at 1173 K and higher [290], and the decomposition of Na_2CO_3 takes place at temperatures of 1124 K [49]. It should be stated that the temperatures presented in different researches should be viewed as the maximal values of the decomposition temperature, since all of these researches were considering mostly a fast and quickly visible decomposition of carbonates. For example research accomplished by Koziol [226] was considering only the decomposition of siderite, magnesite, and siderite-magnesite carbonates within 5–15 minute periods under temperature 743 K; and the research revealed extensive decarbonation for pure siderite while only a tiny amount of its decarbonation was found in other compositions. However, it is clear that tiny decomposition in 5–15 minutes means that over millions and billions of years, the decomposition could be completed.

All this data shows that at the time of the formation of the magma ocean, none of these carbonates had a chance to survive and must have decomposed releasing CO_2, which was joined the other components contained within the early Earth's atmosphere. Estimations by different authors show that Earth currently has enough carbon dioxide bound in carbonate rocks enough to generate an atmospheric partial pressure between approximately 6MPa [176,381] to 21.5 MPa [473]. It is clear that all this carbon dioxide existed in the atmosphere above the magma ocean. It is also obvious from centimeter- to decimeter- or even meter-thick Archean carbonates that significant amounts of carbon dioxide were still present in the Archean atmosphere.

At the same time, the temperatures of the formation of carbonates are significantly lower than those of their stability and decomposition. Kent et al. [214] demonstrated that carbonates on Mars were formed at temperatures no higher than 673 K and most probably less than 473 K. Experimental results of Stoessell et al. [432] show that between 373 and 473 K, the mode of calcite formation changed from direct precipitation into void space produced by the dissolving of dolomite into a pseudomorphic replacement of dolomite, and that dedolomitization textures formed in these experiments are similar to those found in the rock record. Bech et al. [35] estimate the temperature of the formation of carbonates at 375 K. Authors of [345] report that with the increase of the hydrogenation temperature, the temperature of carbonate formation could increase to 653-713 K.

One of the compounds accounted as a marker of anoxic atmosphere environments in the Early and Middle Precambrian is pyrite. Analysis of latest publications on the thermodynamic conditions of the formation and decomposition of pyrite shows that pyrite decomposes at temperatures of 743 K in a CO_2 atmosphere [143]. Since, as it was shown above, the early Earth's atmosphere was rich in CO_2, it is obvious that pyrite could not have formed at the time of the magma ocean, nor until the temperature of the atmosphere and surface of the solidified magma ocean dropped below the maximum temperature of pyrite stability. Different researches show that pyrite decomposes under high temperature conditions such as: 523–1016 K [240], 673-863 K [178], 930 K [91], and ~740 K at Venus [403]. Research also shows temperatures of the formation of pyrite are between 601 and >743 K [270], up to 1015 K [63], 653–773 K [147], and 875 K [425]. The temperature of pyrite formation in gold-bearing veins and gold deposits is ~623 K [139], 563-653 K [466], 623 K [164], ~723 K [330], and ~773 K [253]. In gold deposits, temperatures in the range of 548 K to 623 K are related to the formation of sulphide deposits (including pyrite) [136,154,347]. At the same time research shows that pyrite was not produced experimentally at temperatures in the range of 298 and 373 K [262]. All these data means that elemental sulfur, which has a boiling point as high as 717.75 K [252,286], would exist in the atmosphere above the magma ocean as a gas. It was shown earlier in Eqs. (19)–(21) that during accretion and formation of the magma ocean, some oxygen was released in the transformation of ferric oxide into ferrous oxide and the decomposition of the ferrous oxide. This released oxygen could have reacted with the elemental sulfur forming SO_2 and SO_3 oxides, which then also compose part of the early Earth atmosphere. This is in agreement with the fact of the formation of barite during the Archean before 3.2 Ga [184].

It is known that carbonates are mostly formed in oceanic water, therefore the formation of the water ocean is necessary first. However, formation of a liquid water ocean is under constraints of its boiling point under the corresponding pressure of the atmosphere and its critical point [359]. Some researches accomplished earlier gave different values of estimated temperatures for the formation of the water ocean. Valley et al. [450] report that surface temperatures were below ~473 K at 4.4 Ga and that this was low enough to cause partial or complete condensation of the steam atmosphere to form oceans. They suggested [450] that the volume of today's ocean water would require temperatures above that of the critical point of water (647.2 K, [425]) in order for it to be completely vaporised. Jacobsen and Pimentel-Klose [191] believe in the existence of hot hydrothermal waters (> 648 K) during the Archean. Sauer and Yachandra [395] point to the existence of high-temperature springs (~673 K) and oceans on early Earth. Liu [255] believes that the oceans started to form when the surface temperature of the Earth cooled to below approximately 573–723 K. Nisbet and Sleep [311] also believe that temperatures up to ~623 K are needed to convert the whole ocean into steam. Other research shows [415] that it is possible to form liquid water at temperature over 473 K, and that as soon as the temperature of the atmosphere dropped to perhaps 523 K, water vapour would begin to rain down. Another research states that the Archean ocean temperature was about 623-673 K [391].

Unfortunately, in most of these estimations of the temperature of ocean formation such thermodynamic characteristics as the critical temperature of water (~647.2 K, [425]) and dependence of the boiling point of water on pressure were not taken into account. The boiling point of water for different values of pressure using data [425] and *P-T* conditions for the density of water 958.36 kg/m^3 (its density at the boiling point 373 K and pressure one atmosphere, see [425]) calculated using data of [86] are discussed in [359]. Pilchin and Eppelbaum [359] show that based on the analysis of thermodynamic properties of water and thermodynamic conditions of the early atmosphere, formation of the water ocean took place when the temperature of the atmosphere and that of Earth's surface was between 395 and 615 K.

It is also obvious that at the time of formation and evolution of the magma ocean no liquid water had a chance to exist on the surface, and it was in its gaseous state in the atmosphere.

Taking into account all these facts it is evident that during the formation and evolution of the magma ocean there was a thick and dense atmosphere above it. Estimations of the pressure generated by this atmosphere gave a minimum value of about 35 MPa [359]. This estimation represents the minimal possible pressure of the early Earth's atmosphere, because the amount of water in used for the calculations was that of the present ocean and subsurface underground water. The estimated pressure is a minimum also because some researches show that the content of water on Earth have been that of about 5–6 Earth oceans of today [121,323] and even up to 50 Earth oceans [5,121]. Such a quantity of water would have dramatically increased the atmospheric pressure at the time of the magma ocean. However, even this minimum pressure of the atmosphere is much greater than the critical pressure for water (21.78 MPa). This means that a significant part of the water within the early atmosphere was under supercritical conditions, because since both temperature and pressure within the atmosphere above the magma ocean were higher than those of the critical points of temperature and pressure for water. It is important to state that other key components of the early Earth's atmosphere (SO_2, CO_2, etc.) were also under supercritical conditions.

Since H_2O and CO_2 were by far the primary components of the early Earth's atmosphere, an analysis of the densities of water and carbon dioxide near Earth's surface was conducted. The densities of H_2O and CO_2 with the increase of temperature for values of pressure at 35 MPa were calculated (Figure 3) using the published data [86,215, 252,425].

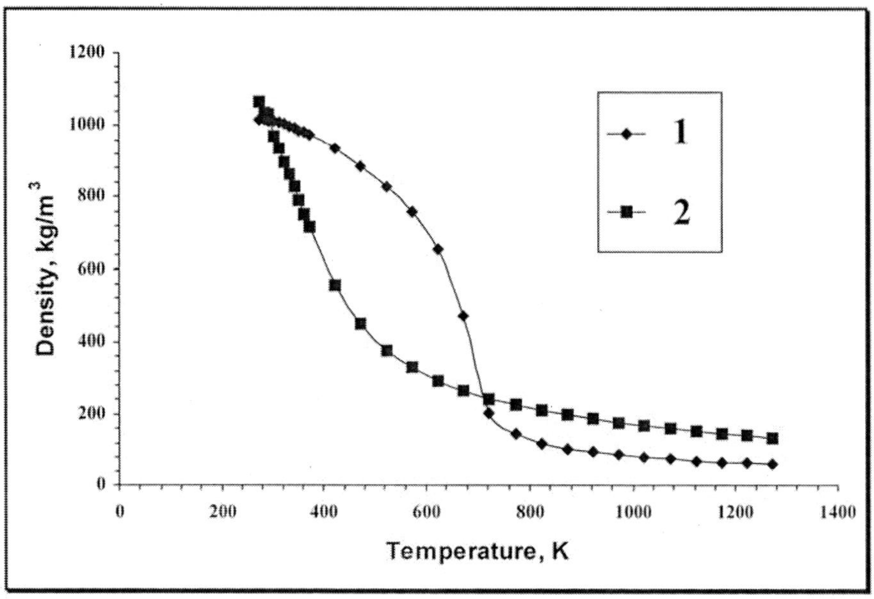

Figure 3. Change of density of water (1) and carbon dioxide (2) with increase of temperature under pressure 35 MPa.

Figure 3 gives the opportunity to compare the densities of the main components of early Earth's atmosphere under different thermodynamic conditions. Analysis of the data in Figure 3 shows that under pressure of 35 MPa, the temperature at which the density of CO_2 becomes greater than that of the density of H_2O is about 723 K. The estimations also assert that for pressure 35 MPa, the density of SO_2 would be greater than the density of H_2O at absolutely all temperatures.

Calculations of the temperatures necessary for maintaining water density at 1000 kg/m^3 (normal density of water (NDW) at 277 K and 0.1 MPa) and 958.36 kg/m^3 (density of water at its boiling point 373.16 K and 0.1 MPa) for atmospheric pressure of 35 MPa, indicate that the values would have been 329 K and 395 K, respectively. The results displayed in Figure 3 show that the density of water reaches a value of about 75.8% of NDW at 573 K, and then decreases very rapidly to 14.4% of NDW at 773 K; and with the further increase of temperature, the density of water decreases much more slowly to a value of about 6% of NDW at 1273 K. Analysis of the density values of these main components of early Earth's atmosphere shows that the difference in their densities at temperatures above 723 K is so significant that it would not permit them to blend easily, since the force of gravity would separate them. This would generate conditions favorable to the formation of different stratified layers within the atmosphere (SO_2-layer with some SO_3, CO_2-layer, H_2O-layer, as well as layers of other fluids). Temperatures in excess of 723 K, there would create conditions where the denser SO_2 and CO_2 would comprise the bottom two layers of the atmosphere. This makes the formation of the water-ocean at temperatures greater than 723 K impossible. In other words, it means that the formation of the water-ocean could be expected to have taken place only at temperatures of below 723 K. Furthermore, for the formation of the water ocean, the SO_2-layer would have to have been reworked somehow. An excellent example of such a reworking of the SO_2-layer is the known relative abundance of bedded barite deposits prior to 3.2Ga [184], as well as the known formation of pyrite in the Archean.

Unfortunately, available data does not allow us to define an exact age of the formation of the water ocean.

The oldest zircon of 4.4 Ga found so far in the Jack Hills area, Yilgarn Craton of Western Australia does not prove that oceans were necessarily present at the time of its formation [340,462], because there could have been other causes for the elevation of $\delta^{18}O$ values in the zircon. The following facts will be found to contradict the statement that $\delta^{18}O$ values within the zoned zircon of 4.4 Ga from the Jack Hills area, Yilgarn Craton of Western Australia indicate the existence of oceans prior to 4.4 Ga:

Since the zircon was grown from the granitic melt, its $\delta^{18}O$ content should not be compared with the standard mantle $\delta^{18}O$ content, since granitic melts are not usually formed in the mantle.

Data of Peck et al. [340] and Wilde et al. [462] show that zircons from the Jack Hills area, Yilgarn Craton of Western Australia are characterized by $\delta^{18}O$ values of about: 7.4 ‰ at 4.353 Ga, 5.7 ‰ at 4.15 Ga, 7.2 ‰ at 4.13 Ga, 6.8‰ at 4.01 Ga, and 6.3 ‰ at 3.6-3.3 Ga. Does it then mean that oceans existed at some periods, but not in other periods of the Hadean–Archean but not in others? Since there is a strong relationship between $\delta^{18}O$ values and the temperature for SiO_2, does it also mean that during the Hadean–Archean there was a drastically fluctuating temperature? According to experimental data on the relationships between $\delta^{18}O$ values and the temperature of quartz [87,492], the temperatures during formation of the zircons of the Jack Hills area, Yilgarn Craton were greater than about 583-623 K.

Based on micro-inclusions of SiO_2 in zircons from the Jack Hills area, Yilgarn Craton [462, Valley et al. [449], and Harrison et al. [165] came to the conclusion that the zircons formed from silica-saturated granitic or granitoid magma. However, silica-saturated magmas are known for high $\delta^{18}O$ content in quartz [222,269], which may have been a possible cause for the elevated $\delta^{18}O$ values of some grains of the zircons.

$\delta^{18}O$ values are higher for quartz formed at relatively high temperatures [141,222,269]. Temperature estimations show that Jack Hills zircons ranging from 4.0 to 4.35 Ga yield crystallization temperature peaks of 670 and 710°C [166].

Research shows that $\delta^{18}O$ values are high for carbonates [123,409], CO_2 [409], and sulfates [42,182]. This means that the zircons of Jack Hills could have become enriched with $\delta^{18}O$ through contact with carbonate rocks or sulfate rocks, or with CO_2 itself from the early Earth atmosphere.

Elevated $\delta^{18}O$ values for barite from Pilbara block, Western Australia [376] and Barberton Mountland of South Africa [159] also point to the possible involvement of sulfates and SO_2 in the elevated $\delta^{18}O$ values of the zircons from the Jack Hills area of the Yilgarn Craton.

Data of Fehlhaber and Bird [131] shows that, in contrast, some minerals in gabbros having been in contact with supercritical water have actually lowered values of $\delta^{18}O$ values.

Taking all of the above into account, it is likely that zircons from the Jack Hills area, Yilgarn Craton of Western Australia represent the first or one of the first attempts at the formation of the crust. It is also clear that the presence of an ocean is not necessary to form a zoned zircon with $\delta^{18}O$ values in some grains a little bit higher than those for the normal mantle.

Conclusion

Physical problems related to modeling the conditions of the formation and evolution of the lithosphere, are discussed. It is shown that none of the main methods (seismic, thermal, or gravitational methods, as well as analysis of the mineral composition of xenoliths and conditions of their stability) for investigating the position and primary physical characteristics of both the lithosphere and asthenosphere are free of ambiguity and uncertainty.

The role of pressure in the analysis of the distribution of density within the lithosphere cannot be ignored, because both the thermal expansion and compressibility of a rock or mineral directly control their density. Ignoring compressibility in calculations leads to a perpetual decrease in the value of the density of a rock with depth; whereas if we take into account both the thermal expansion and compressibility, it could yield results of no change or even an increase in density with depth under the *P-T* conditions within the lithosphere, especially in cases with low temperatures.

Low velocity zones (LVZs) did not necessarily contain partly molten rocks, and could have been formed by specific thermodynamic conditions of the rocks and minerals of the layer (see inequality (15)). LVZs could also be formed at the transition between consecutive layers from the relatively less dense forsterite layer with high seismic velocity to the denser iron-rich layers characterized by much lower values of seismic velocity.

Analysis of the thermal conditions within the solar nebula prior to planetary accretion shows that they were within the conditions of stability of ferric oxide, but changed with the start of accretion to those favorable to the stability of ferrous iron, which must have been the main component of rocks and minerals during accretion.

During accretion and the differentiation of Earth, the planet could have been entirely molten and at some point of its evolution was entirely covered by a magma ocean.

The formation and composition of the early lithosphere were mostly related to processes of the differentiation of matter and the rate of cooling of the magma ocean. The process of differentiation in the magma ocean would start during its formation and continue until its solidification. The stratification of Earth's slabs caused an initial state of separation of rocks and minerals into slabs within the upper mantle and crust, which were strictly regulated by their density, whether solid or melted. The difference between the densities of the felsic, intermediate, mafic, and ultramafic magmatic slabs within the magma ocean was enough to prevent the exchange of matter between them, and therefore mantle-wide convection did not take place.

The solidification of the magma ocean most likely began with the formation of a forsterite (or forsterite-rich peridotite) slab at depths of about 100 km. This was followed by the solidification of Earth's surface, cooled by heat radiation from the surface and the cooling effect of the early atmosphere.

Analysis of the thermodynamic conditions for carbonate rocks shows that under the thermal conditions of the magma ocean, carbonate rocks would have been unstable and decomposed, releasing carbon dioxide into the atmosphere. Water would also be unable to exist in its liquid state at the time of the magma ocean, and would form a thick and dense early atmosphere together with the carbon dioxide. Formation of the water ocean was under the constraints of the boiling point of water at the pressure of the early atmosphere and the critical point of water.

Low geothermal gradients within the magma ocean caused low heat transfer to the surface both by heat conduction and convection, which does not support a fast cooling of the magma ocean and the uppermost slabs of Earth. Analysis of the cooling rates of the mantle and numerous magmatic and/or metamorphic complexes also contradicts a fast cooling of the magma ocean and uppermost layers of Earth after its solidification.

Analysis of magmatic activity during the Early Precambrian points to a temperature maximum in the upper mantle during the main pulse of komatiites appearance and formation of the first large igneous provinces (LIPs) at about 2.8-2.7 Ga; and a maximum of mafic magmatism, related to the formation of LIPs, at about 2.5 Ga. Analysis of the pulses of granulite formation points to the increase of temperature at a depth of about 30 km from ~3.0 Ga to ~2.7 Ga, with its continuing increase to ~2.5 Ga.

At the end of the Archean, the thickness of the lithosphere was $\leq$100 km, including a solid forsterite layer at a depth of about 100 km with possible pockets of magma above it.

Analysis of available data shows that the presence of oceans is not necessary for the enrichment of $\delta^{18}O$ values within 4.4 Ga old zoned zircon from the Jack Hills area, Yilgarn Craton of Western Australia. The enrichment could have been caused by the contact of the forming zircon with quartz, carbonates, CO_2-rich atmosphere, or sulfates known for elevated content of $\delta^{18}O$.

References

[1] Abbott, D., Burgess, L. and Longhi, J., 1994. An empirical thermal history of the Earth's upper mantle. *Jour. of Geophys. Res.*, **99**, 13835-13850.

[2] Abe, Y., 1993. Physical state of the very early Earth. *Lithos*, **30**, No. 3-4, 223-235.

[3] Abe, Y., 1997. Thermal and chemical evolution of the terrestrial magma ocean. *Physics of the Earth and Plan. Inter.*, **100**, No. 1-4, 27-39.

[4] Abe, Y. and Matsui, T., 1986. Early evolution of the Earth: accretion, atmosphere formation, and thermal history. *Jour. of Geoph. Res.*, **91**, No. B13, 291-302.

[5] Abe, Y., Ohtani, E., Okuchi, T., Righter, K., and Drake, M.J., 2000. Water in the early Earth. In: (eds. R. Canup and K. Righter), *Origin of the Earth and Moon*, University of Arizona Press, 413-434.

[6] Airy, G.B., 1855. On the computation of the effect of the attraction of mountain-masses as disturbing the apparent astronomical latitude of stations of geodetic surveys. *Philos. Trans. R. Soc. London*, **145**, 101-104.

[7] Aleinikov, A.L., Belikov, V.T. and Eppelbaum, L.V., 2001. *Some Physical Foundations of Geodynamics* (in Russian, contents and summary in English). Kedem Printing-House, Tel Aviv, Israel, 172 pp.

[8] Allen. P.A. and Allen, J.R. 1995. *Basin Analysis - Principles and applications*. Blackwell Sci. Publ.

[9] Anderson, D.L., 1967. A seismic equation of state. *Geophys. J. R. Astron. Soc.*, **13**,9-30.

[10] Anderson, D.L., 1989. *Theory of the Earth*. Blackwell Sci. Publ., Boston.

[11] Anderson, D.L., 1999. Earth, In: (Beatty, J.K., Petersen, C.C. and Chaikin, A. Eds.), *The New Solar System*, 4th ed., Cambridge University Press, 111-124.

[12] Anderson D.L., 2002a. The case for irreversible chemical stratification of the mantle. *Intern. Geology Review*, **44**, No. 2, 97-116.

[13] Anderson, D.L, 2002b. The inner inner core of Earth. *Proc. of the Nat. Acad Sci USA*, **99**(22), 13966–13968.

[14] Anderson, D.L., 2006. Speculations on the nature and cause of mantle heterogeneity. *Tectonophysics*, **416**, No. 1-4, 7-22.

[15] Anderson, D.L., 2007. *The New Theory of the Earth*. Cambridge University Press, 401 pp.

[16] Arndt, N.T., Francis, D., Hynes, A.J., 1979. The field characteristics and petrology of Archean and Proterozoic komatiites. *Canadian Mineralogist*, **17**, 147-163.

[17] Arndt, N.T. and Nisbet, E.G. (Eds.), 1982. *Komatiites*. George Allen & Unwin, London.

[18] Arndt, N.T., Teixeira N.A. and White, W.M., 1989. Bizarre geochemistry of komatiites from the Crixás greenstone belt, Brazil. *Contrib. to Mineral. and Petrol.*, **101**, No. 2, 187-197.

[19] Artemjev, M.E. and Kaban, M.K., 1994. Density inhomogeneities, isostasy and flexural rigidity of the lithosphere in the Transcaspian region. *Tectonophysics*, **240**, 281-297.

[20] Artemjev, M.E., Kaban, M.K., Kucherinenko, V.A., Demyanov G.V. and Taranov, V.A., 1994. Subcrustal density inhomogeneities of the Northern Eurasia as derived from the gravity data and isostatic models of the lithosphere. *Tectonophysics*, **240**, 248-280.

[21] Artemieva, I.M., 2003. Lithospheric structure, composition, and thermal regime of the East European Craton: implications for the subsidence of the Russian platform. *Earth and Plan. Sci. Lett.*, **213**, 431-446.

[22] Artemieva, I.M., 2006. Global 1°×1° thermal model TC1 for the continental lithosphere: Implications for lithosphere secular evolution. *Tectonophysics*, **416**, 245-277.

[23] Artemieva, I.M. and Mooney, W.D., 2001. Thermal thickness and evolution of Precambrian lithosphere: A global study. *Jour. of Geoph. Research*, **106**, No. B8, 16,387-16,414.

[24] Artyushkov, E.V. Mörner, N.A., Tarling, D.H., 2000. The cause of loss of lithospheric rigidity in areas far from plate tectonic activity. *Geoph. Jour. Intern.*, **143** (3), 752-776.

[25] Ashchepkov, I., Vladykin, N., Ovchinnikov, Y., Anoshin, G., Gerasimov P. and Saprykin, A., 2000. Thermal Gradient, Mantle Layering and Geochemistry Beneath Aldan Shield According to the Kimberlitic Deep Seated Disintegrated Inclusions. Trans. of the Goldschmidt Conf., Oxford, UK. *Journal of Conf. Abst.*, **5**(2), 162-163.

[26] Ashwal, L.D., 1993. *Anorthosites*. Springer-Verlag, Berlin Heidelberg, 422 pp.

[27] Aspler, L.B. and Chiarenzelli, J.R., 1998. Two Neoarchean supercontinents? Evidence from the Paleoproterozoic. *Sedimentary Geol.*, 120, 75-104.

[28] Bada, J.L., 2004. How life began on Earth: a status report. *Earth and Plan. Sci. Lett.*, 226, No. 1-2, 1-15.

[29] Balaganskiy, V.V., Glaznev, B.N. and Osipenko, L.G., 1998. Early Proterozoic evolution of north-east of Baltic Shield: terrain analysis. *Geotectonics*, No.**2,** 16-28 (in Russian).

[30] Baldwin, J.A., Bowring, S.A., Williams M.L. and Williams, I.S., 2004. Eclogites of the Snowbird tectonic zone: petrological and U-Pb geochronological evidence for Paleoproterozoic high-pressure metamorphism in the western Canadian Shield. *Contrib. to Mineral. and Petrol.*, **147**, No. 5, 528-548.

[31] Barley, M.E., Bekker, A. and Krapež, B., 2005. Late Archean to Early Paleoproterozoic global tectonics, environmental change and the rise of atmospheric oxygen. *Earth and Plan. Sci. Lett.*, **238**, 156-171.

[32] Barley, M.E., Doyle, M.G., Kositcin, N., Krapez, B. and Cassidy, K.F., 2004. Late Archaean arc and seamount volcanism in the Kurnalpi Terrane, Eastern Yilgarn Craton, Western Australia. *Predictive Mineral Discovery, Coop. Res. Centre, Geoscience Australia, CRC Conf.*, Extended Abst., 3-7.

[33] Barley, M.E., Kerrich, R., Reudavy, I., Xie, Q., 2000. Late Archaean Ti-rich, Al-depleted komatiites and komatiitic volcaniclastic rocks from the Murchison Terrane in Western Australia. *Australian Jour. of Earth Sci.*, **47**, 873-883.

[34] Beakhouse, G.P. 1985: The Relationship of Supracrustal Sequences to a Basement. Complex in the Western English River Subprovince. *Geol. Assoc. of Canada Spec. Paper* **28**, 69-178.

[35] Bech, L., Onsgaard, J., Hoffmann, S.V. and Godowski, P.J., 2001. CO dissociation on K-modified Cu(112) and Cu(117). *Surface Science*, **482-485**, Part 1, 243-249.

[36] Bedini, R.-M., Blichert-Toft, J., Boyet, M. and Albarède, F., 2004. Isotopic constraints on the cooling of the continental lithosphere. *Earth and Plan. Sci. Lett.*, **223**, No. 1-2, 99-111.

[37] Bégin, N.J. and Pattison, D.R.M., 1994. Metamorphic evolution of granulites in the Minto block, Northern Quebec: extraction of peak P-T conditions taking account of late Fe-Mg exchange. *Jour. of Metamorphic Geol.*, **12**, 411-428.

[38] Bekker, A., Karhu, J.A., Eriksson, K.A. and Kaufman, A.J., 2003. Chemostratigraphy of Paleoproterozoic carbonate successions of the Wyoming Craton: tectonic forcing of biogeochemical change? *Precambrian Research*, **120**, 279-325.

[39] Bennett, J., Donahue, M., Schneider, N. and Voit, M., 2004. *The Cosmic Perspective*. 3rd ed., San Francisco Pearson Education Inc. publ. as Addison Wesley, 844 pp.

[40] Beukes, N.J. and Gutzmer, J., 2003. Aerobic to anaerobic depositional systems tracts in Archean sediments: Implications for the composition of the early hydrosphere and biosphere. *Goldschmidt Conference Abstracts*, **A38**.

[41] Bibikova, E.V., Borisova, E.Yu., Drugova, G.M. and Makarov, V.A., 1997. Metamorphic history and age of aluminous gneisses of the Belomorian Mobil Belt on the Baltic Shield. *Geochemistry Internat.*, **35**, No. 9, 775-784.

[42] Bindeman, I.N., Eiler, J.M., Wing, B.A. and Farquhar, J., 2007. Rare sulfur and triple oxygen isotope geochemistry of volcanogenic sulfate aerosols. *Geochim. et Cosmochim. Acta*, **71**, No. 9, 2326-2343.

[43] Birch, A.B., 1960. The velocity of compressional saves in rocks to 10 kilobars. Part I. *J. of Geophys. Res.*, **65**, 1083-1102.

[44] Birch, A.B., 1961. The velocity of compressional waves in rocks to 10 kilobars. Part 2. *J. of Geophys. Res.*, **66**, 2199-2224.

[45] Birch, F., 1965. Energetics of core formation. *J. of Geophys. Res.*, **70**, 6217-6221.

[46] Birch, F., 1966. Compressibility: Elastic constants. In: (Clark, Jr., S.P., Ed.), *Handbook of Physical Constants*. Revised ed., Geol. Soc. of America. Memoir 97, Washington, D.C., 97-173.

[47] Birch, F., Roy, E.R. and Decker, E.R., 1968. Heat flow and thermal history and New England and New York, In: (E. An-Zen, Ed.), *Studies of Appalachian Geology*, Wiley-Interscience, N.Y.

[48] Birch, F., Schairer, J.F. and Spicer H.C. (eds.), 1942. *Handbook of physical constants*. Geol. Soc. of America. Special papers, No. **36**, 325 pp.

[49] Bird, M.I., Ayliffe, L.K., Fifield, L.K., Turney, C.S.M., Cresswell, R.G., Barrows, T.T. and David, B., 1999. Radiocarbon Dating of "Old" Charcoal Using a Wet Oxidation, Stepped-Combustion Procedure.. *Radiocarbon*, **41**, No. 2, 127-140.

[50] Bjørnerud, M.G. and Austrheim, H., 2004. Inhibited eclogite formation: The key to the rapid growth of strong and buoyant Archean continental crust. *Geology*, **32**, No. 9, 765-768.

[51] Blenkinsop, T.G., Kröner, A. and Chiwara, A., 2004. Single stage, late Archaean exhumation of granulites in the Northern Marginal Zone, Limpopo Belt, Zimbabwe, and relevance to gold mineralization at Renco Mine. *South Afr. J. Geol.*, **107**, 377-396.

[52] Blichert-Toft, J. and Arndt, N.T., 1999. Hf isotope compositions of komatiites. *Earth and Plan. Sci. Lett.*, **171**, No. 3, 439-451.

[53] Bodorkos, S. and Reddy, S. M., 2004. Proterozoic cooling and exhumation of the northern central Halls Creek Orogen, Western Australia: constraints from a reconnaissance $^{40}Ar/^{39}Ar$ study. *Australian Jour. of Earth Sci.*, **51**, No. 4, 591-609.

[54] Bogdanova, S.V. and E. V. Bibikova, 1993. The "Saamian" of the Belomorian Mobile Belt: new geochronological constraints. *Precambrian Research*, **64**, No. 1-4, 131-152.

[55] Bowring, S.A., and Williams, I.S., 1999. Priscoan (4.00-4.03 Ga) orthogneisses from northwestern Canada. *Contrib. Mineral. Petrol.*, **134**, 3-16.

[56] Boyd, F.R., 1973. A pyroxene geotherm. *Geochim. et Cosmochim. Acta*, **37**, 2533–2546.

[57] Boyd, F.R., Gurney, J.J. and Richardson, S.H., 1985. Evidence for a 150–200 km thick Archaean lithosphere from diamond inclusion barometry. *Nature*, **315**, 387–389.

[58] Boyd, F.R. and Mertzman, S.A., 1987. Composition and structure of the Kaapvaal lithosphere, southern Africa. In: Mysen, B. O. (ed.) Magmatic Processes: Physicochemical Principles. *Geochemical Soc. Special Publication* **1**, 13-24.

[59] Boyd, F.R. and Nixon, P.H., 1975. Origins of the ultramafic nodules from some kimberlites of northern Lesotho and the Monastery Mine, South Africa. *Phys. Chem. Earth*, **91**, 431-454.

[60] Boyet, M. and Carlson, R.W., 2005. 142Nd Evidence for Early (>4.53 Ga) Global Differentiation of the Silicate Earth. *Science*, **309**, No. 5734, 576-581.

[61] Bry, M. and White, N., 2007. Reappraising elastic thickness variation at oceanic trenches. *Jour. of Geoph. Res.*, **112**, B08414, 30 pp.

[62] Brey, G.P. and Kohler, T., 1990. Geothermobarometry in four-phase lherzolites. II. New thermobarometers, and practical assessment of existing thermobarometers. *Jour. of Petrology*, **31**, 1353-1378.

[63] Brostigen, G. and Kjekshus, A., 1969. Redetermined crystal structure of FeS_2 (pyrite). *Acta Chem. Scand.*, **23**, 2186-2188.

[64] Brown, G.C. and Mussett, A.E., 1993. *The Inaccessible Earth*. Chapman & Hall, London, U.K., 288 pp.

[65] Buchan, K.L. and Ernst, R.E., 2004. Diabase dyke swarms and related units in Canada and adjacent regions. *Geological Survey of Canada Map* No. **2022A**, scale 1:5,000,000.

[66] Buchan, K.L., Mortensen, J.K., Card, K.D. and Percival, J.A., 1998. Paleomagnetism and U–Pb geochronology of diabase dyke swarms of Minto block, Superior Province, Quebec. Canada. *Canad. Jour. Earth Sci.*, **35**, 1054-1069.

[67] Burov, E.B. and Diament, M., 1995. The effective elastic thickness (Te) of continental lithosphere: What does it really mean? *Jour. of Geoph. Res.*, **100**, No. B3, 3905-3927.

[68] Burov, E. and and Diament, M., 1996. Isostasy, equivalent elastic thickness, and inelastic rheology of continents and oceans. *Geology*, **24**, No. 5, 419-422.

[69] Cameron, A.G.W., 1997. The Origin of the Moon and the Single Impact Hypothesis V. *Icarus*, **126**, No. 1, 126-137.

[70] Card, K.D., 1990. A review of the Superior Province of the Canadian Shield, a product of Archean accretion. *Precambrian Research*, **48**, No. 1-2, 99-156.

[71] Carlson, R.W., Pearson, D.G. and James, D.E., 2005, Physical, chemical, and chronological characteristics of continental mantle. *Rev. Geophys.*, **43**, RG1001.

[72] Carmichael, R.S. (Ed.), 1982. *Handbook of physical properties of rocks*, Volume II, CRC Press Inc., Boca Raton, Florida, 360 pp.

[73] Carslaw, H.S. and Jaeger, J.C., 1959. *Conduction of Heat in Solids*, 2nd ed., Oxford Univ. Press, N.Y.

[74] Carswell, D.A., 1991. The garnet-orthopyroxeneAl barometer: problematic application to natural garnet lherzolite assemblages. *Mineral. Magazine*, **55**, 19-31.

[75] Catling, D.C. and Claire M.W., 2005. How Earth's atmosphere evolved to an oxic state: A status report. *Earth and Plan. Sci. Lett.*, **237**, No. 1-2, 1-20.

[76] Cattin, R., Martelet, G., Henry, P., Avouac, J. P., Diament, M. and Shakya, T.R., 2001. Gravity anomalies, crustal structure and thermo-mechanical support of the Hymalaya of Central Nepal. *Geophys. Jour. Intern.*, **147**, 381-392.

[77] Čermak, V., 1982. A geothermal model of the lithosphere and a map of the thickness of the lithosphere on the territory of the USSR. *Phys. of the Solid Earth*, **18**(1), 18-27.

[78] Čermák, V., 1989. Crustal heat and mantle heat flow in Central and Eastern Europe. *Tectonophysics*, **159**, 195-215.

[79] Čermák, V. and Bodri, L., 1995. Three-dimensional deep temperature modeling along the European geotraverse. *Tectonophysics*, **244**, 1-11.

[80] Chamberlain, K.R. and Frost, B.R., 2005. Granulite metamorphism in the northeastern wind river range at 2705 Ma: evidence for crustal underplating during the stillwater thermal event and plume-style crustal growth. *Proceed. of the Ann. Meet. of the Geol. Soc. of America*, **37**, No. 7, p. 505.

[81] Cheng, S. and Kusky, T. 2007. Komatiites from west Shandong, North China craton: Implications for plume tectonics. *Gondwana Research*, **12**, No. 1-2, 77-83.

[82] Choi, S.H. and Kwon, S.-T., 2005. Mineral chemistry of spinel peridotite xenoliths from Baengnyeong Island, South Korea, and its implications for the paleogeotherm of the uppermost mantle. *Island Arc*, **14** (3), 236-253.

[83] Christensen, N.I. and Mooney, W.D., 1995. Seismic velocity structure and composition of the continental crust: A global view. *J. of Geophys. Res.*, **100B**, 9761-9788.

[84] Christoffel, C.A., Connelly, J.N., and Åhäll, K.-I., 1999. Timing and characterization of recurrent pre-Sveconorwegian metamorphism and deformation in the Varberg–Halmstad region of SW Sweden. *Precambrian Research*, **98**, No. 3-4, 173-195.

[85] Chyba, Ch.F., 1990. Impact delivery and erosion of planetary oceans in the early inner Solar System. *Nature*, **129-133**.

[86] Clark Jr., S.P., (Ed.), 1966. *Handbook of Physical Constants*, Revised ed., The Geol. Soc. of America, Memoir 97.

[87] Clayton, R.N., O'Neil, J.R. and Mayeda, T.K. 1972. Oxygen isotope exchange between quartz and water. *Jour. of Geoph. Res.*, **77**(17), 3057-3066.

[88] Cockell, C.S. and Horneck, G., 2001. The ultraviolet climate of Earth, past and present - theoretical and space-based observations. *Photochem. Photobiol.*, **73**, 447-457.

[89] Collerson, K.D., Jesseau, C.W. and Bridgwater, D., 1976, Crustal development of the Archaean gneiss complex: Eastern Labrador, In: (Windley, B.F., ed.), *The Early History of the Earth*, Wiley, London, 237-253.

[90] Collerson, K.D., McCulloch, M.T. and Nutman, A.P., 1989. Sr and Nd isotope systematics of polymetamorphic Archean gneisses from southern West Greenland and northern Labrador. *Canad. J. of Earth Sci.*, **26**, 446-466.

[91] Conard, B.R., Sridhar, R. and Warner, J.S., 1980, High-temperature thermodynamic properties of chalcopyrite. *Jour. of Chem. Thermodyn.*, **12**, 817-833.

[92] Condie, K.S., 1989a. Origin of the Earth's crust. *Global and Planetary Change*, **1,** No. 1-2, 57-81.

[93] Condie, K.C. 1989b. *Plate Tectonics and Crustal evolution*, 3rd ed., Pergamon Press, N.Y., 476 pp.

[94] Condie, K.C., 2001. *Mantle plumes and their record in Earth history*. Cambridge University Press.

[95] Condie, K.C., 2005. *Earth as an evolving planetary system*. Elsevir Academic Press, Boston.

[96] Condie, K.C., Des Marais, D.J. and Abbot, D., 2000. Geologic evidence for a mantle superplume event at 1.9 Ga. *Geochemistry, Geophysics, Geosystems*, **1**, Paper number 2000GC000095.

[97] Condie, K.C., Kröner, A. and Milisenda, C.C., 1996. Geochemistry and geochronology of the Mkhondo suite, Swaziland: evidence for passive-margin deposition and granulite facies metamorphism in the Late Archean of Southern Africa. *Jour. of African Earth Sci.*, **22**, 483-506.

[98] Cook, F.A. and Turcotte, D.L., 1981. Parameterized convection and the thermal evolution of the Earth. *Tectonophysics*, **75**, No. 1-2, 1-17.

[99] Cooper, C.M., Lenardic, A. and Moresi, L., 2003. Stability of a Chemical Boundary Layer Within a Convecting Mantle. *American Geophysical Union, Fall Meeting*, abstract No. T32C-08.

[100] Cooper, C.M., Lenardic, A. and Moresi, L., 2004. The thermal structure of stable continental lithosphere within a dynamic mantle. *Earth and Plan. Sci. Lett.*, **222**, No. 3-4, 807-817.

[101] Corfu, F., 1988. Differential response of U-Pb systems in coexisting accessory minerals, Winnipeg River Subprovince, Canadian Shield: implications for Archean crustal growth and stabilization. *Contrib. to Mineral. and Petrol.*, **98**, No. 3, 312-325.

[102] Corfu, F., Jackson, S.L. and et Sutcliffe, R.H., 1991. U-Pb ages and tectonic significance of late alkalic magmatism and nonmarine sedimentation: Timiskaming Group, southern Abitibi belt, Ontario. *Canad. Jour. of Earth Sci.*, **28**, 489-503.

[103] Corfu, F., Krogh, T.E., Kwok, Y.Y.,· and Jensen, L.S., 1989, U-Pb zircon geochronology in the southwestern Abitibi greenstone belt, Superior Province. Canad. *Jour. of Earth Sci.*, **26**, 1747-1763.

[104] Corfu, F., Stott, G.M., Breaks, F.W., 1995. U-Pb geochronology and evolution of the English River Subprovince, an Archean low P-high T metasedimentary belt in the Superior Province. *Tectonics*, **14**, No. 5, 1220-1233.

[105] Daly, R., 1940. *Strength and structure of the Earth*. Prentice-Hall, N.Y.

[106] Davaille, A., 1999. Simultaneous generation of hotspots and superswells by convection in a heterogeneous planetary mantle. *Nature*, **402**, 756-760.

[107] Davies, G.F., 1980. Thermal histories of convective Earth models and constraints on radiogenic heat production in the Earth. *J. Geophys. Res.*, **85**, 2517–2530.

[108] Davies, G.F., 1990. Heat and mass transport in the Early Earth. In: (Eds. J.H. Jones and H.E. Newsom), *The Origin of the Earth*. Oxford Univ. Press, N.Y., 151-174.

[109] Davis, W.J., Canil, D., MacKenzie, J.M. and Carbno, G.B., 2003. Petrology and U–Pb geochronology of lower crustal xenoliths and the development of a craton, Slave Province, Canada. *Lithos*, **71**, No. 2-4, 541-573.

[110] Dazhong, S. and Songnian, L. 1985. A subdivision of the Precambrian of China. *Precambrian Research*, **28**, No. 2, 137-162.

[111] Deming, D., 2002. Origin of the Ocean and Continents: A Unified Theory of the Earth. Intern. *Geology Review*, **44**, No. 2, 137-152(16).

[112] Deschamps, F., Trampert. J. and Snieder, R., 2002. Anomalies of temperature and iron in the uppermost mantle inferred from gravity data and tomographic models. *Physics of the Earth and Plan. Inter.*, **129**, No. 3-4, 245-264.

[113] De Smet, J., van den Berg, A.P. and Vlaar, N.J., 2000. Early formation and long-term stability of continents resulting from decompression melting in a convecting mantle. *Tectonophysics*, **322**, No. 1-2, 19-33.

[114] Desmons, J., 1985. Ophiolites through time: An introductions. *Ofioliti*, **10**(2/3), 103-108.

[115] Dickson, M.H. and Fanelli, M., 2004. What is Geothermal Energy? *Istituto di Geoscienze e Georisorse*, CNR, Pisa, Italy, 61 p.

[116] Djomani, Y.H.P., O'Reilly, S.Y.., Griffin, W.L. and Morgan, P., 2001. The density structure of subcontinental lithosphere through time. *Earth Planet. Sci. Lett.*, **184**, 605-621.

[117] Dörr, W., Belka, Z., Marheine, D., Schastok, J., Valverde-Vaquero, P. and Wiszniewska, J., 2002. U–Pb and Ar–Ar geochronology of anorogenic granite magmatism of the Mazury complex, NE Poland. *Precambrian Research*, **119**, 1-4, 101-120.

[118] Doucoure, C.M., de Wit, M.J. and Mushayandebvu, M.F., 1996. Effective elastic thickness of the continental lithosphere in South Africa. *Jour. of Geoph. Res.*, **101**, No. B5, 11,291-11,303.

[119] Doucouré, S.M. and de Wit, 2002. Temporal variation in rigidity and mechanical behaviour of old thick continental lithosphere. *South African Jour.of Geology*, **105**, No. 1, 39-50.

[120] Downes, P.J. and Bevan, A.W.R., 2002. Chrysoberyl, beryl and zincian spinel mineralization in granulite-facies Archaean rocks at Dowerin, Western Australia. *Mineral. Magazine*, **66**, No. 6, 985-1002.

[121] Drake, M.J., 2005. Origin of water in the terrestrial planets. *Meteorit. and Plan. Sci.*, **40**, 519-527.

[122] Drury, M.R., Van Roermund, H.L.M., Carswell, D.A., De Smet, J.H., Van Den Berg, A.P. and Vlaar, N.J., 2001. Emplacement of Deep Upper-Mantle Rocks into Cratonic Lithosphere by Convection and Diapiric Upwelling. *Jour. of Petrology*, **42** No. 1, 131-140.

[123] Ducea, M.N., and Saleeby, J., 2003. Trace element enrichment signatures by slab-derived carbonate fluids in the continental mantle wedge: an example from the Sierra Nevada, California. *Trans. of the Ann. Meet. of the Geol. Soc. of America, Seattle*, **35**, No. 6, p. 138.

[124] Durrheim, R.J. and Mooney, W.D., 1994. Evolution of the Precambrian lithosphere – seismological and geochemical constraints, *J. Geophys. Res*., **99**, 15,359–15,374.

[125] Eppelbaum, L.V., Modelevsky, M. and Pilchin, A.N., 1996. Geothermal investigations in the Dead Sea Rift zone, Israel: Implications for petroleum geology. *Jour. of Petroleum Geol*., **19**(4), 425-444.

[126] Eppelbaum, L.V. and Pilchin, A.N., 2006. Methodology of Curie discontinuity map development for regions with low thermal characteristics: An example from Israel. *Earth and Plan. Sci. Lett*., **243**, No. 3-4, 536-551.

[127] Eppelbaum, L.V., Vaksman, V.L., Kouznetsov, S.V., Sazonova, L.M., Smirnov, S.A., Surkov, A.V., Bezlepkin, B., Katz Y., Korotaeva, N.N. and Belovitskaya, G., 2006. Discovering of microdiamonds and minerals-satellites in Canyon Makhtesh Ramon (Negev desert, Israel). *Doklady Earth Sciences (Springer)*, **407**, No. 2, 202-204.

[128] Eriksson, P.G., Martins-Neto, M.A., Nelson, D.R., Aspler, L.B., Chiarenzelli, J.R., Catuneanu, O., Sarkar, S., Altermann, W. and Rautenbach, C.J. de W., 2001. An introduction to Precambrian basins: their characteristics and genesis. *Sedimentary Geology*, **141-142**, 1-35.

[129] Ernst, R.E., Buchan, K.L. and Campbell, I.H., 2005. Frontiers in Large Igneous Province research. *Lithos*, **79**, 271-297.

[130] Eyles, N. and Young, G.M., 1994. Geodynamic controls on glaciation in Earth history. In: (Deynoux, M., Miller, J. M.G., Domack, E.W., Eyles, N., Fairchild, I.J. & Young, G.M., eds.), *Intern. Geol. Correl. Program Project 260: Earth's Glacial Record*. Cambridge Univ. Press, 1-28.

[131] Fehlhaber, K. and Bird, D.K., 1991. Oxygen-isotope exchange and mineral alteration in gabbros of the Lower Layered Series, Kap Edvard Holm Complex, East Greenland. *Geology*, **19**, No. 8, 819-822.

[132] Finnerty, A.A., 1989. Xenolith-derived mantle geotherms: Whither the inflection? *Contrib. to Mineral. and Petrol*., **102**,367-375.

[133] Finnerty, A.A. and Boyd, F.R., 1987. Thermobarometry for garnet peridotites: Basis for the determination of thermal and compositional structure of the upper mantle. In: (P.H. Nixon, Ed.), *Mantle Xenoliths*, Wiley, N.Y., 381-402.

[134] Franck, S., 1992. Olivine flotation and crystallization of a global magma ocean. *Physics of the Earth and Plan. Inter.*, **74**, No. 1-2.

[135] Friend, C.R.L. and Nutman, A.P., 1994. Two Archaean granulite-facies metamorphic events in the Nuuk–Maniitsoq region, southern West Greenland: correlation with the Saglek block, Labrador. Jour. of the Geol. Society, **151**, No. 3, 421-424.

[136] Fripp, R.E.P., 1976. Stratabound gold deposits in Archean banded iron-formation, Rhodesia. *Economic Geol.*, **71**, 58–75.

[137] Froude, D.O., Ireland, T.R., Kinny, P.D., Williams, I.S., Compston, W., Williams, I.R. and Myers, J.S., 1983, Ion microprobe identi.cation of 4100–4200 Myr-old terrestrial zircons. *Nature*, **304**, 616-618.

[138] Gaál, G. and Gorbatschev, R. 1987. An outline of the Precambrian evolution of the Baltic Shield. *Precambrian Research*, **35**, 15–52.

[139] Gaboury, D., Daigneault, D. and Beaudoin, G., 2000. Volcanogenic-related origin of sulfide-rich quartz veins: evidence from O and S isotopes at the Géant Dormant gold mine, Abitibi belt, Canada. *Mineral. Deposita*, **35**, 21-36.

[140] Galer, S.J.G. and Mezger, K., 1998. Metamorphism, denudation and sea level in the Archean and cooling of the Earth. *Precambrian Research*, **92**, No. 4, 389-412.

[141] Gallagher, V., Feely, M., Högelsberger, H., Jenkin, G.R.T. and Fallick, A.E., 1992. Geological, Fluid inclusion and stable isotope studies of Mo mineralisation, Galway Granite, Ireland. *Mineral. Deposita*, **27**, 314-325.

[142] Gangopadhyay, A., Sproule, R.A., Walker, R.J. and Lesher, C.M., 2005. Re-Os systematics of komatiites and komatiitic basalts at Dundonald Beach, Ontario, Canada: Evidence for a complex alteration history and implications of a late-Archean chondritic mantle source. *Geochim. et Cosmochim. Acta*, **69**, No. 21, 5087-5098.

[143] Golden, D.C., Ming, D.W., Lauer, H.V., (Jr.) and Morris, R.V., 2004. Thermal decomposition of siderite–pyrite assemblages: implications for sulfide mineralogy in Martian meteorite alh84001 carbonate globules. *Lunar and Plan. Sci.*, XXXV, 1396.

[144] Goodwin, A.M., 1977. Archean volcanism in Superior Proince, Canada Shield. *The Geol. Assoc. of Canada*, Special paper **16**, 205-241.

[145] Goodwin, A.M., 1996. *Principles of Precambrian Geology*. Acad. Press Limited, London, 327 pp.

[146] Graham, E.K., Schwab, J.A., Sopkin, S.M. and Takei, H., 1988. The pressure and temperature dependence of the elastic properties of single-crystal fayalite Fe_2SiO_4. *Physics and Chemistry of Minerals*, **16**, No. 2, 186-198.

[147] Grapes, R.H. and Challis, G.A., 1999. Gersdorffite with pentlandite, violarite, pyrrhotite, and pyrite, northwest Nelson, New Zealand. *New Zealand Jour. of Geol. and Geoph.*, **42**, 189-204.

[148] Griffin, W.L., Graham, S., O'Reilly S.Y. and Pearson, N.J., 2004. Highly Siderophile Element Behavior in High Temperature Processes Lithosphere evolution beneath the Kaapvaal Craton: Re–Os systematics of sulfides in mantle-derived peridotites. *Chemical Geology*, **208**, No. 1-4, 89-118.

[149] Griffin, W.L., O'Reilly, S.Y., Abe, N., Aulbach, S., Davies, R.M., Pearson, N.J., Doyle, B.J. and Kivi, K., 2003. The origin and evolution of Archean lithospheric mantle. *Precambrian Research*, **127**, 19-41.

[150] Green, D.H. and Ringwood, A.E., 1967. The stability field of aluminous pyroxene peridotite and garnet peridotite and their relevance in upper mantle structure. *Earth Planet. Sci. Lett.*, **3**, 151-160.

[151] Green, D.H. and Ringwood, A.E., 1972. A comparison of recent experimental data on the gabbro-. garnet granulite-eclogite transition. *Jour. of Geo*logy, **80**, 277-288.

[152] Gregory, R.T., 1991. Oxygen history of seawater revisited: Timescales for boundary event changes in the oxygen isotope composition of seawater, In: (Taylor, H.P. et al., eds.), *Stable isotope geochemistry*, San Antonio, Texas, Lancaster Press, 65-76.

[153] Grotzinger, J.P. and Kasting, J.F., 1993. New constraints on Precambrian ocean composition. *The Journal of Geology*, **101**, 235-243

[154] Groves, D.I., Phillips, G.N., Falconer, L.J., Houstoun, S.M., Ho, S.E., Browning, P., Dahl, N. and Mcnaughton, M.J., 1987. Evidence for an epigenetic origin for BIF-hosted gold deposits in the greenstone belts of the Yilgarn Block, Western Australia. In: (Eds. S.E. Ho and D.I. Groves), *Recent advances in understanding Precambrian gold deposits*, Univ. of Western Australia, Perth, Australia, Publ. 2, 167-179.

[155] Gung, Y. Panning M. and Romanowicz, B., 2003. Global anisotropy and the thickness of continents, *Nature*, **422**, 707-711.

[156] Guo-Wei, Z., Yu-Bao, B., Yong, S., An-Lin, G., Ding-Wu Z. and Tao-Hong, L., 1985. Composition and evolution of the archaean crust in central Henan, China. *Precambrian Research*, **27**, No. 1-3, 7-35.

[157] Gurney, J.J., Moore, R.O., Otter, M.L., Kirkley, M.B., Hops, J.J. and McCandless, T.E. 1991. Southern African kimberlites and their xenoliths. In: (eds. A.B. Kampunzu and R.T. Lubala), *Magmatism in Extensional Structural Settings*. Springer, Berlin, 495-536.

[158] Gutenberg, B., 1959. *Physics of the Earth's Interior*, Academic Press, New York, 240 pp.

[159] Gutzmer, J., Banks, D., de Kock, M.O., McClung, C.R., Strauss, H. and Mezger, K., 2006. The origin and paleoenvironmental significance of stratabound barites from the mesoarchean fig tree group, Barberton mountainland, South Africa. *Trans. of the V South American Symp. on Isotope Geology, contrib*. **311**, 258-262.

[160] Hall, A., 1995. *Igneous petrology*. Longman Scientific & Technical, Singapore.

[161] Halliday, A.N., 2000, Terrestrial accretion rates and the origin of the Moon. *Earth and Plan. Sci. Lett.*, **176**, 17-30.

[162] Hamilton, W.B., 1998. Archean magmatism and deformation were not products of plate tectonics. *Precambrian Research*, **91**, No. 1-2, 143-179.

[163] Hamilton, W.B., 2003. An Alternative Earth. *GSA Today*, **13**, No. 11, 4-12.

[164] Hannington, M.D., Herzig, P.M. and Scott, S.D., 1991. Auriferous hydrothermal precipitates on the modern seafloor. *In*: (eds. Foster, R.P.), *Gold metallogeny andexploration*, Blackie & Son Ltd., London, 249-282.

[165] Harrison, T.M., Blichert-Toft, J., Müller, W., Albarede, F., Holden, P. and Mojzsis, S.J., 2006. Response to Comment on "Heterogeneous Hadean Hafnium: Evidence of Continental Crust at 4.4 to 4.5 Ga". *Science*, **312**, No. 5777, 1139.

[166] Harrison, T.M. and Watson, E.B., 2005. Crystallization temperatures of hadean zircons: possible plate boundary interactions between 4.4 and 4.0 Ga. *Proceed. of the Geol. Soc. of America Ann. Meet., Salt Lake City,* **37**, No. 7, p. 309.

[167] Hawkesworth, C.J. and O'Nions, R.K., 1977. The Petragenesis of some Archaean Volcanic Rocks from Southern Africa. *J. of Petrology*, **18**, 487-520.

[168] Heaman, L.M., 1997. Global mafic volcanism at 2.45 Ga: remnants of an ancient large igneous province? *Geology*, **25**, 299-302.
[169] Heizler, M.T., Ralser, S. and Karlstrom, K.E., 1997. Late Proterozoic (Grenville?) deformation in central New Mexico determined from single-crystal muscovite $^{40}Ar/^{39}Ar$ age spectra. *Precambrian Research*, **84**, No. 1-2, 1-15.
[170] Helmstaedt, H.H. and Scott, D.J., 1992. The proterozoic ophiolite problem, In: (Condie, K.C., Ed.,), *Proterozoic crustal evolution, Development in Precambrian geology*, **10**, Elsevier, 55-95.
[171] Hess, H.H., 1955. Serpentinites, orogeny and epeirogeny. *Geol. Soc. of America Special Paper* **62**, 391-407.
[172] Hessler, A.M., Lowe, D.R., Jones, R.L., and Bird, D.K., 2004. A lower limit for atmospheric carbon dioxide levels 3.2 billion years ago. *Nature*, **428**, 736-738.
[173] Hill, R.E.T., Barnes, S.J., Dowling S.E., and Thordarson, T., 2004. Komatiites and nickel sulphide orebodies of the Black Swan area, Yilgarn Craton, Western Australia. 1. Petrology and volcanology of host rocks. *Mineralium Deposita*, **39**, No. 7, 684-706.
[174] Höckenreiner, M., Söllner, F. and Miller, H., 2003. Dating the TIPA shear zone: an Early Devonian terrane boundary between the Famatinian and Pampean systems (NW Argentina). *Jour. of South Amer. Earth Sci.*, **16**, No. 1, 45-66.
[175] Hokada, T., 2001. Feldspar thermometry in ultrahigh-temperature metamorphic rocks: Evidence of crustal metamorphism attaining ~1100 °C in the Archean Napier Complex, East Antarctica, *Amer. Mineralogist*, **86**, No. 7-8, 932-938.
[176] Holland, D.H., 1978. *The Chemistry of the Atmosphere and Oceans*. Wiley & Sons, Chichester, 351 pp.
[177] Hölttä, P., Huhma, H., Mänttäri, I. and Paavola, J., 2000. P–T–t development of Archaean granulites in Varpaisjärvi, Central Finland II. Dating of high-grade metamorphism with the U–Pb and Sm–Nd methods. *Lithos*, **50**, 121–136.
[178] Hong, Y. and Fegley, B. (Jr.), 1997. Kinetics and Mechanism of Pyrite Thermal Decomposition. *Ber. Bunsenges. Phys. Chem*. **101**, 1870-1881.
[179] Hughes, Patrick J., Frost, B. Ronald, and Swapp, S.M., 2005. High-pressure granulite metamorphism and P-T path of the Archean layered gneiss unit, northern Tetons, Wyoming. Salt Lake City. *Proceed. of the Ann. Meet. of the Geol. Soc. of America*, **37**, No. 7, p. 505.
[180] Haggerty, S.E. and Sautter, V., 1990. Ultradeep (greater than 300 kilometers), ultramafic upper mantle xenoliths. *Science*, **248**, 993-996.
[181] Humayun, M. and Cassen, P., 2000. *Origin of the Earth and Moon* (Eds. R.M. Canup and K. Righter and 69 collaborating authors). Tucson, University of Arizona Press, 3-23.
[182] Humphries, S.A., 2003. *Stable Isotopes of Sulphur and Oxygen in Forested Catchments: Insight from New Techniques into Sulphur Cycling and Dissolved Organic Matter Alteration*. M. Sci. thesis, Univ. of Waterloo, Waterloo, Ontario, Canada, 144 pp.
[183] Huston, D.L. and Large, R.R., 1989. A chemical model for the concentration of gold in volcanogenic. massive sulphide deposits. *Ore Geol. Review*, **4**, 171-200.
[184] Huston, D.L., Logan and Barite, G.A., 2004. BIFs and bugs: evidence for the evolution of the Earth' s early hydrosphere. *Earth Planet. Sci. Lett*. **220**, 41-55.

[185] Ireland, T.R., Rudnick, R.L., Spetsius, Z., 1994. Trace elements in diamond inclusions from eclogites reveal link to Archean granites. *Earth and Plan. Sci. Lett.*, **128**, No. 3-4, 199-213.

[186] Irving, A.J., 1976. On the validity of paleogeotherms determined from xenolith suites in basalts and kimberlite. *Amer. Mineralogist*, **61**, 638-642.

[187] Ito, K. and Kennedy, G.C., 1970. The fine structure of the basalt-eclogite transition. *Mineral. Soc. Amer. Spec. Publ.*, **3**, 77-83.

[188] Iyer, S.S., Choudhuri, A., Vasconcellos, M.B.A. and Cordani, U.G., 1984. Radioactive element distribution in the Archean granulite terrane of Jequié — Bahia, Brazil. *Contrib. to Miner. and Petrol.*, **85**, No. 1, 95-101.

[189] Jacob, D.E., 2004. Nature and origin of eclogite xenoliths from kimberlites. *Lithos*, **77**, No. 1-4, 295-316.

[190] Jacobsen, S.B., 2005. The Hf-W isotopic system and the origin of the Earth and Moon. *Annual Review of Earth and Plan. Sci.*, **33**, 531-570.

[191] Jacobsen, S.B. and Pimentel-Klose, M.R., 1988. A Nd isotopic study of the Hamersley and Michipicoten banded iron formations: the source of REE and Fe in Archean oceans. *Earth and Plan. Sci. Lett.*, **87**, No. 1-2, 29-44.

[192] Jaeger, J.C., 1953. The surface temperature of the Moon. *Austral. J. Phys.*, **6**, No. 1, 10-21.

[193] Jahn, B.M., Auvray, B., Cornichet, J., Bai, Y.L., Shen Q.H. and Liu, D.Y., 1987. 3.5 Ga old amphibolites from eastern Hebei Province, China: Field occurrence, petrography, Sm-Nd isochron age and REE geochemistry. *Precambrian Research*, **34**, No. 3-4, 311-346.

[194] James, D.T., Nunn, G.A.G., Kamo, S., and Kwok, K., 2003. The southeastern Churcill Province revisited: U-Pb geochronology, regional correlations, and the enigmatic Orma domain. Current research, *Newfoundland Dept. of Mines and Energy, Geological Survey*, Report 03-1, 35-45.

[195] Jarvis, G.T. and Campbell, I.H., 1983. Archean komatiites and geotherms - Solution to an apparent contradiction. *Geoph. Res. Lett.*, **10**, 1133-1136.

[196] Jeffreys, H., 1962. *The Earth: Its Origin, History and Physical Constitution.* 4th ed., Cambridge Univ. Press, Cambridge, England.

[197] Johnson, E.A., Rossman, G.R., Dyar, M.D. and Valley, J.W., 2002. Correlation between OH concentration and oxygen isotope diffusion rate in diopsides from the Adirondack Mountains, New York. *Amer. Mineralogist*, **87**, 899-908.

[198] Johnsson, M.J., 1985. Late paleozoic-middle mesozoic uplift rate, cooling rate and geothermal gradient for south-central New York State. *Nuclear Tracks and Radiat. Measur.*, **10**, No. 3, 295-301.

[199] Judge, A.V. and McNutt, M.K., 1991. The relationship between plate curvature and elastic plate thickness: A study of the Peru-Chile trench. *J. of Geophys. Res.*, **96**, 16,625-16,640.

[200] Kaban, M.K., 2001. A gravity model of the North Eurasia crust and upper mantle: 1. Mantle and isostatic residual gravity anomalies. *Russ. Jour. of Earth Sci.*, **3**, No. 2, 143-163.

[201] Kaban, M.K., Schwintzer, P., Artemieva, I.M. and Mooney, W.D., 2003. Density of the continental roots: compositional and thermal contributions. *Earth and Plan. Sci. Lett.*, **209**, 53-69.

[202] Kaban, M.K., Schwintzer, P. and Tikhotsky, S.A., 1999. A global isostatic gravity model of the Earth. *Geoph. Jour. Intern.*, **136** (3), 519–536.

[203] Kaler, J.B., 1994. *Astronomy!*, Harper Collins College Publishers, N.Y., 615 pp.

[204] Kappelmeyer, O. and Haenel, R., 1974. *Geothermics with special reference to application*, Gebruder Borntrargen, Berlin–Stuttgart.

[205] Karato, S.-I. and Karki, B.B., 2001. Origin of lateral variation of seismic wave velocities and density in the deep mantle, *J. Geophys. Res.*, **106**, 21771-21783.

[206] Karg, H., Carter, A., Brix, M.R. and Littke, R., 2005. Late- and post-Variscan cooling and exhumation history of the northern Rhenish massif and the southern Ruhr Basin: new constraints from fission-track analysis. *Intern. Jour. of Earth Sci.*, **94**, No. 2, 180-192.

[207] Kasting, J.F., 1993. Earth's early atmosphere. Special section; evolution of atmospheres. *Science*, **259**, 920–926.

[208] Kasting, J.F. and Ackerman, T.P., 1986. Climate consequences of very high carbon dioxide levels in the Earth's early atmosphere. *Science*, **234**, 1383-1385.

[209] Kasting, J.F. and Donahue, T.M., 1981. Evolution of oxygen and ozone in the earth's atmosphere. In: (J. Billingham, ed.), *Life in the Universe*, MIT Press, Cambridge, MA, 149-162.

[210] Kasting, J.F. and Siefert, J.L., 2002. Life and the evolution of Earth's atmosphere. *Science*, **296**, 1066-1068.

[211] Kaufmann III, W.J., 1994. *Universe*, 4th ed. W.H. Freeman and Co., N.Y., 624 pp.

[212] Kaufmann III, W.J. and Freedman, R.A. 1999. *Universe*, 5th ed. W.H. Freeman and Co, N.Y, 823 pp.

[213] Kaula, W.M., 1979. Thermal evolution of the Earth and Moon growing by planetesimal impacts. *J. of Geophys. Res.*, **84**, 999-1008.

[214] Kent, A.J.R., Hutcheon, I.D., Ryerson, F.J. and Phinney, D.L., 2001. The temperature of formation of carbonate in martian meteorite ALH84001: constraints from cation diffusion. *Geochim. et Cosmochim. Acta*, **65**, No 2, 311-321.

[215] Kerimov, K.M., Pilchin, A.N. and Ibragimov, S.M., 1980. Influence of thermodynamical factor on the overhigh pressure in sedimentary strata. *Azerbaijan Oil Industry*, No. **2**, 6-9 (in Russian).

[216] Kerimov, K.M. and Pilchin A.N., 1986. Some peculiarities of pressure anomaly coefficient alteration in the reservoir rocks with different saturation character, In: *The role of abnormal pressure in oil and gas fields distribution*, Nauka, Moscow, 140-143 (in Russian).

[217] Kerrich, R., Wyman, D., Fan, J. and Bleeker, W., 1998. Boninite series: low Ti-tholeiite associations from the 2.7 Ga Abitibi greenstone belt. *Earth and Plan. Sci. Lett.*, **164**, No. 1-2, 303-316.

[218] Khesin, B.E., Alexeyev, V.V. and Eppelbaum, L.V., 1996. *Interpretation of Geophysical Fields in Complicated Environments*. Kluwer Academic Publishers, Ser.: Modern Approaches in Geophysics, Boston - Dordrecht - London, 368 pp.

[219] Kleine, T., Münker, C., Mezger, K. and Palme, H., 2002. Rapid accretion and early core formation on asteroids and the terrestrial planets from Hf-W chronometry. *Nature*, **418**(6901), 952-955.

[220] Knauth, L.P., 2005. Temperature and salinity history of the Precambrian ocean: implications for the course of microbial evolution. *Palaeogeography, Palaeoclimatology, Palaeoecology*, **219**, No. 1-2, 53-69.

[221] Knauth, L.P. and Lowe, D.R., 2003. High Archean climatic temperature inferred from oxygen isotope geochemistry of cherts in the 3.5 Ga Swaziland Supergroup, South Africa. *Bull. of the Geol. Soc. of Amer.*, **115**, No. 5, 566-580.

[222] Knight, J.T., Ridley, J.R. and Groves, D.I., 2000. The Archean Amphibolite Facies Coolgardie Goldfield, Yilgarn Craton, Western Australia: Nature, Controls, and Gold Field-Scale Patterns of Hydrothermal Wall-Rock Alteration. *Economic Geology*, **95**, No. 1, 49-84.

[223] Knoll, A.H., 2003. *Life on a Young Planet: The First Three Billion Years of Evolution on Earth*. Princeton Univ. Press, 304 pp.

[224] Kozhevnikov, V.N., 2007. Thermal history of the Archean mantle – on prognosis of diamond-bearing cratons. *Trans. of the All-Russian Geol. Confer*. Petrozavodsk, 168-172 (in Russian).

[225] Kozhevnikov, V.N. and Svetov, S.A., 2001. Mantle and crust thermal anomalies in the Archean and the Early Proterozoic: regional analysis, global correlations, metallogenic consequences. Geology and Natural Resources of Karelia, 4, Petrozavodsk, *Karelian Sci. Center of the Russ. Acad. of Sci.*, **3-17** (in Russian).

[226] Koziol, A.M., 2001. A siderite-magnesite decarbonation study. *Proceed. of the Meet. of Geol. Soc. of America*, **33**, Abstract No. 25840.

[227] Koziol, A.M., 2004. Experimental determination of siderite stability and application to Martian Meteorite ALH84001. *Amer. Mineralogist*, **89**, No. 2-3, 294-300.

[228] Krasovskii, S.S., 1989. *Gravity Modeling of the Crust with Implications for Isostasy*. Naukova Dumka, Kiev, Ukraine.

[229] Kröner, A., 1982. Archaean to Early Proterozoic tectonics and crustal evolution: A review. *Revista Brasileira de Geociências*, **12**, 15-31.

[230] Kröner, A., Brown, L., 2005. Structure, composition and evolution of the South Indian and Sri Lankan granulite terrains from deep seismic profiling and other geophysical and geological investigations: A legends initiative. *Gondwana Research*, **8**, 317-335

[231] Kubota, M., Kyaw, K., Watanabe, F., Matsuda H. and Hasatani, M., 2000. Study of Decarbonation of CaCO 3 for High Temperature Thermal Energy Storage. *JCEJAB Jour. of Chemical Engineer. of Japan*, **33**, No. 5, 797-800.

[232] Kuhn, W.R., Walker, J.C. and Marshall, H.G., 1989. The effect on Earth's surface temperature from variations in rotation rate, continent formation, solar luminosity, and carbon dioxide. *J. of Geophys Res.*, **94**(D8), 11129-11136.

[233] Kuskov, O.L. and Kronrod, V.A. 2007. Composition, temperature, and thickness of the lithosphere of the Archean Kaapvaal craton. *Physics of the Solid Earth*, **43**, No. 1, 42-62.

[234] Krogh T.E., 1993. High precision U-Pb ages for granulite metamorphism and deformation in the Archean Kapuskasing structural zone, Ontario: implications for structure and development of the lower crust. *Earth and Plan. Sci. Lett.*, **119**, No. 1-2, 1-18.

[235] Kutas, R.I. and Gordienko, V.V., 1971. Thermal fields of Ukraine. Naukova dumka, Kiev, 138 pp. (in Russian).

[236] Kutasov, I.M., 1999. *Applied Geothermics for Petroleum Engineers*. Elsevier.

[237] Labrosse, S., Hernlund, J.W. and Coltice, N., 2007. A crystallizing dense magma ocean at the base of the Earth's mantle. *Nature*, **450**, 866-869.
[238] Lachenbruch, A.H., 1970. Crustal temperature and heat production: Implications of the linear heat flow relation. *J. Geophys. Res.*, **75**, 3291–3300.
[239] Lambert, I.B., 1982. Early geobiochemical evolution of the Earth. *Revista Brasileira de Geociências*, **12**, 32-38.
[240] Lambert, D.D. Foster, J.G., Frick, L.R., Ripley, E.M. and Zientek, M.L. 1998. Geodynamics of magmatic Cu-Ni-PGE deposits: New insights from the Re-Os isotope system. *Economic Geology*, **93**, 121-136.
[241] Lang, K.R., 1999. The Sun, In: (Beatty, J.K., Petersen, C.C. and Chaikin A., Eds.), *The New Solar System*, 4th ed., Cambridge Univ. Press, 23-38.
[242] Lauretta, D.S. and McSween, H.Y. (eds.), 2006. *Meteorites and the Early Solar System II.* University of Arizona Press, Tucson, AZ, 942 pp.
[243] Le Pichon, X. and Sibuet, J.-C., 1981. Passive margins: a model of formation. *J. Geophys. Res.*, **86**, 3708-3729.
[244] Lee, C.A., 2003. Compositional variation of density and seismic velocities in natural peridotites at STP conditions: Implications for seismic imaging of compositional heterogeneities in the upper mantle. *Jour. of Geoph. Res*, **108**, No. B9, 2441.
[245] Lee, C.A., Brandon, A.D. and Norman, M., 2003. Vanadium in peridotites as a proxy for paleo-fO2 during partial melting: Prospects, limitations, and implications. *Geochim. et Cosmochim. Acta*, **67**, No. 16, 3045-3064.
[246] Lee, C.T. and Rudnick, R.L., 1999. Compositionally stratified cratonic lithosphere: petrology and geochemistry of peridotite xenoliths, In: (J.J. Gurney, J.L. Gurney, M.D. Pascoe and S.H. Richardson, Eds.), *Proc. 7th Int. Kimberlite Conf.*, **2**, 503-521.
[247] Lehrnann, I., 1961. S and the structure of the upper mantle. *Geophys. J. Roy. Astron. Soc.*, **4**, 124-138.
[248] Lenardic, A., Nimmo1, F., and Moresi, L., 2004. Growth of the hemispheric dichotomy and the cessation of plate tectonics on Mars. *Jour. of Geoph. Res.*, **109**, E02003.
[249] Lerner-Lam, A. and Jordan, T., 1983. Earth structure from fundamental and higher-mode waveform analysis. *Geophys. J. R. Astron. Soc.*, **56**, 759-797.
[250] Levin, L., 2006. Structure of the thermal lithosphere and asthenosphere beneath oceans and continents. *Geotectonics*, **40**, No. 5, 357-366.
[251] Li, J. and Agee C.B., 1996. Geochemistry of mantle-core differentiation at high pressure. *Nature*, **381**, 686-689.
[252] Lide, D.R., 2004. (Ed.). *Handbook of Chemistry and Physics*, 85th ed., CRC Press.
[253] Lindblom, S., Broman S. and Martinsson, O., 1996. Magmatic-hydrothermal fluids in the Pahtohavare Cu-Au deposit in greenstone at Kiruna, Sweden. *Mineralium Deposita*, **31**, No. 4, 307-318.
[254] Liu, D.Y., Nutman, A.P., Compston, W., Wu, J.S. and Shen, Q.H., 1992. Remnants of ≥3800 Ma crust in the Chinese part of the Sino-Korean craton. *Geology*, **20**, 339-342.
[255] Liu, L., 2004. The inception of the oceans and CO_2-atmosphere in the early history of the Earth. *Earth and Plan. Sci. Lett.*, **227**, No. 3-4, 179-184.
[256] Lobach-Zhuchenko, S.B., Levchenkov, O.A., Chekulaev, V.P. and Krylov, I.N., 1986. Geological evolution of the Karelian granite-greenstone terrain. *Precambrian Research*, **33**, No. 1-3, 45-65.

[257] Lowe, D.R., 1994. *Early environments: Constraints and opportunities for early evolution, in Early Life on Earth. Nobel Symposium* No.**84** (S. Bengston, Ed.), Columbia Univ. Press, N.Y., 10-23.

[258] Lowe, D.R. and Tice, M.M., 2004. Geologic evidence for Archean atmospheric and climatic evolution: Fluctuating levels of CO_2, CH_4, and O_2 with an overriding tectonic control. *Geology*, **32**, No. 6, 493-496.

[259] Lubimova, E.A., 1968. *Thermics of the Earth and Moon*. Nauka Publ., Moscow, 280 pp. (in Russian).

[260] Lubimova, E.A. and Mayeva, S.V., 1982. Models of the thermal evolution of the Earth. *Izvestiya Akad Sci. SSSR, Fizika Zemli*, No.6, 83-93 (Translated version published in *Izvestiya, Earth Physics*, 1982, 18, No.6, 457-465.

[261] Lubimova, E.A. and Pilchin, A.N., 1978. The thermal parameters of the Earth crust of Kura depression. *Trans. of the All-Union Geothermic Meet., Makhachkala*, **1,** 58-59 (in Russian).

[262] Luther, G.W. III, 1991. Pyrite synthesis via polysulfide compounds. *Geochim. et Cosmochim. Acta*, **55**, 2839-2849.

[263] MacGregor, I.D., 1974. The system MgO-$Al2O_3$-SiO_2: solubility of Al_2O_3, in enstatite for spinel and garnet peridotite compositions. *Amer. Mineralogist*, **59**, 110-119.

[264] MacGregor, I.D. and Basu A., 1974. Thermal structure of the lithosphere: a petrologic model. *Science*, **185**, 1007-l0ll.

[265] Mackenzie, F.T., 1998. *Our Changing Planet: An Introduction to Earth System Science and Global Environmental Change*. Prentice-Hall.

[266] Maclennan, J., McKenzie, D., Gronvöld, K. and Slater, L., 2001. Crustal accretion under northern Iceland. *Earth and Plan. Sci. Lett.*, **191**, No. 3-4, 295-310.

[267] Malkovets, V., Taylor, L., Griffin, W., O'Reilly, S., Pearson, N., Pokhilenko, N., Verichev, E., Golovin, N. and Litasov, K., 2003. Cratonic conditions beneath arkhangelsk, russia: garnet peridotites from the grib kimberlite. *Trans. of the* 8th *International Kimberlite Conf.*.

[268] Maloof, A.C., Halverson, G.P., Kirschvink, J.L., Schrag, D.P., Weiss, B.P. and Hoffman, P.F., 2006. Combined paleomagnetic, isotopic, and stratigraphic evidence for true polar wander from the Neoproterozoic Akademikerbreen Group, Svalbard, Norway. *GSA Bulletin*, **118**, No. 9/10, 1099-1124.

[269] Mariano, G., Sial A.N., and Norman Herz, 1990. Oxygen isotope geochemistry of a potássio porphyritic calc-alkalic composite pluton: the itaporanga batholith, state of Paraíba, northeastern Brazil. *Revista Brasileira de Geociências*, **20**(1-4), 159-164.

[270] Marschik, R. and Fontboté, L., 2001, The Candelaria-Punta del Cobre iron oxide (Cu-Au-Zn-Ag) deposits, Chile. *Economic Geology*, **96**, 1799-1826.

[271] Marshall, J.S., Pounder, E.R. and Stewart, R.W., 1967. *Physics*. Macmillan of Canada, Toronto, 1139 pp.

[272] Martignole, J. and Reynolds, P., 1997. $^{40}Ar/^{39}Ar$ thermochronology along a western Québec transect of the Grenville Province, Canada. *Jour. of Metamorphic Geol.*, **15**, No. 2, 283-296.

[273] Mattisson, T. and Lyngfelt, A., 1999. The reaction between limestone and SO_2 under periodically changing oxidizing and reducing conditions - Effect of temperature and limestone type. *J. of Institute of Energy*, **190-196**.

[274] McGregor, V.R., 1973, The Early Precambrian gneisses of the Godthåb district, West Greenland. *Royal Soc. of London Philos. Trans.*, **A273**, 343-358.

[275] McKenzie, D.P., 1978a. Some remarks on the development of sedimentary basins. *Earth Plan. Sci. Lett.*, **40**, 25-32.

[276] McKenzie, D.P., 1978b. Active tectonics of the Alpine-Himalayan Belt: the Aegean Sea and surrounding regions. *Geophys. J. R. astr. Soc.*, **55**, 217-254.

[277] McKenzie, D. and Fairhead, D., 1997. Estimates of the effective elastic thickness of the continental lithosphere from Bouguer and free air gravity anomalies. *J. of Geoph. Res.*, **102**, No. B12, 27,523-27,552.

[278] McKenzie, D., Jackson J., and Priestley, K., 2005. Thermal structure of oceanic and continental lithosphere. *Earth and Plan. Sci. Lett.*, **233**, No. 3-4, 337-349.

[279] McSween Jr., H.Y., 1993. *Stardust to planets*. St. Martin's Griffin, N.Y, 241 pp.

[280] Meert, J.G., Nédélec, A. and Hall, C., 2003. The stratoid granites of central Madagascar: paleomagnetism and further age constraints on neoproterozoic deformation. *Precambrian Research*, **120**, No. 1-2, 101-129.

[281] Mekhtiev, S.F., Kerimov, K.M. and Pilchin, A.N., 1982. Role of thermal factors in formation and preservation of overhigh stratum pressure in sedimentary cover of Kura depression, *Azerbaijan Oil Industry*, No. **3,** 1-5 (in Russian).

[282] Melezhik, V.A., Fallick, A.E., Rychanchik, D.V. and Kuznetsov, A.B., 2005. Palaeoproterozoic evaporites in Fennoscandia: implications for seawater sulphate, the rise of atmospheric oxygen and local amplification of the $d^{13}C$ excursion. *Terra Nova*, **17**, 141-148.

[283] Menzies, A., Westerlund, K., Gr□tter, H., Gurney, J.., Carlson, J., Fung A. and Nowicki, T., 2004. Peridotitic mantle xenoliths from kimberlites on the Ekati Diamond Mine property, N.W.T., Canada: major element compositions and implications for the lithosphere beneath the central Slave craton. *Lithos*, **77**, No. 1-4, 395-412.

[284] Mercier, J.C. and Carter, N.L., 1975. Pyroxene geotherms. *J. of Geophys. Res.*, **80**, 3349-3362.

[285] Mertanen, S., 2001. Palaeomagnetic Evidences for the Evolution of the Earth during Early Palaeoproterozoic, *Proceed. of the Europ. Union of Geosciences Annual Meet.*, EUG XI, Strasbourg, Cambridge Publ., p.190.

[286] Meyer, B., 1976. Elemental sulfur. *Chemical Reviews*, **76**, No. 3, 367-388.

[287] Mezger, K., Hanson, G.N., and Bohlen, S.R., 1989. High-precision U-Pb ages of metamorphic rutile: application to the cooling history of high-grade terranes. *Earth and Plan. Sci. Lett.*, **96**, No. 1-2, 106-118.

[288] Mezger, K., Bohlen, S.R. and Hanson, G.N., 1990. Metamorphic History of the Archean Pikwitonei Granulite Domain and the Cross Lake Subprovince, Superior Province, Manitoba, Canada. *Jour. of Petrology*, **31**, No. 2, 483-517.

[289] Mezger, K., Rawnsley, C.M., Bohlen, S.R. and Hanson, G.N., 1991. U-Pb garnet, sphene, monazite, and rutile ages: Implications for the duration of high grade metamorphism and cooling histories, Adirondack Mts., New York. *Journal of Geology*, **99**, 415–428.

[290] Misra, M.K., Ragland, K.W. and Baker, A.J., 1993. Wood ash composition as a function of furnace temperature. *Biomass and Bioenergy*, **4,** No. 2, 103-116.

[291] Mitchell, R.H., Carswell D.A. and Clarke, D.B., 1980. Geological implications and validity of calculated equilibration conditions for ultramafic xenoliths from the pipe 200 kimberlite, northern Lesotho. *Contrib. to Mineral. and Petrol.*, **72**, No. 2, 205-217.

[292] Mojzsis, S.J., Coath, C.D., Greenwood, J.P., McKeegan K.D. and Harrison, T.M., 2003. Mass-independent isotope effects in Archean (2.5 to 3.8 Ga) sedimentary sulfides determined by ion microprobe analysis. *Geochim. et Cosmoch. Acta*, **67**, No. 9, 1635-1658.

[293] Mojzsis, S.J., Harrison, T.M. and Pidgeon, R.T., 2001. Oxygen-isotope evidence from ancient zircons for liquid water at the Earth's surface 4.300 Myr ago. *Nature*, **409**, 178-181.

[294] Monié, P., Caby, R. and Arthaud, M.H., 1997. The neoproterozoic Brasiliano orogeny in northeast Brazil: $^{40}Ar/^{39}Ar$ and petrostructural data from Ceará. *Precambrian Research*, **81**, No. 3-4, 241-264.

[295] Mooney, W.D., and Vidale, J.E., 2003. Thermal and chemical variations in subcrustal cratonic lithosphere: evidence from crustal isostasy. *Lithos*, **71**, No. 2-4, 185-193.

[296] Moraes, R., Brown, M., Fuck, R.A., Camargo, M.A. and Lima, T.M., 2002. Characterization and *P–T* Evolution of Melt-bearing Ultrahigh-temperature Granulites: an Example from the Anápolis–Itauçu Complex of the Brasília Fold Belt, Brazil. *Journal of Petrology*, **43**, No. 9, 1673-1705.

[297] Morse, J.W. and Mackenzie, F.T., 1998. Hadean Ocean carbonate geochemistry. *Aquatic Geochemistry*, **4**, No. 3-4, 301-319.

[298] Morris, R.G., Sinclair, H.D. and Yelland, A.J., 1998. Exhumation of the Pyrenean orogen: implications for sediment discharge. *Basin Research*, **10**, 69–85.

[299] Moser, D.E., 1994. The geology and structure of the mid-crustal Wawa gneiss domain: A key to understanding tectonic variation with depth and time in the late Archean Abitibi-Wawa orogen. *Canad. Jour. of Earth Sci.* **31**(7), 1064-1080.

[300] Mueller, P.A., Wooden, J.L., Nutman, A.P. and Mogk, D.W., 1998, Early Archaean crust in the northern Wyoming province: Evidence from U-Pb ages of detrital zircons. *Precambrian Research*, **91**, 295-307.

[301] Mukhin, L.M. and Pimenov, K.Yu., 2002. Impact hot spots on the cold surface of the early Earth. *Planetary and Space Science*, **50**, No. 1, 41-43.

[302] Mutanen, T. and Huhma H., 2004. The 3.5 Ga Siurua trondhjemite gneiss in the Archaean Pudasjärvi Granulite Belt, northern Finland. *Bull. of the Geol. Soc. of Finland*, **75** (1–2), 51-68.

[303] Myers, J.S., 1988, Early Archaean Narryer Gneiss Complex, Yilgarn Craton, Western Australia. *Precambrian Research*, **38**, 297-308.

[304] Nalivkina, A.B., 1978. Archean ophiolite association: on example of Ukraine. *Soviet Geology*, **3**, 93-102.

[305] Nelson, D.R., 1997. Evolution of the Archaean granite-greenstone terranes of the Eastern Goldfields, Western Australia: SHRIMP U-Pb zircon constraints. *Precambrian Research*, **83**, 57-81.

[306] Nelson, D.R., 1998. Granite-greenstone crust formation on the Archaean Earth: a consequence of two superimposed processes. *Earth and Plan. Sci. Lett.*, **158**, 109-119.

[307] Nemchin, A.A., Pidgeon, R.T., Wilde, S.A., 1994. Timing of Late Archaean granulite facies metamorphism in the southwestern Yilgarn Craton of Western Australia:

evidence from U-Pb ages of zircons from mafic granulites. *Precambrian Research*, **68**, No. 3-4, 307-321.

[308] Neugebauer, H.J., 1987. Models of Lithospheric Thinning. *Annual Review of Earth and Plan. Sci.*, **15**, 421-443.

[309] Newton, R.C., 1987. Petrologic aspects of Precambrian granulite facies terrains bearing on their origins. In: (A. Kröner, Ed.), *Proterozoic lithospheric evolution*, Amer. Geoph. Un., Washington, D.C., 11-26.

[310] Ninomiya, Y., Zhang, L., Nagashima, T., Koketsu J. and Sato, A. 2004. Combustion and De–SO_x behavior of high-sulfur coals added with calcium acetate produced from biomass pyroligneous acid. *Fuel*, **83**, No. 16, 2123-2131.

[311] Nisbet, E.G. and Sleep, N.H., 2001. The habitat and nature of early life. *Nature*, **409**, 1083-1091.

[312] Nixon, P.N. (Ed.), 1987. *Mantle xenoliths*. J. Wiley & Sons, 844pp.

[313] Nixon, P.H. and Boyd, F.R., 1973. Petrogenesis of the granular and sheared ultrabasic nodule suite in kimberlites. In: (Nixon, P.H. ed.), Lesotho Kimberlites. *Meseru: Lesotho Nat. Devel. Corporation*, 48-56.

[314] Nunn, J. F., 1998. Evolution of the atmosphere. *Proceed. of Geol Assoc.*, **109**(1), 1-13.

[315] Nutman, A.P., Friend, C.R.L. and Bennett, V.C., 2001. Review of the oldest (4400–3600 Ma) geological and mineralogical record: Glimpses of the beginning. *Episodes*, **24**, No. 2, 93-101.

[316] Nutman, A.P., Kinny, P.D., Compston, W. and Williams, I.S., 1991. SHRIMP U-Pb zircon geochronology of the Narryer Gneiss Complex, Western Australia. *Precambrian Research*, **52**, 275-300.

[317] Nutman, A.P., McGregor, V.R., Friend, C.R.L., Bennett, V.C. and Kinny, P.D., 1996, The Itsaq Gneiss Complex of southern West Greenland; the world's most extensive record of early crustal evolution (3900–3600 Ma). *Precambrian Research*, **78**, 1-39.

[318] Nutman, A.P., Mojzsis, S.J. and Friend, C.R., 1997. Recognition of > or = 3850 Ma water-lain sediments in West Greenland and their significance for the early Archaean Earth. *Geochim. et Cosmochim. Acta*, **61**(12), 2475-2484.

[319] O'Brien, P.J. and Rötzler, J., 2003. High-pressure granulites: formation, recovery of peak conditions and implications for tectonics. *Jour. of Metamorphic Geol.*, **21**, No. 1, 3-20.

[320] Ohmoto, H., 1996. Evidence in pre-2.2 Ga paleosols for the early evolution of atmospheric oxygen and terrestrial biota. *Geology*, **24**(12), 1135-1138.

[321] Ohmoto, H., Watanabe, Y. and Kumazawa, K., 2004. Evidence from massive siderite beds for a CO_2-rich atmosphere before ~ 1.8 billion years ago. *Nature*, **429**, 395-399.

[322] Ohtani, E., 1985. The primordial terrestrial magma ocean and its implication for stratification of the mantle. *Phys. of the Earth and Plan. Inter.*, **38**, 70-80.

[323] Ohtani E., 2005. Water in the mantle. *Elements*, **1**, 25-30.

[324] Okudaira, T., Hayasaka, Y., Himeno, O., Watanabe, K., Sakurai, Y. and Ohtomo, Y., 2001. Cooling and inferred exhumation history of the Ryoke metamorphic belt in the Yanai district, south-west Japan: Constraints from Rb-Sr and fission-track ages of gneissose granitoid and numerical modelling. *The Island Arc*, **10**, No. 2, 98-115.

[325] Omar, G.I., Onstott, T.C. and Hoek, J., 2003. The origin of deep subsurface microbial communities in the Witwatersrand Basin, South Africa as deduced from apatite fission track analyses. *Geofluids*, **3**, No. 1, 69-80.

[326] O'Reilly, S.Y., Griffin, W.L., 2006. Imaging global chemical and thermal heterogeneity in the subcontinental lithospheric mantle with garnets and xenoliths: Geophysical implications. *Tectonophysics*, **416**, 289-309.

[327] Qihan, S., Huifen, X., Zongqing, Z., and Chenglin, J., 1994. Early Precambrian Granulites of China. *Geol. Publ. House*, Beijing, 168 pp.

[328] Papunen, H., Halkoaho, T., Tulenheimo, T. and Liimatainen, J., 1998. Excursion guide to the Kuhmo greenstone belt. *Geol. Surv. of Finland, Special Paper* **26**, 91-106.

[329] Pascal, C., 2006. On the role of heat flow, lithosphere thickness and lithosphere density on gravitational potential stresses. *Tectonophysics*, **425**, No. 1-4, 83-99.

[330] Passier, H.F., de Lange, G.J. and Dekkers, M.J., 2001. Magnetic properties and geochemistry of the active oxidation front and the youngest sapropel in the eastern Mediterranean Sea. *Geoph. Jour. Internat.*, **145**, No. 3, 604-614.

[331] Pavlenkova, N.I., 1995. On a regional seismic boundary in the uppermost mantle. *Physics of the Solid Earth, English Translation*, **31**, No. 12 (July 1996), Russian Edition, 1054-1059.

[332] Pavlenkova, N.I., 1996. General features of the uppermost mantle stratification from long-range seismic profiles. *Tectonophysics*, **264**, No. 1-4, 261-278.

[333] Pavlenkova, N.I., 2006. Long-range profile data on the upper-mantle structure in the Siberian Platform. *Russian Geology and Geophysics*, **47**, No. 5, 626-641.

[334] Pavlenkova, G.A. and Pavlenkova, N.I., 2003. 3-D velocity model of the upper mantle in the northern Eurasia. *Proceed. of the Europ. Geophys. Soc., Geoph. Res. Abst.*, Vol. 5, No. 02666.

[335] Pavlenkova, G.A., Priestley, K. and Cipar, J., 2002. 2D model of the crust and uppermost mantle along rift profile, Siberian craton. *Tectonophysics*, **355**, No. 1-4, 171-186.

[336] Pavlov, A.A. and Kasting, J.F., 2002. Mass-independent fractionation of sulfur isotopes in Archean sediments: Strong evidence for an anoxic Archean atmosphere. *Astrobiology*, **2**(1), 27-41.

[337] Pavlov, A.A., Hurtgen, M.T., Kasting, J.F. and Arthur, M.A., 2003. Methane-rich Proterozoic atmosphere? *Geology*, **31**, No. 1, 87-90.

[338] Pearson, D.G., Canil, D. and Shirey, S.B., 2003. Mantle samples included in volcanic rocks: xenoliths and diamonds. In: (R.W. Carlson, Ed.), *Treatise On Geochmistry*, **2**, The Mantle, Elsevier, N.Y., 171-277.

[339] Peck, W.H., Valley J.W. and Graham, C.M., 2003. Slow oxygen diffusion rates in igneous zircons from metamorphic rocks. *Amer. Mineralogist*, **88**, No. 7, 1003-1014.

[340] Peck, W.H., Valley, J.W., Wilde, S.A. and Graham C.M., 2001. Oxygen isotope ratios and rare earth elements in 3.3 to 4.4 Ga zircons: Ion microprobe evidence for high d18O continental crust and oceans in the Early Archean. *Geochim. et Cosmochim. Acta*, **65**, No. 22, 4215-4229.

[341] Perchuk, L. L., Gerya, T.V., Van Reenen, D.D., Smit, C.A. and Krotov, A.V., 2000. P-T paths and tectonic evolution of shear zones separating high-grade terrains from cratons: examples from Kola Peninsula (Russia) and Limpopo Region (South Africa). *Mineral. and Petrol.*, **69**, No. 1-2, 109-142.

[342] Percival, J.A., 1994. Archean high-grade metamorphism. In: Condie, K. C. (Ed.), Archean crust evolution. *Developm. in Precambrian geol.*, **11**, Elsevier, 357-410.

[343] Percival, J.A. and Skulski, T., 2000. Tectonothermal evolution of the northern Minto block, superior province, Quebec, Canada. *The Canadian Mineralogist*, **38**, 345-378.

[343] Percival, J.A. and Pysklywec, R.N., 2007. Are Archean lithospheric keels inverted? *Earth Planet. Sci. Lett.*, **254**, 393-403.

[344] Percival, J.A. and West, G.F., 1994. Kapuskasing uplift: a geological and geophysical synthesis. *Canad. Jour. Earth Sci.*, **31**, 1256 -1286.

[345] Perry, G.J., Gray A. and Mackay, G.H., 1985. Carbonate formation during hydrogenation of Victorian brown coal. *Fuel Processing Technology*, **10**, No. 3, 285-297.

[346] Petrini, K. and Podladchikov, Y. 2000. Lithospheric pressure-depth relationship in compressive regions of thickened crust. *Jour. of Metamorphic Geol.*, **18**, No. 1, 67-77.

[347] Phillips, G.N., Groves D.I. and Martyn J.E., 1984, An epigenetic origin for Archaean banded iron-formation-hosted gold deposits. *Economic Geology*, **79**, 162-171.

[348] Pilchin, A.N., 1976. On the petrologic composition of the earth at the crust-mantle transition zone. *USSR Acad. of Sci., VINITI Press*, No. **3308**, 22 pp. (in Russian).

[349] Pilchin, A.N., 1978. Analysis of the thermodynamic conditions in sand-shall intervals using change of geothermal gradient with the depth. *Applied Geophysics*, **83**, 112-116 (in Russian).

[350] Pilchin, A.N., 1978. Correction to hydrostatic pressure in the crust of the Middle Kura depression, *In*: *Geophysical investigation of oil, gas and ore deposits in Azerbaijan*, Baku, Azerbaijan, 78-80 (in Russian).

[351] Pilchin, A.N., 1979. On heat absorption in the earth strata (on example of Middle Kura depression). *USSR Acad. of Sci., VINITI Press*, No. **183-79**, 21 pp. (in Russian).

[352] Pilchin, A.N., 1981a. Influence of additional pressure on seismic velocity and density of rocks in Middle-Kura depression. *Applied Geophysics*, **91**, 122-127 (in Russian).

[353] Pilchin, A.N., 1981b. Influence of thermodynamic conditions of different crust strata on local and regional gravity anomalies in Kura depression, In: *Theory and methods of magnetic and gravity fields interpretation*, Kiev, Ukraine, 4 pp. (in Russian).

[354] Pilchin, A.N., 1986. On the role of pressures in tectonic processes. *USSR Acad. of Sci., VINITI Press*, No. 3723-86, 23 pp.

[355] Pilchin, A.N., 2005. The role of serpentinization in exhumation of high- to ultra-high-pressure metamorphic rocks. *Earth and Plan. Sci. Lett.*, **237**, No. 3-4, 815-828.

[356] Pilchin, A.N., and Eppelbaum, L.V., 1997. Determination of the lower edges of magnetized bodies by using geothermal data. *Geophys. Jour. Intern.*, **128**, 167-174.

[357] Pilchin, A.N. and Eppelbaum, L.V., 2002. Some peculiarities of thermodynamic conditions of the Earth crust and upper mantle. *Scientific Israel*, **4**, No. 1–2, 117-142.

[358] Pilchin, A.N. and Eppelbaum L.V., 2004. On the stability of ferrous and ferric iron oxides and its role in rocks and rock-forming minerals stability. *Scientific Israel*, **6**, No. 3-4, 119-135.

[359] Pilchin, A.N. and Eppelbaum L.V., 2005. On the role of thermodynamic conditions in rocks and minerals stability. *Scientific Israel*, **7**, No. 1-2, 88-105.

[360] Pilchin, A.N. and Eppelbaum, L.V., 2006. *Iron and its unique role in Earth evolution.* Monograph, Mexican Geophys. Soc. 9, National Univ. of Mexico, 68 pp.

[361] Pilchin, A.N. and Eppelbaum, L.V., 2007. Stability of iron oxides in the Earth and their role in the formation of rock magnetism. *Acta Geofisica*, **55**, No. 2, 133-153.

[362] Pilchin, A.N. and Eppelbaum, L.V., 2008. Some Causes of Initial Mantle Heterogeneity. *Trans. of the 33rd Intern. Geological Conf.*, Oslo, Norway.

[363] Pilchin, A.N. and Kerimov, K.M., 1980. Peculiarities of thermal regime of the central part of Kura depression. *Oil&Gas Geology and Geophysics*, No. **4**, 23-25 (in Russian).

[364] Pilchin, A.N. and Kerimov, K.M., 1982. Device for measurements of pressure and temperature in a bore hole, Russian patent No. 953195. *Bulletin of inventions*, No. **31** (in Russian).

[365] Pilchin, A.N. and Kerimov K.M., 1987. Prognosis of temperatures and pressures in not drilled section intervals in Azerbaijan oil and gas fields by using seismic data, In: *Prognosis of oil and gas deposits by geophysical methods* (in Russian), Baku, 55-61.

[366] Pollack, H.N., 1997. Thermal characteristics of the Archaean, In: (de Wit, M.J., and Ashwal, M.D., eds.), *Greenstone belts*, Oxford, UK, Clarendon Press, 223-232.

[367] Potrel, A., Peucat J.J. and Fanning, C.M., 1998. Archean crustal evolution of the West African Craton: example of the Amsaga Area (Reguibat Rise). U--Pb and Sm--Nd evidence for crustal growth and recycling. *Precambrian Research*, **90**, No. 3-4, 107-117.

[368] Poulton, S.W., Fralick, P.W., Canfield, D.E., 2004. The transition to a sulphidic ocean approximately 1.84 billion years ago. *Nature*, **431** (7005), 173-177.

[369] Pratt, J.H., 1855. On the attraction of the Himalaya Mountains and of the elevated regions beyond them, upon the plumb line in India. *Philos. Trans. R. Soc., London*, **145**, 53-100.

[370] Priestley, K. and Debayle, E., 2003. Seismic evidence for a moderately thick lithosphere beneath the Siberian Platform. *Geoph. Res. Lett.*, **30**, No. 3, 1118.

[371] Puchtel, I.S., Hofmann, A.W., Mezger, K., Shchipansky, A.A., Kulikov, V.S. and Kulikova, V.V., 1996. Petrology of a 2.41 Ga remarkably fresh komatiitic basalt lava lake in Lion Hills, Central Vetreny Belt, Baltic Shield. *Contrib. Mineral. Petrol.*, **124**, 273-290.

[372] Puchtel, I.S. and Humayun, M., 2005. Highly siderophile element geochemistry of 187Os-enriched 2.8-Ga Kostomuksha komatiites Baltic Shield. *Geochim. et Cosmochim. Acta*, **69**, 1607-1618.

[373] Puchtel, I.S., Humayun, M., Campbell, A.J., Sproule, R.A., Lesher, C.M. (2004) Platinum group element geochemistry of komatiites from the Alexo and Pyke Hill areas, Ontario, Canada. *Geochim. et Cosmochim. Acta*, **68**, 1361-1383.

[374] Ramachandra, H.M. and Roy, A., 2001. Evolution of the Bhandara-Balaghat granulite belt along the southern margin of the Sausar Mobile Belt of central India. *Proceed. of the Indian Acad. of Sci., Earth and Plan. Sci.*, **110**, No. 4, 351-368.

[375] Renner, R., Nisbet, E.G., Cheadle, M.J., Arndt, N.T., Bickle M.J. and Cameron, W.E., 1994. Komatiite Flows from the Reliance Formation, Belingwe Belt, Zimbabwe: I. Petrography and Mineralogy. *Jour. of Petrology*, **35**, No. 2, 361-400.

[376] Richards, I.J., Gregory, R.T., Ferguson, K.M. and Douthitt, C.B., 2001. Archean hydrothermal alteration and metamorphism of the Pilbara block, Western Australia. *Proceed. of the GSA Ann. Meet., Paper* No. **165-0.**

[377] Richter, F.M., 1985. Models for the Archean thermal regime. *Earth and Plan. Sci. Lett.*, **73**, 350-360.

[378] Righter, K. and Drake, M.J., 1997. Metal-silicate equilibrium in a homogeneously accreting earth: new results for Re. *Earth and Plan. Sci. Lett.*, **146**, No. 3-4, 541-553.

[379] Ringwood, A.E., 1960. Some aspects of the thermal evolution of the Earth. *Geochim. et Cosmochim. Acta*, **20**, No. 3-4, 241-259.

[380] Ringwood, A.E. and Green, D.H., 1966. An experimental investigation of the gabbro-eclogite transformation and some geophysical implications. *Tectonophysics*, **3**, 383-427.

[381] Ronov, A.B. and Yaroshevsky, A.A., 1967. Chemical structure of the Earth's crust. *Geochemistry*, **11**, 1285-1309 (in Russian).

[382] Rosen, O.M., 1989. Two geochemically different types of Precambrian crust in the Anabar Shield, North Siberia. *Precambrian Research*, **45**, No. 1-3, 129-142.

[383] Roya, A., Kagamib, H., Yoshidac, M., Royd, A., Bandyopadhyaye, B.K., Chattopadhyayf, A., Khand, A.S., Huing A.K. and Pale, T., 2006. Rb–Sr and Sm–Nd dating of different metamorphic events from the Sausar Mobile Belt, central India: implications for Proterozoic crustal evolution. *Jour.of Asian Earth Sci.*, **26**, No. 1, 61-76.

[384] Rubie, D.C., Gessmann C.K. and Frost, D.J., 2004. Partitioning of oxygen during core formation on the Earth and Mars. *Nature*, **429**, 58-61.

[385] Rubin, A.E., and Mittlefehldt, D.W., 1993. Evolutionary History of the Mesosiderite Asteroid: A Chronologic and Petrologic Synthesis. *Icarus*, **101**, No. 2, 201-212.

[386] Rudge, J.F., Champion, M.E.S., White, N., McKenzie, D., Lovell, B., 2008. A plume model of transient diachronous uplift at the Earth's surface. *Earth Planet. Sci. Lett.*, **267**, No. 1-2, 146-160.

[387] Rudnick, R.L., 1997. Thermal structure of Archean cratons: a new look at conductive geotherms and xenolith P-T arrays. *Proceed. of the 7th Ann. V.M. Goldschmidt Conf.*, No. **2281**.

[388] Rudnick, R.L., McDonough W.F. and O'Connell, R.J., 1998. Thermal structure, thickness and composition of continental lithosphere. *Chemical Geology*, **145**, No. 3-4, 395-411.

[389] Rye, R. and Holland, H.D., 1998. Paleosols and the evolution of atmospheric oxygen: a critical review. *Amer. J. Sci.*, **298**(8), 621-72.

[390] Sabitova, T.M., Lesik, O.M. and Adamova, A.A., 1998. Velocity and density heterogeneities of the Tien-Shan lithosphere. *Pure and Appl. Geoph.*, **151**, No. 2-4, 539-548.

[391] Sáenzl, J.P., Foustoukos, D.I. and Seyfried, W.E., 2003. Experimental Constraints on Hydrothermal Fe-flux in Archean-Like Sea Water: Implications for Fe Transport in the Precambrian Seas and Precipitation of Banded Iron Formation. *AGU Fall Meet. Suppl., Eos Trans. AGU,* **84**(46), Abstract PP21C-1178.

[392] Safronov, V.S., 1969. Evolution of Protoplanetary cloud and the formation of the Earth and planets. Nauka, Moscow, (in Russian), *NASA Tech. Transl.* **F-677**, Washington, D.C. 1972.

[393] Safronov, V.S., 1978. The heating of the earth during its formation. *Icarus*, **33**, No. 1, 3-12.

[394] Sankaran, A.V., 2004. The oxygen state of the oceans of Proterozoic eon. *Current Science*, **87**, No. 3, 280-281.

[395] Saucr, K. and Yachandra, V.K., 2002. A possible evolutionary origin for the Mn_4 cluster of the photosynthetic water oxidation complex from natural MnO_2 precipitates in the early ocean. *Proced. of the Nat., Acad. Sci. USA*, **99**, No. 13, 8631-8636.

[396] Schiøtte, L., Compston, W., and Bridgwater, D., 1989, Ion probe U-Th-Pb zircon dating of polymetamorphic orthogneisses from northern Labrador, Canada. Canad. *Jour. Earth Sci.*, **26**, 1533-1556.

[397] Schiøtte, L., Nutman, A.P. and Bridgwater, D., 1992. U-Pb ages of single zircons within "Upernavik" metasediments from the Archaean Nain Province: regional implications. *Canad. Jour. of Earth Sci.*, **29**, 260-276.

[398] Schiøtte, L., Hansen, B.T., Shirey, S.B. and Bridgwater, D., 1993. Petrological and whole rock isotopic characteristics of tectonically juxtaposed Archaean gneisses in the Okak area of the Nain Province, Labrador: relevance for terrane models. *Precambrian Research*, **63**, No. 3-4, 293-323.

[399] Schmidberger, S.S. and Francis, D., 1999. Nature of the mantle roots beneath the North American craton: mantle xenolith evidence from Somerset Island kimberlites. *Lithos*, **48**, No. 1-4, 195-216.

[400] Schmitz, M.D. and Bowring, S.A., 2003. Constraints on the thermal evolution of continental lithosphere from U-Pb accessory mineral thermochronometry of lower crustal xenoliths, southern Africa. *Contrib. to Mineral. and Petrol.*, **144**, No. 5, 592-618.

[401] Schulze, D.J., 1989. Constraints on the abundance of eclogite in the upper mantle. *Jour. of Geophys.Res.*, **94**, No. B4, 4205-4212.

[402] Scott, H.P., Hemley, R.J., Mao, H., Herschbach, D.R., Fried, L.E., Howard, W.M. and Bastea, S., 2004. Generation of methane in the Earth's mantle: In situ high pressure–temperature measurements of carbonate reduction. *Proced. of the Nat. Acad. Sci. USA*, **101**, No. 39, 14023-14026.

[403] Seiff, A., Shofield, J.T., Kliore, A.J., Taylor, F.W., Limaye, S.S., Revercomb, H.E., Sromovsky, L.A., Kerzhanovich, V.I., Moroz, V.I., Marov, M.Ya., 1986. Models of the structure of the atmosphere of Venus from the surface to 100 km altitude. In: (Kliore, A.J., Moroz, V.I., Keating, G.M., Eds.), *The Venus International Reference Atmosphere*. Pergamon, Oxford, 3-58.

[404] Sen, G., Keshav, S. and Bizimis, M., 2005. Hawaiian mantle xenoliths and magmas: Composition and thermal character of the lithosphere. *Amer. Mineralogist*, **90**, 871-887.

[405] Sengupta, P., Sen, J., Dasgupta, S., Raith, M., Bhui, U.K. and Ehl, J., 1999. Ultra-high Temperature Metamorphism of Metapelitic Granulites from Kondapalle, Eastern Ghats Belt: Implications for the Indo-Antarctic Correlation. *Jour. of Petrology*, **1065-1087**

[406] Sharkov, E.V., and Bogatikov, O.A., 2001. Early Stages of the Tectonic and Magmatic Development of the Earth and Moon: Similarities and Differences. *Petrology*, **9**, No. 2, 97-118.

[407] Sharma, K.K., Bal, K.D. Parshad, R., Lal, N. and Nagpaul, K.K., 1980. *Tectonophysics*, **70**, No. 1-2, 135-158.

[408] Sharma, T. and Clayton, R.N., 1965. Measurement of 0-18/0-16 ratios of total oxygen of carbonates. *Geochim. et Cosmochim. Acta*, **29**, 1347–1353.

[409] Sharp, Z.D., Papike, J.J. and Durakiewicz, T., 2003. The effect of thermal decarbonation on stable isotope compositions of carbonates. *Amer. Mineralogist*, **88**, No. 1, 87-92.

[410] Shaw, R.K. and Arima, M., 1998. A corundum–quartz assemblage from the Eastern Ghats Granulite Belt, India: evidence for high P–T metamorphism? *Jour. of Metamorphic Geol.*, **16**, No. 2, 189-196.

[411] Shen, J.J., Papanastassiou, D.A. and Wasserburg, G.J., 1996. Precise Re-Os determinations and systematics of iron meteorites. *Geochim. et Cosmochim. Acta*, **60**, No. 15, 2887–2900.

[412] Shimizu, K., Nakamura, E. and Maruyama, S., 2005. The Geochemistry of Ultramafic to Mafic Volcanics from the Belingwe Greenstone Belt, Zimbabwe: Magmatism in an Archean Continental Large Igneous Province. *Jour. of Petrology*, **46**(11), 2367-2394.

[413] Shumlyanska, L., Tsvetkova, T., Bugaenko, I. and Zayets, L., 2006. 3-D velocity model of the upper mantle of the south-western part of the East-European Platform. *Geoph. Res. Abst.*, Vol. 8, 00762.

[414] Shuldiner, V.I., 1982. *Precambrian basement of the Circumpacific Belt and neighboring platforms*, Moscow, Nedra, 226 pp. (in Russian).

[415] Sleep, N.H., Zahnle, K. and Neuhoff, P.S., 2001. Initiation of clement surface conditions on the earliest Earth. *Proc. Natl. Acad. Sci. US*, **98**, No. 7, 3666-3672.

[416] Smelov, A.P., Timofeev, V.F. and Zaitsev, A.I., 2003. A geodynamic model for the formation of the north Asian craton in the Early Precambrian. *Proceed. of the 8th Intern. Kimberlite Conf.*, Victoria, British Columbia, Canada, Extended Abstracts Volume, Abst. 9.P12

[417] Smith, D., 1999. Temperatures and pressures of mineral equilibration in peridotite xenoliths: review, discussion, and implications. In (Y. Fei, C.M. Bertka and B.O. Mysen, Eds.). Mantle Petrology: Field Observations and High Pressure Experimentation: A Tribute to Francis R. (Joe) Boyd. *The Geochem.Soc. Spec. Publ.* **6**, 171-198.

[418] Smit, C.A., Van Reenen, D.D., Gerya, T.V. and Perchuk, L.L., 2001. *P–T* conditions of decompression of the Limpopo high-grade terrane: record from shear zones. *Jour. of Metamorphic Geol.*, **19**, No 3, 249-268.

[419] Smyth, J.R., Jacobsen, S.D. and Hazen, R.M., 2001. Comparative crystal chemistry of orthosilicate minerals. In: (Hazen, R.M., Downs, R.T. Eds.), High-pressure, high-temperature crystal chemistry. *Reviews in Mineral. and Geochem.*, **41**, 187-209.

[420] Snoeyenbos, D.R., Williams, M.L. and Hanmer, S., 1995. Archean high-pressure metamorphism in the western Canadian Shield. *Europ. Jour. of Mineralogy*, **7**, 1251-1272.

[421] Sobolev, S.V. and Babeyko, A.Y., 1994. Modeling of mineralogical composition, density and elastic wave velocities in anhydrous magmatic rocks. *Surveys in Geophysics*, **15**, No. 5, 515-544.

[422] Solomatov, V.S., 2000. Fluid dynamics of a terrestrial magma ocean. In: *Origin of the Earth and Moon* (R. Canup and K. Righter, eds.), University of Arizona Press, Tucson, Arizona, 323-338.

[423] Solomon, M. and Groves, D.I., 1994. *The geology and origin of Australia's mineral deposits*. Clarendon Press, Oxford, 951 pp.

[424] Song, B., Nutman, A.P., Liu, D., and Wu, J., 1996. 3800 to 2500 Ma crustal evolution in the Anshan area of Liaoning Province, northeastern China. *Precambrian Research*, **78**, 79–94.

[425] Speight, J.G., 2005. *Lange's handbook of Chemistry*. 16th ed., McGaw-Hill, N.Y.

[426] Speziale, S., Duffy, T.S. and Angel R.J., 2004. Singlecrystal elasticity of fayalite to 12 Gpa. *J. of Geophys. Res.*, **109**, B12, B12202.

[427] Speziale, S., Jiang, F. and Duffy, T.S., 2005. Compositional Dependence of the Elastic Wave Velocities of Mantle Minerals: Implications for Seismic Properties of Mantle Rocks. In: (van der Hilst, R.D., Bass, J., Matas, J., Trampert, J., Eds.), *Earth's Deep Mantle: Structure, Composition, and Evolution, AGU, Geophysical Monograph Series* **160**, 303-322.

[428] Spohn, T. and Schubert, G., 1991, Thermal equilibration of the Earth following a giant impact. *Geoph. Jour. Internat.*, **107**, 163-170.

[429] Spooner, E.T.C., Beckinsa, R.D., Fyfe, W.S. and Smewing, J.D., 1974. O^{18} Enriched ophiolitic metabasic rocks from E Liguria (Italy), Pindos (Greece), and Troodos (Cyprus). *Contributions to Mineralogy and Petrology*, **47** (1), 41-62.

[430] Stein, C.A., 1995. Heat Flow of the Earth. Global Earth Physics. *A Handbook of Physical Constants AGU Reference Shelf 1*, American Geophysical Union, 144-158.

[431] Stern, R.A. and Bleeker, W., 1997, Age of the world's oldest rocks re.ned using Canada's SHRIMP: the Acasta Gneiss Complex, Northwest Territories, Canada. *Geoscience Canada*, **25**, 27–33.

[432] Stoessell, R.K., Klimentidis, R.E. and Prezbindowski, D.R., 1987. Dedolomitization in Na – Ca – Cl brines from 100° to 200°C at 300 bars. *Geochim. et Cosmochim. Acta*, **51**, No. 4, 847-855.

[433] St-onge, M.R. and Lucas, S.B., 1995. Large-scale fluid infiltration, metasomatism and re-equilibration of Archaean basement granulites during Palaeoproterozoic thrust belt construction, Ungava Orogen, Canada. *Jour. of Metamorphocal Geol.*, **13**, 509-535.

[434] Storm, L.C. and Spear, F.S., 2005. Pressure, temperature and cooling rates of granulite facies migmatitic pelites from the southern Adirondack Highlands, New York. *Jour. of Metamorphic Geol.*, **23**, No. 2, 107-130.

[435] Streepey, M.M., van der Pluijm, B.A., Essene, E.J., Hall, C.M. and Magloughlin, J.F., 2000. Late Proterozoic (ca. 930 Ma) extension in eastern Laurentia. *GSA Bulletin*, **112**, No. 10, 1522-1530.

[436] Sukumaran, P.V., 2001. Early Planetary Environments and the origin of life. *Resonance*, **6**, No. 10, 16-28.

[437] Sumner, D.Y., 2000. Late Archean Carbonate Facies Models. *Trans. of the 31st IGC Meeting*, Rio de Janeiro.

[438] Sumner, D.Y. and Grotzinger, J.P., 2000. Late Archean aragonite precipitation: Petrography, facies associations, and environmental significance. In: Carbonate Sedimentation and Diagenesis in the Evolving Precambrian World (J.P. Grotzinger and N.P. James, eds). *SEPM Special Publication* No. **67**, 123-144.

[439] Svetov, S.A. and Smolkin, V.F., 2003. Model *P–T* conditions of high-magnesia magma generation in the Precambrian of the Fennoscandian shield. *Geochemistry Internat.*, **41**, 799-811.

[440] Tackley, P.J., 1998. Three dimensional simulations of mantle convection with a thermo-chemical basal boundary layer. In: (Gurnis, M., Wysession, M.E., Knittle, E. and Buffett, B.A., eds.), *The core-mantle boundary region*, Washington, DC, American Geophysical Union, *Geodynamics Series*, **28**, 231-253.

[441] Tackley, P.J., 2000, Mantle convection and plate tectonics: Toward an integrated physical and chemical theory. *Science*, **288**, 2002-2007.

[442] Takahashi, E., 1990. Speculations on the Archean mantle: Missing link between komatiite and depleted garnet peridotite. *J. Geophys. Res.*, **95**(B10), 15941-15954.

[443] Tonks, W.B. and Melosh, H.J., 1993. Magma ocean formation due to giant impacts. *J. of Geophys. Res.*, **98**, 5319-5333.

[444] Trikkel, A. and Kuusik, R., 2003. Modeling of decomposition and sulphation of oil shale carbonates on the basis of natural limestone. *Estonian Acad. Publ., Oil Shale*, **20**, No. 4, 491-500.

[445] Tugarinov, A.I. and Bibikova, E.V., 1975.Geochronology of ancient formations of the Kola Peninsula. *Geochemistry*, No **9,** 1275-1281 (in Russian).

[446] Turcotte, D.L. 1980. On the thermal evolution of the Earth. *Earth and Plan. Sci. Lett.*, **48**, No. 1, 53-58.

[447] Turcotte, D.L. and Schubert, G., 2002. *Geodynamics: Applications of continuum physics to geological problems* (2nd ed.), Cambridge University Press.

[448] Utsunomiya, S., Murakami, T., Nakada, M. and Kasama, T., 2003. Iron oxidation state of a 2.45-Byr-old paleosol developed on mafic volcanics. *Geochim. et Cosmochim. Acta*, **67**, No. 2, 213-221.

[449] Valley, J.W., Cavosie, A.J., Fu, B., Peck, W.H. and Wilde, S.A., 2006. Comment on "Heterogeneous Hadean Hafnium: Evidence of Continental Crust at 4.4 to 4.5 Ga". *Science*, **312**, No. 5777, p. 1139.

[450] Valley, J.W., Peck, W.H., King, E.M. and Wilde, S.A., 2002. A cool early Earth. *Geology*, **30**, No. 4, 351-354.

[451] van den Berg, A.P. and Yuen, D.A., 2002. Delayed cooling of the Earth's mantle due to variable thermal conductivity and the formation of a low conductivity zone. *Earth and Plan. Sci. Lett.*, **199**, No. 3-4, 403-413.

[452] van Thienen, P., Vlaar, N.J., and van den Berg, A.P., 2005. Assessment of the cooling capacity of plate tectonics and flood volcanism in the evolution of Earth, Mars and Venus. *Physics of the Earth and Plan. Inter.*, **150**, 287–315.

[453] Veizer, J., Hoefs, J., Ridler, R.H., Jensen, L.S. and Lowe, D.R., 1989. Geochemistry of Precambrian carbonates: I. Archean hydrothermal systems. *Geochimica et Cosmochimica Acta*, **53**, 845-857.

[454] Vening-Meinesz, F.A., 1939. Tables fondamentales pour la reduction isostatique regionale. Bull. Geod., **63**, 711-776.

[455] Verzhbitsky, E.V., 1996. *The Geothermal Regime and Seafloor Tectonics of Marine Basins along the Alpine–Himalayan Belt*. Nauka, Moscow (in Russian).

[456] Walter, M.J. and Trønnes, R.G., 2004. Early Earth differentiation. *Earth and Plan. Sci. Lett.*, **225**, 253-269.

[457] Walker, J.C.J., 1985. Carbon dioxide on the early Earth. *Origins of Life*, **16**, 117-127.

[458] Watts, A.B., 2001. *Isostasy and flexure of the lithosphere*. Cambridge University Press.

[459] Wentzcovitch, R.M. and Stixrude, L., 1997. Crystal chemistry of forsterite: A first-principles study. *American Mineralogist*, **82**, 663-671.

[460] Wessel, P., 1993. A Reexamination of the Flexural Deformation Beneath the Hawaiian Islands, *J. Geophys. Res.*, **98**, 12,177-12,190.

[461] Wilde, S.A., Middleton, M.F. and Evans, B.J., 1996. Terrane accretion in the southwestern Yilgarn Craton: evidence from a deep seismic crustal profile. *Precambrian Research*, **78**, No. 1-3, 179-196.

[462] Wilde, S.A., Valley, J.W., Peck, W.H., and Graham, C.M., 2001, Evidence from detrital zircons for the existance of continental crust and oceans on Earth 4.4 Gyr ago. *Nature*, **409**, 175-178.

[463] Wood, R.J., Walter, M.J. and Wade, J., 2006. Accretion of the Earth and segregation of its core. *Nature*, **441**, 825-833.

[464] Xie, Q., Kerrich, R., and Fan, J., 1993. HFSE/REE fractionations recorded in three komatiite-basalt sequences, Archean Abitibi greenstone belt: Implications for multiple plume sources and depths. *Geochimica et Cosmochimica Acta*, **57**, 4111–4118.

[465] Yakubchuk, A.S., Nikishin, A.M., and Ishiwatari, A., 1994. A Late Proterozoic ophiolite pulse, In: Circum - Pacific Ophiolites (A. Ishiwatari, J. Malpas and H. Ishizuka, Eds.), *Proceedings of the 29th Intern. Geol. Congress, Part D, Kyoto* **92**, Utrecht, The Netherlands, 273-286.

[466] Yao, Y., Morteani, G. and Trumbull, R.B., 1999. Fluid inclusion microthermometry and the P–T evolution of gold-bearing hydrothermal fluids in the Niuxinshan gold deposit, eastern Hebei province, NE China. *Mineral. Deposita*, **34**, 348-365.

[467] Yin, Q., Jacobsen, S.B., Yamashita, K., Blichert-Toft, J., Télouk, P. and Albarède, F., 2002. A short timescale for terrestrial planet formation from Hf-W chronometry of meteorites. *Nature*, **418**(6901), 949-952.

[468] Yuan, H. and Dueker, K.G., 2005. Upper mantle tomographic V_p and V_s images of the Rocky Mountains in Wyoming, Colorado and New Mexico: evidence for a thick heterogeneous chemical lithosphere. In: (Karlstrom, K.E., and Keller, G.R., eds), The Rocky Mountain region: an evolving lithosphere. *Tectonics, Geochemistry, and Geophysics. Geophys. Monogr. Ser.* **154**, 329-146.

[469] Yukutake, T., 2000. The inner core and the surface heat flow as clues to estimating the initial temperature of the Earth's core. *Physics of the Earth and Plan. Inter.*, **121**, No. 1-2, 103-137.

[470] Zahnle, K.J., Kasting, J.F., Pollack, J.B., 1988. Evolution of a steam atmosphere during Earth's accretion. *Icarus*, **74**, 62-97.

[471] Zaun, P.E. and Wagner, G.A., 1985. Fission-track stability in zircons under geological conditions. *Nuclear Tracks and Radiation Measurements*, **10**, No. 3, 303-307.

[472] Zeck, H.P., 1996. Betic-Rif orogeny: subduction of Mesozoic Tethys lithosphere under eastward drifting Iberia, slab detachment shortly before 22 Ma, and subsequent uplift and extensional tectonics. *Tectonophysics*, **254**, No. 1-2, 1-16.

[473] Zhang, Y. and Zindler, A., 1993. Distribution and evolution of carbon and nitrogen in Earth. *Earth and Planet. Sci. Lett.*, **117**, 331-345.

[474] Zhao, Z.-F., Zheng, Y.-F., Wei, C.-S. and Gong, B., 2004. Temporal relationship between granite cooling and hydrothermal uranium mineralization at Dalongshan in China: a combined radiometric and oxygen isotopic study. *Ore Geology Reviews*, **25**, No. 3-4, 221-236.

[475] Zheng, Y. and Arkani-Hamed, J., 2002. Rigidity of the Atlantic oceanic lithosphere beneath New England seamounts. *Tectonophysics*, **359**, No. 3-4, 359-369.

In: The Lithosphere: Geochemistry, Geology and Geophysics ISBN: 978-1-60456-903-2
Editors: J.E. Anderson et al, pp. 69-121

Chapter 2

CONTINENTAL AND OCEANIC LITHOSPHERE STRUCTURE FROM THE LONG-RANGE SEISMIC PROFILING

N.I. Pavlenkova
Institute Physics of the Earth, Moscow

Abstract

During the last decades of the XX century several long-range seismic profiles with large chemical and Peaceful Nuclear Explosions (PNE) were carried out by Russian institutions in the Atlantic Ocean and in the Eurasia continent. They are the Angola-Brazil geotraverse with the investigation depth of 100 km and a system of PNE profiles in Russia with the wave penetrating depth of 700 km.

2-D lithosphere velocity models were constructed for all these profiles using a common methodology for the wave field interpretation. They show that along the geotraverse the oceanic basin lithosphere is of 60-70 km thickness and it is underlined by a low velocity layer (the asthenosphere). Beneath the mid-oceanic ridge instead of the asthenosphere uplift, several local low velocity zones (asthenolites) are revealed at depth of 20-50 km. The seismic velocities between these zones are too high (up to 8.5 km/s) for such high heat flow area, they may be explained by the anisotropy effects.

In Eurasia structural peculiarities of the upper mantle are difficult to describe in the classical lithosphere-asthenosphere system. The asthenosphere can not be traced as a low velocity layer, on the contrary the lithosphere is rheologically stratified. It follows from the data of 25 PNEs which were used to compile a 3-D upper mantle velocity model for the central part of the continent. Five basic boundaries were traced over the study area: N1 and N2 boundaries at a depth around 100 km, L boundary at a depth of 180-240 km and H boundary at 300-330 km. The depth maps for each boundary and the velocity distribution map in the uppermost mantle were compiled.

In general, the old and cold cratons have higher velocities in the lithosphere than the young platforms with higher heat flows. Mostly horizontal inhomogeneity is observed in the uppermost mantle: the velocities change from the average 8.0-8.1 km/s to 8.3-8.4 km/s in some blocks of the Siberian Craton and the Urals. At the depth of 100-120 km the local high velocity blocks disappear and low velocity layers are often observed. The velocity inversions are characterised by higher electrical conductivity and it means that many of them may be a result of fluids concentration. These structural features propose that the depth of 100-120 km is a bottom of a brittle part of the lithosphere.

The visible changes of the matter plasticity are observed also at the depths of around 250 km (beneath the L boundary) where the mantle structural pattern is changed too. The H boundary has a mirror form relative to the upper boundaries (its depths are greater beneath L boundary uplifts). At these depths the Q-factor is decreased. Both these features indicate an increasing of the matter plasticity, which makes an isostatic equilibrium of the upper mantle. Thus, the L boundary may be considered as a bottom of the continental lithosphere which agrees with the heat flow data.

The rheological stratification of the lithosphere follows also from the regional boundary structure. All the boundaries are not simple discontinuities, they are heterogeneous (thin layering) zones which generate multiphase reflections. Many of them may be a result of fluids concentration at some critical PT levels. The fluids change mechanical properties of the matter, they initiate partly melting and metasomatism of the mantle material. The matter flow along weak layers results in origin of the seismic anisotropy and variation of low and high velocities.

1. Introduction

The lithosphere studies are important for understanding the evolution of different tectonic units and the nature of geophysical fields. It is becoming clear that the interaction between the crust and upper mantle is a key process influencing the development of geological structure. In order to understand this interaction the upper mantle structure has to be studied in high detail. Such studies have recently become possible due to the prevalence of long-range seismic profiles. Among them the deepest studies are the extensive seismic exploration, conducted in Russia and in the Atlantic. In Russia several long-range seismic profiles were carried out with large chemical and Peaceful Nuclear Explosions (PNE) with depth of investigation of 700 km. In the South Atlantic, the Angola-Brazil geotraverse was made with the investigation depth of 100 km.

The goal task of this study is to use the wave analysis and the 2-D and 3-D velocity models for the determination of the principal regularities in the crust and upper mantle velocity structure and their correlations with geological features or geophysical fields. It is interesting also to answer such questions as: are there regional seismic boundaries in the upper mantle and what does the lithosphere-asthenosphere system look like beneath continents and oceans? The important problem is the origin of the velocity inhomogeneity and seismic boundaries.

2. Lithosphere Structure of the Northern Eurasia

2.1. Long-Range Profiling in the Northern Eurasia

The Northern Eurasia is crossed by several long-range seismic profiles (Fig. 1), which were shot with chemical and Peace Nuclear Explosions (PNE). The studies were made by the GEON Center of the USSR Ministry of Geology (now Ministry of Natural Resources of Russia). It was a large program of seismic profiling carried out in 1970-1980 in order to study comprehensively the structures of the upper mantle and the mantle transition zone to a depth of 700 km (Benz eta l, 1992).

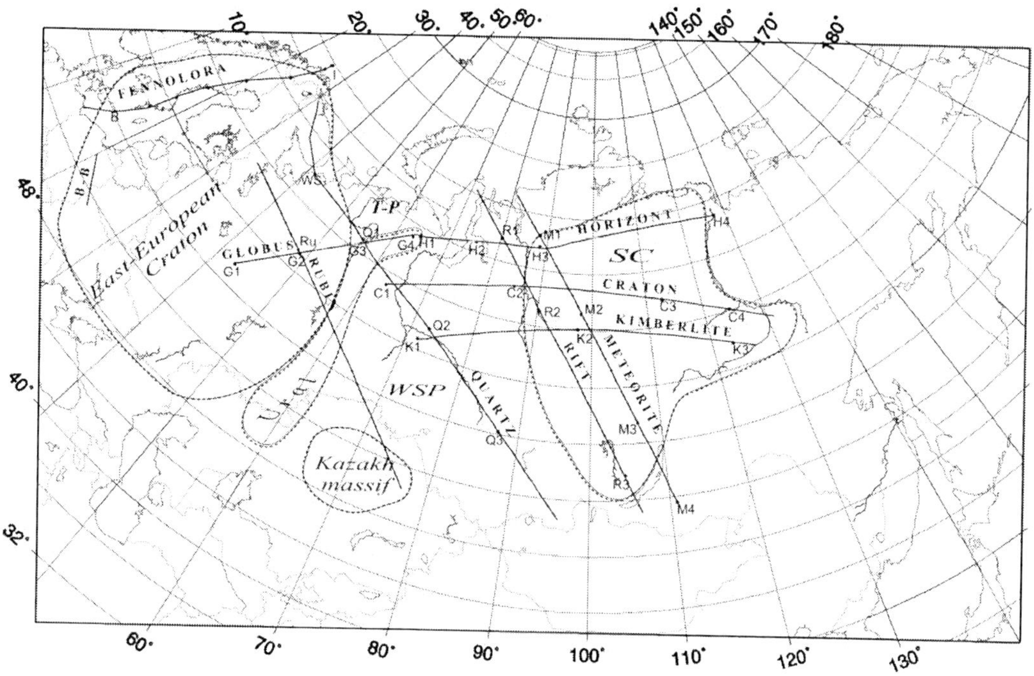

Figure 1. Map showing the location of the long-range seismic profiles and main tectonic structures in the Northern Eurasia. SC - Siberian Craton, WSP – West-Siberian Plate, T-P – Timan-Pechora Plate. Letters indicate locations of the Peace Nuclear Explosion (PNE): Q1, Q2, Q3 and WS (White Sea) along the profile QUARTZ, G1, G2, G3 and G4 – along the profile GLOBUS; H1, H2, H3 and H4 – along the profile HORIZONT; C1, C2, C3 and C4 – along the profile CRATON; K1, K2 and K3 along the profile KIMBERLITE; R1, R2 and R3 along the profile RIFT; M1, M2, M3 and M4 along the profile METEORITE; Ru along the profile RUBIN; large chemical explosions B and I al ong the profiles FENNOLORA and BB.

Three-component analog seismic stations equipped with short period seismometers (1-2 Hz) were deployed along the profile, recording all events (explosions and earthquakes) during one week. The distance between the stations was 10 km. The length of the profiles varies from 1500 to 3200 km, shot points (SP) intervals between the chemical explosions are 100-150 km. The number of PNEs varies from one on the RUBIN profile (SP Ru in Fig.1) to four on the CRATON (SPs C1-C4) and GLOBUS (SPs G1-G4) profiles. The chemical explosions provided recordings up to 300-600 km offsets, the PNE - up to 3200 km offsets. Both types of explosions provided only high quality P- wave recording.

The profiles cross several large geostructures of the Northern Eurasia: the East European-Craton, the Siberian Craton, the Urals, the West-Siberian Plate and Timan-Pechora Plate (Fig.1). The geostructures iffer in age, in geological history, in crustal structure and geophysical fields (Beloussov et al, 1991; Pavlenkova, 1996 a; Pollack et al., 1993).

The East-European Craton (EEC) is of the Archean-Proterozoic ages. Two large geostructures are distinguished in the northern part of the craton: the Baltic Shield and the Russian platform. Crustal thickness changes from 40 to 50 km and the average velocity in the consolidated crust is around 6.5 km/s. The craton is distinguished by strongly differentiated mosaic potential fields characterizing a complex structure of the consolidated crust. The heat flow distribution is rather smooth (40–50 mW/m^2).

The *Timan-Pechora Plate*, of the Caledonian (?) age, has a fairly thick, particularly in the pre-Ural region, sedimentary cover. The crustal thickness is nearly the same as that of the craton, but potential fields are much less disturbed and generally reflect the topography of the basement surface. The heat flow is elevated (up to 60 mW/m^2).

The Urals is a Paleozoic orogenic belt with its own specific features of the crustal structure: the thickness of the crust is high (up to 55 km) and its upper part contains higher density rocks whose occurrence is delineated by significant magnetic and gravity anomalies all along the belt. The heat flow is lowered to 40 mW/m^2.

The West-Siberian Plate (WSP), of the Caledonian–Hercynian age, is covered by Mesozoic sediments 3–15 km thick. The sedimentary cover is thickest (more than 15 km) in the northern part of the plate. The magnetic and gravitational fields and the crustal structure differ little from those of the EEC. In contrast, the heat flow is strongly differentiated, attaining values of 60–70 mW/m^2.

The Siberian Craton is of Archean-Proterozoic age. Two shields and two large depressions are distinguished in the craton area. The depression in the western part of the craton, the Tunguss Basin, is of 8-10 km deep and filled with high density sediments and with plato-basalts. In the eastern part of the craton there is 12 km deep Vilyui Basin filled with younger sediments. The crustal thickness is 40-45 km, the average velocity is 6.5-6.6 km/s. The magnetic field has sharply differentiated anomalies with different orientations. It is an effect of the plateau-basalts. The heat-flow is much lower than in the East-European Craton: 30 mW/m^2.

The first interpretation of the long-range profile data was made by GEON Centre and crustal models were constructed from the chemical explosion records for all profiles. The PNE data was processed in the GEON Center mainly during the last decades (Egorkin, 1980; Egorkin, Pavlenkova, 1981; Egorkin and Chernyshov, 1983; Egorkin et al., 1987; Egorkin, 1997, 1999). Some PNE data were also interpreted at the Institute of Physics of the Earth of the Russian Academy of Science (Pavlenkova, 1996 a, b; Pavlenkova et al., 1996).

During the 1990s the PNE records were digitised and some of the data became available for international groups. Three profiles were discussed in many publications: QUARTZ (Mechie et al., 1993; Riberg et al., 1996; Morozov et al., 1999; 2000), RIFT (Cipar et al., 1993; Priestly et al., 1994; Pavlenkova et al., 2002) and CRATON (Egorkin, 1997; Nielsen et al., 1999). Some profiles: GLOBUS, RUBIN, HORIZONT, have been interpreted only in the last years (Pavlenkova N.I and Pavlenkova G.A., 2006). General analysis of the data and their comparison with the other long-range seismic profiles were made in (Fuchs, 1997).

Comparison of the obtained models showed no serious difference between them in the regional plan. In all models a significant difference in velocities between the cratons and young platforms is observed and in the mantle transition zone three boundaries are recognized approximately at the same average depth. The models differ, however, in other features of the upper mantle structure. It is due to the ambiguous solutions of the inverse seismic problems, the different volumes of information used and to different methods for interpretation of wave fields. For example, some researchers obtained only ID mantle velocity models (Cipar et al., 1993; Priestly et al., 1994) or used only the first waves for seismic tomographic inversion (Nielsen et al., 1999), others, on the contrary, applied velocity filtering to pick as more secondary waves as possible (Egorkin, 1999). Most researchers determined the mantle structure from nuclear-explosion data only, not invoking chemical-explosion records. All velocity cross-sections were constructed at different detail levels. Some authors show many boundaries and many inversion zones (Egorkin and Chernyshov, 1983; Egorkin et al., 1987; Morozov et al., 1999), and other authors present very simple models (Nielsen et al., 1999). Besides, the models are presented in different forms and it is difficult to use them for the compilation of maps or 3-D velocity models.

To gain deeper insights into this ambiguity and to determine the most reliable features of resulting models, the experimental data from the all long-range profiles were compared and they were processed in terms of a unified approach, and combined solutions for both chemical and nuclear explosions were obtained. This allowed us to refine the general structure of the upper mantle and to make all resulting models agree with each other at intersection points of profiles, thereby increasing the reliability of the models. The authors have been reprocessing these data over many years in order to construct crust and upper-mantle velocity models following the same technique for all profiles. The long-range profile FENNOLORA, a part of the European Geotraverse (Mueller and Ansorge, 1988) was also included in these studies. This combined processing helped us to recognize the common features of the crust and upper mantle structure throughout the whole investigated area, in particular, to trace the reference boundaries, and thus increase the total validity of the constructions. Some generalized results are given in (Pavlenkova, 1996 a, b; Pavlenkova et.al., 2002; Pavlenkova G.A and Pavlenkova N.I., 2006). All models are presented in the common form. They characterise the crust and upper mantle structure. Regarding the mantle transition zone, there are not enough data to compile 2-D velocity models, because the zone boundaries are traced only in the narrow central portions of the profiles. These data are not considered in this paper.

2.2. The Observed Wave Fields

The average distances of wave recording from chemical explosions is 250-300 (Fig. 2), only some shots were recorded up to 500-600 km. With the distance between the seismic stations of 10 km, reliable records were obtained only for major groups of P waves characterizing the crust structure. The first arrivals are refractions from the sedimentary cover (Psed), from the consolidated crust (Pg) and from the upper mantle (Pn). The secondary arrivals are the Moho reflections (PmP) and some reflections off the crustal and upper mantle boundaries. When the sediment cover is thin, Pg forms first arrival with a velocity of 6.0 km/s at 0—50 km from the shot point. At a distance of 100-200 km, its velocity increases to 6.5-6.6 km/s and, sometimes, even 6.8-7.0 km/s. In basins, the travel times of first arrivals are characterized by velocities of 3.5—5.0 km/s (Psed) at distances of 10-40 km from the shots (offsets) and 6.0-6.4 km/s (Pg) at 40-140 km. Wave Pn forms first arrival with a velocity of 8.0—8.4 km/s at offsets of 170-300 km. Although the long (up to 600 km from the shots) records were obtained for Pn from some chemical explosions.

Record sections from the nuclear explosion (Figs. 3 and 4) show the complex structure of the mantle wave fields with drastically varying apparent velocities and amplitudes of first arrivals and numerous later high-amplitude arrivals. The most records, however, show that observed waves may be approximated by several regular wave groups: Pn, P_{N1}, P_{N2}, P_L and P_H

The apparent velocity of Pn varies over a broad range of values (7.8—8.5 km/s). Therefore, its travel times from different shots differ by 2—3 s. The highest velocities (up to 8.5 km/s) are observed in the Siberian Craton and in the Urals. The velocities of 8.1—8.2 km/s are specific for Pn waves in the other regions.

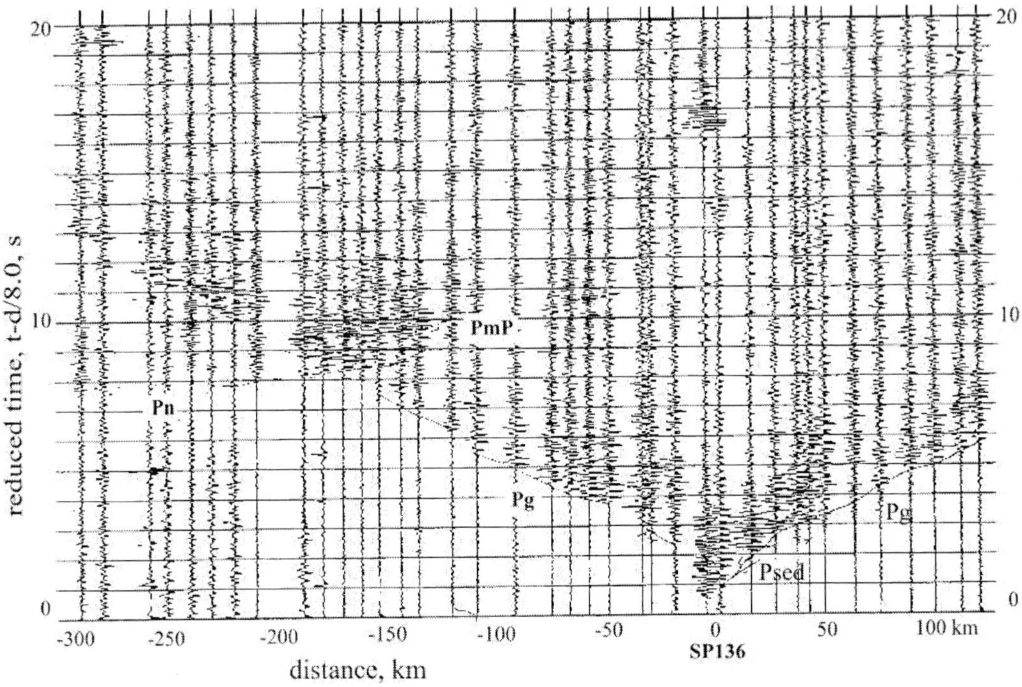

Figure 2. Record-section of the crustal and upper mantle waves from the chemical explosion shot point (SP) 136 (QUARTZ profile). The redaction velocity is 8.0 km/s. Psed is refraction from the sediments, Pg is refraction from the basement, PmP is reflection from the Moho and Pn is mantle refraction. The records show that at the trace intervals of 10 km only the first arrivals and the Moho reflections are possible to correlate.

The waves Pn and P_{N1} have often step-like travel-times because of the alteration of layers with high and low velocities in the uppermost mantle. For example, the records from SP Cl show its attenuation at an offset of 700—800 km (Fig. 3). This suggests the existence of a velocity inversion zone or a decrease in the velocity gradient at a depth of about 100 km.

At 800-1600 km offsets the first arrivals have velocities of 8.4 –8.5 km/s, these are P_{N1} and P_{N2} waves. At first these waves were interpreted as one wave P_N reflected from the N boundary at the depth of around 100 km. This boundary was identified in many other regions and was considered to be a principal boundary in the upper mantle (Pavlenkova, 1996 b). It corresponds to the 8° boundary by (Thybo and Perchuc., 1997). More detailed studies of the PNE wave fields show, however, that the P_N wave may be divided into two waves: P_{N1} and P_{N2} (Fig.3 and 4) which correspond to N1 and N2 boundaries. The boundary velocities are close (8.35 – 8.4 km/s) and their depths differ only in 20-30 km, that is why it was difficult to separate them. A characteristic feature of the P_{N1} waves is that they are often recorded after sharp attenuating (shadow zone) of Pn waves, suggesting the N1 boundary is at the bottom of the low velocity layer.

The P_L wave is observed at the first arrivals at distances of 1500—1800 km. It has a higher apparent velocity (8.6-8.7 km/s) than P_N waves. At the first arrivals this wave is weak, whereas in the later ones its intensity increases at offsets of 800-1300 km. Sometimes, the records of P_L continue the multiphase record of P_{N2}; therefore, these waves are difficult to separate (Fig. 3 and 4). P_L waves are interpreted as reflections from the L boundary at the depths of 200-250 km. This boundary is known from seismology as the Lehmann discontinuity (Lehmann, 1959).

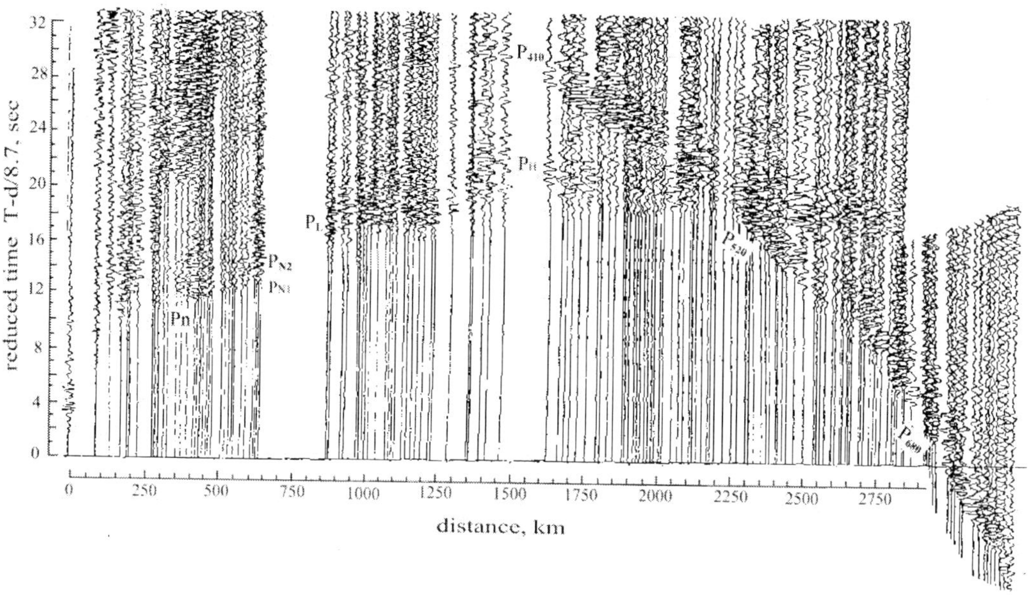

Figure 3. Record-section of the upper mantle waves from the PNE C1 along CRATON profile (the reduction velocity is 8.7 km/s). Several regular mantle waves may be traced in the records: Pn – refraction in the uppermost mantle (apparent velocities Va = 8.0-8.5 km/s), P_{N1} (Va = 8.2-8.5 km/s), P_{N2} (Va = 8.3-8.6 km/s), P_L (Va = 8.6-8.8 km/s) and P_H (Va = 8.8- 9.0 km/s) are the reflections and the refractions from the corresponding seismic boundaries N1, N2, L and H. The waves P_{410}, P_{520} and P_{680} are from the boundaries of the mantle transition zone at the depths of around 410, 520 and 680 km. The shadow zone observed at the first arrivals at distances of 750—1250 km corresponds to the low velocity layer at depth of 100-120 km.

The P_H waves are recorded mainly as secondary arrivals at the distances of 1700-2200 km. It is wave P_{350} in (Thybo at al., 1997). Due to the length, many-phases records of these waves, it is difficult to determine depths of the correspondent reflector, that is why we labelled the wave as P_H. (At first a boundary near the upper-lower mantle transition zone were revealed by A.Hales, 1969). The P_H waves are weak in most record-sections. However, sometimes, the reflections from H boundary are so strong that they can be traced to the distances of 2700 km, crossing P_{520} wave (Fig. 3).

Intense seismic records were obtained at the first and secondary arrivals at offsets of 1500-3000 km (Fig. 3), they are the waves from the upper-lower mantle transition zone. The most distinct waves are well known in seismology P_{410},. P_{520} and P_{680} waves (the wave index marks the depth of corresponding boundary, km). These waves differ significantly in apparent velocity (Va): P_{410} (Va = 10 km/s), P_{520} (Va = 10.5 km/s), and P_{680} (Va = 11 km/s).

The apparent velocities and the intensity of the described upper mantle waves change from profile to profile. To determine main tendencies in the lateral variation of the velocities in the whole area of study we used comprehensive analysis of the reversed and overlapping travel times of all profiles together. It shows that they may be divided into three regional travel-time curves which correlate with the tectonic structure: they are East-European, the Siberian Craton, and West-Siberian curves (Fig. 5).

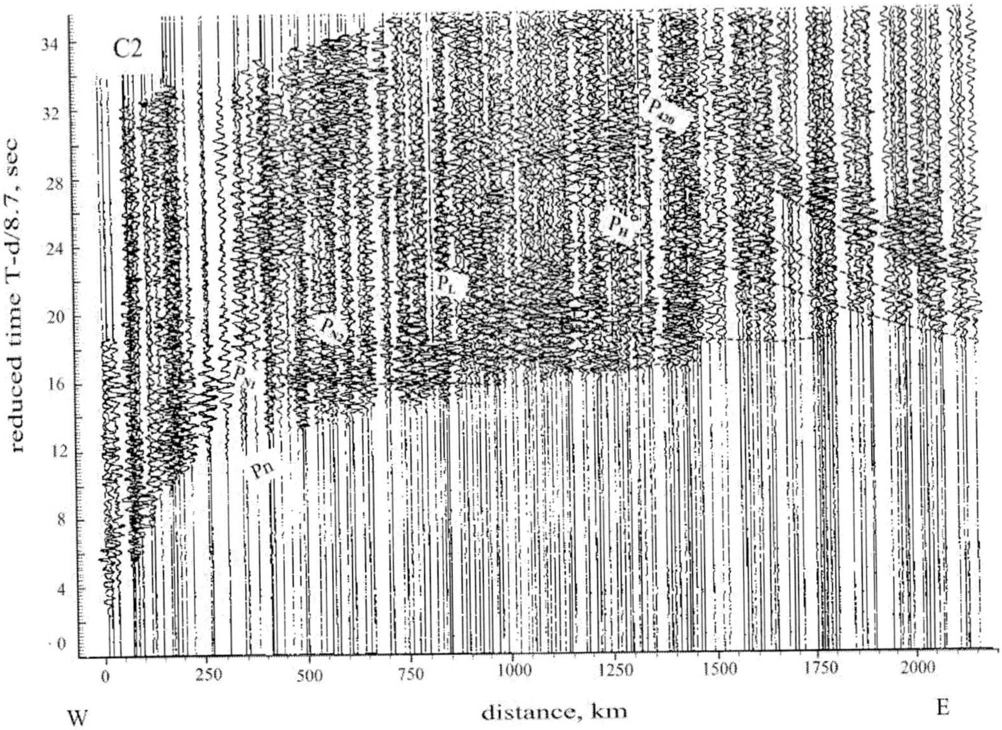

Figure 4. Record-section of the upper mantle waves from the PNE C2 along CRATON profile. All regular waves have long codas and may be separated only at the first arrivals. The wave indications are in Figure 3.

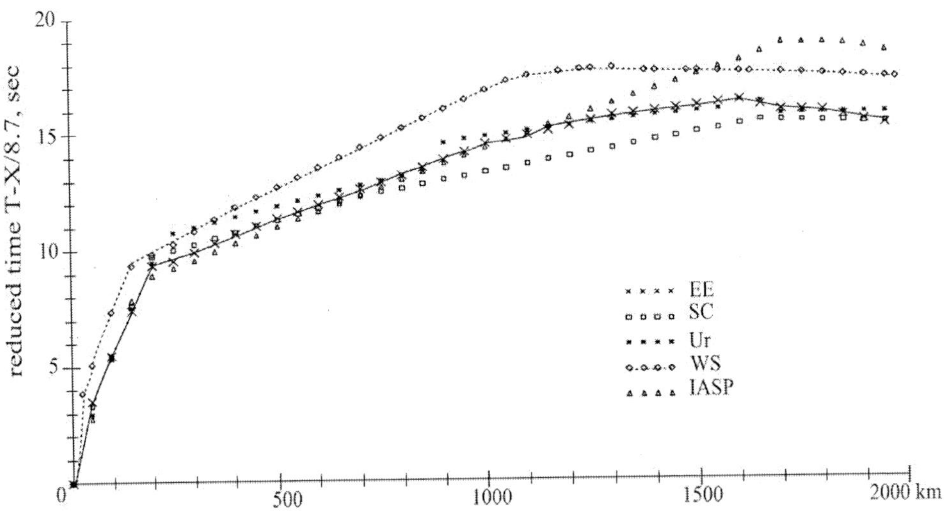

Figure 5. Regional travel-times of the first arrivals for the East-European Craton (EE), West-Siberian Plate (WS), the Urals and the Siberian Craton (SC) in comparison with IASP-91 times. Reduction velocity is 8.7 km/s. The EE regional travel-times were constructed as average from the travel-times, recorded along FENNOLORA and GLOBUS profiles, from SP Q1 with observation to the north and from the White Sea Shot. The WS travel-times were determined from PNEs Q2, Q3, H1, H2, C1, C2 and C3 data. The Siberian Craton regional travel-times (SC) were determined from the profiles RIFT, METEORITE and KIMBERLITE data.

The East-European regional travel-times (EE) were constructed as average from the travel-times, recorded along the FENNOLORA and GLOBUS profiles, from SP Q1 with observation to the north and from the White Sea Shot (WS). All these profiles are located inside the East-European Craton and they form a compact group of the observed times which indicates similarities in the crust and upper mantle structure of the craton at different parts.

The West-Siberian regional travel-times (WS) were determined at first as average times from SPs Q2 and Q3 of the QUARTZ profile and from reversed shots H1 and H2 of the HORIZONT. They characterise two young platforms: the West-Siberian and the Timan-Pechora. Later the WS times were completed by the data from SPs C1, C2 and C3 of the profile CRATON which indicate differences in the velocity structure between the northern and southern parts of the craton.

The WS travel-times differ from that of the EE one by 1.0-1.5 sec delay of the first arrivals at offsets of 200-1500 km and by corresponding shifting of P_N and P_L waves to the larger times. This time delay is mainly a result of the low velocity sediments in the West Siberia. The relative WS time increase, observed at distances of 200-700 km, occurs due to a decrease in first arrival velocities of 8.1-8.2 km/s in comparison with that of 8.1-8.3 km/s for the EE group.

The Siberian Craton regional travel-times (SC) were determined from the profiles RIFT, METEORITE and KIMBERLITE. They differ from the EE and WS travel-times by higher velocities (8.3-8.5 km/s) of the first arrivals within offset ranges of 200-1700 km.

The IASP-91 travel-times differ from all regional travel-times, mainly, in the lower velocities of P_N waves. As a result time delays of 2-5 sec are observed between IASP-91 and all regional travel times at the distances of 1000-2000 km (Fig.5).

2.3. Methodology of the Wave Field Interpretation and 2-D Velocity Modeling

The present-day system of interpretation of deep seismic sounding (DSS) data includes several steps: (1) processing of seismic records with different kinds of filtering, (2) correlation of coherent waves, (3) solution of inverse problems, (4) mathematical modeling (ray tracing) and construction of an optimal velocity model.

Filtering of seismic records is aimed to improve their resolution and tracing coherent waves. To the data discussed, two kinds of filtering were applied: frequency and velocity filtering. The potentialities of frequency filtering for picking crustal and mantle reflections are very high (Carbonell et al., 2000) and that is an important stage of the record processing. As regards the velocity filtering, its results are different for the chemical and the PNE records.

Applying this filtering to the chemical explosion data with traces spaced 10 km and more apart, when the reflections are obtained from a maximum of 7-8 traces (Fig. 2), seems unjustified. Velocity filtering applied to such records created many crustal reflections which did not correlate with original records. As a result the crustal sections constructed after the filtering are a complex system of small blocks with numerous intracrustal reflections (Egorkin, 1980; Egorkin et al., 1987). Due to the large interval between seismic stations many of these reflections might be artificial, producing unreliable details of the crustal structure. The models constructed from the primary records of the first waves and from Moho reflections are the more schematic (Pavlenkova, 1996 a).

Applying velocity filtering to correlation of mantle waves (Egorkin, 1999) is more justified because these waves are observed at larger intervals including 20-30 station records.

Thus, during our interpretation, much emphasis was placed on indicating which features of the crustal and upper mantle structure were the most reliable and could be determined from the observed data without extensive computer processing.

The wave correlation is the most complicated and unique stage of the seismic data interpretation. The correlation of the secondary waves was extremely ambiguous because of their intricate record, long coda and interference. Phase correlation is impossible in this field; usually, one can peak only the fronts of high-amplitude wave packets which are traceable for long distances and can be continued in the first arrivals (Fig. 3 and 4).

An important stage of the record analysis was determination of the regular waves which may be traced in most record-sections. Primarily the first arrivals were analysed separately for every shot-point and they were divided in branches with different apparent velocities. Strong changes in amplitudes and, especially, time delays of the first arrivals were marked. Then, all visible secondary phases were traced. The main attention was paid to the phases which had long intervals of correlation and which could be extended as first arrivals. They were several crustal and upper mantle P-waves described above. S-waves were not distinguished on the original records and could not be included in our consideration.

For some oldest profiles (GLOBUS, HORIZONT) it was difficult to perform wave correlation on the records digitized from magnetic types due to poor quality of the types after many years of storage in bad conditions. In those cases, we used the times taken from original paper record-sections.

The next important step in interpreting the data was analysis of the travel times of coherent waves and determination of their nature using all known techniques.

At this step it was important to use all information on the recorded waves from both chemical explosions and PNEs. The combination of the PNE records together with the chemical ones increase the amount of data on the uppermost mantle structure which is difficult to get from the PNE records alone. Many chemical explosions gave the mantle refraction Pn records up to 500-600 km offsets, these records may be used for detail sub-Moho velocity studies.

The reflections from Moho (PmP waves), recorded from the chemical explosions help also to determine the PNE first wave origin at 200-300 km offset because PmP waves are usually weak on PNE records and Pn refractions sometimes also have low amplitude and are not visible within the high amplitude reflections from the uppermost mantle boundaries (Fig.3 and 4).

Velocity modeling. Different methods are applied to construct velocity models. As soon as correlation of first arrivals seems more reliable, some researchers prefer the inverse solutions using these waves (Zelt and Smith, 1992). But DSS disproved the assumed reliability. The first waves are often of low amplitude and thus are not visible on seismograms. Later arrivals are more intense and can be picked as apparent first arrivals, which leads to errors in solving classical inverse problems.

These limitations are typical of seismic tomographic inversions (Nelson et al., 1999). The models obtained by these methods are restricted by a velocity function monotonously increasing with depth, without low-velocity layers and sharp boundaries. These solutions are also limited by the volume of information used: the constructions are based mainly on the travel times of first arrivals. The secondary arrivals have important information on the seismic boundaries (reflectors) which are often the main interest of the study. The reflections decrease the non-uniqueness of the velocity model construction and increase accuracy of depth determination. The velocity inversions are also important features of the mantle structure and there is no reason to exclude them from the interpretation.

The velocity modeling is the major method for constructing detail 2-D cross-sections. It consists in ray tracing for some starting models and comparison of the results with the observed fields. In case of their discordance, the model is corrected, and new calculations are continued out until the calculated and observed times agree within 0.1-0.2 s and the synthetic seismograms reflect the true relationship among the intensities of dominant waves. Most programs for the solution of a forward 2D problem are based on the ray method (Cerveny et al., 1977), widely used owing to its simplicity and obviousness. The method is convenient for mass-volume calculations and, particularly, wave field analysis. In this work we used the program described in (Cerveny and Psencik, 1983).

The accuracy and reliability of velocity modeling are determined not only by the convergence of calculated and observed fields but also by all previous interpretation steps. They depend particularly on the quality of starting models. To begin modeling with "a clean slate" is an ungracious and long work. In practice, the closer the starting model to the final solution and the less minor unjustified details it includes, the more reliable is this solution.

2D starting velocity models are often compiled from 1D solutions obtained for each shot point. As a result, the model is usually complicated by numerous small-scale heterogeneities which hamper distinguishing the main, most reliable peculiarities of velocity cross-section. This is because experimental travel times are usually distorted by subsurface heterogeneities, which are ascribed to deeper layers on 1D modeling. In minor-detail observation systems, it is difficult to get rid of such false heterogeneities during the velocity modeling. Very often additional heterogeneities compensating the effect of the false ones are introduced into the model, which complicates it yet more. An example of such a model is the lithosphere cross-section along the FENNOLORA profile (Guggisberg and Berthelsen, 1987).

A useful method for the travel time analysis and for determination of the reliable starting model is the construction of the time cross-sections. For this the intercept-time method (Pavlenkova, 1982; Pavlenkova et al., 2002) may be used. In this method the observed refraction and wide-angle reflection travel times are reduced with different reduction velocities Vr and transformed to the source-receiver mid-points. The envelopes of such travel-times are the intercept time curves $t_i = f(x, Vr)$ which may be recalculated into the depth to the velocity level V = Vr.

In Fig. 6b the travel-times of the mantle waves recorded from PNEs along the CRATON profile are reduced with Vr = 8.5 km/s and transformed to the mid-points. The t_i curve shows that the corresponding velocity boundary N1 is not horizontal. There is a regular change in the arrival times of P_{N1} in transition from the Siberian Craton to the West Siberian Plate: The envelope of the first-wave travel times, $t_i(x, V = 8.5$ km/s), corresponds to a time of 10 s in the eastern part of the Siberian Craton and 13 s within the West-Siberian Plate. This is not related to the crustal structure or lower velocity in the sedimentary cover of the West-Siberian Plate, because the time of the Pn first arrivals is nearly the same for all shot points: about 9-10 s.

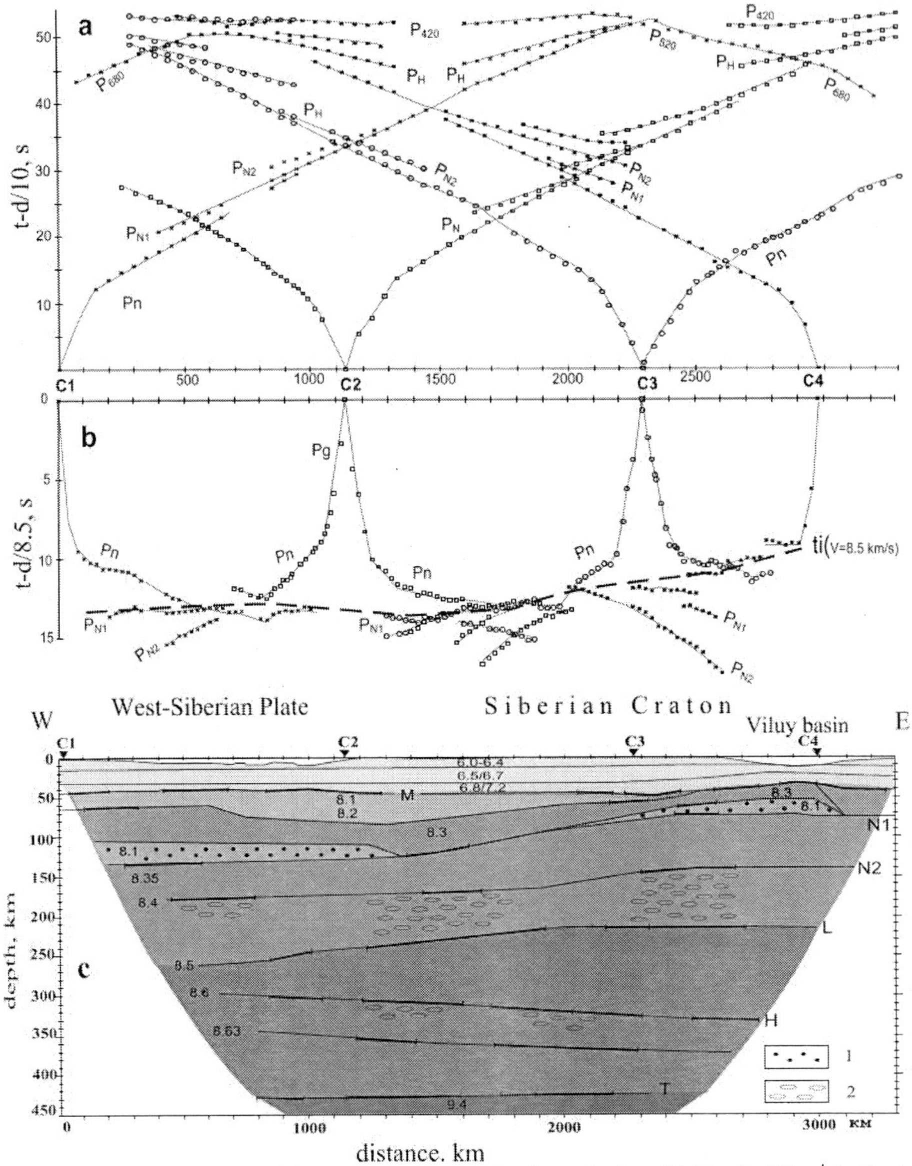

a - Direct and reverse travel times of the mantle waves. Reduction velocity =10.0 km/s. (Wave indications are in Figure 3)

b - The reduced and transformed to mid-points travel times of the basic mantle waves. Reduction velocity Vr=8.5 km/s. Dashed line t_i (8.5) is the intercept time line for the velocity level V = 8.5 km/s. These travel-times demonstrate the method of the seismic boundary form determination from the intercept time curve t_i (8.5). The latter shows that the seismic boundary with velocity 8.5 km/sec uplifts beneath the eastern part of the Siberian Craton

b - Velocity cross-section. The seismic boundaries N1, N2, L, H are the upper mantle basic boundaries with constant velocity (the velocities increase linearly between the boundaries). T boundary is the top of the upper/lower mantle transition zone. The reflectors are shown by the thick lines. 1 – lower velocity layer, 2 – high reflectivity zone

Figure 6. Observed and resulting data along the profile CRATON. C1, C2, C3 and C4 are PNE locations.

A similar pattern is observed along the parallel KIMBERLITE profile, but a strong change in ti(x, V = 8.5 km/s) is observed immediately at the boundary between the West-Siberian Plate and the Siberian Craton. Along the CRATON profile (Fig.6), the change is gradual and occurs in the central part of the platform rather than at the boundary of its structures of different ages.

The intercept time method helps also to determine the nature of the secondary arrivals: the reflection travel-times which correspond to the boundary with the velocity V = Vr, coincide with the curve t_i (Vr,x) at the critical points.

At the last stage the starting models were improved by ray-tracing. To control how reliable the main boundaries are and how stable the velocities between them are, the velocity cross-sections were determined with ray tracing, independently, along each profile. The boundaries were modelled without changing the velocities along them. Only sometimes the N boundary velocities were also changed but not more than in 0.05 km/s. If the main boundaries were not enough to describe the velocity structure and to match the observed travel-times, some additional boundaries were included in the models. The discrepancy of calculated and observed travel times for the first arrivals is within 0.1 s. For the later arrivals, whose correlation is often ambiguous because of the multiphase recording, the discrepancy between the observed and calculated times sometimes reaches 0.2 s.

The analysis of the waves lasted up to the end of data interpretation. Only numerous calculations of rays and travel times for a series of different models give insight into the nature of recorded waves and their correlation.

2.4. 2-D Velocity Models

2.4.1. QUARTZ Profile

The QUARTZ profile crosses the East-European Craton (EEC), Timan-Pechora Plate, the Northern Urals, the West-Siberian Plate, and Altai. Forty-eight chemical explosions and three PNEs Q1, Q2, and Q3 (Fig. 1) were fired on the profile. The entire profile is characterized by high quality of P wave records and the most extended system of direct and reverse observations. 400 stations recorded waves at offsets of 3200 km from the PNEs and up to 300–600 km from the chemical explosions.

A few international working groups participated in processing and interpretation of QUARTZ data. The chemical sources records were first analyzed and a crustal model was constructed (Egorkin and Chernyshov, 1983). The next stage of the interpretation was related to the activities of a German working group headed by K. Fuchs. The researchers processed PNE records. Preliminary (Mechie et al., 1993) and more detailed subsequent (Ryberg et al., 1996) models were constructed. In both cases, data from chemical explosions were not processed and the model of the crust was incorporated as a priori information. Later the profile data were processed in the United States and a crustal model based on data from chemical explosions was constructed by tomography methods (Schueller et al., 1997) and using PNE records, a velocity model of the upper mantle was constructed to a depth of 200 km (Morozova et al., 1997; Morozov et al., 1999; 2000).

The models of these authors differ in structure of both crust and mantle. The differences between the crustal models are due to different methods of their construction. Egorkin (1999) widely applied velocity filtering for processing of records and, therefore, obtained complex models with numerous small blocks. Schueller et al. [1997] used only the first arrivals and the Moho reflections for the construction of their 2-D model of the crust. The distinctions in the upper mantle models constructed by different authors are also due to the volume of the information used.

In the reconstructions presented below, the records from 15 chemical sources were incorporated into the general solution, and the NW segment of the profile was complemented by the WS PNE data (Fig. 1). The WS nuclear source was located on the QUARTZ profile, but records were obtained on the White Sea-Vorkuta profile, nearly parallel to QUARTZ and located a few kilometers to the north. The comparison of the WS records with the reverse PNE Q1 at reciprocal points showed that it can be used for construction of a velocity model along QUARTZ.

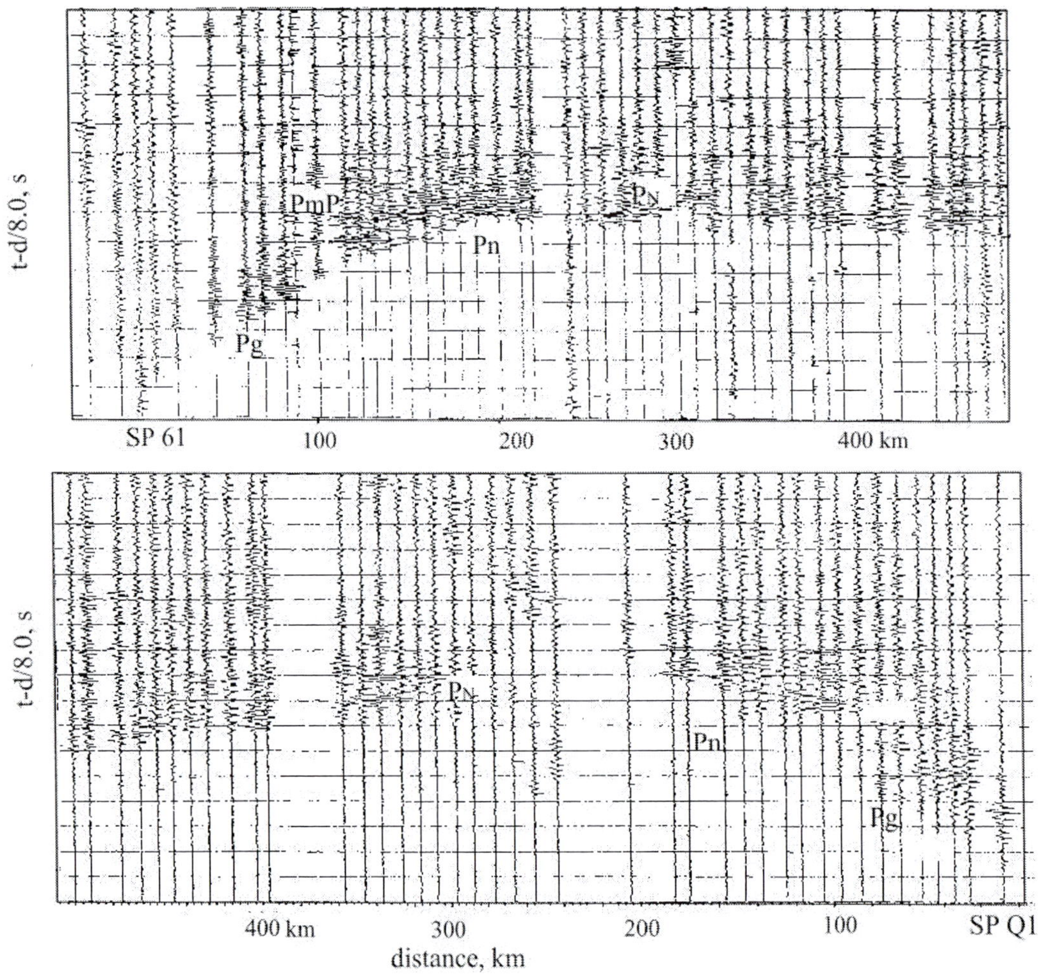

Figure 7. The reversed record sections from the chemical explosion SP 61 and PNE Q1 (QUARTZ profile). P_N waves have different apparent velocities in the reversed records and that means the corresponded reflector would be inclined to the east. The other wave indications are in Figure 2.

The crustal structure. The four main waves are traceable on records from chemical sources along the QUARTZ: Psed, waves refracted in the sedimentary cover; Pg, waves refracted in the crystalline crust; PmP, Moho reflections; and Pn, Moho refractions (Fig. 2). The mantle waves P_N arriving after Pn could be also traced from some shot-points (Fig. 7). Travel-time data from the waves served as a basis for the construction of a velocity model of the crust and uppermost mantle. The crustal reflections of intracrustal boundaries were also used.

Apparent velocities, amplitudes, and the intercept time t_i of crustal waves vary significantly from one geostructure to another. The sedimentary thickness within the Baltic Shield and Russian platform is either zero or very small, and Pg records start here from distances of 0–30 km. They consist of three branches with velocities of about 6.2, 6.5, and 6.8 km/s. The intercept time of the Pn wave amounts to 6–7 s and the Pn velocity is 8.1 km/s. Within the Timan-Pechora Plate, Psed branches are long and their apparent velocity varies from 3.5 to 5.5 km/s; Pg velocities (6.0 to 6.6 km/s) are lower compared to the Russian Plate. The Pn value of t_i is 7–8 s, and the apparent velocity is 7.9–8.0 km/s. Very poor records obtained for the Urals do not resolve even the PmP waves. An approximate Pn value of t_i is 11–12 s.

Due to the thick sedimentary cover of the West-Siberian Plate (WSP), the Psed waves with velocities of 3.5–4.5 km/s are observed in the most records up to a distance of 20-30 km. First arrivals of Pg waves yield velocities of 6.2–6.4 km/s. High velocity (6.8–7.0 km/s) reflections from the lower crust are traceable at the secondary arrivals. According to the apparent velocities of PmP waves, mean velocities in the WSP crust are higher compared to the Timan-Pechora Plate. The value of t_i for Pn is about 8 s and the apparent velocity is 8.0 km/s.

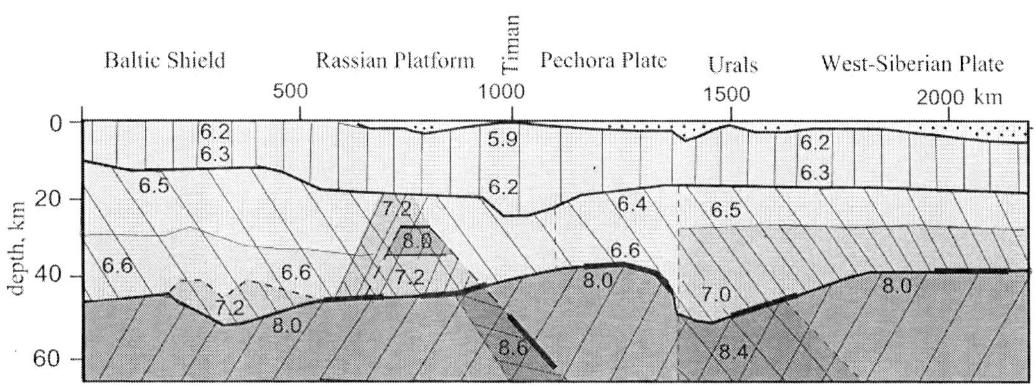

Figure 8. Seismic cross-section of the crust along the northern part of the profile QUARTZ. The most differences in the crustal and uppermost mantle structure are observed at the boundaries between the Russian platform (East-European Craton), the Timan-Pechora Plate and the Urals. The inclined reflector beneath the Moho corresponds to the P_N waves in Figure 7.

The crustal velocity model for the NW part of the profile is shown in Fig. 8. The most prominent horizontal heterogeneities in the crustal structure are seen in the transitional zones between the Russian platform, the Timan-Pechora Plate and the Urals. The velocities at the basement are 6.2 km/s beneath the Russian platform and 5.9–6.0 km/s beneath the Timan-Pechora Plate, where the high velocity (6.8–7.0 km/s) layer of the lower crust observed in all remaining parts of the profile is absent. The depth to Moho varies along the profile from 40–45 km beneath the EEC to 55 km beneath the Urals. A relatively thin crust is characteristic of the Timan-Pechora Plate.

A high velocity body is recognizable in the middle crust beneath the Russian platform edge. This indicates that the crust at the EEC edge was significantly reworked by mantle intrusions, which is generally untypical of the northern margin of the platform. In previously obtained velocity models (Egorkin and Chernyshov, 1983; Schueller et al., 1997), a significant rise of Moho was revealed instead of the given high velocity body. The detailed analysis of wave fields by the reduced travel-time method has demonstrated that this is a body with mantle velocities separated from the mantle by a velocity inversion zone.

A characteristic feature of the crustal structures is in the Russian platform – the Urals transition zone is the presence of well-expressed fault zones separating the platform from the Timan Ridge and the Pechora Plate from the Urals. The type of the crust and the Moho structure changes across these zones. Likewise, the upper mantle velocities abruptly change here from 8.2 to 7.9 km/s. This change takes place at an inclined reflector dipping eastward from Moho to a depth of 80 km. This reflector was reconstructed from reverse records of chemical SP 61 and PNE Q1 (Fig. 7). The wave Pn from PNE Q1 is very weak but, at offsets of 400 km, its first arrivals are succeeded by a very high amplitude wave P_N also traceable at the secondary arrivals with a very high apparent velocity (up to 9 km/s). The reverse SP 61 record section displays no waves with high apparent velocities. The Pn and the subsequent P_N waves are recorded as first arrivals with velocities typical of the upper mantle (about 8.0–8.1 km/s). In separate treatments of chemical and nuclear explosions, this important feature was not resolved because P_N from SP 61 was considered to continue the Pn wave, and the data from SP Q1 were interpreted in terms of the variability of the Moho velocities.

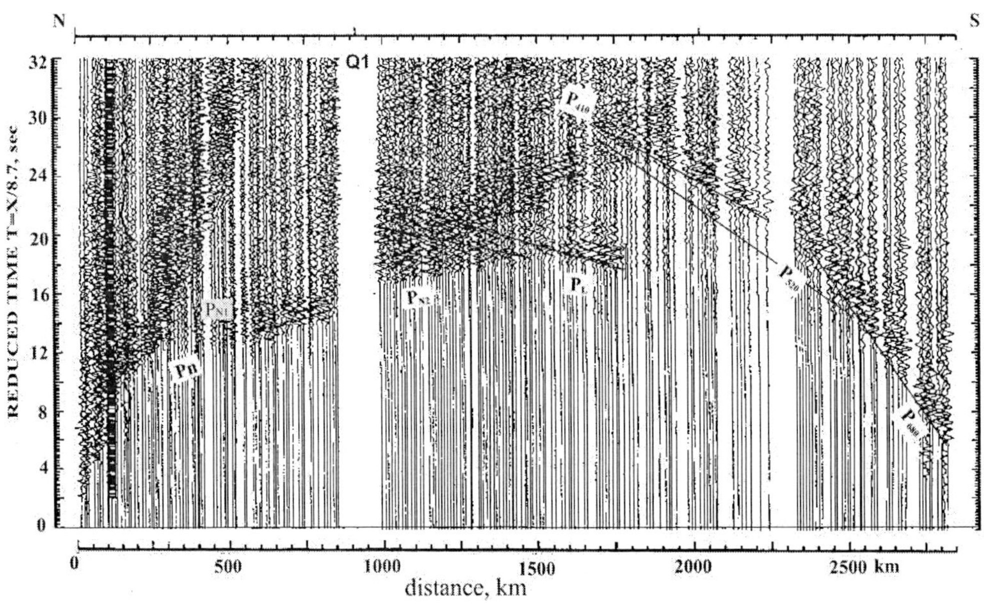

Figure 9 Record-section of the upper mantle waves from the PNE Q1 along QUARTZ profile. Legend is in Figure 3. The peculiar features of the records are the high amplitude secondary arrivals with lower velocities (the multiple refractions) and the sharp attenuation of P_L wave. The latter may indicate decreasing of the velocity gradient beneath the L boundary.

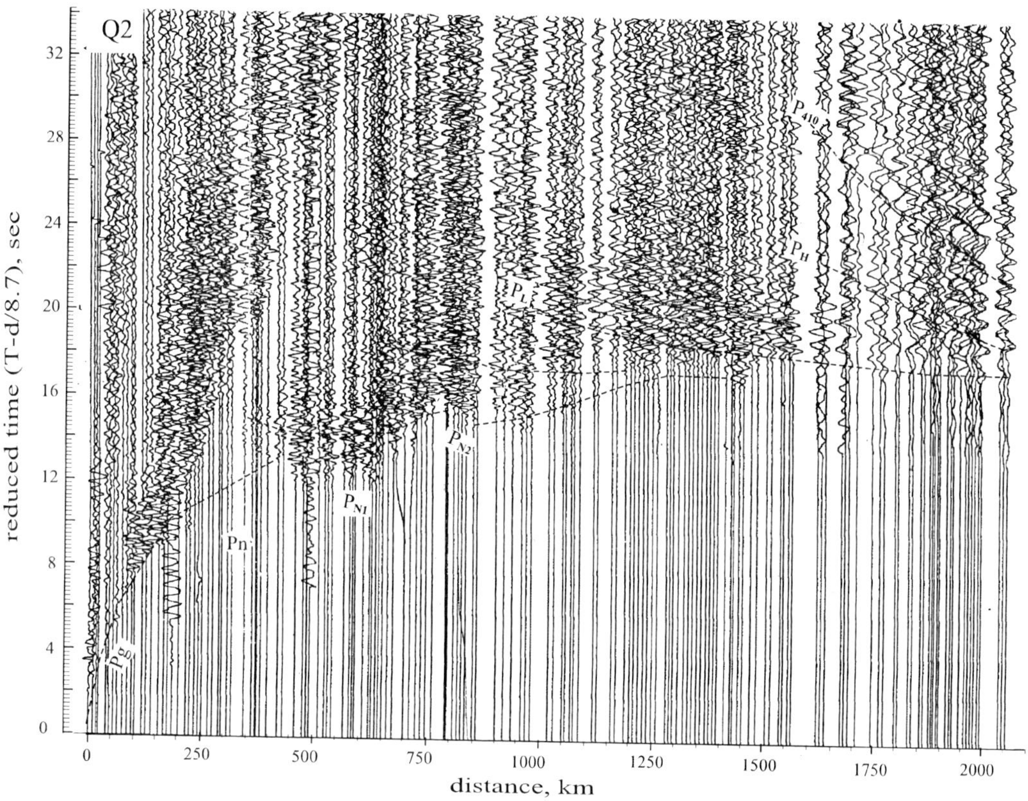

Figure 10. Record-section of the upper mantle waves from the PNE Q2 along QUARTZ profile with observations to the north-west. Legend is in Figure 3. The section demonstrates the complicated and unreliable records of Pn, P_{N1} and P_{N2} waves and the dominant P_L wave.

The structure of the Urals is asymmetric. Beneath its western slope, the Moho abruptly dips, forming a bench, while beneath the eastern slope it rises rather smoothly toward the WSP. The upper mantle is also very heterogeneous here. A layer of anomalously high velocities (more than 8.4 km/s) underlying Moho is revealed beneath the Urals.

The structure of the crust and upper mantle in the remaining part of the profile is less disturbed. The crustal thickness gradually increases from 40 km under the WSP to 50 km under Altai, and abrupt changes in the crust type are not observed.

The upper mantle structure. Records from nuclear sources obtained on the profile reflect a complex pattern of the wave fields with sharp variations in apparent velocities and amplitudes of mantle waves recorded as first arrivals and as high amplitude secondary arrivals (Figs. 9-11).

Differences between wave fields from different shot points are observed not only in arrival times of reference waves but also in their arrival patterns, relative intensities, and the presence of additional waves. With observations in SE azimuths, the wave field from SP Q1 (Fig. 9) is characterized by a distinct splitting of the first arrivals into two branches. The first branch, Pn, with velocities of 8.3–8.4 km/s is observed at offsets of 400–900 km; farther, it rapidly decays and, with a certain time delay, the wave P_{N1} having the same intensity and apparent velocity is identified at distances of up to 1500 km. The difference between these travel-times is 1-2 s (Fig. 12 a). Such a wave pattern is characteristic of a velocity inversion zone. At a source–receiver distance of 1500, first arrivals are the wave P_L with an anomalously high velocity (8.8 km/s), but the interval of its observation is very short (up to 1800 km); at greater distances, the wave rapidly attenuates. All these waves are reliably identified only as first arrivals because subsequent parts of records are filled with multiples with low apparent velocities (less than 8.0 km/s). A specific feature of 20^0 waves from SP Q1 is their distinct division into waves P_{410} and P_{680} that have here records typical of reflections. The weak P_{520} refractions are also observed but their correlation is unreliable.

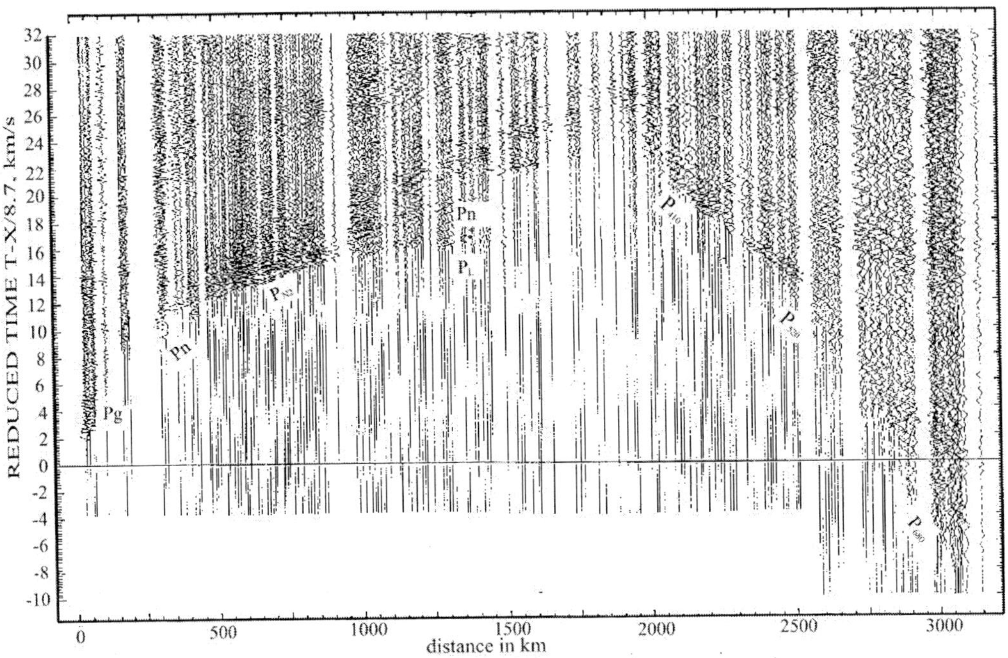

Figure 11. Record section of the upper mantle waves from the PNE Q3 along QUARTZ profile. Legend is in Figure 3. A specific feature of the records is the wave Pn recorded at the secondary arrivals up to distances of 1500 km. The wave was considered as a telesismic wave of a complicate origin (Ryberg et al., 1995; Morosov et al., 1998). It might be, however, the ordinary refraction Pn with high intensity due to focusing effect of the upper mantle inclined boundaries at PK 2700-3200 km (see Figure 12).

A different pattern is observed in SP Q2 observations in NW azimuths (Fig. 10). The Pn wave is not recorded here, and distinct first arrivals appear only at offsets of 500 km and persist to 1200 km; the arrivals are the P_N waves with high velocities (8.6–8.7 km/s). They are separated into two waves P_{N1} and P_{N2} by steep benches. The secondary arrivals clearly define the P_L wave. Very weak arrivals are also recognizable in the subsequent part of the record, at distances of 1600–2000 km; this is the P_H wave, which was rather reliably identified on other long-range profiles, but here its presence can only be suggested.

The wave field from SP Q3 is more stable (Fig.11). As in the preceding cases, the wave Pn is very weak. At distances of 300–1000 km, distinct first arrivals are formed by P_N waves, followed by the wave P_L that is less expressed here and decays at a distance of 1500 km from the source. The secondary arrivals are a low velocity (8 km/s or less) wave that was called the teleseismic wave Pn in (Ryberg et al., 1995; 1998; Tittgemeyer et al., 1996; Morozov et al.,

1998; Nielsen et al., 2002). Its origin, widely discussed in the literature, is still not defined unambiguously. According to our calculations, this is an ordinary refracted wave from the upper mantle originating above the boundary N2. Their high intensity at the large offsets is explained by focus effect of the inclined velocity isolines in area of the ray incoming in the upper mantle (PK 2700-3000 km in Fig.12b).

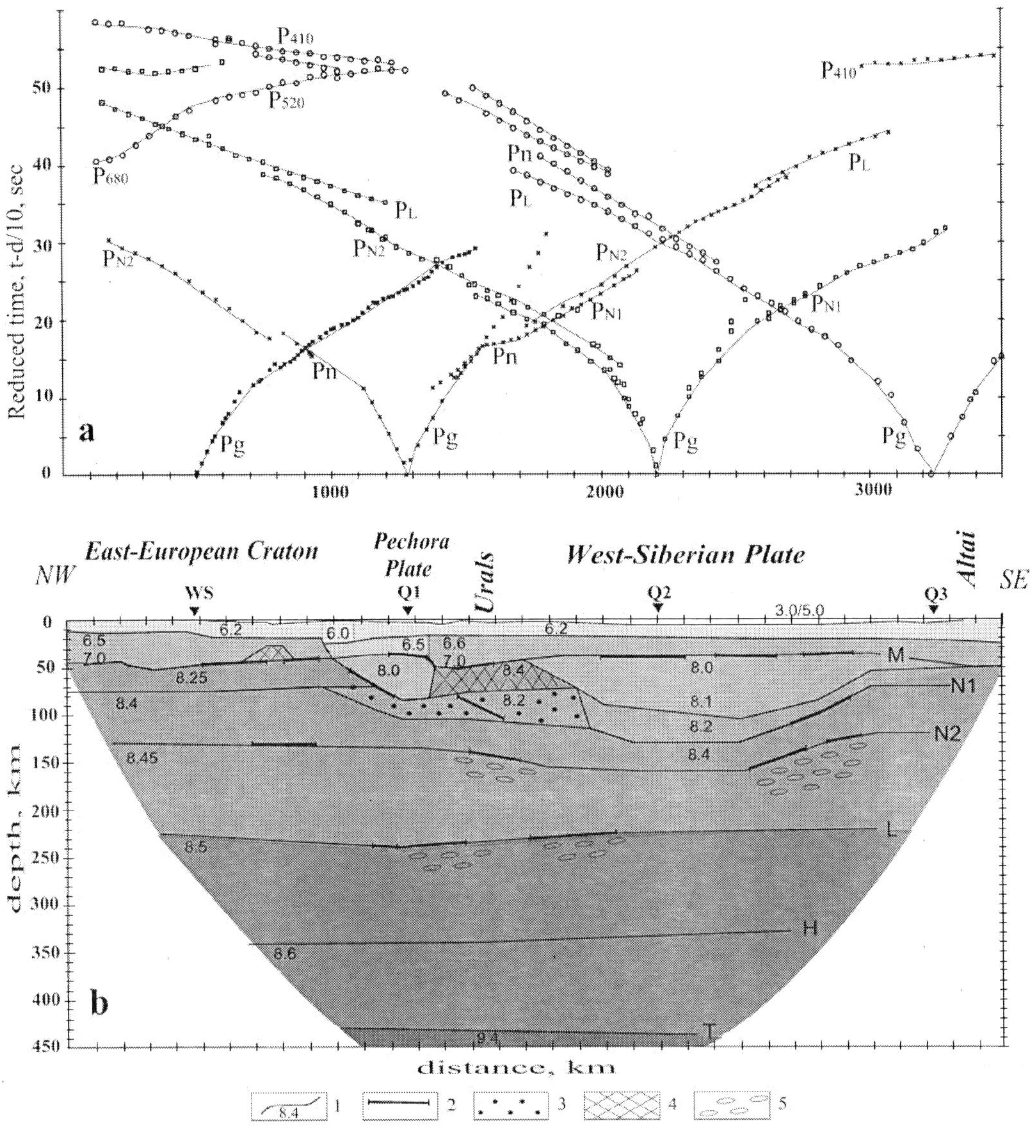

Figure 12. Observed travel times (a) and velocity cross-section (b) along the profile QUARTZ. 1 – refraction boundary, 2 – reflectors, 3 – low velocity zone, 4 – high velocity block, 5 – high reflectivity zone. The most differences in the upper mantle structure are characteristic for the old East-European Craton and the young Pechora and West-Siberian platforms. The differences are observed only up to depth of 100-120 km and that determines the ridge part of the lithosphere. No indication is revealed for the "thermal" lithosphere bottom (depth of 250-300 km).

The main feature of the upper mantle structure along the QUARTZ profile is its significant heterogeneity in the upper 150 km correlating well with large tectonic structures. Higher velocities (over 8.2 km/s) in the upper mantle are characteristic of the EEC, the Urals, and the Altai, while smaller values (8.0–8.1 km/s) are observed beneath the Timan-Pechora Plate and WSP. Higher velocities (8.3 km/s) are observed in the Urals immediately under Moho. Overall, the sharpest variations in the upper mantle structure are typical of the Timan–Urals region. Here, velocity variations are significant in both the crust and the mantle, the crustal thickness also varies, and the reflectors dipping eastward are resolved. These reflectors could be related to the subduction zone at the NE margin of the ancient platform, if they dipped in an opposite direction, under the East-European Craton. Their dip angle evidently reflects the general tendency of tectonic movements in this region, with thrusts overriding the EEC margin.

Another characteristic feature of the upper mantle structure is the presence of a lower velocity zone at depths of 70–100 km. Its reliable identification is based on the reverse records from PNEs Q1 and Q2 (Figs. 9, 10 and 12 a), where a shadow zone and displacements of first arrivals by nearly 2 s at an offset of 800 km are observed.

For the Western Siberia two models are possible, with or without a lower velocity layer at depths of 80–100 km, but we chose the second variant because the observed fields do not provide direct evidence for a velocity inversion.

Another layer of lower velocities could be expected at a depth of about 200 km, where the asthenosphere should be present according to heat flow data (Cermak, 1985). Possibly the vertical gradient of velocity decreases here under the boundary L, resulting in a decay of first waves at distances greater than 1500 km. On the whole, horizontal heterogeneity of the mantle is insignificant at depths below 200 km.

The regional boundaries identified in the upper mantle differ in their characteristics. The boundary N1 is often the base of a lower velocity layer. The strongest reflections are obtained from the boundaries N2 and L. However, their records are of a complex multiphase type, indicating a complex structure of these boundaries represented by heterogeneous groups of layers.

Data on the structure of the upper–lower mantle transition zone are obtained only from the central part of the profile (distances of 1800–2500 km along the profile). Three main boundaries are identified in this zone. All of them slightly dip southeastward: the top of the transition zone (T boundary) dips from 410 to 420 km, and an inner boundary, from 520 to 530 km. The depth obtained for the base of the transition zone is significantly smaller than the value obtained from world data: 640–650 km rather than 680 km.

2.4.2. CRATON Profile

The CRATON profile crosses the Siberian Craton and the eastern part of the West-Siberian Platform (Fig.1). Four PNEs and around 30 chemical explosions were recorded along the profile providing information of the whole crust, upper mantle and the mantle transition zone. The PNE records are characterised by the stable wave fields with clear correlation of the regular waves (Fig. 3, 4 and 13). In the eastern part of the profile, several chemical explosions were recorded up to 600-700 km offsets and gave important information on the uppermost mantle structure (Fig. 14).

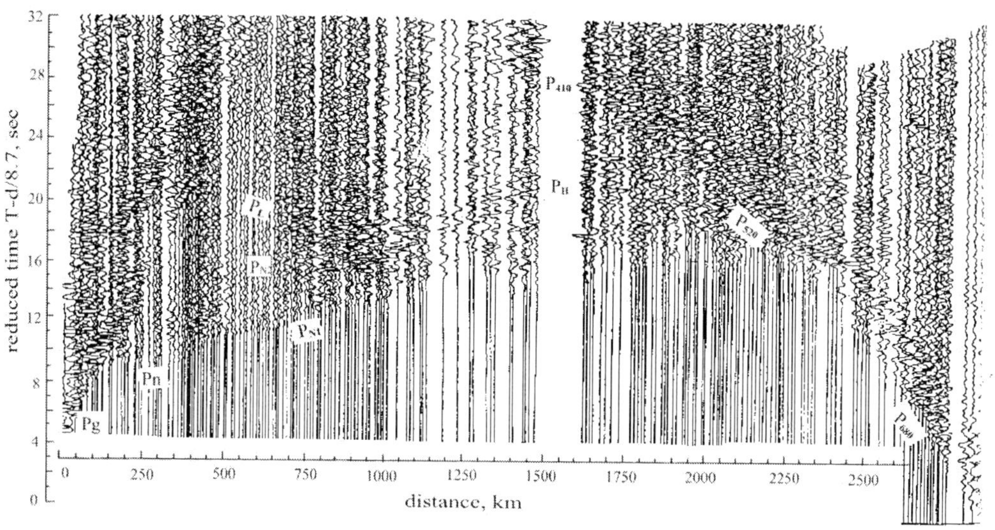

Figure 13. Record-section of the upper mantle waves from the PNE C4 along CRATON profile. Legend is in Figure 3.

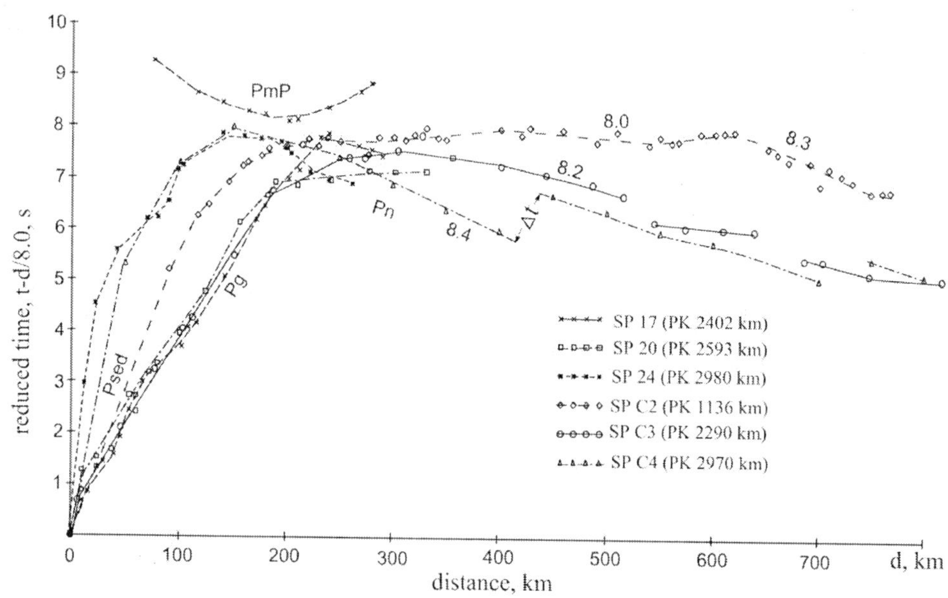

Figure 14. Comparison of the travel-times of crustal and mantle waves from different shot points (SP) of the profile CRATON which demonstrate the velocity variation in the uppermost mantle. Legend is in Figure 2. Numbers are the apparent velocities.

The constructed cross-section (Fig.6) reflects the major specific structure of the crust and upper mantle of the Siberian Craton and West-Siberian Plate. The velocity model of the crust includes four layers: sedimentary cover and three layers of consolidated part with velocities of 6.0-6.4 km/s (upper crust), 6.5-6.7 km/s (the middle crust), and 6.8-7.2 km/s (the lower crust). The thickness of sediments along most of the profile is small (0-3 km), but it increases to 10 km in the area of the Vilyui basin, where the sedimentary cover is characterized by velocities of 3.5-5.0 km/s. In the other parts of the platform, the velocities in sediments are, on the contrary, high, 4.5-5.0 km/s.

The structure of the crystalline crust also significantly varies mainly in the area of the Vilyui basin. The upper crust is 10 km thick in the platform area, whereas beneath the basin it is lacking. The thickness of the middle crust also varies along the profile: it is 20 km in the platform area and as small as 10 km beneath the basin. The lower crust is 15-17 km throughout the profile.

The average depth to the Moho varies along the profile within 40-45 km. decreasing to 36 km beneath the Vilyui basin. The velocities beneath the Moho average 8.1—8.2 km/s, except the profile interval 2500-3000 km where they increase to 8.3 km/s. This anomalous high-velocity block is localized in the western part of the basin, and its eastern margin coincides with the area of the sharpest subsidence of the basement (Fig. 6c).

The obtained upper mantle model shows that the highest heterogeneity of the velocity section is specific for the upper 130 km of the mantle. In addition to a high-velocity block, low-velocity layers have been revealed at depths of 100-130 km in the western part and at depths of 60-70 km in the eastern part of the profile. In the central part no velocity inversion is observed. The depth to the boundary N1 changes from 130 km in the west to 70 km in the east of the profile.

The boundaries N1, N2 and L are of similar structure. They rise from west to east, but the rise amplitude decreases with depth. The depth of N2 is 180 km in the west and 140 km in the east, and L boundary occurs at a depth of 260 km in the west and rises to a depth of 220 km in the eastern part of the profile. The boundary H is nearly horizontal but slightly tends to subsidence from west to east in contrast to the rise of the upper boundaries.

All the boundaries are not simple first-order discontinuities but a multilayered heterogeneous zone, which generates complicated multiphase reflections. In Fig. 6 these domains are marked by ovals.

The structure of the transition zone is not shown in the figure as it was studied only in a small part of the profile in the interval 1000-2000 km. All the boundaries in this zone including the top of the transition zone (T boundary in Fig. 6) are near horizontal (in the depth ranges 410-430, 500-520, and 660-680 km respectively).

2.4.3. KIMBERLITE Profile

The KIMBERLITE profile also crosses the Siberian Craton and the eastern part of the West-Siberian Plate (Fig.1). This profile features three PNEs. The upper mantle waves are not so strong on the Kimberlite profile (Fig.15); we determined, however, all waves described above: P_{N1}, P_{N2}, P_L and P_H. On the records of the chemical explosions there are also mantle arrivals up to 400-600 km offsets (Fig.16).

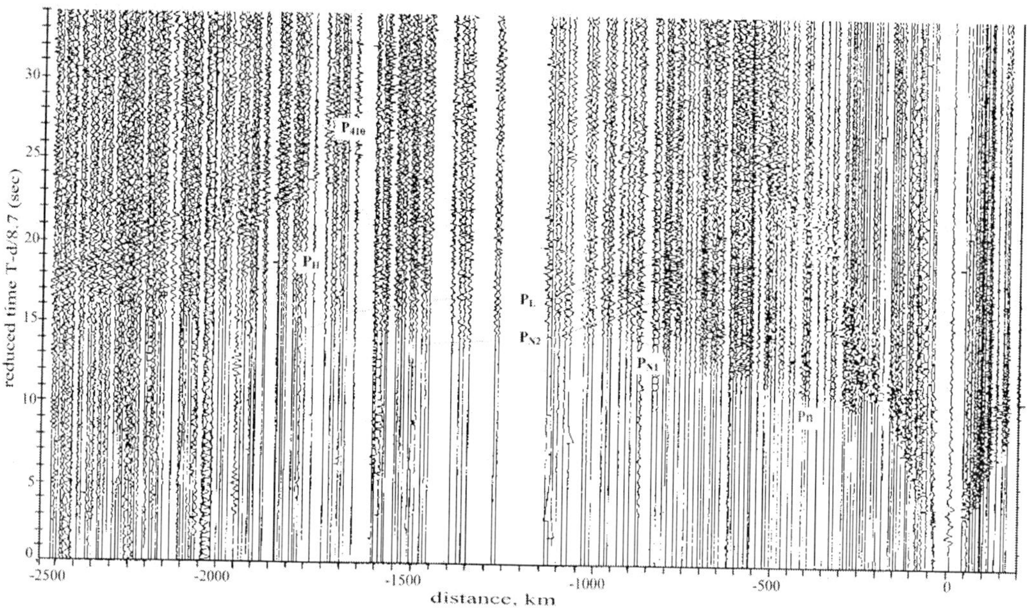

Figure 15. Record-section of the upper mantle waves from the PNE K3 along KIMBERLITE profile. Legend is in Figure 3. The records of the P_{410} wave are unusual in the section: two separate reflections are observed from the transition zone top.

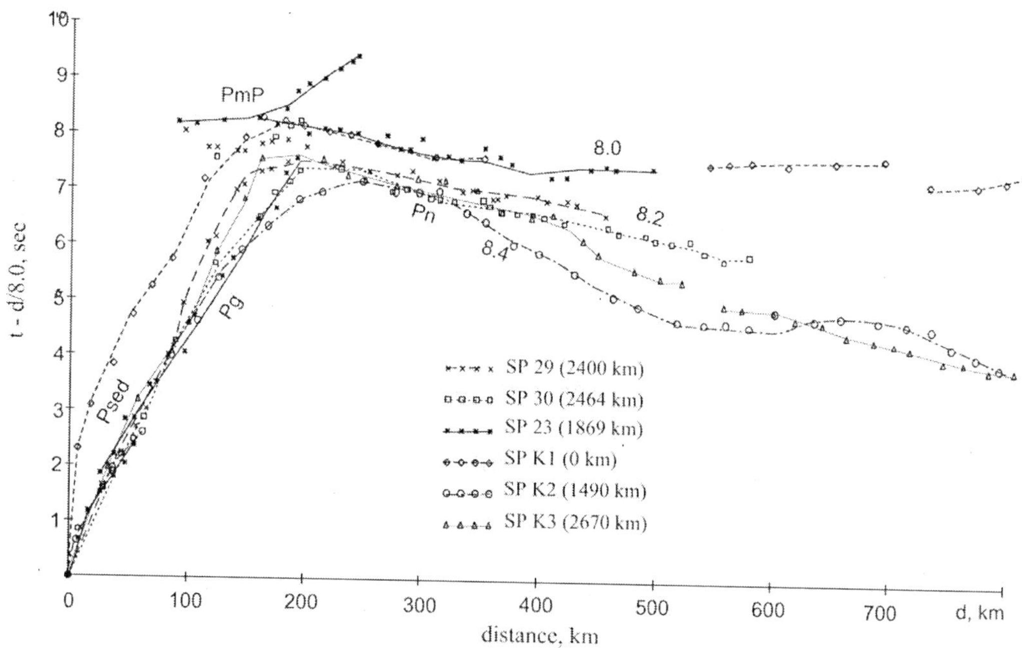

Figure 16. Comparison of the travel-times of crustal and uppermost mantle waves from different shot points (SP) of the profile KIMBERLITE which demonstrate the velocity variation in the uppermost mantle. Legend is in Figure 14.

The crustal structure along the profile is similar to the CRATON profile: three layered consolidated crust with the average thickness of 40-45 km is observed in the Siberian Craton and the West-Siberian Platform region and it changes in the Vilyui Basin.

The first arrivals of the mantle phases show higher velocities than in the CRATON profile. The largest velocity variations are observed in the upper 150 km of the mantle (Fig. 17). They change from the average 8.1- 8.2 km/s to 8.3 km/s in two high velocity blocks: at the boundary between the Siberian craton and the West-Siberian Platform (Tunguss block) and in the Vilyui Basin area (Vilyui block). In the western part of the profile at the depth of 100-120 km a low velocity zone is determined. The velocity inversion is also outlined beneath the Vilyui high velocity block.

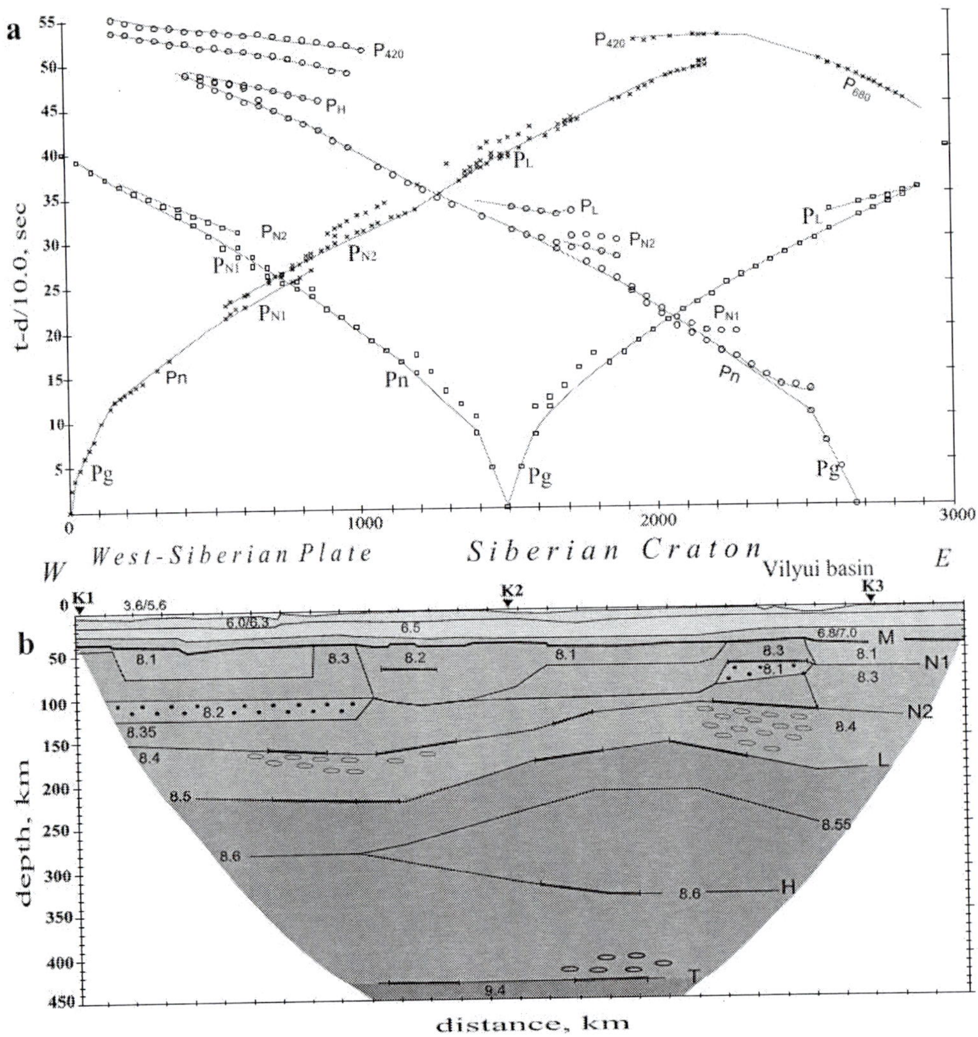

Figure 17. Observed travel times (a) and velocity cross-section (b) along the profile KIMBERLITE. Legends are in Figure 3 and 12. The most inhomogeneous is the upper 120 km of the lithosphere with the high velocity blocks beneath the Moho and with low velocity zones at depth of 70-120 km. The differences in the velocities between the West-Siberian Plate and the Siberian Craton are characteristic for the whole upper mantle up to depth of 400 km.

At the depths of 100-250 km a clear difference is observed in the mantle structure between the Siberian Cratonove the Siberian Craton and the West-Siberian Platform. As was mentioned before; the envelope of the reduced travel-times that is the intercept-time curve for the velocity level 8.5 km/s (the N2 boundary), shows a time decrease from west to east. This is imaged by an uplift of the N2 boundary directly at the western margin. The similar structure is typical of the N1 and L boundaries (Fig.17b). At the depth of 300 km the situation changes: the H boundary depth increases beneath the craton. To describe this structural change an additional boundary is included in the model with V= 8.55 km/s.

Thus, the velocity cross-section along the KIMBERLITE profile in general is similar to that of the CRATON profile. There are, however, some differences. The Tunguss high velocity block is not distinguished on the CRATON profile. On the contrary in this part of the profile, the uppermost mantle velocities are lower than those in the other area of the craton. That indicates the mantle structure of the northern part of the Siberian Craton to change on the CRATON profile not beneath the craton margins as on the KIMBERLITE profile but in its central part.

2.4.4. RIFT Profile

The Rift profile crosses the northern part of the West-Siberian Platform in the area of deep Pur-Gidan Basin, the western part of the Siberian Craton and the Baikal Rift Zone (Fig.1). Three PNEs and around 35 chemical explosions were shot on the profile. The experimental data and some results of their interpretation are described in (Cipar et al., 1993; Priestly et al., 1994; Pavlenkova et al., 2002).

The most complicated wave pattern and crustal structure is observed in the Pur-Gedan Basin (Fig. 18). Here, the wave field is defined by low velocities in the sediments (from 2.7 to 5.3 km/s) and by extremely large values of Pg intercept-times t_i (5 sec in reduced scale with a velocity reduction of 8.0 km/s). It corresponds to a basement depth of around 15 km. The Pg velocities are high - 6.3-6.4 km/s. The refractions from the lower crust with velocities more than 7 km/s are clearly traced at the first arrivals. The Moho reflections are not stable beneath the basin and sometimes they are difficult to separate from the reflections at the lower crustal boundaries. In the southern margin of the Pur-Gedan basin the Pn first arrivals consist of several branches with high apparent velocities. They correspond to high velocity intrusions in the crust (Fig.18).

In the Siberian Craton, the velocity of Psed is approximately 6 km/s. This high velocity is due to the plateo-basalts in the sediments. However, as follows from the time delay between the Psed and Pg waves, the basalt layers are underlain by lower velocity sediments (5.5 km/s). The velocities in the basement (6.2-6.4 km/s) are lower than beneath the Pur-Gedan basin. The refractions with higher velocities (6.8-7.0 km/s) characterizing the lower crust have been observed here at the first arrivals. The Moho reflections and refractions are stable, they were recorded at the time around 8 sec (at the reduction velocity of 8.0 km/s), Pn velocity is 7.9-8.8.2 km/s.

The mantle waves observed on the profile also show some peculiarities (Fig. 19 and 20). The first wave, Pn, is recorded at distances of 200-800 km. Its intensity attenuates rather sharply with distance, and apparent velocities vary from 8.0 to 8.6 km/s. The next wave group (P_N) was recorded as a first arrival at distances of 800-1600 km with dominant average velocities of 8.4-8.6 km/s. This wave group is observed as the second arrival as well. The shadow zone observed between the Pn and P_{N1} on PNE R2 records (Fig. 20) requires a velocity inversion at about 80 km depth.

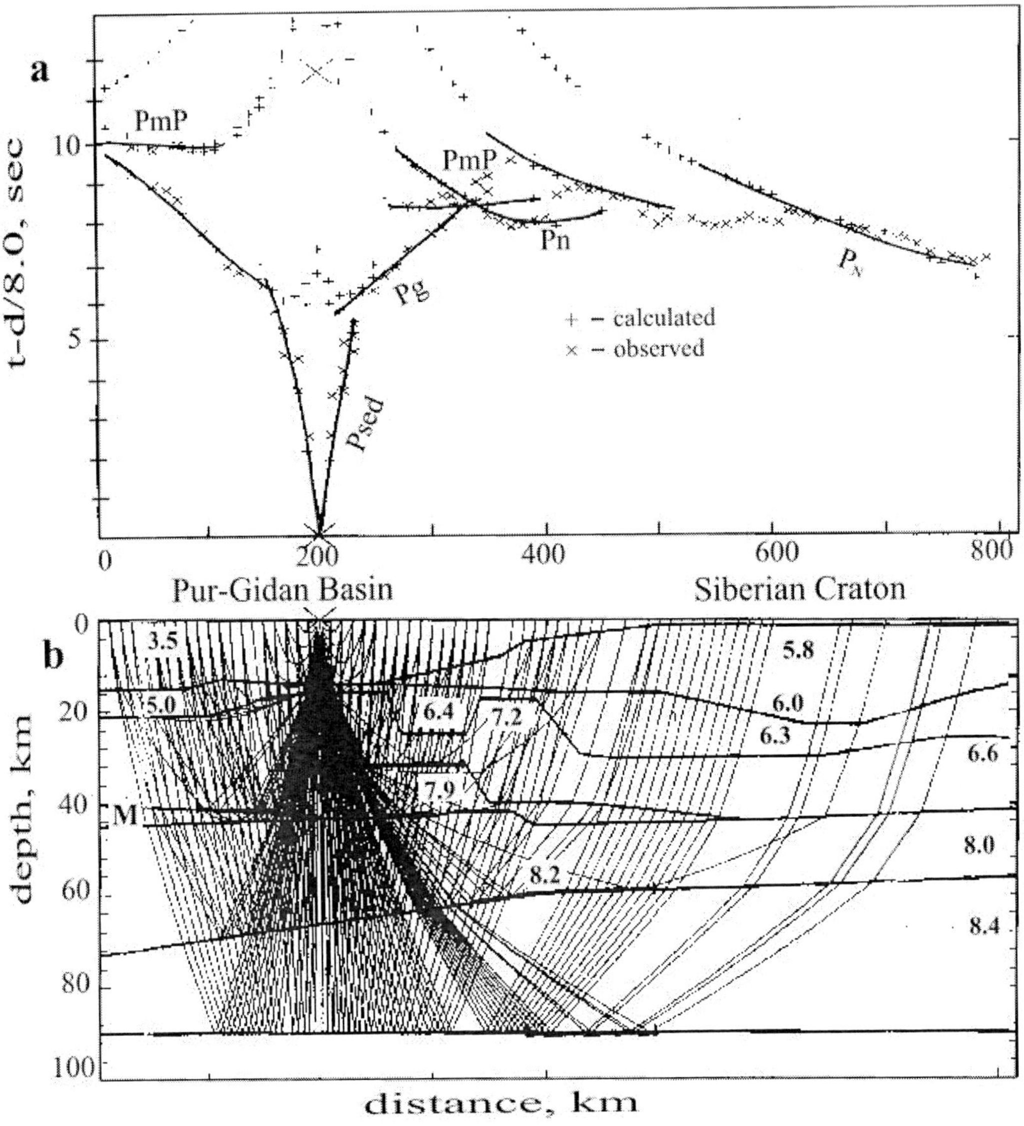

Figure 18. Ray-tracing results for the chemical explosion SP 245 in the northern part of the profile RIFT: (a) the observed and calculated travel-times, (b) the velocity model and the ray pattern. Numbers are the apparent velocities. The travel-times and the model demonstrate complicate block structure of the crust at the northern boundary of the Siberian Craton

The relation between P_{N1} and P_{N2} waves changes from one shot to another. From PNE R1 there are no strong differences in their apparent velocity or arrival time, and they interfere at the first arrivals. SP R2 records show a clear shadow zone and time delay after P_N wave attenuation. The wave P_L was recorded as the first and secondary arrivals at distances from 1200 to 2000 km.

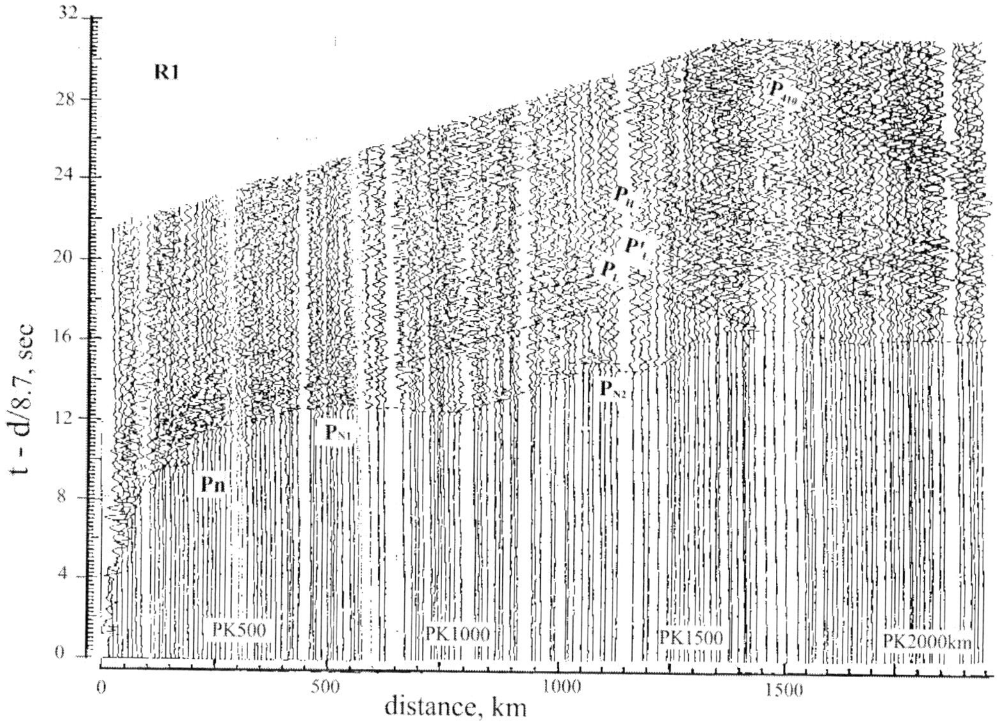

Figure 19. Record-section of the upper mantle waves from the PNE R1 along RIFT profile. Legend is in Figure 3. Specific features of the records are the anomalous high velocities of the first arrivals at distances of 400-750 km showing the increasing of the upper mantle velocities beneath the Siberian Craton, and multi-phases high amplitude group around the P_H wave.

The PNE R3 display different upper mantle wave properties (apparent velocity and arrival time) and the wave patterns are difficult to divide into separate waves (Fig.21 a). The first events are very weak and are followed by an intensive group of waves with velocities of around 8.4 km/s. However, waves P_N and P_L were distinguished at this background.

The wave pattern changes greatly in the mid-craton area (100-1500 km on Fig.21): the times from both shots R1 and R2 increase here and to the south high apparent velocities (up to 8.7 km/sec) are characteristic for both P_N and P_L waves.

Record-sections for R1 and R2 also display multi-phased secondary arrivals with lower apparent velocities than the first arrivals. In some cases these waves appear to be a continuation of P_N waves at larger offsets but more obviously they are multiple refractions in the uppermost mantle.

The upper mantle 2-D velocity model (Fig. 21 b) shows strong horizontal inhomogeneity, particularly in the upper 100 km of the upper mantle. Two blocks (500 km long) of anomalously high velocity (8.4 - 8.6 km/s) are imaged, one in the central part of the Tunguss basin (Tunguss block) and one in the southern part of the craton near the Baikal Rift (Pre-Baikal block). In the Tunguss block (located between 800-1500 km in Fig.21), the reversed travel-time curves show first arrival apparent velocities of 8.4 km/s. The P_N wave from SPs R1, R2 and R3 has similar high velocities in this region.

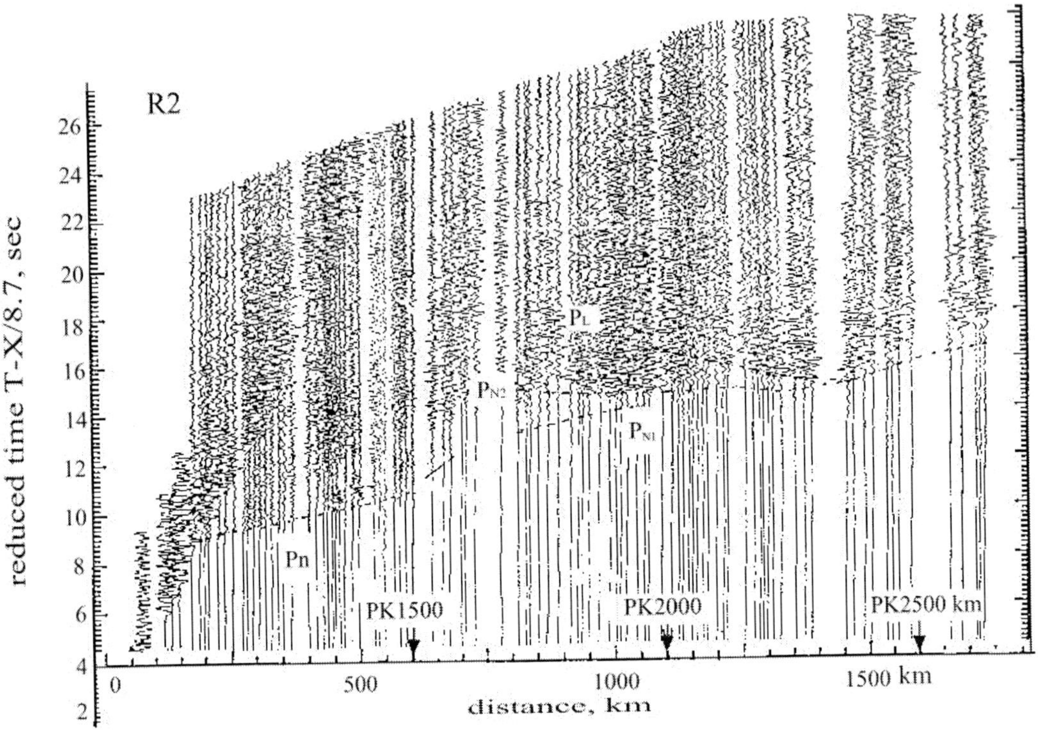

Figure 20. Record-section of the upper mantle waves from the PNE R2 along RIFT profile with observations to the east. Legend is in Figure 3. The bright shadow zone with the first arrivals time delay of 2 sec is observed at distances of 600-750 km. This wave pattern corresponds to the low velocity layer at depth of 80-100 km (Figure 21).

The velocity inhomogeneity is characteristic mainly in the uppermost mantle. Deeper the velocity inversion zone is distinguished at the depth of 100 km. The velocities beneath the boundary are 8.4-8.5 km/s. We supposed the velocity in the inversion zone to be 7.9 km/s; consequently, its thickness is approximately 30-40 km. The most reliable determination of the zone was obtained from R2 records (Fig. 20), which show clear shadow zone and a delay time at the first arrival. The records from R1 and R2 confirm or at least do not contradict this velocity inversion: the R1 first arrivals attenuate at the same offsets as from R2. R3 records do not show such attenuation but strong secondary arrivals with the same velocity as the first ones agree with the inversion model.

The mantle boundary L is traced at a depth of 180-230 km. The reflections P_L are not of such intensity as along the other profile but they were recorded by all PNEs. At a depth of 250-300 km several reflection boundaries were also found from SP R1 and R3.

The upper mantle structure beneath the low velocity zone is difficult to describe using only the basic layers: several intermediate boundaries are added and the velocities vary in the layers. The most complex velocity structure is observed beneath the Tunguss block (Fig. 21b): the N2 and L boundaries uplift here and higher (in 0.1 km/s) velocities are typical of the boundaries in comparison with the basic starting model. It suggests that in this area (depth 100-250 km, profile interval 1000-1500 km) there is a fault zone with several reflectors dipping to the north. The zone generates reflections with anomalous high apparent velocities and a complex wave pattern.

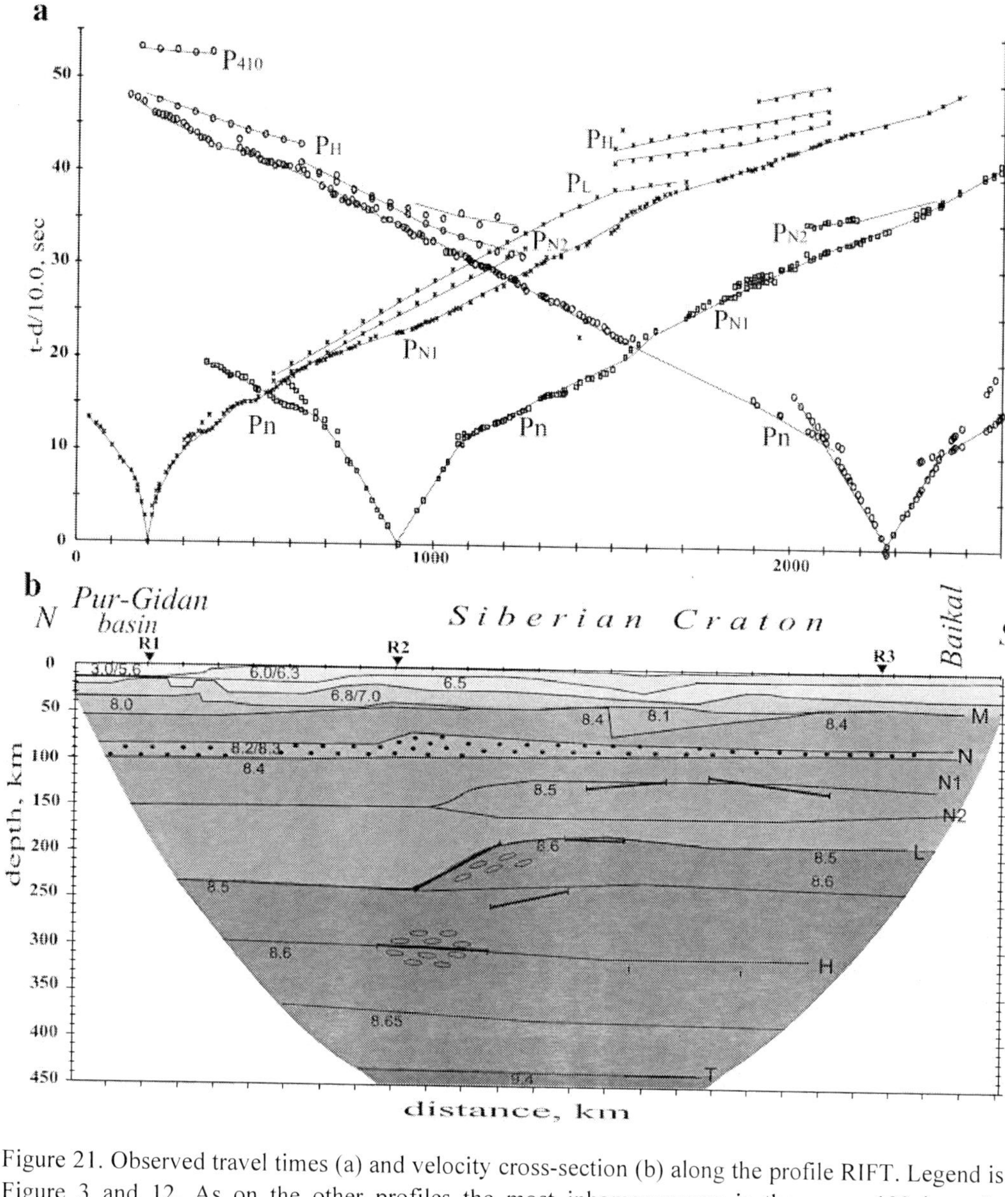

Figure 21. Observed travel times (a) and velocity cross-section (b) along the profile RIFT. Legend is in Figure 3 and 12. As on the other profiles the most inhomogeneous is the upper 100 km of the lithosphere with the high velocity blocks and low velocity zone at the bottom. The strong increasing of the velocities is observed in the centre of the Siberian Craton (PK 900-1100 km) in the depth interval 120-240 km.

2.4.5. METEORITE Profile

The METEORITE profile crosses the Siberian Craton from north to south, from the Taimir Orogen Belt to the Baikal Rift Zone (Fig.1). Four PNEs but no chemical explosion were shot along the profile. The crustal model was compiled from other seismic data obtained in this region.

On this profile only the Pn waves are recorded as stable first arrivals. The apparent velocities of these waves in average are higher than those in the other profiles: 8.2- 8.4 km/s.

At offsets of 700-1000 km Pn attenuates and at the larger distances the first arrivals are weak (Fig.22). The secondary arrivals are complex and the basic wave correlation is more difficult than on the other profiles.

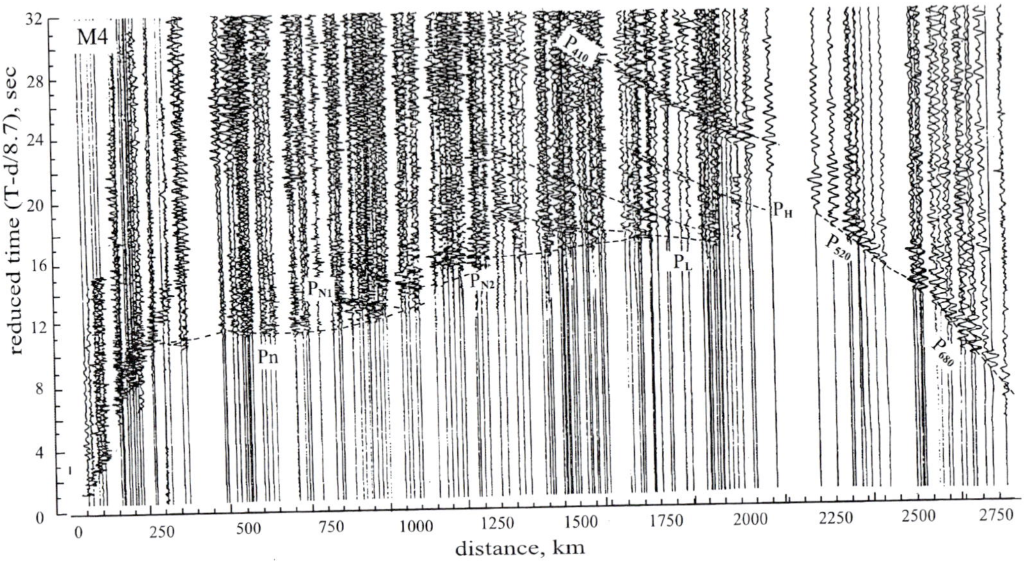

Figure 22. Record-section of the upper mantle waves from the PNE M4 along METEORITE profile. Legend is in Figure 3. The high velocity first arrivals at distances of 300-750 km correspond to high velocity Pre-Baikal block (Figure 23).

As on the other profiles the uppermost mantle is the most heterogeneous; it is of block structure, with variations in block velocities from 8.1 to 8.3 km/s (Fig. 23b). High-velocity blocks are distinctly recognized from the observed travel-times: from the regular variations in the apparent velocities of different waves in the same profile intervals. For example, in the interval 2100-2400 km, the apparent velocities in all direct and reversed refraction travel-time curves increase (Fig.22 a). That points to the existence of a high-velocity mantle block immediately beneath the Moho. Another high-velocity block was recognized in the upper mantle in the interval 1000-1600 km. Its roof, however, comes onto the Moho surface only in the interval 1500-1600 km.

In the northern part of the profile, a low-velocity layer has been recognized at depths of 80-100 km. This layer causes a decrease in the apparent velocity of first waves from PNE M2 in the interval 300-600 km and attenuation of the waves from PNE M3 at PK 900 km.

The topography of the mantle boundaries is much more intricate along this profile as compared with the other profiles. The boundaries N1, N2, and L are characterized by close wave velocities; therefore, it is often difficult to distinguish between the travel-time curves of their waves because of the strong variations in the apparent velocities of first arrivals and in the uppermost mantle heterogeneities. The boundaries have an inverse topography. The boundary N2 occurs at a depth of 150 km in the northern part of the profile; then it rises to 120 km at PK 2000 km and, finally, lowers beneath the Baikal Rift Zone. The boundary L also rises from north to south, particularly at PK = 1100-1400 km. But this rise is traceable only to PK 1600 km; then the boundary downwards beneath the rise of the above-lying boundaries (PK 1600-2200 km). The depth of the boundary H is virtually constant along the profile, 310 km. Besides these reference boundaries, there are reflecting horizons in the bottom of the upper mantle, which complicates the section even more.

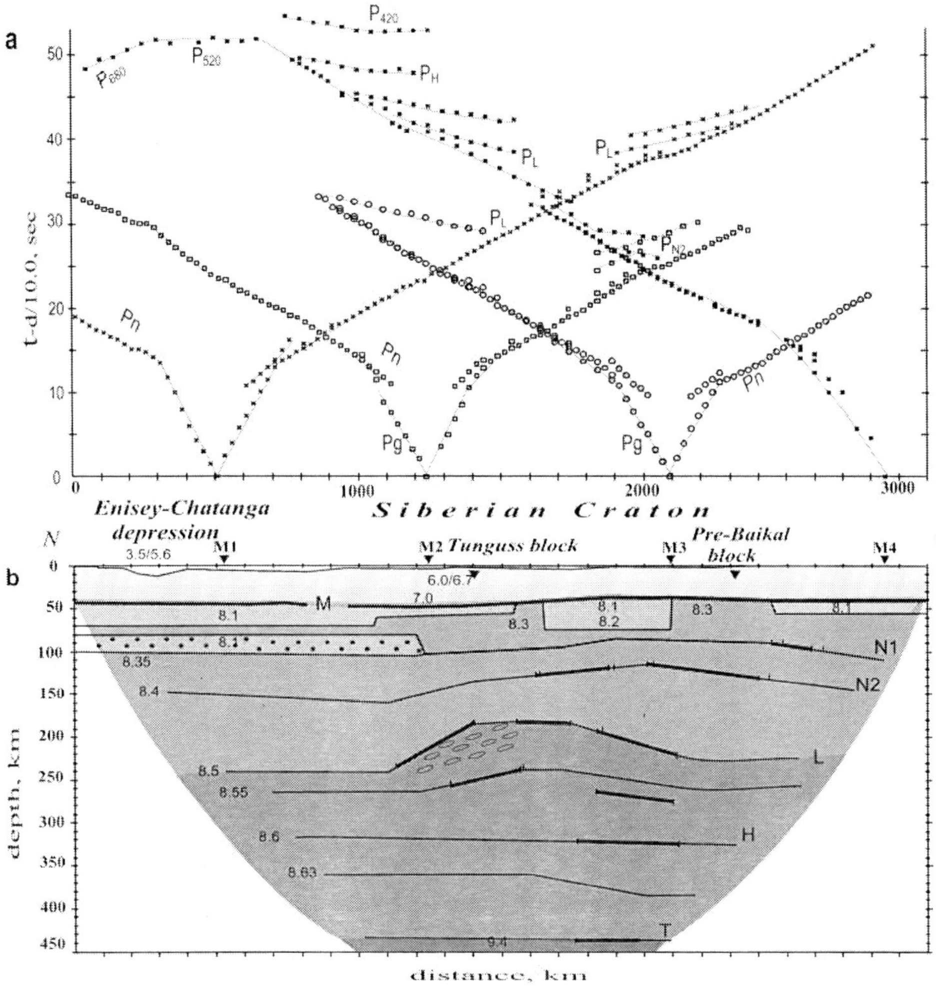

Figure 23. Observed travel times (a) and velocity cross-section (b) along the profile METEORITE. Legend in Figure 3 and 12. Note the structural features of the upper mantle to be changed in the central part of the Siberian Craton but at the boundary with the West-Siberian Plate.

2.4.6. RUBIN Profile

The RUBIN profile has a dense system of the chemical explosions but only one PNE (Ru in Fig.1). The mantle waves from the explosion are recorded up to a 1500 km distance. The large differences are observed in the wave pattern between the RUBIN and the other profiles. Weak Pn waves are not visible in the records, instead of the high amplitude reflections from the N1, N2 and L boundaries are characteristic for this profile (Fig. 24). A peculiarity of the RUBIN wave-field is also the anomalously high apparent velocities of P_{N1} and P_{N2} waves: more than 9.0 km/s. As a result the first arrival travel-times show a large negative anomaly at distances 700-1200 km relative to the other profile data.

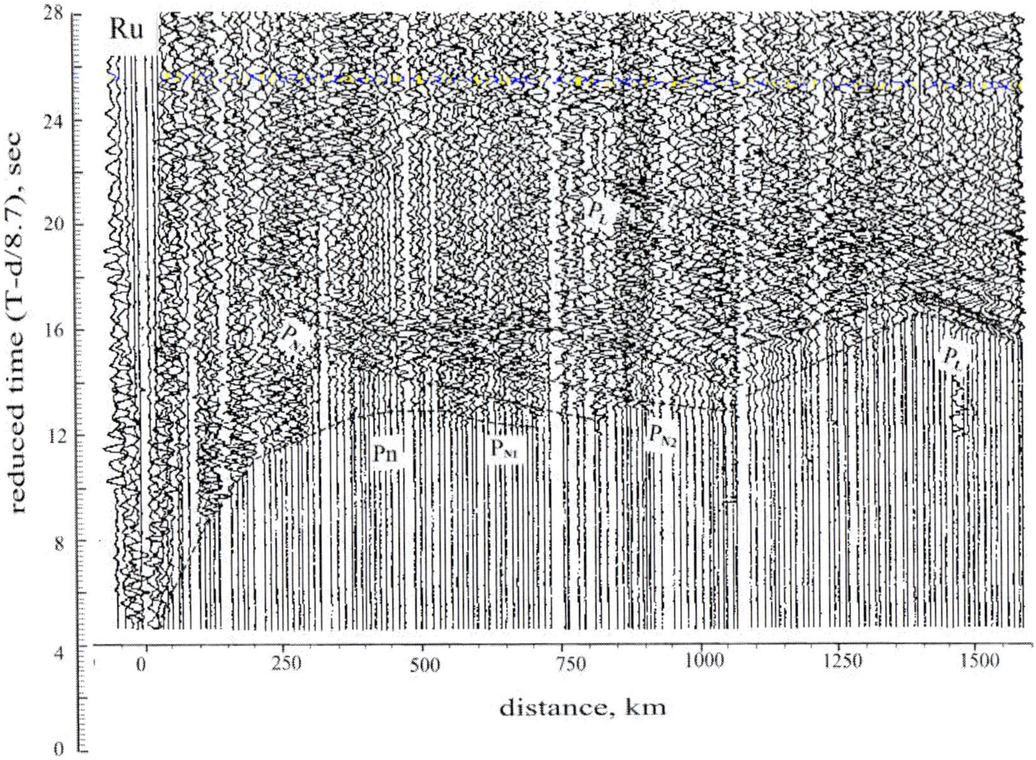

Figure 24. Record-section of the upper mantle waves from the PNE Ru along RUBIN profile. Legend is in Figure 3. Anomalous high apparent velocities of the P_{N1} and P_{N1} waves are characteristic of the records. They reveal the upper mantle reflectors dipping from the Urals to the East-European Craton (Figure 25).

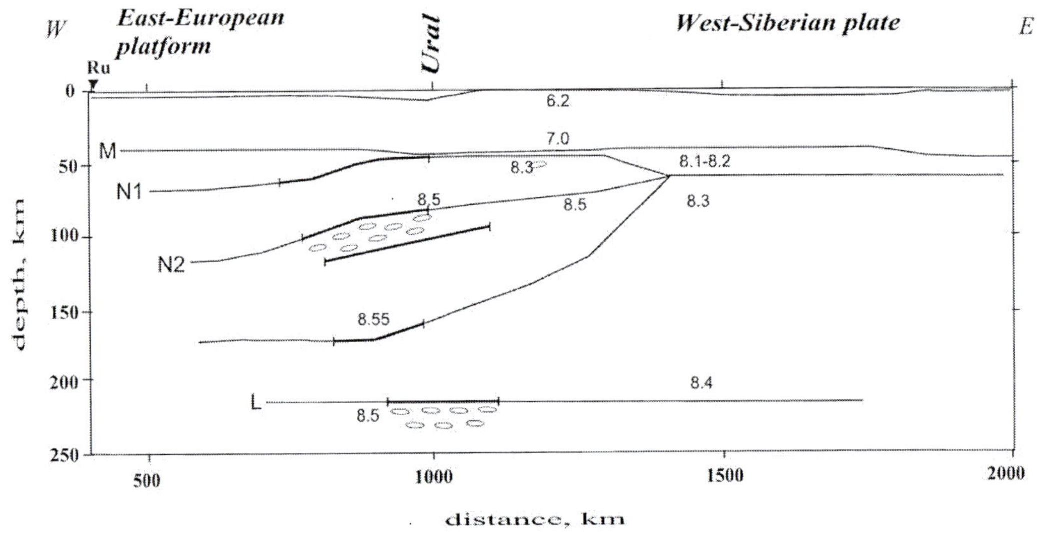

Figure 25. Velocity cross-section along the profile RUBIN. Legend is in Figure 12.

To explain the observed wave pattern two possible models were considered: the anomalous apparent velocities of P_N waves are caused (1) by increasing of the mantle velocities in the central part of the profile and (2) by inclinations of the reflectors to the west. The ray-tracing has shown that the second version is more reliable. Due to the first version unrealistic high velocities should be suggested in the uppermost mantle. The reflections from inclined boundaries may have very high apparent velocities. The inclined N1 and N2 boundaries (Fig. 25) generate the reflections with apparent velocities up to 9 km/s. Both boundaries are shallow beneath the Urals. The L boundary is near horizontal.

2.4.6. GLOBUS Profile

The profile crosses the eastern part of the East-European platform and partly the Timan-Pechora Plate (Fig. 1). It is 1500 km long. Four PNEs were recorded with the intervals of 400 km between the shot-points (Fig. 26a). No chemical explosions were made along the profile.

A characteristic feature of the GLOBUS wave fields is poor amplitudes of the Pn waves and recording of the P_{N1} as apparent first arrivals at 200-600 km offsets. The upper mantle velocity structure does not change strongly along the profile (Fig. 26b). The observed times of the mantle waves change mainly due to the crustal structure variations. A regular increase of the velocities with depth is clearly observed from 8.15 beneath the Moho up to 8.3 km/s at the N1 boundary (depth of 70 km) and up to 8.4-8.5 km/s at the N2 boundary (depth of 120 km).

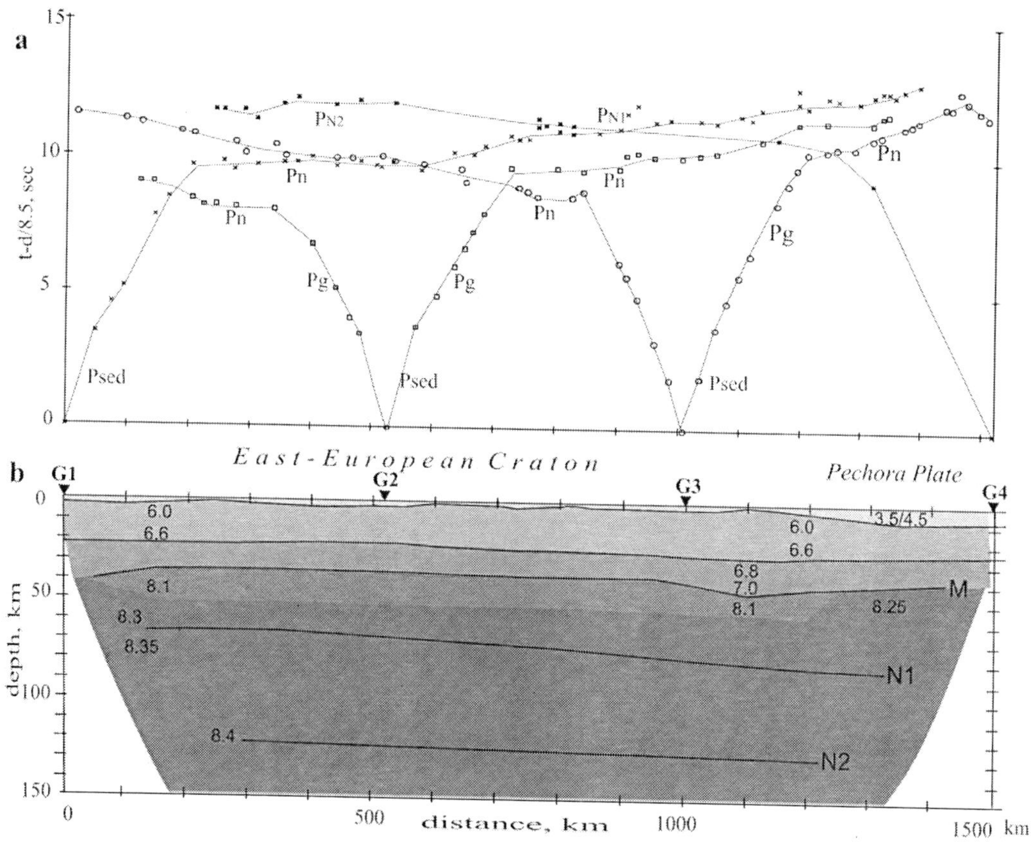

Figure 26. Velocity cross-section along the profile GLOBUS. The upper mantle structure of the East-European Craton looks much more homogeneous than the Siberian Craton. Legend is in Figure 12.

2.4.7. HORIZONT Profile

The HORIZONT profile crosses the northern parts of the Urals, of the West-Siberian Plate and of the Siberian Craton (Fig1). Four PNEs were made here and only some chemical explosions (Fig.27a). The latter are irregularly spared, not covering the eastern part of the profile.

The longest observations were carried out from the SP H1 and H4 with offsets up to 1100 km. They served to penetrate the upper mantle to the depths of 140 km. Travel-times of the mantle waves vary mainly due to the changes of the sediments thickness and sometimes it is difficult to understand which apparent velocities show the mantle structure or the basement relief. The three basic waves, however, were determined: the Pn, P_{N1} and P_{N2}.

In the western part of the profile the uppermost mantle velocities are 8.0 –8.1 km/s which are lower than those beneath the craton (Fig. 27b). The N1 boundary with velocity of 8.4 km/s is determined at the depth of 100 km with the uplift up to 70 km beneath the northern edge of the Urals. The N2 boundary with the velocity of 8.4 km/s is observed at the depth of 110-130 km. A low velocity zone is distinguished above the N1 boundary in the cratonic area.

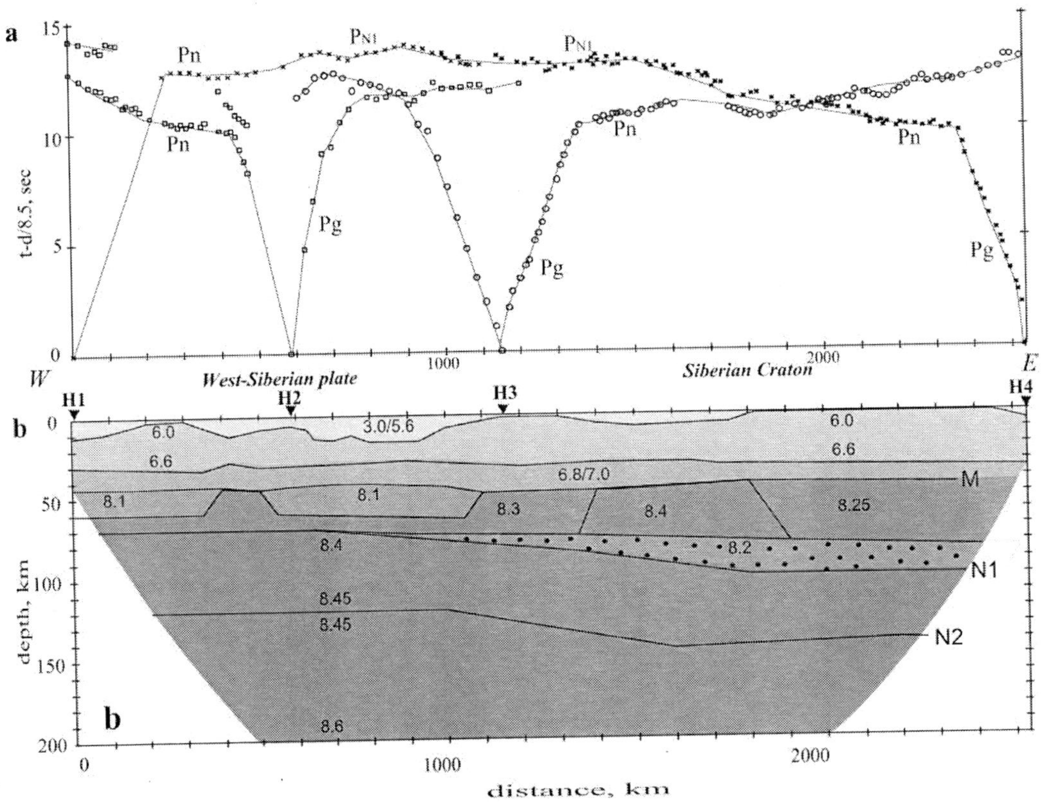

Figure 27. Velocity cross-section along the profile HORIZONT. The typical for the Siberian Craton uppermost mantle structures (high velocity block and the velocity inversion zone) are observed along the profile. Legend is in Figure 12.

2.4.8. FENNOLORA Profile

The profile crosses the Baltic Shield from the Barents Sea to the Baltic Sea (Fig. 1). This profile is a part of the European geotraverse measured within the framework of one of the largest international projects [Mueller and Ansorge, 1988]. The upper mantle waves were recorded at distances from a source up to 2000 km. This result could be attained due to explosions in the Baltic and Barents seas. The stations interval was 3-5 km; the shot spacing was 200-300 km.

Several international groups interpreted the FENNOLORA profile data. The well known interpretation was made by Gugesberg and Berthelsen (1987). Their results yield a complex distribution of velocities highly variable both laterally and vertically (from 8.0 to 8.6 km./s). A characteristic feature of the upper mantle model is thin layering with alternation of low velocity and high velocity zones. On the background of this complicated structure it is difficult to see if there are some regular changes of the velocities along the profile or with depth. To answer these questions all the experimental data were analysed and a new cross-section was constructed in the form of the basic model described above (Pavlenkova, 2005).

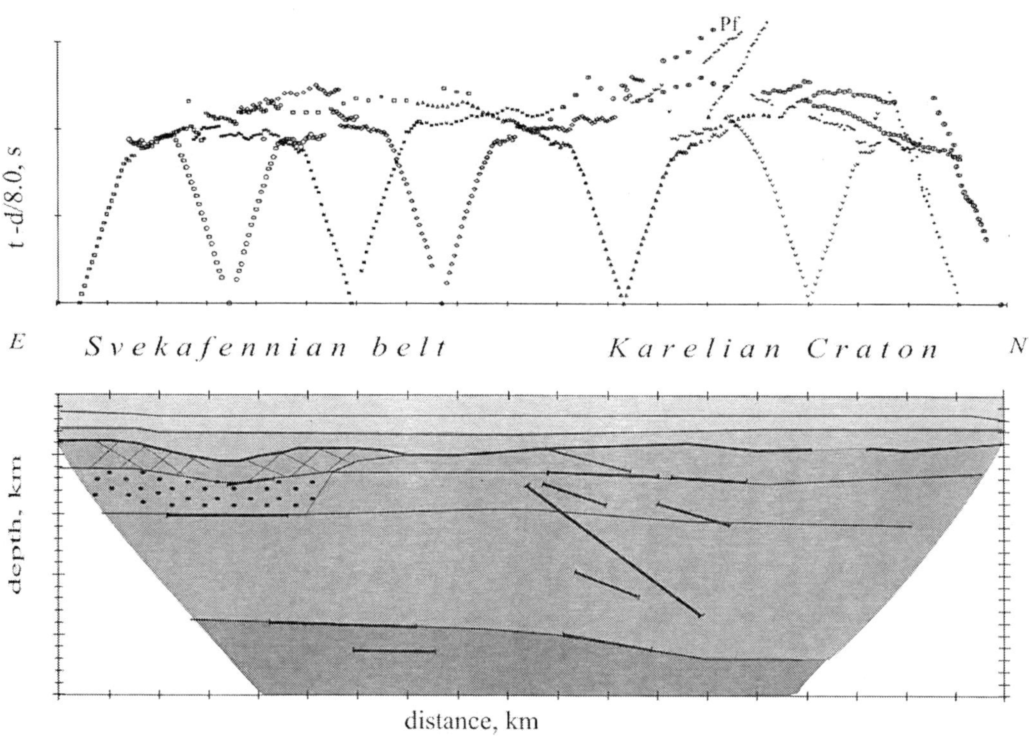

Figure 28. (a) The reduced and transformed to mid-points travel times of the basic waves. Reduction velocity Vr=8.0 km/s. Pf are the reflections from the inclined boundaries in the upper mantle. (b) Velocity cross-section along the FENNOLORA profile. Legend is in Figure 12. A clear difference is observed in the uppermost mantle structure between the two tectonic domains of the Baltic Shield. The difference follows from two types of the observed wave fields (Figure 29).

A variable mantle wave field is typical of the FENNOLORA profile. Changes of wave types, shadow zones, and abrupt transitions from a regular waveform to a chaotic one and vice versa are commonly observed in first arrivals. Two types of the wave fields are observed along the profile. In the southern part, the waves Pn with apparent velocities of 8.2-8.3 km/s are well traced (even phase correlation is possible) up to a distance of 400 km from the source, farther, they decay abruptly and P_N waves are traceable as visible first arrivals (Fig. 28a). In the northern part of the profile no regular waves are observed, and even intensities and arrival times of first events vary continuously, whereas local wave trains of an indefinite form are observed at the secondary arrivals (Fig. 28 b). This field can be called chaotic. Wave structures with a sufficiently clear shape of their first phase or, rather, the high-intensity front are noted only in some cases.

The travel-times imply that a low velocity layer bounded by velocities of about 8.3- 8.4 km/s at its upper and lower boundaries is present only in the southern part of the profile at depths of 80 to 100 km. In the northern part of the profile the first arrivals with the apparent velocities of 8.0-8.2 km/s are observed without significant breaks in the times. This means that the upper mantle velocities increase smoothly with depth in the northern part of the profile.

In order to gain more detailed constraints on lateral variations in velocities, the observed travel-times were reduced with the reduction velocity Vr = 8.5 km and transformed to source-receiver midpoints (Fig. 29a). These travel-times reveal an anomalous wave pattern at the center of the profile. Here, inclined travel-time curves from SPs G, I, and H disagree with the general tendency and are characterized by anomalous apparent velocities. This means that these waves arrived from inclined reflectors steeply dipping northward.

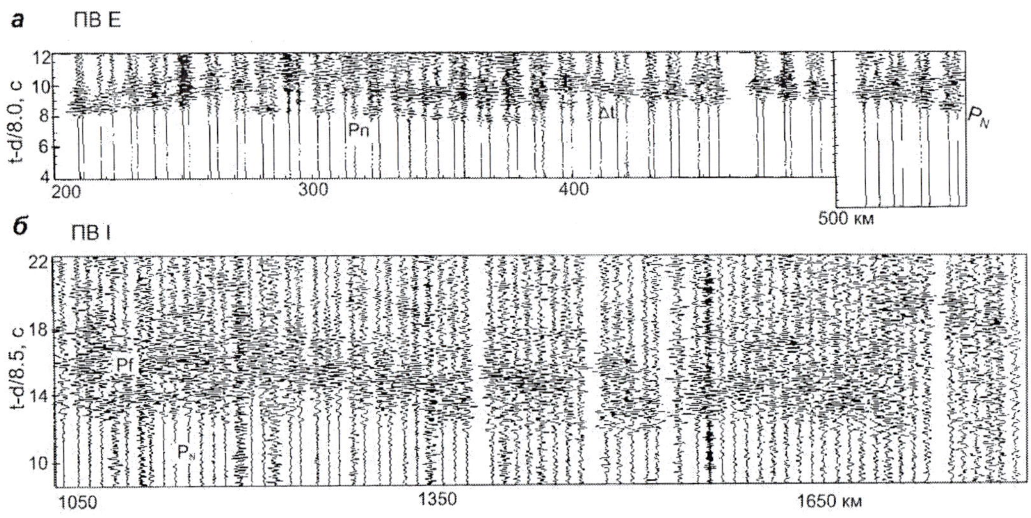

Figure 29. Typical record sections of the mantle waves for the southern (a) and northern (b) parts of the profile FENNOLORA: Records from SP E show the high velocity first arrivals at offsets of 200-400 km, then the shadow zone with time delay Δt at offsets of 400 km. These wave patterns correspond to high and low velocity layers in the lithosphere of Svekafennian Belt (Figure 28 b) Records from SP I show chaotic first arrivals and the reflection Pf from the inclined boundary in the mantle.

The obtained model differs significantly from the model by (Gugesberg and Berthelsen, 1987). The most significant distinction of the new model is the absence of several lower velocity layers extending all along the profile. These layers were identified by solving l-D problems for each individual shot point, i.e., without regard for the lateral heterogeneity of the medium. The reliability of such a great number of inversion zones and local heterogeneities cannot be proved.

The simplicity of the new velocity model does not imply a loss of information. The following structural features of the upper mantle are discovered. The division of the mantle into two (northern and southern) blocks is seen. The deepest subsidence of the Moho and the highest velocities (8.3 km/s) under this boundary are noted in the southern part of the profile. The mantle structure in its northern part is smoother and velocities in the uppermost mantle are lower (8.1 km/s). This velocity difference persists up to a depth of 80 km. A lower velocity layer observed at greater depths in the southern part virtually smooth out the lateral velocity variation in the mantle at a depth of about 100 km. A series of northward-dipping reflectors is observed in the central part of the profile at depths of 40 to 200 km. The inferred two blocks with different upper mantle structure correspond to two geologic structures of different ages. The southern block is associated with the Svecofennian province, and the northern block, with the Karelian Craton.

2.5. 3-D Velocity Model of the Upper Mantle

The regular waves described above constrain a basic model for the upper mantle of the study area. The model includes five layers divided by the boundaries: N1 with velocity V = 8.35 km/s and depth interval D = 70-120 km, N2 (V = 8.4 km/s, D = 100-130 km), L (V = 8.5 km/s, D = 200-230 km) and H (V = 8.6 km/s, D = 300-330 km). Characteristic parameters of the boundaries are the velocities along them. (Note, the boundary velocities differ from the corresponding wave apparent velocities due to the effect of the Earth surface curvature).

The depths to the boundaries at the most profile intersection points differ not more than by 10 km. That is good agreement which showed good enough stability in the principal layer determination. In some cross-points, however, the depth differences have reached 20-30 km. They were observed at the ends of the profiles where the velocity structure cannot be determined correctly. Another cause of the depth differences in the intersection points are complicated many phase records of the basic waves. At the first stage of the wave correlation ,it is impossible to determine the same phases along all profiles. Time differences between the phases are 0.4-0.6 sec and ray tracing shows that the corresponding depth differences may be up to 20 km. To exclude such cases, the records and cross-sections were analysed once more and additional ray tracing was made to get agreement between the models at their cross-points.

The seismic cross-sections constructed along all the profiles may be used to compile a 3-D upper mantle model. To describe the complicated velocity structure of the uppermost mantle, the map of the velocity distribution at the depth of 60 km was constructed (Fig.30). For deeper parts of the mantle the 3-D model is presented in the form of depth maps to the basic boundaries (Fig. 31). As the velocities are constant along the boundaries and they change linearly between them, such maps describe the velocity model in each point of the 3-D space. The maps were compiled directly from the profile cross-sections. It became possible because the cross-section is presented in the common form using the basic velocity model. The only problem of the mapping is drawing depth isolines between the profiles. For that we used the correlations between the mantle structure, tectonic and geophysical fields.

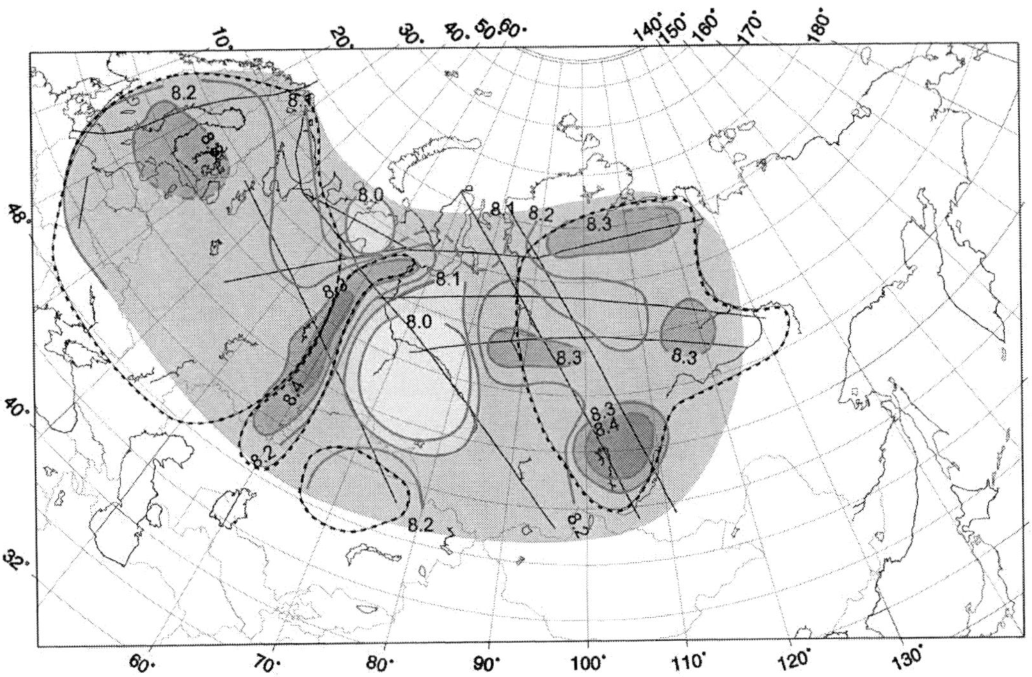

Figure 30. Scheme of the uppermost mantle velocity distribution at depth of 60 km. The thin lines are the seismic profiles, the dotted lines show the main tectonic units (Figure 1). The velocities change from 8.0-8.1 km/s beneath the young West Siberian and Timan-Pechora platforms to the average 8.2 km/s in the East-European Craton and to 8.3-8.4 km/s in some blocks of the Siberian Craton.

The resulting models characterise the mantle structure in the following way (Figs. 30 and 31). Beneath the Moho the velocities change from 8.0-8.2 km/s in the West Siberia to 8.3-8.4 km/s in some blocks of the Siberian Craton and the Urals. Four high velocity blocks are outlined in the Siberian Craton. Three of them are determined from the reliable data of the reversed profiles: the Tunguss block in the area of the profile METEORITE, RIFT and KIMBERLITE, the Vilyui block in the western part of the Vilyui Basin from the CRATON and KIMBERLITE data and Pere-Baikal block in the southern part of the craton from the RIFT and METEORITE data. The high velocities are distinguished also along the HORIZONT profile and in the Urals along QUARTZ and RUBIN profiles.

The East-European Craton mantle looks more homogeneous than that of the Siberian Craton. A local high velocity anomaly is observed only in the Baltic Shield. The lowest velocities (8.0-8.1 km/s) are characteristic of the central part of the West Siberia and for the Timan-Pechora Plate. In all the other regions the normal velocities 8.1-8.2 km/s are observed.

At the depth of 100 km the velocity distribution looks otherwise (Fig. 31). The most local high velocity anomalies disappear and only two large anomalies are observed: lower velocities in the central part of the West Siberia and the higher velocities in the Middle Urals. These structural features follow from the N1 boundary map. They are near 130 km in the central part of the West-Siberian Plate and only 60 km in the Middle Urals.

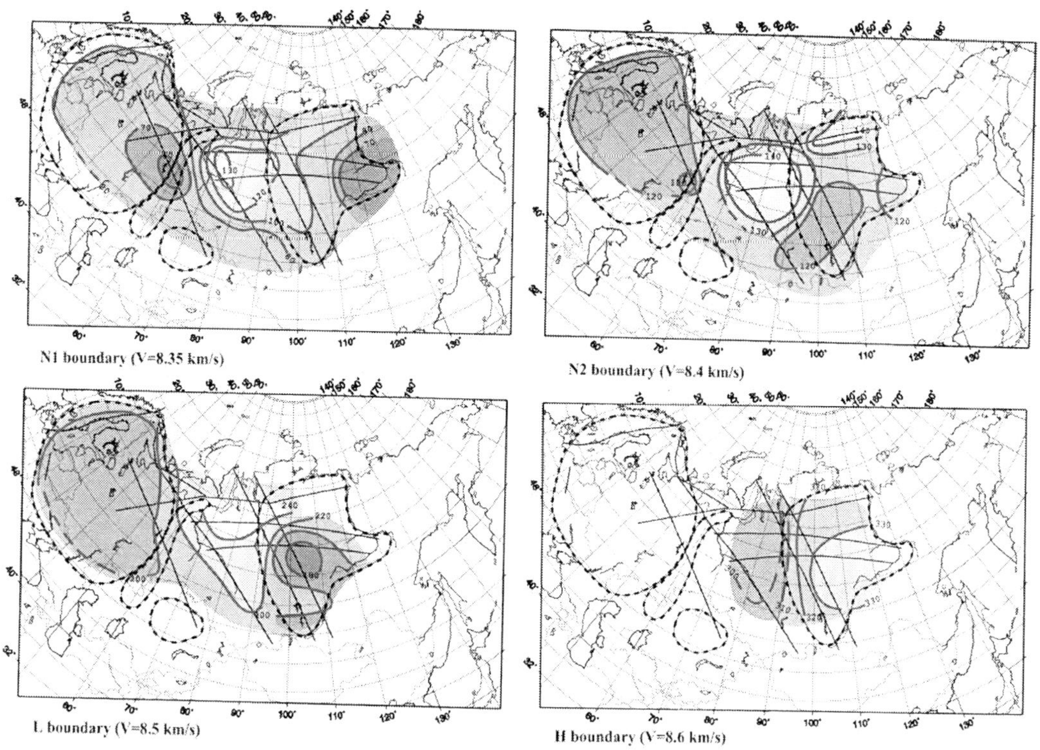

Figure 31. Depth maps of the mantle boundary N1 (boundary velocity is 8.35 km/s), N2 (8.4 km/s), L (8.5 km/s), H (8.6 km/s). The correlations of the mantle velocities with tectonics are observed to the depth of 250 km: in the West -Siberian and Timan-Pechora young platforms, the velocities are lower than in the adjacent ancient platforms. But for the deeper mantle a reverse correlation is revealed. The general regularity is violated in the northern Siberian Craton where the upper mantle structure is closer to the young platforms.

The depth maps of the N2 and L boundaries show the pictures similar to the map of N1 boundary. The West Siberia region is characterised with the lower velocities: depths to the N2 boundary reach here 140 km as regards to the high velocity zone in the eastern part of the Siberian craton where the N2 boundary depths is 120 km. There are, however, some differences: L boundary clearly shows that the higher velocity area is concentrated in the central part of the craton. The depth map of the H boundary reveals the opposite pictures: higher depths beneath the eastern part of the craton (330 km) and the depths of 300 km beneath the West-Siberian platform.

Some correlations of the uppermost mantle velocities with tectonics and with geophysical fields are noted. The lower velocities are observed in the areas of higher heat flow and in tectonic active areas. In the West -Siberian and Timan-Pechora Plates, likewise associated with young platforms, the mantle velocities are lower than in the adjacent ancient platforms. This general regularity, however, is violated in the northern Siberian Craton where the upper mantle is characterized by lower velocities although the heat flow is low as in other parts of the craton. Possibly, the present-day mantle there undergoes a high thermal activity, but the deep heat has not reached the surface yet. The thermal activity might be related to the proximity of an Arctic shelf, which agrees with the general regularity of decrease in the upper-mantle velocities from inner to peripheral areas of continents. This might also be due to the influence of the Yenisei-Khatanga active area, though the no less active Baikal Rift Zone does not exert such a serious effect on the mantle velocities.But the described correlation is not so visible for the deeper parts of the upper mantle, where there is a reverse correlation between the shallow and the deep mantle boundaries.

3. Deep Seismic Research along the Angola-Brazilian Geotraverse

3.1. Introduction

For a long time the upper mantle structure of the Atlantic has been studied on the basis of the extremely limited seismological observations or gravimetric data. Over large areas of oceans there are virtually no seismological stations. Earthquakes, usually occurring in the near mid-oceanic ridges, are not sufficiently strong for their registration on the land. As a result, the surface waves data used for the structural model are considerably averaged over great areas of the ocean. With such averaging, it was possible to determine only the most general features of the mantle structure. The velocity inversion zone (waveguide) at a depth from 40 to 200 km was established (Cristensen et a!. 1980; Qkal, 1977). It was shown that the depth to this waveguide regularly increases with movement from mid-oceanic ridges to oceanic basins. These absolute depths have been determined approximately by different authors and different values have been obtained because the computation results are considerably dependent on the type of model used: anisotropic or isotropic. As an average it was assumed that the lithospheric thickness in the Atlantic varies from 20 km in the axial zone of the ridge to 50-80 km in the abyssal basins.

These determinations completely explain the nature of the gravity field with a large minimum in the ridge region and are consistent with the fundamental principles the plate tectonics and, therefore, a model of the oceanic lithosphere, gradually thickening from the ridge to the basins has already become classic (Okal, 1977). In the model it also was assumed that the structure of the crust and upper mantle is quite symmetric relative to the axial zone of the ridge.

In 1970-1980s the seismic studies of the uppermost mantle were carried out in oceans (Reit et al., 1969; Steinmeiz et al., 1977; Orcutt, Dorman, 1979; Asada end Shimamura, 1979; Shimamura et al., 1983; LADLE working group, 1983). In the Atlantic the depth of the investigation was limited by 20-30 km. Much deeper data were obtained during 1980-1986, when the 'Sevmorgeologiya' Trust of the USSR Ministry of Geology and the Institute of Physic of the Earth the USSR Academy of Sciences, carried out geophysical research on the crust and upper mantle in the South Atlantic along the Angola-Brazilian geotraverse (Margin and Pogrebetskiy, 1986; Zverev, Tulina, 1996). It was a complex of the reflected wave method, deep seismic sounding (DSS) and magnetic-gravity surveys.

Seven DSS profiles were worked along the geotraverse which shed light on the structure of all the principal oceanic structures: central part of the Mid-Atlantic Ridge, its western and eastern slopes, Brazilian and Angola Basins (Fig. 32 a). A system of direct and reversed observations with a recording to distance of 600 km was carried out for each profile. It corresponds to a depth of penetration of seismic rays up to 80 km. In world experience there is no such research in the ocean equivalent with respect to density of observations and their effective depth. Accordingly, the results of DSS along this geotraverse are of special value.

They provide factual data on upper mantle structure characterizing all the principal oceanic structures.

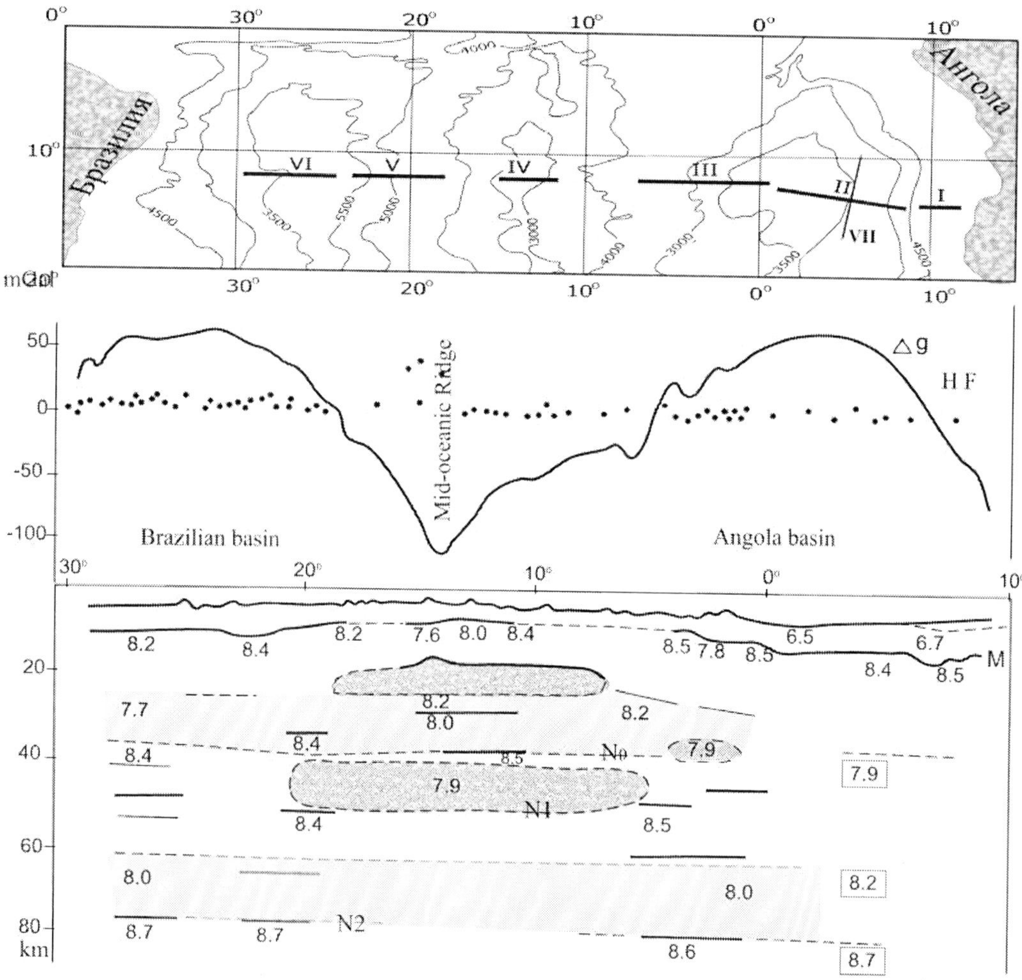

Figure 32. Geophysical cross-section along the Angola-Brazilian Geotraverse: (a) Location of DSS profiles I-VII along the geotraverse, (b) Gravity (Δg) and heat flow (HF) data (c) Seismic cross-section. The thick lines are the seismic boundaries, the numbers are seismic velocities along the geotraverse, the numbers in squares are the seismic velocities observed in Angola basin along the profile VII. The data shows that instead of the unified asthenosphere rising toward the ridge axis, several low velocity layers are traced beneath the ridge. Another special feature is asymmetric block structure: there are the Angola basin block with homogeneous anisotropic upper mantle and the mid-oceanic - Brazilian block with several velocity inversion zones.

Preliminary results of the DSS data for the geotraverse were presented in (Zverev, Tulina, 1996). The processing of the experimental data for individual DSS profiles has been carried out by different interpreters using different methods. These differences do not change the principal conclusions on the upper mantle structure, but nevertheless they complicate the construction of a composite seismic section for the whole geotraverse and its use for other geophysical data interpretation. An attempt, however, was made to generalize DSS materials using a common approach and on this basis to construct a combined seismic model of crust and upper mantle along the entire geotraverse (Pavlenkova et al., 1993).

3.2. Experimental Data

Fig. 32 shows that deep structure along the geotraverse has been studied nonuniformly. The largest gaps between the DSS profiles exist on either side of the axial zone of the ridge between 8^0 and 11^0 W and 15^0 and 18^0 W. In addition, there are gaps in tracing the deep boundaries.

In order to construct a composite section and to identify seismic boundaries at the places where there were gaps it was necessary to formulate some general criteria for comparing the observed wave fields and determining the nature of the inhomogeneities.

A comparison of the wave fields along the profiles was made using the dynamic and kinematic characteristics of several reference waves. Principal differences in the wave pattern are observed between the Angola Basin and other parts of the geotraverse. In the Angola Basin (profile II-III) only one mantle wave Pn is traced as the first arrivals at long distances (up to 500 km from the source) and it has an anomalous high velocity: 8.5 km/s at offsets of 30-300 km and 8.8 km/s at larger offsets (Fig. 33 b).

In the mid-oceanic ridge region and in the Brazilian Basin among the mantle waves two principal groups were discriminated: Pn wave associated with the Moho and the top of the upper mantle and wave group P_N, corresponding to deeper parts of the mantle (Fig. 33 a). Waves Pn are usually traced at the first arrivals at distances of 30 -150 km from the source; their apparent velocities are 8.2 - 8.5 km/s.

Waves of the P_N group are discriminated at the secondary arrivals after attenuation of the Pn waves, forming a series of almost parallel travel-time curves with velocities 8.4 - 8.6 km/s. Such a character of the wave pattern shows that they were formed in layers with approximately identical velocities separated by zones of lower velocities. Three principal waves are traced: N_0, N_1 and N_2. The most stable is the N_2 wave.

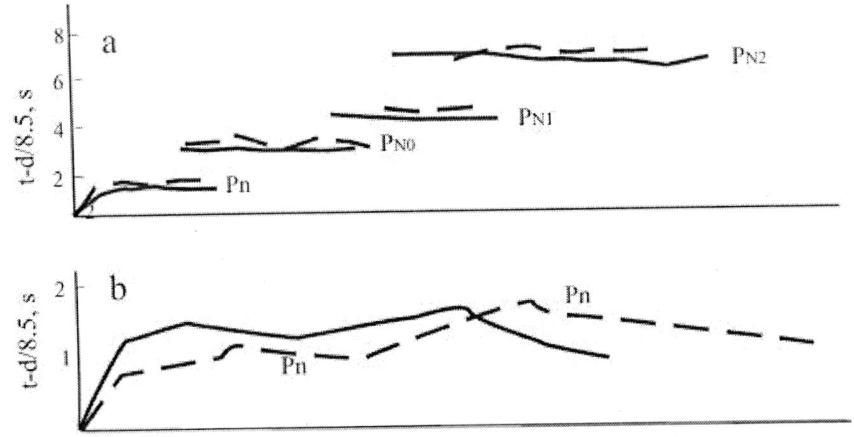

Figure 33. Travel-times of reference waves for: (a) the Mid- Atlantic Ridge and Brazilian Basin, (b) Angola Basin. Pn are waves from top of mantle, P_{N0}, P_{N1} and P_{N2} are the waves from boundaries within mantle at the bottoms of the lower velocity layers (Figure 32 c).

The depths of the N boundaries vary from profile to profile. But this is not always attributable to their actual differences. Sometimes this is due to a general uncertainty in the seismic problem solution in the case of velocity inversion zones. In such a case the velocity section was constructed in the following way. Using the distance to the point of wave attenuation at the top of the low velocity zone, the depth to this top was determined. This is an approximate estimate, substantially dependent on the adopted velocity gradient in the covering medium. Then the first arrivals time delays (Fig. 34) were used in determining the "inversion intensity," that is, the product of thickness of the inversion zone and the velocity decrease in it. In constructing the composite section it was natural to apply identical assumptions concerning velocities in the inversion zones in order not to introduce discrepancies in the boundary depths. The velocities of 7.9-8.0 km/s were resumed in the inversion zones.

The resulting cross-sections became more comparable with respect to the unity of the form of their representation. However, the difficulties in combining them into a single model, attributable to the gaps between the profiles remained. The gravity data were used for combining the N boundaries and inversion layers into some system. For example, the form of the principal minimum in the Mid-Atlantic Ridge region was used for outlining two zones with low velocity in the region of this minimum. The first zone, lying at a depth of about 20 km, was bounded by 17^0 and 10^0 W; the second (depth about 50 km) - by 6^0 and 18^0W. In the latter case allowance also was made for the anomalous properties of the M discontinuity in these sectors and the presence of lower velocity blocks beneath it.

3.3. The Composite Seismic Section along the Geotraverse

A considerable inhomogeneity of the uppermost mantle was reliably established on all DSS profiles: the velocities along the M discontinuity vary from 7.8 to 8.5 km/s; their regional decrease in the direction of the Mid-Atlantic Ridge also is noted (Fig. 32 c).

In the Angolan Basin at a depth 20-80 km there is a block with anomalously high velocities up to 8.8 km/s, no analogue of which is observed in the Brazilian Basin. The reason for such high velocities is difficult to explain from the petrology point of view, that is why this area was specially investigated. Spatial observations were made on meridian profile VII for the purpose of studying seismic anisotropy (Zverev, Tulina, 1996). It was determined that the velocities of 8.8 km/s are observed only along the geotraverse; in perpendicular direction (profile VII) the velocity section is close to the section in the Brazilian Basin.

The difference in the mantle structure of the two basins in general is not manifested in the gravity and heat fields and therefore there is basis for assuming that the existence of an anisotropic block in the Angola Basin is related to its special stressed state and not a difference in the petrology or thermal state of mantle matter.

Thus, the DSS data along the Angola-Brasilian Geotraverse shows that instead of the unified asthenosphere rising toward the ridge axis, several low velocity layers are traced (Fig. 32 c). Only in the narrow axial zone of the ridge there is a block with anomalously low velocities, which corresponds to the classic concepts of an uplift asthenosphere.

The constructed velocity model does not contradict surface wave data. Due to their low frequency, surface waves cannot discriminate the thin layers traced by DSS. However, it is not difficult to visualize that the overall effect of the series of such layers, limited the ridge region, will be identical for surface waves to the effect of a unified asthenosphere rising in the region of the mid-oceanic ridge. The gravitational effect of such models also is identical.

The other special features of the upper mantle along the Angola-Brazilian geotraverse are a structural asymmetry and block structure. The largest blocks are of the mid-oceanic ridge and of the oceanic basins. A clearly expressed block is outlined also in the axial zone of the ridge. The boundaries of those blocks are manifested clearly in all geophysical fields. For example, the wave pattern changes sharply at the boundaries of the block associated with the ridge. At its eastern boundary (38^0W) the high velocity branches of Pn wave characteristic for the Angola Basin is no longer traced and the velocities at the Moho decrease from 8.5 to 8.3 km/s. A similar picture also is observed on the western slope of the ridge (21^0W) which is supplemented here by a sharp cutoff of the N boundaries. The sharpest decrease in the gravity field also is observed precisely in these regions.

The block of the axial zone of the ridge is characterized by low velocities along the M discontinuity: 7.5 km/s against a background of 8.5 km/s typical for the geotraverse. This occurs in a rather narrow zone which creates a local gravity minimum in the region 14^0W. The boundaries of this anomalous zone agree well with the distribution of the heat flow, which is almost constant along the entire geotraverse and which has a distinct maximum only in this zone (Fig. 32 b).

However, it must be noted that the block structure of the upper mantle does not always correspond to the principal structures of the ocean. For example, the western boundary of the central block (21^0W) is located somewhat to the east of the boundary between the ridge and the Brazilian Basin (24^0 W). It most likely corresponds to a change in the character of the gravity and magnetic fields, along which Yu.Ye. Pogrebitskiy draws the boundary between the African and the American plates of the South Atlantic (Margin and Pogrebelskiy, 1986). According to their model, the axis of an ancient rift separating these plates ran in the region 21^0W.

The idea of the South Atlantic formation at the boundary between two different plates possibly is confirmed in the upper mantle structural asymmetry relative to the present-day axial zone of the ridge. For example, the width of the eastern slope of the ridge is far greater than that of the western slope. But the principal asymmetry is associated with a fundamentally different structure of the mantle in the Brazilian and Angolan Basins. In the Brazilian Basin a distinct alternation of layers with lower and higher velocities has been established. The velocities observed here near the Moho do not exceed 8.2—8.4 km/s. The mantle of the Angola Basin is characterized, as mentioned above, by structural anisotropy and no velocity inversion zones were discriminated.

4. Conclusion

This chapter presents the results of collection and analysis of seismic data received from the long-range seismic profiles during the last 25-30 years in the Northern Eurasia and in the South Atlantic. In the Northern Eurasia the profiles were carried out with large chemical and Peace Nuclear Explosions (PNE). Nine seismic profiles with total length of more than 20,000 km including 25 PNEs and many large chemical explosions were interpreted using the common methodology, and 2-D velocity models of the upper mantle were constructed along all profiles. The results were used for the compilation of five structural maps describing 3-D upper mantle structure of the Northern Eurasia down to the depth of 400 km.

The geophysical research along the Angola-Brazilian geotraverse made it also possible to introduce a number of important corrections into general concepts concerning the deep structure of the basic oceanic domains: the mid-oceanic ridge and the oceanic basins.

The comprehensive analysis of all the obtained data enable us to answer many questions which are important not only for regional studying but for global geodynamics as well. The key questions are the following:

What are the general changes of the upper mantle velocity structure and how do they correlate with tectonics and geophysical fields?

Are there some regional seismic boundaries in the upper mantle and what are their origins?

What is the lithosphere-asthenosphere system beneath different tectonic domains of the Northern Eurasia and of the South Atlantic?

Are there any problems in understanding the origin of the upper mantle seismic heterogeneity?

Answers to these questions might be the following.

1. The long-range seismic profile data have shown that the upper mantle of the Northern Eurasia and the South Atlantic is highly heterogeneous both vertically and laterally. This is expressed as variations in seismic velocities, topography of seismic boundaries, degree of layering, and local heterogeneity (Figs. 6, 12,17, 21 and 23).

The crust and uppermost mantle structure along the most profiles might be described by a block-layered model which suggests a combination of near horizontal stratification (layering) divided by several blocks. The block structure usually reflects the tectonics. For instance, the East-European and the Siberian Cratons have higher velocities in the uppermost mantle as regards the young West-Siberian and Timan-Pechora platforms. The Urals is also characterised by high velocities beneath the Moho but its deeper structure is closer to the West-Siberian platform. The velocity differences between the small geostructures are observed down to 100-120 km and between the large tectonic domains - down to 200-250 km (Fig.30 and 31).

The close correlation is also determined between the uppermost mantle velocities and the heat flow. In the areas with low heat flow the upper mantle velocities are the higher, the higher heat flow regions are correlated with the lower mantle velocities.

In the South Atlantic an asymmetry of the crust and upper mantle structure relative to the axial zone of the ridge was established. The latter is manifested in a different extension of the western and eastern slopes of the ridge and a different seismic structure of the crust and upper mantle in the Angolan and Brazilian Basins (Fig.32). This block structure reflects not only the present day division of the ocean into the ridge and the basins, but possibly more ancient structures.

2. Another important conclusion following from the presented data, is that the upper mantle is clearly stratified. In the Northern Eurasia besides velocity layering the regional reflecting boundaries are traced in the large areas. They are characterised by stable boundary velocities. Those are N1 boundary (the boundary velocity 8.35 km/s), N2 (8.4 km/s), L (8.5

km/s) and H (8.6 km/s) boundaries. The mantle reflections are distributed irregularly in the time-distance scales and it might suggest irregular space distribution of correspondent reflectors. The most reflectors, however, coincide with concrete velocity discontinuities.

In the South Atlantic the upper mantle is also considerably stratified: it is represented by alternation of the lower and higher velocities with a considerable difference in their values (from 7.8 to 8.5 km/s). The most stable boundaries separating their layers are at depths of 40 and 80 km.

All boundaries recognized in the upper mantle are not of the first order. They are thick zones with alternating high- and low-velocity layers. Both the first and secondary waves from these boundaries are complicated by many phase groups with long coda (Figs. 2, 3, 9-11, 3 and 15) which characterise the corresponding boundaries as high reflective zones (Morozov and Smithson, 2000; Nielsen et al., 2002). Average velocity contrasts are not high at the boundaries (0.0- 0.1 km/s) but they are high within the thin layering.

The existence in the mantle lithosphere regional seismic boundaries is an unexpected result because it looks unrealistic to find the regular and strong enough velocity contrasts in the upper mantle of the old platforms. No phase transitions were revealed at the depths where the boundaries are traced (Griffin et al, 1998; Kukkonen et al., 2002). Though, Solov'eva et al. (1989) note that xenoliths from the Siberian Platform kimberlites taken from the depths of occurrence of seismic boundaries have indications of film melting. It is difficult to explain the partial melting in the dry mantle lithosphere at its low temperatures. Hence, these layers might be confined to areas enriched in water or other fluids decreasing the melting temperature. The boundary structure confirms this proposal: as mentioned above they are not simple first order discontinuities but the zones with alternation of the low and high velocity thin layers.

3. The observed stratification of the continental upper mantle is difficult to present in the traditional lithosphere-asthenosphere model (Fuch and Froidevauxs, 1987). From the heat flow data the Siberian "thermal" lithosphere is outlined by a zone of possible solidus in the dry mantle at around 200-250 km depth (Cermak, 1985). The upper-mantle model shows the absence of a distinct low-velocities which could be related to this zone.

In contrast the presented data show the "thermal" lithosphere to be rheologically stratified. Strong changes of the rheology may be supposed at depths of 70-120 km at N boundaries. The most important features of the boundaries are that they are good reflectors or reflectivity zones and they divide the lithosphere into two portions with different inner structure. Above the N boundaries the sub-Moho lithosphere is complex, laterally inhomogeneous, beneath them its structure appears to be less complex.

The N boundaries have been distinguished in many other regions as well. They were called as the 8° boundary by Thybo and Perchuc (1996) and may be considered to have a global significance.

Another specific feature of the N boundaries is that they often underline low velocity layers.

Why there exist the low velocity zones within the continental lithosphere at a depth of 100 km is not easy to understand. The partial melting is difficult to propose at these depths in the low temperature lithosphere. As in the case of the seismic boundaries, it may be suggested that at a depth of 100-150 km a weak zone appears due to the concentration of mantle fluids and corresponding metamorphic processes (Pavlenkova, 1996; Perchuc and Thybo, 1996). In

Baltic Shield these velocity inversions are characterized by higher electrical conductivity (Kovtun et al., 1994) favoring the existence of fluids at this depth interval. The fluids change mechanical properties of the matter, they initiate partial melting and metasomatism of the mantle material at low temperature which results in the seismic velocity changes.

The formation of thin layers of partial melting in the upper mantle agrees with the laws of mechanics. In a two-phase medium a uniform distribution of the fluid component is impossible. It must be concentrated at some pressure and temperature levels.

The described characteristics of the N boundaries (change of the upper mantle general heterogeneity and existence of the lower velocity zones) indicate that they may be bottoms of the rigid parts of the lithosphere (the mechanical but thermal lithosphere) and beneath them the mantle material is more plastic and can not preserve its own inhomogeneity.

A change of the rheology is visible also at the L boundary. In many record-sections the waves from this boundary attenuate, and low velocity gradient zones may be proposed beneath the boundary. The Q factor which has been determined from the mantle wave spectrums (Egorkin, Kun, 1978) decreases at the depths of 200-250 km. And finally the increasing of the plasticity at these depths follows from structural features of the L and H boundaries. H boundary has a form reverse to L boundary (Fig. 17), and it means that the matter between the boundaries can flow to create an isostatic equilibrium of the upper mantle.

Thus, in the Northern Eurasia the upper mantle should be described not by the system lithosphere-asthenosphere but by a system of layers of different rheology separated by thin weak zones.

4. In the ocean it is also very difficult to make a clear delimitation between the lithosphere and the asthenoshere against a background of general stratification. In general in the South Atlantic the velocity inversion zones may be interpreted in an ordinary way as partial melting zones existing over a prolonged period of time. According to the heat flow data, the partial melting in the ridge region is possible at a depth of 20-30 km, in the basins at a depth not less than 60-80 km. The proposal that in the ridge region the low velocity layers may be represented by molten matter has been also confirmed by the results of density modeling. This modeling shows that the density in the velocity inversion zones beneath the ridge is the minimum possible for the limitations on the density/velocity ratio and on the average is 0.1 g/cm^3 less than in these same layers, but in the region of basins (Pavlenkova et al., 1993).

However, the presented above Angola-Brazilian geotraverse data show that the transition zone from the lithosphere to the asthenosphere looks far more complex than in the classic lithosphere-asthenosphere model (Fig. 32 c). This is not simply a boundary rising toward the mid-oceanic ridge, but a series of layers with reduced velocities and densities (asthenospheric lenses) permeating the lithosphere. They are best developed in the ridge region. The lithosphere itself also is inhomogeneous: it is considerably stratified maybe due to relict of asthenospheric lenses.

Thus, the classic model of the oceanic lithosphere with a gradually increasing thickness with increasing the distance from the axial zone of the ridge must be replaced by a stratified lithosphere permeated by low velocity layers in the mid-oceanic ridge zone.

5. The presented data reveal some problems in understanding the nature of abnormally high velocities (up to 8.5 km/s) in the uppermost part of the mantle.

For instance, the specific feature of the Siberian Craton uppermost mantle is the existence of blocks with velocities up to 8.4 km/s (Fig.30). The velocities in some local blocks reach the values of 8.5-8.6 km/s (in the cross-sections and at the mapping the average values of 8.3-8.4 km/s are shown for large domains). Similar velocity were distinguished in other parts of the craton as well (Suvorov et.al., 1999). The nature of these blocks can not be explained in a conventional way as a change in composition because they are not expressed in geophysical fields. As shown in (Christensen, 1984; Kern, 1993; Sobolev and Fuchs, 1993) all known mantle rock varieties are characterized by close seismic velocities but differ in density.

Two possible explanations of this phenomenon may be suggested: (1) The mantle matter in high-velocity blocks is depleted in iron (Deshamps et al., 2002) or (2) the blocks are characterized by velocity anisotropy, i.e., the lateral velocities are much higher than the vertical ones.

The low content of iron in the upper mantle might be interpreted by its removal into the Earth's crust during plato-basalt eruptions. But in this case, it is unclear why the high-velocity blocks are localized along the periphery of the craton rather than in its western part, where most plateau basalts occur.

The most plausible explanation for these high velocity zones is seismic anisotropy which is typical of the continental upper mantle and is based on the velocity anisotropy of olivine crystals, the main component of mantle rocks (Fuchs, 1983; Babushka et al., 1984). In the Siberian Craton case, however, it is not the asymuthal anisotropy. In the Tunguss block the high velocities are observed on the crossed profiles METEORITE and KIMBERLITE. The same results are obtained for the other blocks (Suvorov et al., 1999). It might be the depth anisotropy when in the horizontal directions the velocities are high, and they are lower in the vertical (depth) direction.

On the Angola-Brazilian geotraverse the velocities of 8.5 km/s look also unrealistic for the high heat flow region, and they might be explained by asymuthal anisotropy. The observations along the meridian profile VII crossed the geotraverse (Fig. 32) confirm the seismic anisotropy in the uppermost mantle of the Angola basin: the velocities below the M boundary on the profile VII are lower than along the geotraverse at least in 0.3 km/s (Zverev and Tulina, 1996). The same origin of the high velocities in the oceanic uppermost mantle was determined in other regions (Shimamura et al., 1983; Shearer and Orcutt, 1985; Ekstrom and Dziewonski, 1998).

In the mid-oceanic ridge area it is also possible to assume that the high velocity boundaries were formed at the bottom of zones of partial melting as a result of prolonged differentiation of mantle matter and the dropping out of dense material with a high melting point. The boundary structure possibly reflects the sequence of this differentiation and the high boundary velocities confirm the idea of their association with the densest components of mantle matter.

Thus, the long-range seismic studies made in the Northern Eurasia and in the South Atlantic reveal many new details of the upper mantle structure which are important for understanding the geodynamic processes and tectonic evolution of the principal geostructures of the continents and oceans. The profiling revealed a fine stratification of the crust and upper mantle: high velocity layers alternate with lower velocity ones and strong reflection boundaries often separate the layers. Several such boundaries have a global significance and suggest rheological stratification of the lithosphere (they separate brittle and weak layers). The nature of these boundaries and of velocity inversion zones may be explained by fluids concentration at some critical depths. They may play important roles at any geodynamic processes: to be detachment zones at the lithosphere blocks moving or channels for flowing of the mantle matter for the isostatic equilibrium of the lithosphere. Together with deep faults the thin layering zones form a channel system for the mantle fluids and matter transportation. During tectonic activation these weak layers were transformed in asthenolites by partial melting and provoked plume tectonics. In general the rheological stratification can better explain origin of the observed tectonic pattern than other models.

References

Asada, T. and Shimamura, H., 1979. Long-range refraction experiments in deep oceans // *Tectonophysics*, vol.56. P. 67-82

Babuska V., Plomerova J., Sileny I., 1984. Spatial variations of P-residuals and deep structure of the European lithosphere // *Geophys. J. R. Astron. Soc.* **79**. P.363-383

Beloussov V.V., Pavlenkova N.I. and Kvjatkovskaya G.N.(Eds.), 1991. *Deep structure of the USSR territory (in Russian).* Nauka, Moscow: 234 pp.

Benz, H.M., Huger, J.D., Leith, W.S., Mooney, W.D., Solodilov, L.N., Egorkin, A.V. & Ryaboy, V.S.,1992. *Deep seismic sounding in Northern Eurasia. EoS*, **73**, 297-300.

Carbonell,R., J.Gallart, A.Perez-Estaun, J.Diaz, S.Kashubin, J.Mechie, F.Wenzel and J.Knapp, 2000. Seismic wide-angle constraints on the crust of the southern Urals // *J.Geoph.Res*., vol.105, NO B6. P.13755-13777.

Cermak, V., 1985. Thickness of the lithosphere of the USSR territory from geothermal data. *Geology and Geophysics*, N **5**, 33-40.

Cerveny,V., Molotkov, I.A., Psencik, I., 1977. Ray method in seismology., Univerzita Karlovo, Praha.

Cerveny V. and I. Psencik, 1983. "SEIS83-Numerical Modelling of Seismic Wave Fields in 2-D Laterally Varying Layered Structure by the Ray Method," in: Documentation of Earthquake Algorithms, World Data Center A, for *Solid Earth Geophys*, Ad. by E. R. Engdahl (Boulder,Rep. SE-35, 1983) pp. 36–40.

Cipar, J.J., Priestley, K., Egorkin, A. & Pavlenkova, N., 1993. From rift to rift: the Yamal Peninsula - Lake Baikal deep seismic sounding profile. *Geophys.Res.Lett*., **20**, 1631-1634.

Cristensen, D.H., Kimbal, J.K. and Mank F.J, 1980. Reylaigh wave groupvelocity dispersion in the North and South Atlantic Oceans. *Bull. of Seismol. Soc. of America*, **70**, No 5, 1787-1809

Christensen, N.I , 1984. The Magnitude, Symmetry and Origin of Upper Mantle Anisotropy Based on Fabric Analyses of Ultramafic Tectonics, Geophys. *J. R. Astron. Soc.***76**, 89–111.

Deschamps F., J. Trampert, and R. Snieder, "Anomalies of Temperature and Iron in the Uppermost Mantle Inferred from Gravity Data and Tomographic Models." *Phys. Earth Planet. Inter*. **129**, 245—264 (2002).

Egorkin A.V., 1980. The upper mantle. In: Zverev S.M. and Kosminskaja I.P. (Eds.) Seismic models of the lithosphere of the major geostructures of the territory of the USSR. *Nauka*, Moscow, 145-153 (In Russ.)

Egorkin A.V., 1997. Evidence for 520-km discontinuity. In: *Upper mantle heterogeneities from active and passive seismology*, Ed. K.Fuchs. Dordrecht, Kluwer Acad.Pub. 51-61

Egorkin A.V., 1999. Study of the mantle by superlong geotraverses. *Physics of the Earth*, N **7-8**, 114-130.

Egorkin, A.V. and Chernyshov, N.M., 1983. Peculiarities of mantle waves from long-range profiles. *J.Geophys*., **54**: 30-34.

Egorkin A.V. and V.V.Kun, 1978. P-wave attenuation in the upper mantle of the Earth. *Physics of the Earth*, **4**, 25-36

Egorkin, A.V., Zuganov, S.K., Pavlenkova, N.A. & Chernyshev, N.M., 1987. Results of lithosphere studies from long-range profiles in Siberia. *Tectonophysics*, **140**., p.29-47.

Ekstrom, G., Dziewonski, A.M., 1998. The unique anisotropy of the Pacific upper mantle. *Nature* **394**, 168-172.

Fuchs K. (Ed.), 1997. Upper mantle heterogeneities from active and passive seismology, NATO ASI Series (1.Disarmament Technologies - vol.17), Contribution №336, International Lithosphere Program, Kluwer Academic Publishers. Dordrecht. 366 pp.

Fuchs, K., 1983. Recently formed elastic anisotropy and petrological model for the continental subcrustal lithosphere in southern Germany // *Phys.Earth Planet.Inter*., **31**. P. 93-118.

Fuchs, K. and C. Froidevaux (Eds., 1987.) Composition, Structure and Dynamics of the lithosphere-Asthenosphere Systems. *Geodynamic Series* , **16**, International Lithosphere Program Contribution. Washington. D.C. 327 pp.

Griffin W.L., O'Reilly S.Y., Ryan C.G., Gaul O., Ionov D., 1998. Secular variation in the composition of subcontinental lithospheric mantle// J. Braun et al. (Ed.), Structure and evolution of the Australian continent. *AGU Geodynam. Ser*., v. 26, P. 1-25.

Guggisberg B. and Berthelsen K., 1987. A two-dimensional velocity model for the lithosphere beneath the Baltic Shield and its possible tectonic significance // *Terra Cognita*, **7**. P. 631-638

Hales, A.L.,1969. A seismic discontinuity in the lithosphere, Earth and *Planet.Sci.Lett.,* **7**, 44-46

Kern, H.M., 1993. Physical properties of crustal and upper mantle rocks with regards to lithosphere dynamics and high pressure mineralogy.// *Physics of the Earth and Planetary Interiors*, **79**, 113-136.

Kovtun, A.A., Vagin, S.A., Vardanyants, I.L., Kovkina E.L., Uspenskii N.I., 1994. Magnetotelluric sounding of the crust and mantle structure in the eastern Baltic Shield, *Physics of the Earth*, 1994, no. 3, 32-36

Kukkonen, I.T., K.A. Kinnunen, and P. Peltonen, 2003. Mantle Xenoliths and Thick Lithosphere in the Fennoscandian Shield, *Phys. Chem. Earth*, **28**, 349—360.

Lehmann, I., 1959. Velocities of longitudinal waves in the upper part of the Earth's mantle, *Ann. Geophys*., **15**, 93-118.

Margin, I.M. and Yu.Ye.Pogrebetskiy (Eds.), 1986. Lithosphere of the Angolin Basin and Eastern Slope of the South-Atlantic Ridge. PGO "Sevmorgeologija", *Leningrad*, **167** pp.(In Russ.)

Mechie, J., A.V.Egorkin, K.Fuchs, T.Riberg, L.N.Solodilov & F.Wenzel, 1993. P-wave mantle velocity structure beneath Northern Eurasia from long-range recordings along the profile QUARTZ, *Phys. Earth planet. Inter*.,**79**, 269-286.

Morozov, I.B., Morozova, E.A., Smithson, S.B., Solodilov, L.N., 1998. On the nature of the teleseismic Pn phase observed on the ultralong-range profile "Quartz," Russia. *Bulletin of the Seismo-logical Society of America* **88**, 62-73.

Morozov, I.B., Morozova, E.A., Smithson, S.B., & Solodilov L., 1999. Heterogeneity of the uppermost mantle beneath Russian Eurasia from the ultralong-range profile Quartz, *J. geophys Res.,* **104**(B9), 20 329-20 348.

Morozov, I.B., Smithson, S.B., 2000. Coda of long-range arrivals from nuclear explosions. *Bulletin of the Seismological Society of America* **90**, 929-939.

Morozova, E.A., I. B. Morozov, and S. B. Smithson,1997 "Heterogeneity of the Uppermost Eurasian Mantle along the DSS Profile 'Quartz', Russia," In *Upper Mantle Heterogeneities from Active and Passive Seismology,* Ed. By K. Fuchs Kluwer, Dordrecht, 139–146.

Morozova, E.A., Morozov, I.B., Smithson, S.B., Solodilov, L., 2000. Lithospheric boundaries and upper mantle heterogeneity beneath Russian Eurasia; evidence from the DSS profile QUARTZ. In: Carbonell, R., Gallart, J., Tome, M. (Eds.), Deep Seismic Profiling of the Continents and their Margins. *Tectonophysics*, vol. 329, pp. 333-344.

Mueller, St. and Ansorge,J. Deep seismic sounding of the mantle lithosphere // G.Nolet and B.Dost (Eds.), *European Geotraverse (EGT) Project, the upper mantle*. ESF, Strasbourg, 1988. P. 63-76

Nielsen, L., Thybo, H. & Solodilov, L., 1999. Seismic tomographic invertiön of Russian PNE data along profile Kraton, *Geophys. Res. Lett.*, **26**, 3413-3416.

Nielsen, L., Thybo, H., Egorkin, A.V., 2002. Implications of seismic scattering below the 8° discontinuity along PNE profile Kraton. *Tectonophysics* **358**, 135-150.

Nielsen, L., Thybo, H., Morozov, I.B., Smithson, S.B., Solodilov, L., 2003. Teleseismic Pn arrivals: influence of upper mantle velocity gradient and crustal scattering. *Geophysical Journal International* **153**, F1-F7.

Okal, E.A., 1977. The effect of intrinsic oceanic upper mantle heterogeneity regionalization of long period Reylaigh wave phase velocity. *Geophys. J. R. Astron. Soc.,* **49**, 357-370.

Orcutt, J.A. and Dorman, L.M., 1979. Results of Scripps long range profiles in the Pacific // *Tectonophysics*, vol. 56. P. 83

Pavlenkova G.A., Pavlenkova N.I. Upper mantle structure of the Northern Eurasia from peaceful nuclear explosion data // *Tectonophysics* **416**. 2006. P. 33-52

Pavlenkova G.A., Priestly K. and Cipar, 2002. 2-D model of the crust and uppermost mantle along Rift profile, Siberian craton. *Tectonophysics*, **355**, 171-186.

Pavlenkova, N.I., 1982. The intercept-time method: possibilities and limitations. *J.Geophysics*., **51**, 85-95.

Pavlenkova, N.I., 1988. The nature of seismic boundaries in the continental lithosphere, *Tectonophysics*, **154**, 211-255.

Pavlenkova, N.I., 1996a. Crust and upper mantle structure in Northern Eurasia from seismic data. In: *Advances in Geophysics*, Academic Press, Inc.(Eds. R.Dmowska, B.Saltzmann), Vol.37..

Pavlenkova, N.I., 1996b.General features of the upper mantle stratification from long-range seismic profiles.// *Tectonophysics*, **264**, 261-278.

Pavlenkova, N.I., 1996c. Role of fluids in the formation of the crustal seismic layering // Isvestija, *Physics of the Solid Earth,* № **4**. 1996. P. 51-61.

Pavlenkova N.I., 2005. Ambiguity of the interpretation of DSS data: a case of the Fennolora profile (the Baltic Shield) // Isvestija, *Physics of the Solid Earth*, vol.41, №2. pp.132-141

Pavlenkova, N.I., Pavlenkova G.A. and Solodilov L.N., 1996. High seismic velocities in the uppermost mantle of the Siberian craton. *Tectonophysics*, **262**, 51-65.

Pavlenkova, N.I., Pogrebitsky, Yu.E. and Romanjuk,T.V., 1993. Seismic-density model of the crust and upper mantle of the South Athlantic along Angola-Brasil geotraverse, *Physics of the Earth,* **N10**, 27-38.

Perchuc E.& Thybo H., 1996. A new model of upper mantle P-wave velocity below the Baltic Shield: indication of partial melt in the 95 to 160 km depth range. *Tectonophysics* **253**, 227-245.

Pollack H.N., S.J. Hurter, J.R. Johnson, 1993. Heat flow from the Earth's interior: analysis of the global data set, *Rev. Geophys*., **31**, 267-280.

Priestly K., Cipar, J., Egorkin A.V. and Pavlenkova, N.I.,1994. Upper mantle velocity structure beneath the Siberian Platform. *Geophys.J.Ins*., **118**, 364-378

Ryberg, T., K.Fuchs, A.V. Egorkin and L.Solodilov, 1995. Observation of high-frequency teleseismic Pn on the long-range Quartz profile across northern Eurasia. *J.Geophys.Res*., vol.100, N9, 18151-18163.

Raitt, R.W., Shor, G.G., Francis, T.J.G. and Morris, G.B., 1969. Anisotropy of the Pacific upper mantle // *J.Geophys.Res*., vol.74. P. 3095-3109

Ryberg T., Wenzel F., Egorkin A.V., Solodilov L., 1997. Short-period observation of the 520 km discontinuity in northern Eurasia // *J. Geophys. Res*., V.102. № 3. P.5413-5422.

Ryberg T., Wenzel F., Egorkin A.V., Solodilov L., 1998. Properties of the mantle transition zone in northern Eurasia // *J. Geophys. Res*., V.103. № B1. P.811-822.

Ryberg, T., F. Wenzel, J. Mechie, A. Egorkin, K. Fuchs and L. Solodilov, 1996. Two-Dimensional Velocity Structure beneath Northern Eurasia Derived from the Super Long-Range Seismic Profile Quartz, *Bull. Seismol. Soc. Am*., **86**, P. 857-867.

Shearer, P., Orcutt, J., 1985. Anisotropy in the oceanic lithosphere – theoty and observations from the Ngendei seismic refraction experiment in the south-west Pasific // *Geophys. J. Roy. Astron. Soc*., vol. 80. P. 493-526

Shimamura, H., Asada, T., Suyehiro, K., et al., 1983. Longshot experiments to study velocity anisotropy in the oceanic lithosphere of the North Western Pacific //*Phys.Earth.Planet.Inter*., vol.31. P.384-392

Schueller, W, I. B. Morozov, and S. B. Smithson, 1997. Crustal and Uppermost Mantle Velocity Structure of Northern Eurasia along the Profile 'Quartz', *Bull. Seismol. Soc. Am.* **87**, 414–426.

Sobolev, S.V. & K.Fuchs, 1993. Seismic velocities and density in the deep continental lithosphere from the composition of xenoliths // *Terra Nova*, **5**, Abstract suppl.1 EUG V11, Strasbourg. 333-334.

Solov'eva, L.V., Vladimirov, B.M., Kiselev, A.I. and Zavijalov, L.L.,1989. Two stages of mantle metasamatites of deep xenoliths from Yakutia kimberlites and their relation to lithosphere processes. In: Precambrian metasamotites and their ore deposits. *Nauka*, Moscow, p.3-17./In Russ./

Steinmetz, J., R.B .Whitmarsh and V. S .Moreira, 1977. Upper mantle structure beneath Mid-Atlantic Ridge north of the Azores based on observations compressional waves. Geophys. *J. R. Astron. Soc*., **50**, 350-380.

Suvorov V.D., Parasotka B.S. and Cherny S.D., 1999. Deep seismic studies in Yakutuya.// *Physics of the Earth*, **7/8,** 94-113.

The LADLE study group, 1983. A lithospheric seismic refraction profile in the western North Atlantic Ocean. Geophys. *J. R. Astron. Soc.*, **75**, 23-69

Thybo H., Perchuc E., 1997. The seismic 8^0 discontinuity and partial melting in continental mantle // *Science*. 1997. V. 275. P.1626-1629.

Thybo Y., Perchuc E., Pavlenkova N.I., 1997. Two reflectors in the 400 km depth range revealed from Peace Nuclear Explosion seismic sections // K.Fuchs (Ed.) Upper mantle heterogeneities from active and passive seismology. *NATO ASI Series*, vol.17, 1997. Kluwer Academic Publishers. P.97-104

Tittgemeyer M., F.Wenzel, K.Fuchs and T.Ryberg, 1996. Wave propagation in the multiple-scattering upper mantle - observation and modeling. *Geophys.J.Int.,* **127**, 492-502

Yegorkin, A.V.and Pavlenkova, N.I., 1981. Studies of mantle structure of USSR territory on long-range seismic profiles. *Physics of the Earth and Planetary Interior*, **25**, pp.12-26

Zelt C.A, Smith R.B. Seismic traveltime inversion for 2-D crustal velocity structure // *Geophys. J. Int.* 1992. 108. P. 16-34

Zverev, S.M. and Tulina Ju.V.(Eds.),1996. Deep seismic sounding of the lithosphere on the Angola-Brazil Geotraverse. Geophysical Centre RAN, Moscow, 190 pp, (in Russ.)

In: The Lithosphere: Geochemistry, Geology and Geophysics
Editors: J.E. Anderson et al, pp. 123-140
ISBN: 978-1-60456-903-2

Chapter 3

THE FATE OF SUBDUCTED OCEANIC CRUST AND THE ORIGIN OF INTRAPLATE VOLCANISM

Alan D. Smith
Department of Earth Sciences, University of Durham, Durham, DH1 3LE, UK

Abstract

Standard geodynamic models envisage subducted oceanic crust to be part isolated in thermal boundary layers and part remixed into the depleted mantle. The two fates for such material result from combining remixing models based on observation from orogenic lherzolites, with theoretical models for the generation of intraplate volcanism by mantle plumes. The concepts were combined because high $^{3}He/^{4}He$ ratios in intraplate basalts were interpreted to require a primitive source component and the convecting mantle was considered unable to retain ^{3}He on melting. However, high $^{3}He/^{4}He$ ratios may reflect low U+Th sources, and differences in $^{3}He/^{4}He$ between MORB and OIB can be explained by sampling of the convecting mantle. Interpretations of high $^{186}Os/^{188}Os$ in intraplate lavas as evidence for interaction with the core are likewise tenuous, as the signatures can be explained by pyroxenites or mantle sulphides. Instead, the remixing models should have been combined with models for the tapping of shallow mantle sources by plate tectonic processes, to give an explanation for the origin of intralate volcanism from the convecting mantle without plumes. Pyroxenitic sources for intraplate volcanism may be generated at convergent margins if subducted oceanic crust undergoes melting in the back-arc region, or along the flanks of convective upwellings beneath ocean ridge systems as melts from altered eclogite or sediment components of recycled crust react with peridotites of the depleted mantle. Generation of intraplate melts occurs in off-axis regions as a result of fluxing of the pyroxenite-veined mantle with fluids derived from dehydration or decarbonation of later generations of subducted slabs in the shallow mantle.

Introduction

The development of plate tectonic models in the 1960's led to the question of what is the fate of subducted oceanic crust? Early models (Armstrong 1968) suggested such material is recycled back toward mid ocean ridge systems by convection within the upper mantle.

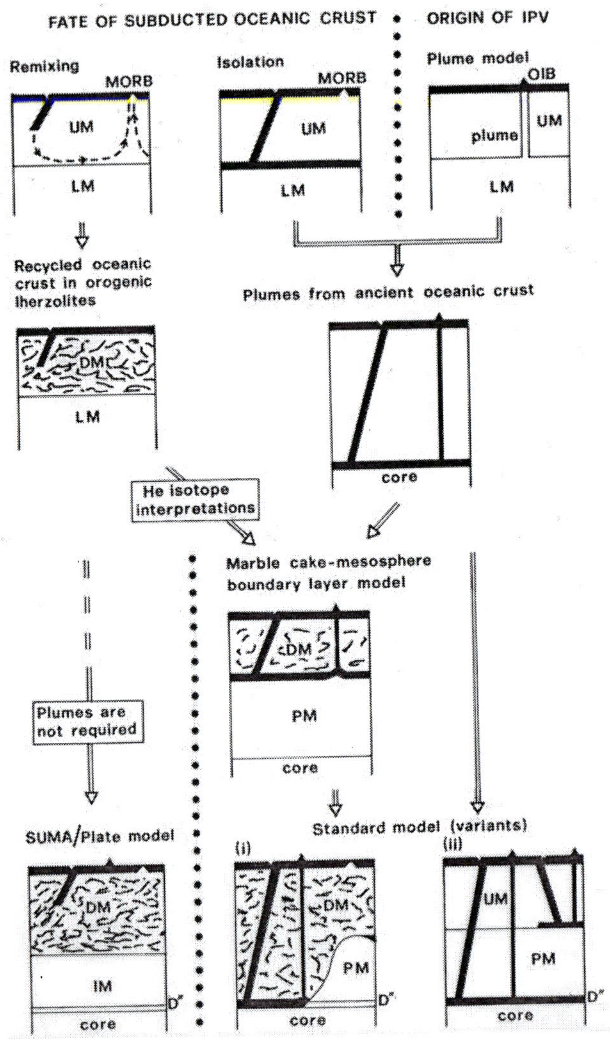

Figure 1. Evolution of concepts concerning the fate of subducted oceanic crust. Remixing (Armstrong 1968) and isolation (Dickinson and Luth 1971) of subducted material were proposed in parallel with development of the mantle plume model for the origin of intraplate volcanism (IPV) (Morgan 1971). The remixing model was supported by interpretation of pyroxenites in orogenic lherzolite massifs as recycled oceanic crust (Polvé and Allègre 1980), whereas the concept of isolating subducted crust was incorporated into the plume model of Hofmann and White (1982). The remixing and plume concepts were subsequently combined in the mesosphere boundary layer and marble-cake models of Allègre and Turcotte (1985, 1986), to give the standard geodynamic model on the basis of helium isotope interpretations. Several variants of the standard model exist (e.g. (i) Kellogg et al. 1999, (ii) Courtillot et al. 2003) because of difficulties in reconciling the requirements for a deep primitive mantle layer with geophysical models of mantle convection. Doubts about the existence of plumes and the ability of the plume model to provide a comprehensive explanation for intraplate volcanism, have resulted in the remixing model being revisited in the SUMA model of Meibom and Anderson (2004), which now forms an important component of the plate model where plumes are not required for the generation of intraplate volcanism. Abbreviations: DM = depleted mantle MORB source, IM = isolated mantle (may be of primitive composition or contain deeply subducted slabs which can not be entrained by mantle convection), LM = lower mantle, PM primitive mantle, UM = upper mantle.

Evidence for such remixing came from the interpretation of pyroxenite bands in orogenic lherzolite massifs as subducted oceanic crust that had been stretched out and thinned by several orders of magnitude, by convection within the mantle over periods of several hundred million years (Polvé and Allègre 1980). The structure of the centimetre-metre thick bands within a predominantly lherzolite matrix found in such massifs was labelled as 'marble-cake mantle' by Allègre and Turcotte (1986). Other models, however, suggested that subducted oceanic crust was isolated at depth rather than being remixed with the mantle (Dickinson and Luth 1971). This latter suggestion became an integral part of models for the origin of intraplate volcanism which had been developed in parallel with models for the fate of subducted oceanic crust, with the suggestion by Hofmann and White (1982) that subducted material collected in a thermal boundary layer at the base of the mantle before becoming buoyant and rising as mantle plumes. Both the concepts of remixing and isolating subducted oceanic crust are now part of the standard geodynamic model for the Earth, where it is envisaged that remixed oceanic crust is responsible for geochemical heterogeneity in the convecting mantle, whilst oceanic crust stored in thermal boundary layers serves as the source of intraplate volcanism (Fig. 1).

The number of studies based on the standard model, however, belies the number of unresolved problems with the model. These include uncertainties as to whether oceanic crust should be stored at the core-mantle boundary or base of the upper mantle, and the number of plumes (e.g. Anderson 2005a). Advocates of the plume model have argued that the uncertainties reflect a theory in its early stages of development (Sheridan 1994; Sleep 2007). However, it remains that plume model cannot readily account for intraplate volcanism away from where plumes impact beneath the lithosphere (e.g. Natland and Winterer 2005, Hirano et al. 2006), such that the standard model must involve two distinct origins for intraplate volcanism. Rather than seeking ad hoc mechanisms to make the plume model fit, it should be asked whether the duplicate origins for intraplate volcanism are a consequence of unnecessary incorporation of concepts during development of the standard model. The reasoning for combining two opposing models for the fate of subducted oceanic crust should therefore be re-evaluated. The argument presented in this study is that the problems with the standard model result from inclusion of the plume model, and that remixing subducted oceanic crust with the convecting mantle as in the plate model (Foulger 2002, 2007), can adequately explain the heterogeneity found in ocean-ridge (MORB) and intraplate (OIB) basalts.

Development of the Standard Model

Why Were Two Contrasting Models for the Fate of Subducted Oceanic Crust Included?

The concepts of remixing and isolating subducted oceanic crust were combined in the geodynamic models of Allègre and Turcotte (1985, 1986) on the basis of the interpretation of Allègre et al. (1983) regarding He isotope signatures in mantle-derived rocks. The isotope ^{3}He is primordial, whereas ^{4}He is primordial and radiogenic from the decay of ^{238}U, ^{235}U, ^{232}Th. The isotope ratios of helium are expressed as R/Ra where Ra is the atmospheric ratio of $1.38x10^{-6}$. Early analyses of rock and gas samples indicated higher $^{3}He/^{4}He$ ratios in Hawaiian samples compared to MORB (e.g. Craig and Lupton 1976, Kurz et al. 1983).

Although it was noted that low ratios could reflect high U+Th in the source region, the principal focus was on interpretation of high $^{3}He/^{4}He$ as indicating an excess of primordial ^{3}He. The concept of mantle beneath the 660 km seismic discontinuity having a primitive composition had been suggested from crust-mantle evolution models in which the upper mantle MORB-source was considered the residue from formation of the continental crust (e.g. Allègre 1982). A primitive lower mantle had also been suggested from the chondritic Nd-Sr isotope compositions measured in continental basalts (Wasserburg and DePaolo 1979), and a deep primitive reservoir was suggested by Allègre et al. (1983) as the source of apparent excess primordial ^{3}He in Hawaiian basalts.

Although $^{3}He/^{4}He$ ratios ratios appeared to be higher in the source of intraplate volcanic rocks, the ratios in MORB indicated the convecting mantle was not devoid of ^{3}He. The latter feature was considered incompatible with degassing at ocean ridges, as it was considered that re-mixing with subducted oceanic crust would not be able to replenish the ^{3}He budget (Allègre and Turcotte 1985). Subducted oceanic crust, along with delaminated continental lithosphere following the model of McKenzie and O'Nions (1983), was thus suggested to collect at the base the upper mantle with ^{3}He introduced into the upper mantle by diffusion through, or perturbation of, the mesospheric layer of subducted material (Allègre and Turcotte 1985). Allègre and Turcotte (1986) elaborated on recycling subducted crust into the convecting mantle following Polvé and Allègre (1980), but as in Allègre and Turcotte (1985), allowed collection of a proportion of subducted oceanic crust in plume sources at the base of the upper mantle.

Recycling crust into the convecting mantle was resisted in some models on the basis that trace element signatures in Atlantic and Pacific MORB did not indicate the involvement of a crustal component (e.g. Rehkämper and Hofmann 1997, Hofmann 1997). However, other studies embraced the concept of isolating a fraction of subducted oceanic crust, estimated at 13% by Christensen and Hofmann (1994), in plume sources, whilst remixing the larger fraction with the convecting mantle as an explanation for heterogeneity in the depleted mantle MORB-source (e.g. Saunders et al. 1988). The notion of high $^{3}He/^{4}He$ ratios requiring a primordial component has remained one of the cornerstones of the plume model (e.g. Kellogg and Turcotte 1990a, van Keken et al. 2002), but the problem of reconciling the location of such a layer with geophysical evidence against layered mantle convection has been a persistent problem which has lead to a multitude of variations in the distribution of plume sources (e.g. Hofmann 1997, Albarède and van der Hilst 1999, Kellogg et al. 1999, Courtillot et al. 2003) (Fig. 1). Despite the variations, this model is referred to here as the standard model because of the widespread, largely unquestioned adoption of the concepts concerned.

A Core-Signature in Intraplate Volcanism?

The Re-Os isotopic system (^{187}Re decays to ^{187}Os) was refined as a tool for studying mantle evolution in the early 1990's. Suprachondritic $^{187}Os/^{188}Os$ ratios found in OIB were interpreted within the framework of the standard model to result from recycling of oceanic crust into plume sources (e.g. Hauri and Hart 1993). However, it was also suggested that the Os isotopic signatures could be generated by interaction of plume sources with the outer core (Walker et al. 1995). Support for the latter model, and hence isolation of plume sources at the core-mantle boundary, came with development of the Pt-Os isotope system (^{190}Pt decays to

^{186}Os) and the demonstration of coupled enrichments in $^{186}Os/^{188}Os$ – $^{187}Os/^{188}Os$ in picrites from Hawaii and komatiites from Gorgona Island (Brandon et al. 1999, 2003). Because ^{190}Pt is a low abundance isotope with a long half life, and recycled crust has low Os content, such material is unsuitable for explaining elevated $^{186}Os/^{188}Os$ signatures, which instead were used to infer the presence of up to 1% outer core material in plumes (Brandon et al. 1998).

The core interaction model was supported by Humayun et al. (2004) who suggested that high Fe contents of intraplate basalts relative to MORB, also resulted from interaction with the core. Coupled with suggestions that ^{3}He might be incorporated into plume sources from the core (Porcelli and Halliday 2001), placement of plume-sources at the core-mantle boundary appeared to present a solution to the paradox that whilst $^{3}He/^{4}He$ ratios could be interpreted as a primordial mantle signature, there was no evidence (e.g. Hofmann et al. 1986) for mixing with a primitive mantle reservoir in trace element signatures in intraplate lavas.

Melting Regimes in Plumes

The plume model explains the large volumes of some intraplate volcanic provinces by virtue of the thermal anomaly inherent in invoking a source deep in the mantle. The thermal anomalies allowed for plumes are typically 100-200^{0}C above the convecting mantle adiabat. However, low temperature plumes have been suggested (Cordery et al. 1997, Takahashi et al. 1998) and the possibility of a temperature range of 200^{0}C within the convecting mantle has been interpreted against there being any thermal anomaly involved in the generation of intraplate volcanism (Anderson 2000). Melting models have considered plumes to be composed of fertile peridotite or eclogite fragments embedded within a depleted peridotite matrix (e.g. Cordery et al. 1997, Kogiso et al. 1998, Kogiso and Hirschmann 2006). The amount of eclogitic in a plume has typically been estimated at 10-30% (e.g. Sobolev et al. 2007), although some models based on Os isotope calculations have required up to 90% recycled crust in a plume (Becker et al. 2000). Pyroxenitic sources have recently been suggested as more suitable than olivine-bearing sources in accounting for the high Ni and Si contents of OIB (Sobolev et al. 2005).

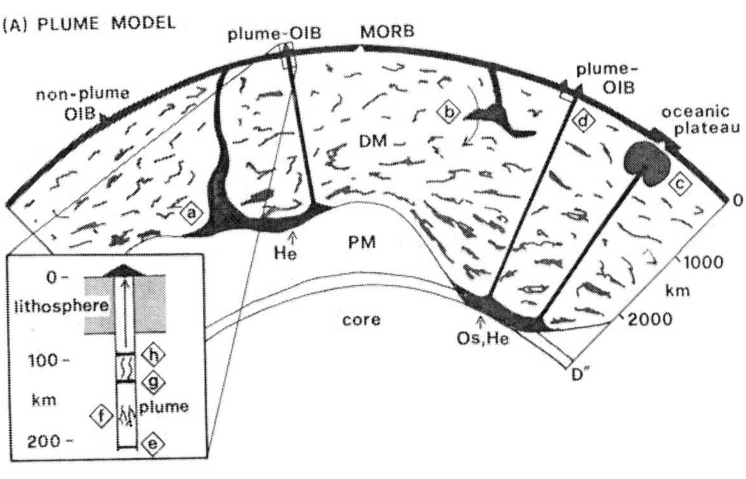

Figure 2. Continued on next page.

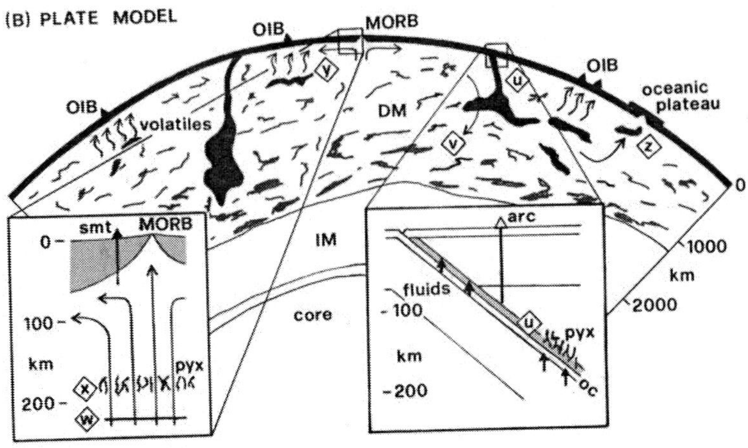

Figure 2. Comparison of the fate of subducted oceanic lithosphere in the plume (variant of Kellogg et al. 1999 illustrated) and plate models. In the plume model, subducted material is both isolated in thermal boundary layer plume sources [a] and remixed with the convecting mantle [b]. Material in plume sources becomes buoyant after 1-2 Gyr, rising as plumes. In the oceanic domain, plateau volcanism is considered to result from the impact of plume heads (albeit in conjunction with ocean ridge systems) [c], whereas plume tails form ocean island chains [d]. Detail inset illustrates melting regimes within a plume: [e] melting of eclogite [f] formation of pyroxenites, [g] melting of pyroxenites, [h] melting of peridotite (see also figure 3a). In the plate model, pyroxenites (pyx) may be formed as part of a layer of refertilized peridotite (shaded, inset) in the hanging wall of the mantle wedge at convergent margins [u], before remixing of slabs with the convecting mantle [v]. In the forearc region, the refertilized peridotite is generated from fluids/melts from the crustal layer (oc) of the slab, whereas in the back arc region, fluids/melts may be generated as a result of serpentinite dehydration in the slab as indicated by arrows. Pyroxenites may also be formed along the flanks of ocean ridge upwellings [x] as a result of melting of altered recycled eclogite [w]. Pyroxenites in the latter region may then be incorporated into the shallow sub-lithospheric mantle where they may undergo melting as a result of lowering of the solidus by invasion of CO_2- or H_2O- rich fluids [y] from thermal equilibration of younger subducted slabs at depths of 300-400 km in the mantle (see also figure 3b). Voluminous melting, such as involved in the generation of oceanic plateaus, is explained by entrinment of large fragments of subducted oceanic crust into ocean ridge upwelling [z].

The standard model is thus summarised in Figure 2 based on the model of Kellogg et al. (1999) that arranges primitive mantle and plume sources to reconcile interpretations of He isotope systematics with tomographic studies of the mantle. Some subducted slabs are deflected at the base of the upper mantle and undergo remixing with the convecting mantle, whereas others sink into the deep mantle to be isolated as plume sources at the core-mantle boundary or around 1600 km depth where geophysical models have indicated an increase in viscosity suggesting a change in mantle composition. Subducted material in the thermal boundary layers is usually envisaged to become buoyant after 1 to 2 Gyrs, with oceanic plateaus or continental flood basalt provinces formed by plume heads and ocean island chains formed by plume tails (Hofmann and White 1982, Hofmann 1997). Melting in a plume begins at approximately 200 km depth on crossing of the eclogite solidus (Sobolev et al. 2005) (Fig. 3a). The resulting melts react with the overlying mantle peridotite to form pyroxenites which subsequently undergo melting at shallower depths accompanied to varying extents by the surrounding peridotite, to form a range of intraplate melt compositions (Sobolev et al. 2005).

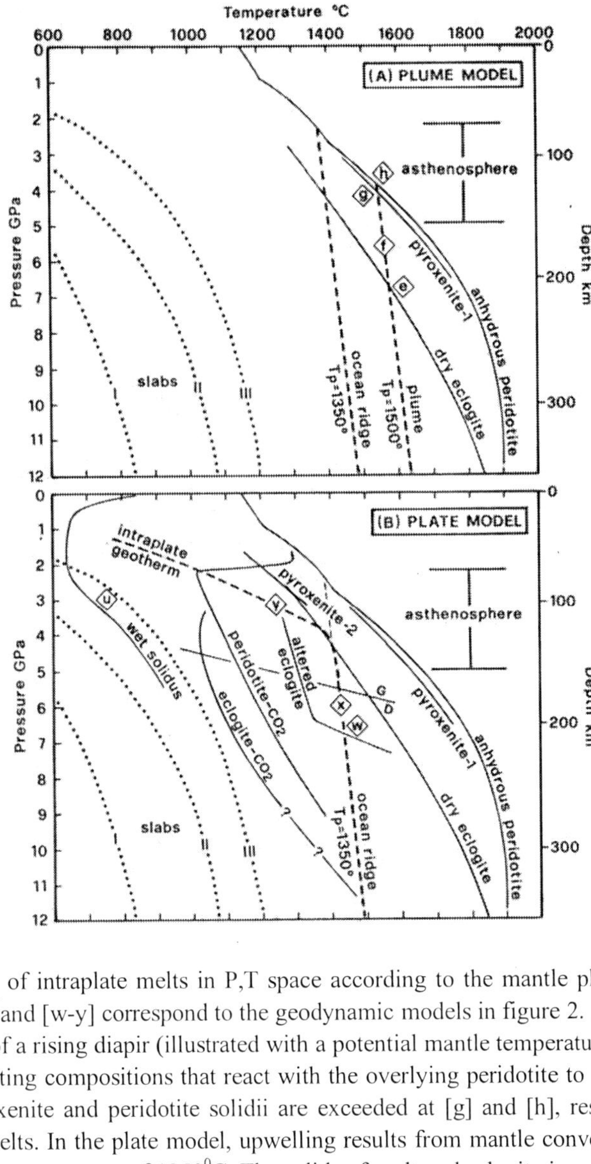

Figure 3. Generation of intraplate melts in P,T space according to the mantle plume and plate models. Stages outlined [e-h] and [w-y] correspond to the geodynamic models in figure 2. In the plume model, the eclogite component of a rising diapir (illustrated with a potential mantle temperature T_p of 1500^0C) begins to melt at [e], generating compositions that react with the overlying peridotite to form pyroxenites in the region [f]. The pyroxenite and peridotite solidii are exceeded at [g] and [h], respectively, generating a range of intraplate melts. In the plate model, upwelling results from mantle convection, and is illustrated for a potential mantle temperature of 1350^0C. The solidus for altered eclogite is reached at [w], generating melts which react with the overlying peridotite in the region [x]. Along the flanks of an ocean ridge system, the solidus would not be reached, allow incorporation of pyroxenites into the shallow sub-lithospheric mantle. However, generation of intraplate melts could be achieved along the intraplate geotherm at [y] by fluxing with volatiles such as CO_2 from the devolatilization of slabs at depth (note intersection of eclogite-CO_2 solidus with adiabat at ~ 11 GPa), in which case the pyroxenite solidus should be suppressed similar to the solidus of perdiotite-CO_2 and eclogite-CO_2 relative to volatile-free compositions. Solidii: anhydrous peridotite (Green and Falloon 2005), pyroxenite-1 (MIX1G of Hirshmann et al. 2003, Kogiso et al. 2003), pyroxenite-2 (MORB-like pyroxenite of Pertermann and Hirschmann 2003), dry eclogite (Yasuda et al. 1994), altered eclogite (Spandler et al. 2007), peridotite-CO_2 (Presnall and Gudfinnsson 2005, Dasgupta et al. 2007), eclogite-CO_2 (Dasgupta et al. 2004), wet solidus (basalt, greywacke and pelite compositions; Schmidt et al. 2004). P,T profiles for top of subducting slabs: I = old lithosphere, fast subduction, II = old lithosphere, slow subduction, III = young lithosphere, slow subduction, are from Kincaid and Sacks (1997). G-D is the graphite-diamond transition.

Consequence: Duplication of Mechanisms

The trend over the last two decades has been to assume a plume origin for intraplate volcanism, but despite its widespread application, significant weaknesses have been pointed out in the plume model. Few examples of intraplate volcanism conform to the predictions of the plume model (e.g. Sheth 1999, Anderson 2005a), but perhaps the greatest problem is the plurality of mechanisms, which applies not only to the fate of subducted oceanic crust, but also to the origin of the intraplate volcanic rocks the model was devised to explain in the first place. The plume model was originally developed as an explanation for linear age-progressive ocean island chains (Morgan 1971). However, much of the intraplate volcanism in the Pacific basin is neither linear nor age progressive (Natland and Winterer 2005). Likewise, much of the intraplate volcanism in regions such as Africa and Asia shows little temporal or spatial variation that could be linked to plumes, but shows a strong relationship with lithospheric architecture and can be readily explained by plate tectonic processes (Flower et al. 1998, Smith 1998, Bailey and Woolley 2005, Liégeois et al. 2005).

To explain all examples of intraplate volcanism by the plume model would also require a large number of plumes. Estimates of the number of plumes have ranged up to 5240 (Malamud and Turcotte 1999), but most are between 7 (Courtillot et al. 2003) and 42 (Crough and Jurdy 1980), which would require a non-plume origin for most examples of intraplate volcanism. Proponents of the plume model have appealed to "flow" of plume material for thousands of kilometres through the asthenosphere (e.g. Ebbinger and Sleep 1998, Niu et al. 2002) or considered the upper mantle to be entirely composed of plume residues (e.g. Morgan et al. 1995). But even with such modifications it remains that non-plume processes would still be required to tap the migrating plume material or residues.

What if the Marble-Cake and Plume Models Had not Been Combined?

A Path Followed Two Decades Later

Non-plume mechanisms which have recently been applied to various examples of intraplate volcanism include lithospheric loading (Hieronymus and Bercovici 2000, Got et al. 2008), shallow mantle convection (Ballmer et al. 2007), and propagating fractures (Stuart et al. 2007). While such mechanism could be used to alleviate difficulties with the plume model, it should be remembered that these are modern versions of concepts which were suggested as explanations for intraplate volcanism (e.g. Jackson and Shaw 1975, Bonatti and Harrison 1976, Walcott 1976) before geodynamic thinking became dominated by the plume model. Had the emphasis not switched to exploring the plume model, it is tempting to postulate that derivation of both MORB and OIB from the convecting mantle could have been explored in conjunction with such models in the 1980's. At that time, preferential melting of fertile heterogeneities in the ocean ridge environment had been proposed for the origin of alkalic axial seamounts which show many similar geochemical features to OIB (Zindler et al. 1984). Generation of OIB from a veined mantle without plume influence had also been explored (Fitton and James 1986).

The possibility of generating MORB and OIB from a common reservoir was also suggested by the study of Hamelin and Allègre (1988), which showed the pyroxenite layers in orogenic lherzolite massifs to cover the range of Pb isotopic composition found in both MORB and OIB. Petrological studies in the 1990's on the orogenic lherzolites, however, indicated two population of pyroxenite to be present in the massifs, with only a small proportion of the pyroxenite layers having the geochemical characteristics of recycled oceanic crust (Kornprobst et al. 1990). The majority of pyroxenites were interpreted to have formed as high pressure cumulates of melts that had intruded the peridotites, and a lack of relationship between radiometric ages and thickness of the layers was cited against the marble-cake concept (Pearson et al. 1993, Pearson and Nowell 2004). However, subducted materials are likely to undergo melting at some stage during remixing with the mantle (Anderson 2006), and the complexity observed in the lherzolite massifs can be considered evidence that such a process has happened for a large proportion of the recycled material.

Instead, it was not until the statistical mantle assemblage (SUMA) model of Meibom and Anderson (2004) that generation of both MORB and OIB from oceanic crust remixed with the convecting mantle was re-considered. In the SUMA model, subducted oceanic crust is remixed entirely with the convecting mantle, with a greater proportion of recycled oceanic crust involved in the generation of OIB relative to MORB. Voluminous examples of intraplate volcanism are explained by invoking a greater size range in recycled materials than in the marble-cake model. MORB and OIB do not form simple endmembers in isotopic space, but this can be resolved by invoking either the recycling of additional materials such as delaminated lower crust or continental mantle (e.g. Anderson 2005b, Lustrino 2005, Ishikawa et al. 2007), or modifications to subducted material such as slab melting in addition to the slab dehydration processes usually considered (Smith 2005). Development of the SUMA model accompanied renewed interest in stress fields and lithospheric architecture as controls on intraplate volcanism (e.g. Favela and Anderson 2000, Smith, 2003a), and the concepts became integral parts of what is now the plate model where plumes are not required to explain any aspect of intraplate volcanism (Foulger 2002, 2007, Foulger and Natland 2003).

Helium Isotopes Re-visited

The standard model was founded on beliefs that the convecting mantle should have been degassed of ^{3}He which could not be replaced by remixing of oceanic crust, and that high ^{3}He/^{4}He ratios indicated an excess of ^{3}He. Experimental studies have now shown that He may reside in olivine and be more compatible on melting than U or Th as the bulk distribution coefficients for the latter decline rapidly as clinopyroxene and garnet are consumed (Parman et al. 2005). Harzburgitic and dunitic residues from melting at ocean ridges could therefore retain ^{3}He in addition to being characterised by high ^{3}He/^{4}He, and impart such signatures to the convecting mantle upon recycling. Introduction of ^{3}He into the convecting mantle by subduction of cosmogenic dust could further supplement the He budget of this reservoir (Anderson 1993). The interplanetary dust particles were suggested to retain some volatiles to temperatures of up to 950^{0}C, which would allow survival to depths of 400 km depending on slab geotherm. Such replenishment was generally dismissed following the work of Staudacher and Allègre (1988) that suggested subducting slabs are degassed of noble gases in the subduction zone, but recycled atmospheric noble gas signatures have been found in Alpine ultramafic rocks of the Horoman peridotite (Matsumoto et al. 2001), indicating that noble gases can survive the subduction barrier.

The interpretation of high $^3He/^4He$ ratios indicating a primitive source is problematic because He contents are higher in MORB than OIB: the converse would be expected if OIB were derived from a primitive and therefore relatively undegassed source (Anderson 1998a,b). Rather than reflecting an abundance of ^3He, high $^3He/^4He$ ratios can be interpreted as a deficit in ^4He from a source with low U+Th (Anderson 1998a,b). If the latter model is correct, there should also be a correlation between low $^{206}Pb/^{204}Pb$ and high $^3He/^4He$, as ^{206}Pb is the end product of one of the principal decay series producing ^4He. Although the inverse relationship, high $^{206}Pb/^{204}Pb$ with low $^3He/^4He$ has been demonstrated for high-μ (HIMU) OIB (e.g. Hanyu and Kaneoka, 1997), OIB with low $^{206}Pb/^{204}Pb$ have long been considered to show both high (Iceland) and low (Gough, Tristan da Cunha) $^3He/^4He$ (e.g. Zindler and Hart 1986). However, in a recent study by Class and Goldstein (2005), a significant correlation was noted between low $^{206}Pb/^{204}Pb$ and high $^3He/^4He$ when OIB with similar neodymium isotopic composition were compared.

Class and Goldstein (2005) interpreted their results relative to a convecting mantle that underwent a reduction in $^3He/^4He$ ratio through time as a result of recycling of relatively He-deficient, but U-rich recycled crust over time. High $^3He/^4He$ -low $^{206}Pb/^{204}Pb$ signatures were suggested to be maintained in ancient mantle that had been isolated from remixing with the convecting mantle by storage in plume sources. However, there is no reason, other than conformity with the standard model, why the older reservoir should be equated with plume sources instead of unmelted streaks of recycled oceanic crust in the convecting mantle. The range of $^3He/^4He$ ratios in MORB (9$\pm$4 unfiltered; Anderson 2000) lies within the range shown by intraplate basalts (5 to >40). The correlations observed by Class and Goldstein (2005) are therefore compatible with sampling of a common source as proposed to explain the He isotopic variation in MORB and OIB by Anderson (2000) and Meibom et al. (2003). The narrow range in MORB may thus only reflect homogenisation during more extensive melting than involved in the generation of intraplate basalts.

Osmium Isotopes and Heterogeneity in the Convecting Mantle

The concept of a core Os signature in some intraplate lavas has been a key piece of evidence for an ultra-deep origin for the sources of such volcanism, but the interpretations rest on a series of assumptions, detailed counter-arguments to which are given in Meibom et al. (2004). Essentially, formation of the inner core has to take place shortly after accretion of the Earth, whereas thermal modelling (Labrosse et al. 2001) has indicated much later formation. Addition of outer core material should also impart tungsten isotopic variations that are not observed in intraplate basalts (Scherstén et al. 2004). The tungsten and osmium isotope signatures were suggested to have been decoupled by percolation of FeO-saturated melts through the lowermost mantle during the early history of the Earth (Humayun et al. 2004), but non-plume mechanisms involving remixing of sediments, and isotopic evolution in pyroxenites and sulphides have also been suggested as means of generating high $^{186}Os/^{188}Os$ in OIB.

Most crustal materials have too low Os content and Pt/Os ratio to produce the high $^{186}Os/^{188}Os$ ratios found in Hawaiian lavas. However, metalliferous sediments are an exception and it has been suggested that recycling of a few percent of such material into the source of intraplate melts could explain the $^{186}Os/^{188}Os$ – $^{187}Os/^{188}Os$ compositions of Hawaiian lavas (Ravizza et al. 2001, Baker and Jensen 2004, Schersten et al. 2004). Nielsen et al. (2006) also noted that remixing of ferromanganese sediments into the mantle could explain the Tl-isotope systematics of Hawaiian lavas, but as there was no correlation between Tl and $^{186}Os/^{188}Os$, these authors preferred the plume model to explain Os isotope variations and the sediment recycling model to explain Tl isotopic variations. Humayun et al. (2004) discounted a ferro-manganese sediment component for generating the Os isotope variation, however, on the basis of there being no complementary enrichment in Mn in the Gorgona komatiites.

Combined Pt, Os and Re analyses on pyroxenites are rare, but samples from ophiolite suites (Bay of Islands and Urals belt; Edwards 1990; Garuti et al. 1997) have high Pt/Os and Re/Os ratios that are expected to result in generation of supra-chondritic $^{186}Os/^{188}Os$ – $^{187}Os/^{188}Os$ ratios (Smith 2003b). High Pt/Os ratios have also been reported for pyroxenites from the Beni Bousera orogenic lherzolite massif by Luguet et al. (2008), who estimated the samples to show a comparable range of $^{186}Os/^{188}Os$ ratios to Hawaiian lavas by calculating their isotopic composition from an assumed age and depleted mantle initial ratio. The calculated signatures were supported by two measured $^{186}Os/^{188}Os$ ratios that bracketed the variation found in the Hawaiian lavas. The Pt, Os abundances in the Beni Bousera pyroxenites reported by Luguet et al. (2008) are also similar to those of Hawaiian picrites analysed by Brandon et al. (1998), which would be consistent with derivation of the lavas by large degrees of melting of a pyroxenitic source. Nonetheless, Os contents in pyroxenites are still around a thousand times lower than in mantle sulphides, suggesting the latter are more likely to control Os isotope compositions in the mantle (Luguet et al. 2008). Sulphides from ophiolite suites display a much wider range in $^{186}Os/^{188}Os$ than the Hawaiian lavas, and the isotopic composition of a melt could be controlled by only a small amount of such minerals in its source (Meibom et al. 2004). Such interpretations suggest the range of $^{186}Os/^{188}Os$ estimated for the depleted mantle by Brandon et al. (1999) was too restricted, and that sufficient heterogeneity exists within the convecting mantle to explain the $^{186}Os/^{188}Os$ signatures of intraplate volcanic rocks without need to invoke interaction with the core (Luguet et al. 2008, Meibom 2008).

Production of Pyroxenitic Sources in the Convecting Mantle

The crustal recycling process envisaged in the plate model is outlined in figure 2b. After termination of subduction, slabs are entrained and thinned by mantle convection as in the marble-cake model, although inefficient remixing or stagnation of deeply subducted slabs in the lower mantle (Hirose et al. 1999) will not affect the following model as plume processes are not invoked. The time for thermal equilibration of subducted material has been estimated at 100-200 Myrs (Stein and Stein 1997), hence slabs will be thermally equilibrated with the mantle within timeframes for convective stirring (240-960 Myrs depending on layered or whole mantle convection; Kellogg and Turcotte, 1990b). The ongoing subduction of oceanic lithosphere will result in recycled crust with varying thickness and state of thermal equilibration being present within the mantle (Anderson 2006). Such material is unlikely to be homogeneously distributed, as localisation of subduction regimes for tens of millions of years may produce large-scale mantle domains characterised by hydrous or enriched mantle interspersed with large accumulations of slabs as proposed to underlie the western Pacific region (Komiya and Maruyama 2007).

At convergent margins, the subducted material first undergoes modification as a result of dehydration or melting of the slab (Poli and Schmidt 2002, Smith 2005). The wet solidus, which is similar for basalt and sediment compositions, is intersected by the geotherm for hot slabs at depths of 80-100 km, suggesting that slow subduction of young slabs will result in melting (Schmidt et al. 2004) (Fig. 3b). Slab melting may have been more common in the Precambrian, but the majority of modern slabs likely undergo dehydration. The latter process takes place in the fore-arc region, such that the basaltic and sedimentary layers of the slab are essentially anhydrous by the time the slab lies beneath the arc region (e.g. Tatsumi and Eggins 1995). Water driven off the slab produces volatile-bearing minerals in the hanging wall of the mantle wedge. Convection induced in the mantle wedge by descent of the slab, drags the volatile-bearing assemblages to greater depth, whereupon the breakdown of amphibole to phlogopite, followed by phlogopite to K-richterite, releases fluids that lead to the generation of arc volcanism (e.g. Tatsumi and Eggins 1995). K-richterite is stable to depths of 300-400 km above cold slabs and formation of this mineral is a possible means of introduction of water into the deep mantle (Konzett et al. 2000).

Pyroxenite-rich peridotites in the Solomon Islands (Berly et al. 2006) and Cabo Ortegal in Spain (Santos et al. 2002) have been suggested to form by fluid metasomatism of the shallow mantle, but their compositions are too low in Al_2O_3 to constitute a source for intraplate volcanism. Pyroxenites may also be formed at greater depth in the mantle wedge as a result of metasomatism by slab-derived fluids/melts generated by fluxing of the crustal layers of the slab by fluids from the breakdown of hydrous minerals in the peridotitic layers of the slab (Ringwood 1990) (Fig. 2b). Such a model has been suggested for formation of the sources of E-MORB (Donnelly et al. 2004), and a mantle wedge setting has been suggested for formation of pyroxenites in the Beni Bousera massif (Davies et al. 1993, Pearson et al. 1993). However, other studies (Sánchez-Rodriguez and Gebauer 2000) have suggested formation of pyroxenites in the Beni Bousera massif during mantle upwelling associated with the opening of the Tethys Ocean, which leads to an alternative mechanism of generation of pyroxenites along the flanks of ocean ridge upwellings.

The sequence of melting for upwelling convecting mantle containing streaks of recycled eclogite in a peridotite matrix has similarities with the plume model of Sobolev et al. (2005): the ocean ridge adiabat intersects the dry eclogite solidus around 120 km depth (Fig. 3b), generating melts which react with the overlying mantle to form pyroxenites which may subsequently play a role in the genesis of MORB (Hirschmann and Stolper, 1996). However, the eclogite solidus is dependent on the K content of the recycled material, and is suppressed between 2 and 6 GPa in altered eclogite compositions (Spandler et al. 2007). Recycled sediments are also likely to have high K content, and are postulated to show a similar melting behaviour to altered eclogite at such pressures. Melts generated from altered eclogite or recycled sediments could therefore react with mantle peridotites to form pyroxenites at around 180 km depth. It is postulated here that the pyroxenites thus formed would have a similar composition to the compositions reported by Pearson et al. (1993) which crystallized in the stability field for diamond and show isotopic evidence for a sediment component in their source. Such pyroxenites have MgO contents intermediate between the MORB-like and refractory compositions previously investigated in melting experiments (Hirschmann et al.

2003, Pertermann and Hirschmann 2003, Kogiso et al. 2003) and their solidii are predicted to lie between the curves for such compositions in figure 3b. The average intraplate geotherm would not intersect the solidus for such pyroxenites, which if formed along flanks of ocean ridge upwellings, could then be incorporated into the shallow mantle in the off-axis region (Fig. 2b). Because of the curvature of the intraplate geotherm away from the pyroxenite solidus at pressures around 4 GPa, this model is insensitive to the potential mantle temperature of the upwelling mantle.

The off-axis geotherm may be raised a few tens of degrees Celsius by shear heating in the asthenosphere (Smith and Lewis 1999, Doglioni et al. 2005), but the principal cause of melting is likely to be lowering of the solidus by the introduction of volatiles. Very cold slabs may be capable of transporting H_2O to the transition region (Ivanov and Litasov 2008). Similarly, CO_2 may be transported to the transition region in carbonated eclogite as the solidus for such material is not intersected by any of the potential slab geotherms (Dasgupta et al. 2004) (Fig. 3b). As such slabs undergo thermal equilibration with the mantle, however, they may release volatiles which would then migrate upwards following the intraplate geotherm in thermal regime (Dasgupta et al. 2004). The presence of volatiles markedly reduces the solidii for peridotite and eclogite compositions (e.g. Green and Falloon 2005, Presnall and Gudfinnsson 2005), and is expected to produce a similar effect on the pyroxenite solidus. Undersaturated intraplate melt compositions are thus suggested to be produced along the intraplate geotherm at temperatures of between 1200-1300^{0}C and approximately 3 GPa pressure in figure 3b, in agreement with the P,T estimates of Green and Falloon (1998) for the generation of intraplate melts and the observations of Frezzotti and Peccerillo (2007) for the presence of CO_2- rich fluids in the mantle beneath Hawaii. The composition of sources for intraplate volcanism may thus be controlled by processes at convergent margins, with generation of melts controlled by the volatile content of the shallow mantle.

Conclusion

Two competing models for the fate of subducted oceanic crust were proposed following the acceptance of plate tectonics: remixing with the convecting mantle and isolation at depth in the mantle. The former was supported by evidence from orogenic lherzolite massifs and became the marble-cake mantle model. The latter, which remains speculative, was developed into the mantle plume hypothesis for the origin of intraplate volcanism. The marble-cake model was combined with the plume model, thereby giving rise to modern geodynamic models, on the basis of interpretations of rare gas isotope systematics in the 1980's. The model that resulted was subsequently supported by interpretations of Os isotope systematics as indicating a component from the Earth's core in the source of intraplate volcanic rocks.

Helium isotope models developed over the last decade have shown the original interpretations to be non-unique. Likewise, Os isotope systematics in intraplate volcanic rocks can be interpreted as evidence for a pyroxenite or sulphide rather than core signature. Instead of incorporating both marble-cake and plume models, it is suggested a geodynamic model based on remixing subducted crust solely into the convecting mantle can account for the geochemical features of ocean ridge and intraplate volcanism. The duplication of mechanisms that must otherwise result to explain plume- and non-plume intraplate volcanism in the standard model is, in effect, a consequence of unnecessary duplication of fates for subducted oceanic crust.

In the proposed model, pyroxenitic sources for intraplate volcanism are formed at depth in the mantle wedge at convergent margins, or along the flanks of ocean ridge upwellings as melts from subducted oceanic crust react with mantle peridotites. The pyroxenites are subsolidus in the P,T regimes of the shallow off-axis mantle, but may undergo melting to produce intraplate melts in the presence of volatiles, for example, if the shallow mantle is refluxed with CO_2 from the breakdown of subducted carbonates in the transition region. The control on the location of intraplate volcanism in the model is plate tectonic, not the result of deep-seated plumes.

References

Albarède, F.; van der Hilst, R.D. *Trans. Am. Geophys. Union Eos* 1999, 80, 535-539.

Allègre, C.J. *Tectonophysics* 1982, 81, 109-132.

Allègre, C.J.; Turcotte, D.L. *Nature* 1986, 323, 123-127.

Allègre, C.J.; Turcotte, D.L. *Geophys. Res. Lett.* 1985, 207-210.

Allègre, C.J.; Staudacher, T.; Sarda, P.; Kunz, M. *Nature* 1983, 303, 762-766.

Anderson, D.L. *Tectonophysics* 2006, 416, 7-22.

Anderson, D.L. In *Plates, Plumes, and Paradigms;* Foulger, G.R.; Natland, J.H.; Presnall D.C.; Anderson D.L.; Eds.; Special Paper 388; Geological Society of America; Boulder, CO, 2005a; 31-54.

Anderson, D.L. *Elements* 2005b, 1, 271-275.

Anderson D.L. *Geophys. Res. Lett.* 2000, 27, 3623-3626.

Anderson, D.L. *Science* 1993, 261, 170-176.

Anderson, D.L. *Internat. Geol. Rev.* 2000, 42, 289-311.

Anderson, D.L. *Proc. Natl. Acad. Sci.* 1998a, 95, 4822-4827.

Anderson, D.L. *Proc. Natl. Acad. Sci.* 1998b, 95, 9087-9092.

Armstrong, R.L. *Rev. Geophys.*1968, 6, 175-199.

Bailey, D.K.; Woolley, A.R. In *Plates, Plumes, and Paradigms;* Foulger, G.R.; Natland, J.H.; Presnall D.C.; Anderson D.L.; Eds.; Special Paper 388; Geological Society of America; Boulder, CO, 2005; 365-377.

Baker, J.A.; Jensen, K.K. *Earth Planet. Sci. Lett.* 2004, 277-286.

Ballmer, M.D.; van Hunen, J.; Ito, G.; Tackley, P.J.; Bianco, T.A. *Geophys. Res. Lett.* 2007, 34, L23310, doi:10.1029/2007GL031636.

Becker, H.; Jochum, K.P.; Carlson, R.W. *Chem. Geol.* 2000, 163, 65-99.

Berly, T.J.; Hermann, J.; Arculus, R.J.; Lapierre, H. *J. Petrol.* 2006, 47, 1531-1555.

Bonatti, E.; Harrison, C.G.A. *Nature* 1976, 263, 402-404.

Brandon, A.D.; Walker, R.J.; Puchtell, I.S.; Becker, H.; Humayun, M.; Revillon, S. *Earth Planet. Sci. Lett.* 2003, 206, 411-426.

Brandon, A.D.; Norman, M.D.; Walker, R.J.; Morgan, J.W. *Earth Planet. Sci. Lett.* 1999, 174, 25-42.

Brandon, A.D.; Walker, R.J.; Morgan, J.W.; Norman, M.D.; Prichard, H.M. *Science* 1998, 280, 1570-1573.

Christensen, U.R.; Hofmann, A.W. *J. Geophys. Res.* 1994, 99, 19867-19884.

Class, C.; Goldstein, S.L. *Nature* 2005, 436, 1107-1112.
Cordery, M.J.; Davies, G.F.; Campbell, I.H. *J. Geophys. Res.* 1997, 102, 20179-20197.
Courtillot, V.; Davaille, A.; Besse, J.; Stock, J. *Earth Planet. Sci. Lett.* 2003, 205, 295-308.
Craig, H.; Lupton, J.E. *Earth Planet. Sci. Lett.* 1976, 31, 369-385.
Crough, S.T.; Jurdy, D.M. *Earth Planet. Sci. Lett.* 1980, 44, 224-230.
Dasgupta, R.; Hirschmann, M.M.; Smith, N.D. *J. Petrol.* 2007, 48, 2093-2124.
Dasgupta, R.; Hirschmann, M.M.; Withers, A.C. *Earth Planet. Sci. Lett.* 2004, 227, 73-85.
Davies, G.R.; Nixon, P.H.; Pearson, D.G.; Obata, M. *Geology* 1993, 21, 471-474.
Dickinson, W.R.; Luth, W.C. *Science* 1971, 174, 400-404.
Doglioni, C.; Green, D.H.; Mongelli, F. In *Plates, Plumes, and Paradigms;* Foulger, G.R.; Natland, J.H.; Presnall D.C.; Anderson D.L.; Eds.; Special Paper 388; Geological Society of America; Boulder, CO, 2005; 735-749.
Donnelly, K.E.; Goldstein, S.L.; Langmuir, C.H.; Spiegelman, M. *Earth Planet. Sci. Lett.* 2004, 226, 347-366.
Ebbinger, C.J.; Sleep, N.H. *Nature* 1998, 395, 1788-1791.
Edwards, S.J. *Can. Mineral.* 1990, 28, 537-552.
Favela, J.; Anderson, D.L. In *Problems in Geophysics for the new Millenium;* Boschi, E.; Ekström, G.; Morelli, A.; Eds.; Editrice Compositori, Bologna, 2000; 463-498.
Fitton, J.G.; James, D. *Philos. Trans. Roy. Soc. Lond.* 1986, A317, 253-266.
Flower, M.F.J.; Tamaki, K.; Hoang, N. In *Mantle Dynamics and Plate Interactions in East Asia;* Flower, M.F.J.; Chung, S.-L.; Lo, C.-H.; Lee, T.-Y.; Eds.; Geophysical Monograph 27; American Geophysical Union, Washington, D.C., 1998; 67-88.
Foulger, G.R. In *Plates, Plumes, and Planetary Processes*; Foulger, G.R; Jurdy, D.M.; Eds.; Special Paper 430; Geological Society of America; Boulder, CO; 2007, 1-25.
Foulger, G.R. *Astron. Geophys.* 2002, 43, 6.19-6.23.
Foulger, G.R.; Natland, J.H. *Science,* 2003, 300, 921-922.
Frezzotti, M.-L.; Peccerillo, A. *Earth Planet. Sci. Lett.* 2007, 262, 273-283.
Garuti, G.; Fershtater, G.; Bea, F.; Montero, P.; Pushkarev, E.V.; Zaccarini, F. *Tectonophysics* 1997, 276, 181-194.
Got, J.-L.; Monteiller, V.; Monteux, J., Hassani, R.; Okubo, P. *Nature* 2008, 451, 453-456.
Green, D.H.; Falloon, T.J. In *Plates, Plumes, and Paradigms;* Foulger, G.R.; Natland, J.H.; Presnall D.C.; Anderson D.L.; Eds.; Special Paper 388; Geological Society of America; Boulder, CO, 2005; 217-247.
Green, D.H.; Falloon, T.J. In *The Earth's Mantle: Composition, Struture, and Evolution;* Jackson, I.; Ed.; Cambridge University Press, 1998; 311-378.
Hamelin, B.; Allègre, C.J. *Earth Planet. Sci. Lett.* 1988, 91, 117-131.
Hanyu, T.; Kaneoka, I. *Nature* 1997, 390, 273-276.
Hauri, E.H.; Hart, S.R. *Earth Planet. Sci. Lett.* 1993, 114, 353-371.
Hieronymus, C.F.; Bercovici, D. *Earth Planet. Sci. Lett.* 2000, 181, 539-554.
Hirano, N.; Takahashi, E.; Yamamoto, J.; Abe, N.; Ingle, S.P.; Kaneoka, I.; Hirata, T.; Kimura, J.-I.; Ishii, T.; Ogawa, Y.; Machida, S.; Suyehiro, K. *Science* 2006, 313, 1426-1428.
Hirose, K.; Fei, Y.; Ma, Y.; Mao, H.-K. *Nature* 1999, 397, 53-56.
Hirschmann, M.M.; Stolper, E.M. *Contrib. Mineral. Petrol.* 1996, 124, 185-208.
Hirschmann, M.M.; Kogiso, T.; Baker, M.B.; Stolper, E.M. *Geology* 2003, 31, 481-484.
Hofmann, A.W. *Nature* 1997, 385, 219-229.

Hofmann, A.W.; White, W.M. *Earth Planet. Sci. Lett.* 1982, 57, 421-436.
Hofmann, A.W.; Jochum, K.P.; Seifert, M.; White, W.M. *Earth Planet. Sci. Lett.* 1986, 79, 33-45.
Humayun, M.; Qin, L.; Norman, M.D. *Nature* 2004, 306, 91-94.
Ishikawa, A.; Kuritani, T.; Makishima, A.; Nakamura, E. *Earth Planet. Sci. Lett.* 2007, 259, 134-148.
Ivanov, A.V.; Litasov, K.D. *Nature Precedings* 2008, <http://hdl.handle.net/10101/npre.2008.1489.1>
Jackson, E.D.; Shaw, H.R. *J. Geophys. Res.* 1975, 80, 1861-1874.
Kellogg, L.H.; Hager, B.H.; van der Hilst, R.D. *Science* 1999, 283, 1881-1884.
Kellogg, L.H.; Turcotte, D.L. *Earth Planet. Sci. Lett.* 1990a, 99, 276-289.
Kellogg, L.H.; Turcotte, D.L. *J. Geophys. Res.* 1990b, 95, 421-432.
Kincaid, C.; Sacks, I.S. *J. Geophys. Res.* 1997, 102, 12295-12315.
Kogiso, T.; Hirschmann, M.M. *Earth Planet. Sci. Lett.* 2006, 249, 188-199.
Kogiso, T.; Hirschmann, M.M.; Frost, D.J. *Earth Planet. Sci. Lett.* 2003, 216, 603-617.
Kogiso, T.; Hirose, K.; Takahashi, E. *Earth Planet. Sci. Lett.* 1998, 162, 45-61.
Komiya, T.; Maruyama, S. *Gondwana Res.* 2007, 11, 132-147.
Konzett, J.; Yang, H.; Frost, D.J. *J. Petrol.* 2000, 41, 583-603.
Kornprobst, J.; Piboule, M.; Roden, M.; Tabit, A. *J. Petrol.* 1990, 31, 717-745.
Kurz, M.D.; Jenkins, W.J.; Hart, S.R.; Clague, D. *Earth. Planet. Sci. Lett.* 1983, 66, 388-406.
Labrosse, S.; Poirier, J.-P.; Le Mouël, J.-L. *Earth Planet. Sci. Lett.* 2001, 190, 111-123.
Liégeois, J.-P.; Benhallou, A.; Azzouni-Sekkal, A.; Yahiaoui, R.; Bonin, B. In *Plates, Plumes, and Paradigms;* Foulger, G.R.; Natland, J.H.; Presnall D.C.; Anderson D.L.; Eds.; Special Paper 388; Geological Society of America, Boulder, CO, 2005; 379-400.
Luguet, A.; Pearson, D.G.; Nowell, G.M.; Dreher, S.T.; Coggon, J.A.; Spetsius, Z.V.; Parman, S.W. *Science* 2008, 319, 453-456.
Lustrino, M. *Earth Sci. Rev.* 2005, 72, 21-38.
Malamud, B.D.; Turcotte, D.L. *Earth Planet Sci. Lett.* 1999, 174, 113-124.
Matsumoto, T.; Chen, Y.; Matsuda, J.-I. *Earth Planet Sci. Lett.* 2001, 185, 35-47.
McKenzie, D.; O'Nions, R.K. *Nature* 1983, 301, 229-231.
Meibom, A. *Science* 2008, 319, 418-419.
Meibom, A.; Anderson, D.L. *Earth Planet. Sci. Lett.* 2004, 217, 123-139.
Meibom, A.; Frei, R.; Sleep, N.H. *J. Geophys. Res.* 2004, B02203 10.1029/2003JB002602.
Meibom, A.; Anderson, D.L.; Sleep, N.H.; Frei, R.; Chamberlain, C.P.; Hren, M.T.; Wooden, J.L. *Earth Planet. Sci. Lett.* 2003, 197-204.
Morgan, W.J. *Nature* 1971, 230, 42-43.
Morgan, J.P.; Morgan, W.J.; Zhang, Y.-S.; Smith, W.H.F. *J. Geophys. Res.* 1995, 100, 12753-12767.
Natland, J.H.; Winterer, E.L. In *Plates, Plumes, and Paradigms;* Foulger, G.R.; Natland, J.H.; Presnall D.C.; Anderson D.L.; Eds.; Special Paper 388; Geological Society of America; Boulder, CO, 2005; 687-710.
Nielsen, S.G.; Rehkämper, M.; Norman, M.D.; Halliday, A.N.; Harrison, D. *Nature* 2006, 439, 314-317.
Niu, Y.; Regelous, M.; Wendt, I.J.; Batiza, R.; O'Hara, M.J. *Earth Planet. Sci. Lett.* 2002, 199, 327-345.
Parman, S.W.; Kurz, M.D.; Hart, S.R.; Grove, T.L. *Nature* 2005, 437, 1140-1143.

Pearson, D.G.; Nowell, G.M. *J. Petrol.* 2004, 45, 439-455.
Pearson, D.G.; Davies, G.R.; Nixon, P.H. *J. Petrol.* 1993, 34, 125-172.
Pertermann, M.; Hirschmann, M.M. *J. Geophys. Res.* 2003, 108, doi:10.1029/2000JB000118.
Poli, S.; Schmidt, M.W. *Ann. Rev. Earth Planet. Sci.* 2002, 30, 207-235.
Polvé, M.; Allègre, C.J. *Earth Planet. Sci. Lett.* 1980, 51, 71-93.
Porcelli, D.; Halliday, A.N. *Earth Planet. Sci. Lett.* 2001, 192, 45-56.
Presnall, D.C.; Gudfinnsson, G.H. In *Plates, Plumes, and Paradigms;* Foulger, G.R.; Natland, J.H.; Presnall D.C.; Anderson D.L.; Eds.; Special Paper 388; Geological Society of America; Boulder, CO, 2005; 207-216.
Ravizza, G.; Blusztajn, J.; Pritchard, H.M. *Earth Planet. Sci. Lett.* 2001, 188, 369-381.
Rehkämper, M.; Hofmann, A.W. *Earth Planet. Sci. Lett.* 1997, 147, 93-106.
Ringwood, A.E. *Chem. Geol.* 1990, 82, 187-207.
Sánchez-Rodríguez, L.; Gebauer, D. *Tectonophysics* 2000, 316, 19-44.
Santos, J.F.; Schärer, U.; Gil Ibarguchi, J.I.; Girardeau, J. *J. Petrol.* 2002, 43, 17-43.
Saunders, A.D.; Norry, M.J.; Tarney, J. *J. Petrol.* 1988, Special Lithosphere Issue, 415-445.
Scherstén, A.; Elliot, T.; Hawkesworth, C.; Norman, M. *Nature* 2004, 427, 234-237.
Schmidt, M.W.; Vielzeuf, D.; Auzanneau, E. *Earth Planet. Sci. Lett.* 2004, 228, 65-84.
Sheridan, R.E. *Geology* 1994, 22, 763-764.
Sheth, H.C. *Tectonophysics* 1999, 311, 1-29.
Sleep, N.H. In *Plates, Plumes, and Planetary Processes*; Foulger, G.R; Jurdy, D.M.; Eds.; Special Paper 430; Geological Society of America, Boulder, CO, 2007; 29-42.
Smith, A.D. In *Plates, Plumes, and Paradigms;* Foulger, G.R.; Natland, J.H.; Presnall D.C.; Anderson D.L.; Eds.; Special Paper 388; Geological Society of America, Boulder, CO, 2005; 303-325.
Smith, A.D. *Internatl. Geol. Rev.* 2003a, 45, 287-302.
Smith, A.D. *J. Geodynam.* 2003b, 36, 469-484.
Smith, A.D.; Lewis, C. *Earth Sci. Rev.* 1999, 48, 135-182.
Smith, A.D. In *Mantle Dynamics and Plate Interactions in East Asia;* Flower, M.F.J.; Chung, S.-L.; Lo, C.-H.; Lee, T.-Y.; Eds.; Geophysical Monograph 27; American Geophysical Union; Washington, D.C., 1998; 89-105.
Sobolev, A.V. and 19 others *Science* 2007, 316, 412-417.
Sobolev, A.V.; Hofmann, A.W.; Sobolev, S.V.; Nikogostan, I.K. *Nature* 2005, 434, 590-597.
Spandler, C.; Yaxley, G.; Green, D.H.; Rosenthal, A. *J. Petrol.* 2007, doi: 10.1093/petrology/egm039
Staudacher, T.; Allègre, C.J. *Earth Planet. Sci. Lett.* 1988, 89, 173-183.
Stein, S.; Stein, C.A. *Science* 1997, 275, 1613-1614.
Stuart, W.D.; Foulger, G.R.; Barall, M. In *Plates, Plumes, and Planetary Processes*; Foulger, G.R; Jurdy, D.M.; Eds., Special Paper 430; Geological Society of America; Boulder, CO, 2007; 497-506.
Takahashi, E.; Nakajima, K.; Wright, T.L. *Earth Planet. Sci. Lett.* 1998, 162, 63-80.
Tatsumi, Y.; Eggins, S. *Subduction Zone Magmatism;* Blackwell Science: Cambridge, MA, 1995, 211p.
van Keken; P.E.; Hauri, E.H.; Ballentine, C.J. *Ann. Rev. Earth Planet. Sci.* 2002, 30, 493-525.
Walcott, R.I. In *The Geophysics of the Pacific Ocean Basin and its Margin;* Sutton, G.H.; Manghnani, M.H.; Moberly, R.; Eds.; Geophysical Monograph 19; American Geophysical Union; Washington, D.C., 1976; 431-438.

Walker, R.J.; Morgan, J.W., Horan, M.F. *Science* 1995, 269, 819-822.

Wasserburg, G.J.; DePaolo, D.J. *Proc. Natl. Acad. Sci. USA* 1979, 76, 3594-3598.

Yasuda, A.; Fujii, T.; Kurita, K. *J. Geophys. Res.* 1994, 99, 9401-9414.

Zindler, A.; Hart, S.R. *Ann. Rev. Earth Planet.Sci.* 1986, 14, 493-571.

Zindler, A.; Staudigel, H.; Batiza, R. *Earth Planet. Sci. Lett.* 1984, 70, 175-195.

In: The Lithosphere: Geochemistry, Geology and Geophysics ISBN: 978-1-60456-903-2
Editors: J.E. Anderson et al, pp. 141-169

Chapter 4

HELIUM ISOTOPE VARIATIONS ALONG THE NIIGATA-KOBE TECTONIC ZONE, CENTRAL JAPAN

Koji Umeda[1,*], Atusi Ninomiya[1], Koji Shimada[1] and Junichi Nakajima[2]

[1] Tono Geoscientific Research Unit, Geological Isolation Research and Development Directorate, Japan Atomic Energy Agency, Toki 509-5102, Japan.
[2] Research Center for Prediction of Earthquakes and Volcanic Eruptions, Graduate School of Science, Tohoku University, Sendai 980-8578, Japan.

Abstract

A linear zone with high strain rates along the Sea of Japan coast, the Nigata-Kobe Tectonic Zone (NKTZ), is considered to be associated with rheological heterogeneity in the lower crust and/or upper mantle, which may be caused by the upwelling of aqueous fluid and/or melt related to subduction of the Philippine Sea and Pacific Plates. In order to elucidate the geographic distribution of $^3He/^4He$ ratios along the NKTZ, new helium isotope data from hot spring gases and water samples were determined. In the southern NKTZ, $^3He/^4He$ ratios lower than the atmospheric value indicate that radiogenic helium dominates over any mantle helium input from aqueous fluids generated during the dehydration of the subducting Philippine Sea slab because a mantle wedge, the potential source of mantle helium, appears to be absent. Higher $^3He/^4He$ ratios are observed in the central NKTZ where active volcanoes are concentrated, suggesting the existence of magmatic fluids in the lower crust and upper mantle. The $^3He/^4He$ ratios of most hot springs in the northern NKTZ, a non-volcanic region, can be interpreted as a three-component mixture of mantle helium associated with magmatism of Middle Miocene age, radiogenic crustal helium and atmospheric helium. However, the $^3He/^4He$ ratios of gases close to active faults in the northern NKTZ are similar to those near active volcanoes in the central NKTZ, suggesting that active faults may facilitate transfer of mantle helium carried by aqueous fluids derived from the subducting Pacific Plate slab from the lower crust to the Earth's surface.

* E-mail address: umeda.koji@jaea.go.jp

1. Introduction

Observational data on the surface deformation field obtained from the nationwide GPS earth observation network (GEONET) of the Geographical Survey Institute of Japan have provided a great deal of information related to temporal and spatial variations of interplate coupling at plate boundaries. One of the most remarkable observations made using GEONET is the presence of a zone, known as the Niigata-Kobe Tectonic Zone (NKTZ), with high strain rates along the Japan Sea coast (Sagiya et al., 2000). This high strain rate zone, ~ 500 km long in the NE-SW direction and ~ 100 km wide (Figure 1), undergoes contraction in the WNW-ESE direction (~10^{-7}/yr.), which is a few times larger than in the surrounding regions (Sagiya et al., 2000). Previous studies proposed that the NKTZ is a plate boundary between the Eurasian (or Amurian) Plate and the North American (or Okhotsk) Plate (Shimazaki and Zhao, 2000; Heki and Miyazaki, 2001).

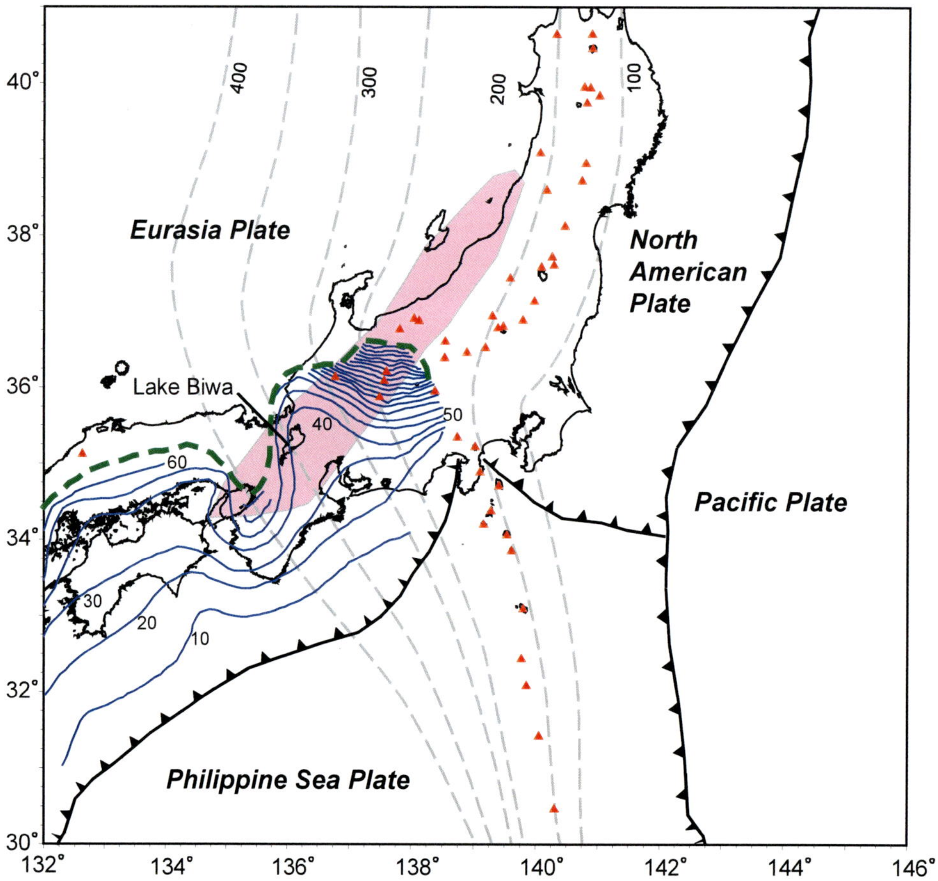

Figure 1. Generalized structure of central Japan. The NKTZ is denoted by a pink belt. The red triangles represent active volcanoes. The iso-depth contours of the Pacific Plate (Zhao and Hasegawa, 1993) and the Philippine Sea Plate (Nakajima and Hasegawa, 2007b) are shown by broken curves with an interval of 50 km and blue curves with an interval of 10 km, respectively. A green broken curve represents the leading edge of the Philippine Sea slab estimated by high-resolution 3-D seismic tomography (Nakajima and Hasegawa, 2007b).

On the other hand, Iio et al. (2002) and Yamasaki and Seno (2005) regarded the NKTZ as an intraplate deformation zone near the eastern margin of the Eurasian plate. Iio et al. (2002) proposed a model in which a water-weakened zone, with lower viscosity caused by dehydration of the subducting Pacific Plate, exists in the lower crust. Yamasaki and Seno (2005) evaluated the effects of viscosity heterogeneities in the lower crust and upper mantle along the NKTZ, indicating that low velocity in the uppermost mantle trenchward of the NKTZ accounts for the observed high strain rates. They suggested that the upwelling of aqueous fluid generated by dehydration of the subducting Philippine Sea and Pacific Plates, partial melting of mantle above the Pacific Plate, and serpentinization in the mantle above the Philippine Sea Plate are possible origins of the viscosity heterogeneities in the upper mantle. Therefore, the strain accumulation process within the NKTZ is associated with rheological heterogeneity in the lower crust and/or upper mantle, which might be attributed to aqueous fluid and melt related to the subducting Philippine Sea and Pacific Plate.

Noble gases and their isotopes are excellent natural tracers for elucidating mantle-crust interactions in different geotectonic provinces because they are chemically inert and thus conserved in crustal rock-water systems. The helium isotopes are of particular interest as they can provide unequivocal evidence for the presence in the crust of mantle-derived materials. ^{3}He is essentially primordial and retained in the Earth's interior, whereas ^{4}He is mainly produced in the crust by the decay of U and Th, so any ^{3}He/^{4}He ratio at the Earth's surface larger than the local and crustal production rates indicates degassing of mantle helium (Ozima and Podosek, 2002). Thus, the helium isotope variations in a subduction zones can be used for evaluating the upward movement of aqueous fluids originating from the dehydration of a subducting slab (Matsumoto et al., 2003; Umeda et al., 2006a) and the presence of partial melting of the mantle (Italiano et al., 2000; Umeda et al., 2007a).

Although abundant helium isotope data from natural gas and oil wells in a sedimentary basin of Middle Miocene age and fumarolic and hot spring gases around active volcanoes have been reported in the central part of the NKTZ (e.g., Wakita et al., 1990; Kusakabe et al., 2003), few gas samples have been measured previously in other regions. In order to provide geochemical insights into factors controlling the strain accumulation process associated with the NKTZ, we present new helium isotope data from the northern and southern NKTZ, and clarify the geographic distribution of the ^{3}He/^{4}He ratios in throughout the NKTZ, using both the new and existing data. Additionally, we inspect the relationship of the helium isotope ratios to geophysical structures, which could image the presence of aqueous fluids and melt related to the subducting slabs from the Philippine Sea and Pacific Plates beneath the NKTZ, especially with regards to seismic velocity and resistivity structure of the crust and mantle.

2. Analytical Procedures and Results

Analytical samples were collected from 30 hot springs in and around the NKTZ (Figure 3). The sampled hot springs are distributed in the Miocene to Pliocene sedimentary and volcanic rocks (#1 - 19) and the Jurassic to early Cretaceous sedimentary and granitic rocks (#20 - 30). Details of sample collection methods can be found in Nagao et al. (1981). In the cases when there were no visible bubbles, water samples were also collected to measure the isotopic ratios of dissolved gases in water. The dissolved gases were expelled from the solution by ultrasonic agitation and collected in a glass bottle. The isotopic ratios of He and Ne were determined using the VG5400 system at the Laboratory for Earthquake Chemistry, University of Tokyo. Blank levels for He and Ne determination were lower than 1×10^{-9} cm^3 STP. They are less than 0.1 % of the amount of sample gases, so blank correction was not necessary. Corrections of interference peaks of $^{40}Ar^{++}$ and CO_2^{++} on $^{20}Ne^+$ and $^{22}Ne^+$ were not needed because the amounts of background Ar and CO_2 are negligible as compared to the sample neon. Mass spectrometry details, including purification procedures, can be found in Aka et al. (2001).

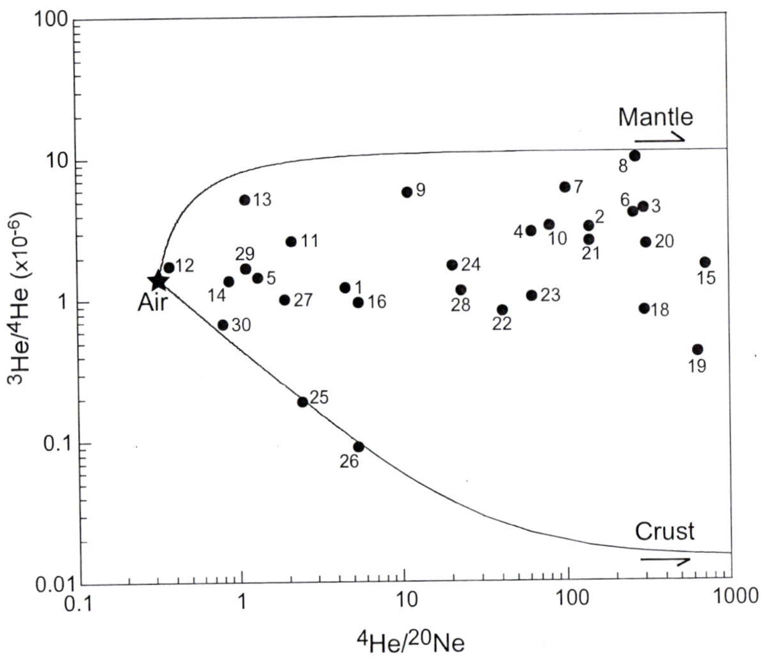

Figure 2. $^3He/^4He$ versus $^4He/^{20}Ne$ diagram for the hot spring gases obtained in this study (filled circles with sample numbers). The mixing curves assume atmospheric components have $^3He/^4He = 1.4 \times 10^{-6}$ and $^4He/^{20}Ne = 0.32$; mantle components have $^3He/^4He = \sim 11 \times 10^{-6}$ (e.g., Graham et al., 1992) and $^4He/^{20}Ne = 1000$ (Sano and Wakita, 1985); and crustal radiogenic components have $^3He/^4He = 1.5 \times 10^{-8}$ (Mamyrin and Tolstikhin, 1984) and $^4He/^{20}Ne = 1000$ (Sano and Wakita, 1985). .

The $^3He/^4He$ ratios of the samples shown in Table 1, range from 0.065 to 7.1 RA (RA denotes the atmospheric $^3He/^4He$ ratio of 1.4×10^{-6}). This variation is not related to the helium concentration in each gas sample. Figure 2 shows the relationship between the measured $^3He/^4He$ and $^4He/^{20}Ne$ rations. The mixing curves assume atmospheric components have $^3He/^4He = 1.4 \times 10^{-6}$ and $^4He/^{20}Ne = 0.32$; mantle components have $^3He/^4He = \sim 11 \times 10^{-6}$ (e.g., Graham et al., 1992) and $^4He/^{20}Ne = 1000$ (Sano and Wakita, 1985); and crustal radiogenic components have $^3He/^4He = 1.5 \times 10^{-8}$ (Mamyrin and Tolstikhin, 1984) and $^4He/^{20}Ne = 1000$ (Sano and Wakita, 1985). We calculated the contribution of atmospheric, radiogenic and mantle-derived helium on the basis of the analytical $^3He/^4He$ and $^4He/^{20}Ne$ ratios (Sano and Wakita, 1985). The mantle helium content relative to total helium was estimated to range widely from 0 to 90.4 % (Table 1).

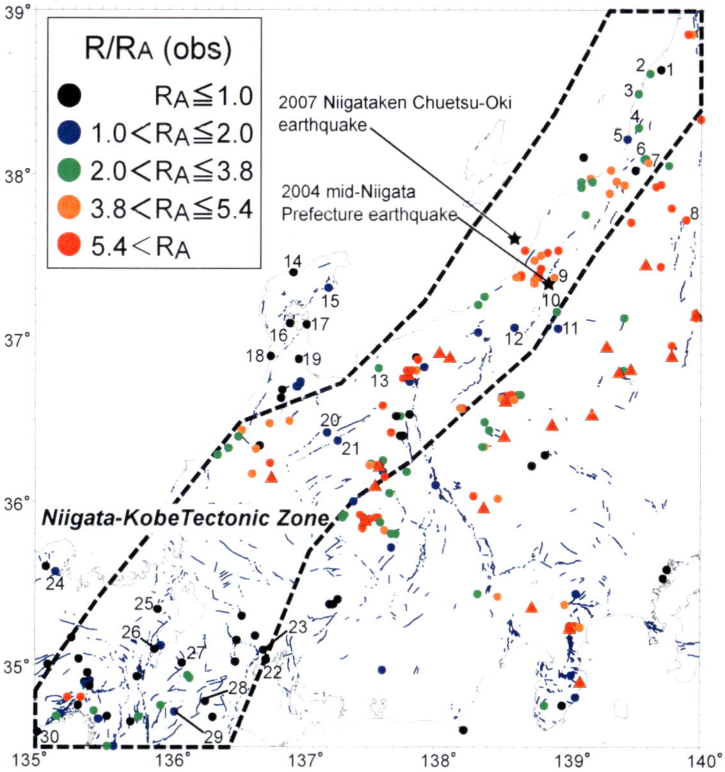

Figure 3. Geographic distribution of $^{3}He/^{4}He$ ratios in and around the NKTZ with localities of hot springs sampled in this study indicated with sample numbers (See Table 1). Also shown in this figure are active volcanoes (solid red triangles), active faults (dark blue lines). Stars denote the epicenters of recent large earthquakes in the northern NKTZ.

3. Variations in the $^{3}HE/^{4}HE$ Ratios along the Niigata-Kobe Tectonic Zone

Several previous investigations have reported helium isotope data in and around the NKTZ. The published helium isotope data used in the following discussion were compiled from previous reports (Table 2). The majority of samples used in the studies were collected from fumaroles, natural gas wells and bubbling hot springs, the remainder are from natural spring waters. The synthesized data were compiled from a total of 457 samples including data obtained from this study.

The geographic distribution of $^{3}He/^{4}He$ ratios in and around the NKTZ is shown in Figure 3. Also shown are active volcanoes (Committee for Catalogue of Quaternary Volcanoes in Japan, 1999) and active faults (Working Group for Compilation of 1:2,000,000 Active Faults Map of Japan, 2000) in this figure. The NKTZ is divided into three parts on the basis of geological setting: the southern part consisting of the Jurassic to early Cretaceous sedimentary and granitic rocks, the central part where volcanoes are concentrated, and the northern part corresponding to the Miocene sedimentary basin formed as a back-arc rift basin associated with the opening of the Sea of Japan.

Table 1. Isotopic compositions of the hot spring gases sampled in and around the Niigata-Kobe Tectonic Zone. The analytical error for $^{4}He/^{20}Ne$ is ~15% of the values given.

No.	Site name	Sample Type	$^{3}He/^{4}He$ (x 10^{-6}) [±1σ]			$^{3}He/^{4}He$ (R/RA)	$^{4}He/^{20}Ne$	Temp. (°C)	pH	Contribution of relative mantle helium (%)
1	Yunose	Water	1.22	±	0.04	0.87	4.5	42	9.2	10
2	Atsumi	Water	3.23	±	0.04	2.3	140	53	7.4	29
3	Gatsugi-yurihana	Water	4.35	±	0.10	3.1	300	55	8.0	40
4	Mahoroba	Water	2.99	±	0.05	2.1	62	68	7.5	27
5	Senami	Water	1.45	±	0.04	1.0	1.3	92	8.3	10
6	Kira	Water	4.03	±	0.06	2.9	260	61	7.1	37
7	Takanosu	Water	6.04	±	0.07	4.3	100	52	7.1	55
8	Atsushio	Gas	9.95	±	0.11	7.1	270	65	6.8	90
9	Yomogihira	Gas	5.71	±	0.07	4.1	11	15	9.2	52
10	Urasa	Gas	3.29	±	0.05	2.4	80	18	-	30
11	Ikasawa	Gas	2.62	±	0.04	1.9	2.1	53	9.1	22
12	Matsunoyama	Gas	1.75	±	0.03	1.2	0.37	98	7.9	5.0
13	Unazuki	Gas	5.21	±	0.07	3.7	1.1	98	8.0	44
14	Nebuta	Gas	1.38	±	0.04	1.0	0.87	25	9.3	7.8
15	Jomon-mawaki	Gas	1.73	±	0.02	1.2	720	47	-	16
16	Wakura	Gas	0.96	±	0.02	0.68	5.4	95	8.0	7.9
17	Yugawa	Gas	0.47	±	0.02	0.34	16000	15	6.9	4.1
18	Chirihama	Gas	0.83	±	0.02	0.59	300	52	8.1	7.4
19	Eiwa	Gas	0.42	±	0.02	0.30	640	60	7.7	3.7
20	Wariishi	Gas	2.45	±	0.04	1.8	310	46	9.7	22.2
21	Miyagawa	Gas	2.57	±	0.08	1.8	140	34	8.3	23
22	Nagashima R3	Gas	0.83	±	0.03	0.60	41	60	8.4	7.4
23	Nagashima R15	Gas	1.04	±	0.03	0.74	62	58	8.0	9.2
24	Iwadaki	Water	1.74	±	0.06	1.2	20.4	58	7.1	16
25	Kutsuki	Water	0.19	±	0.01	0.13	2.4	26	8.4	<0
26	Ogoto	Water	0.091	±	0.004	0.065	5.3	33	9.5	<0
27	Junibou	Water	1.01	±	0.02	0.72	1.9	27	8.4	7.0
28	Sarubino	Water	1.16	±	0.02	0.83	23	34	8.2	10
29	Tsukigase	Water	1.69	±	0.03	1.2	1.1	30	7.8	12
30	Iwaya	Gas	0.68	±	0.04	0.49	0.79	20	7.5	1.0

Table 2. Compiled helium isotope data for fumarolic gases, hot spring gases and hot spring waters from Central Japan.

No.	Site name	Location		Sample	$^3He/^4He$ (x 10^{-6})			$^4He/^{20}Ne$	Ref.
		Latitude	Longitude	Type	[±1σ]				
1	Higashi Amarume	38.853	139.930	Gas	7.77	±	0.15	400	a
2	Higashi Amarume	38.853	139.930	Gas	7.14	±	0.22	120	b
3	Amarume (1060m;82_3)	38.853	139.897	Gas	6.39	±	0.13	43	a
4	Amarume (1060m;83_6)	38.853	139.897	Gas	7.53	±	0.13	290	b
5	Amarume (1000m;83_6)	38.853	139.897	Gas	7.53	±	0.32	710	b
6	Amarume (600m;83_6)	38.853	139.897	Gas	7.9	±	0.1	1100	b
7	Yunose	38.638	139.693	Water	1.22	±	0.04	4.5	c
8	Atsumi	38.615	139.609	Water	3.228	±	0.043	140	c
9	Nogyoshikenjo	38.333	139.997	Water	8.7	±	0.45	98.1	d
10	Oguni	38.053	139.757	Gas	3.76	±	0.08	210	e
11	Awanoyu	37.936	139.697	Gas	9.01	±	0.09	170	f
12	Iide (Oguni)	37.923	139.657	Gas	11.12	±	0.11	2.1	e
13	Iide (Yamato)	37.793	139.777	Gas	8.05	±	0.13	2.1	e
14	Atsushio	37.723	139.887	Gas	9.95	±	0.11	270	c
15	Okuaidu Chinetusei	37.435	139.694	Gas	10.42	±		20	g
16	Katugi-Yuribana	38.492	139.521	Water	4.35	±	0.1	300	c
17	Mahoroba	38.283	139.522	Water	2.99	±	0.05	62	c
18	Senami	38.213	139.438	Water	1.45	±	0.04	1.3	c
19	Aga Oki 71	38.103	139.113	Gas	1.9	±	0.02	140	a
20	Aga Oki 4s	38.103	139.113	Gas	0.746	±	0.042	19	b
21	Aga Oki 11	38.103	139.113	Gas	0.341	±	0.02	20	b
22	Aga Oki 2s	38.103	139.113	Gas	0.138	±	0.005	110	b
23	Aga Oki 11s	38.103	139.113	Gas	0.762	±	0.18	100	b
24	Aga Oki 13s	38.103	139.113	Gas	1.62	±	0.08	260	b
25	Aga Oki 14s	38.103	139.113	Gas	0.93	±	0.057	48	b
26	Yuzawa	38.096	139.578	Gas	3.42	±	0.13	111	e
27	Katusranoseki	38.092	139.566	Gas	3.28	±	0.1	65	e
28	Kira	38.083	139.577	Water	4.03	±	0.06	260	c
29	Takanosu	38.070	136.089	Water	6.041	±	0.073	100	c
30	Shiunji SK-23	38.023	135.020	Gas	6.7	±		200	h

Table 2. Continued

No.	Site name	Location		Sample	$^3He/^4He$ (x 10^{-6})			$^4He/^{20}Ne$	Ref.
		Latitude	Longitude	Type	[±1σ]				
31	Shintainai	38.023	139.498	Gas	0.31	±	0.01	48	e
32	Higashi Niigata MS-24	37.973	139.167	Gas	3	±		4.3	i
33	Higashi Niigata NG-5	37.973	139.167	Gas	2.2	±		7.8	i
34	Higashi Niigata NS-13	37.973	139.167	Gas	4.4	±		15	i
35	Higashi Niigata MG-22	37.973	139.167	Gas	1.3	±		4	j
36	Higashi Niigata NS-13	37.973	139.167	Gas	4.2	±		6.6	j
37	Higashi Niigata MS-22	37.973	139.167	Gas	5.9	±		63	h
38	Shibata	37.955	139.358	Gas	7.39	±	0.23	595	e
39	Matsuzaki 1	37.953	139.180	Gas	3.22	±	0.09	91	a
40	Matsuzaki AB-4	37.953	139.097	Gas	4.3	±		96	h
41	Matsuzaki 7+10	37.953	139.097	Gas	3.51	±	0.06	110	b
42	Matsuzaki 7+10	37.953	139.097	Gas	3.85	±	0.09	22	b
43	Ninoji	37.933	139.417	Gas	6.72	±	0.1	66	e
44	Matsuzaki 2	37.920	139.097	Gas	3.23	±	0.03	77	a
45	Tsukioka	37.883	139.307	Gas	6.12	±	0.11	330	e
46	Minami Aga AD-31	37.753	139.130	Gas	3.76	±	0.08	63	a
47	Minami Aga Al-30	37.753	139.130	Gas	3.08	±	0.02	15	b
48	Kanose (1000m)	37.707	139.466	Gas	9.13	±	0.29	194	e
49	Mitsuke SK-24a	37.533	138.927	Gas	10.5	±		42	j
50	Mitsuke SK-40D	37.533	138.927	Gas	10.9	±		810	h
51	Amaze R-1	37.533	138.677	Gas	9.1	±		38	h
52	Mitsuke SK-11 82_9	37.520	138.847	Gas	8.61	±	0.29	1000	a
53	Mitsuke SK-11 83_6	37.520	138.847	Gas	8.67	±	0.28	>2500	b
54	Mitsuke SK-38	37.520	138.847	Gas	8.16	±	0.19	1700	b
55	Mishima (5652-5708m)	37.503	138.797	Gas	0.28	±	0.01	468	k
56	Mishima (5652-5708m)	37.503	138.797	Gas	0.25	±	0.02	22	k
57	Mishima (4745.5-4751.5m)	37.503	138.797	Gas	7	±	0.1	79	k
58	Mishima (4745.5-4751.5m)	37.503	138.797	Gas	7.1	±	0.1	76	k
59	Kumoide SK-20	37.473	138.747	Gas	7.1	±		85	h
60	Minami Nagaoka AD-5	37.423	138.797	Gas	9.4	±		68	i

Table 2. Continued

No.	Site name	Location		Sample	$^3He/^4He$ (x 10^{-6})			$^4He/^{20}Ne$	Ref.
		Latitude	Longitude	Type	[±1σ]				
61	Minami Nagaoka Jurakuji-1	37.423	138.797	Gas	9.5	±		1100	h
62	Kita Katakai SK-1 82_6	37.386	138.780	Gas	8.65	±	0.08	240	a
63	Kita Katakai SK-1 83_6	37.386	138.780	Gas	8.57	±	0.26	2500	b
64	Nakadoori T-H	37.386	138.647	Gas	7.69	±	0.08	590	a
65	Nakadoori T-L	37.386	138.647	Gas	6.98	±	0.1	130	b
66	Nakadoori 3	37.386	138.647	Gas	7.2	±	0.23	2000	b
67	N-Katakai SK-1	37.373	138.767	Gas	8.45	±	0.28	2500	b
68	Yoshii SK-7D	37.373	138.647	Gas	9.2	±		220	j
69	Higashi Kashiwazaki	37.373	138.647	Gas	9	±		2000	h
70	Katakai SK-6	37.370	138.780	Gas	5.59	±	0.08	77	a
71	Katakai SK-1	37.370	138.780	Gas	8.54	±	0.39	670	b
72	Hirai T-H	37.370	138.613	Gas	7.33	±	0.09	1100	a
73	Hirai T-L	37.370	138.613	Gas	6.9	±	0.1	130	b
74	Hirai 1	37.370	138.613	Gas	6.93	±	0.14	830	b
75	Yomogihira	37.368	138.897	Gas	5.71	±	0.07	11	c
76	Katakai SK-22	37.353	138.747	Gas	9.2	±		15	i
77	M-Katakai SK-2	37.353	138.747	Gas	5.41	±	0.06	370	b
78	S-Katakai SK-1	37.333	138.747	Gas	6.9	±	0.13	830	b
79	Kubiki CH-93	37.253	138.377	Gas	4.1	±		13	i
80	Meiji I-18 82_3	37.203	138.330	Gas	3.1	±	0.06	110	a
81	Meiji I-18 83_6	37.203	138.330	Gas	3.31	±	0.1	480	b
82	Urasa	37.161	138.916	Gas	3.29	±	0.05	80	c
83	Shintainai MS-17	37.123	139.417	Gas	4.2	±		12	h
84	Matsunoyama	37.065	138.601	Gas	1.75	±	0.03	0.37	c
85	Ikazawa	37.059	138.926	Gas	2.62	±	0.04	2.1	c
86	Bessho 2 82_3	37.036	138.330	Gas	2.09	±	0.08	63	a
87	Bessho 2 83_6	37.036	138.330	Gas	1.99	±	0.07	2000	b
88	Bessho 5	37.036	138.330	Gas	2.28	±	0.07	330	b
89	Renge (97_7)	36.803	137.797	Gas	8.35	±	0.22	31.7	l
90	Renge	36.803	137.797	Gas	9.24	±	0.16	3.6	m

Table 2. Continued

No.	Site name	Location		Sample	$^3He/^4He$ (x 10^{-6})			$^4He/^{20}Ne$	Ref.
		Latitude	Longitude	Type	[±1σ]				
91	Renge	36.803	137.797	Gas	9.72	±	0.15	14	m
92	Yunota	36.653	138.647	Gas	4.47	±	0.1	23.8	n
93	Hanashiki (83_8)	36.653	138.647	Gas	5.16	±	0.03	2.8	n
94	Hanashiki	36.653	138.647	Gas	5	±		1.53	o
95	Shiriyaki (83_8)	36.653	138.630	Gas	5.3	±	0.09	1.7	n
96	Shiriyaki	36.653	138.630	Gas	5.13	±			p
97	Shiriyaki	36.653	138.630	Gas	5.28	±		0.97	o
98	Ana Jigoku A	36.653	138.580	Gas	9.02	±		2.03	o
99	Ana Jigoku B	36.653	138.580	Gas	9.08	±		2.89	o
100	Kitagawa fumarole (83_3)	36.648	138.537	Gas	9.58	±	0.12	500	n
101	Kitagawa fumarole Fukin (83_3)	36.648	138.537	Gas	9.06	±	0.2	500	n
102	Kitagawa fumarole	36.648	138.537	Gas	10	±		2.54	o
103	Yugama Kako	36.643	138.537	Gas	10.5	±		12	o
104	Manza Yubatake	36.641	138.510	Gas	2.1	±		0.35	o
105	Oku Manza	36.633	138.520	Gas	9.33	±		1.89	o
106	Karabuki fumarole	36.633	138.512	Gas	8.06	±	0.25	30.9	n
107	Karabuki fumarole	36.633	138.512	Gas	8.69	±		1.27	o
108	Shakunage	36.628	138.505	Gas	5.7	±		0.62	o
109	Sesshogawara fumarole (83_1)	36.626	138.562	Gas	9.34	±	0.12	20.4	n
110	Sesshogawara fumarole (83_3)	36.626	138.562	Gas	9.15	±	0.4	250	n
111	Sesshogawara fumarole (83_5)	36.626	138.562	Gas	9.55	±	0.26	250	n
112	Sesshogawara fumarole (83_8)	36.626	138.562	Gas	8.12	±	0.14	58.8	n
113	Sesshogawara fumarole	36.626	138.562	Gas	10.5	±			p
114	Sesshogawara fumarole	36.626	138.562	Gas	10.2	±		11.8	o
115	Kusatsu Jizo	36.623	138.599	Gas	7.69	±		1.27	o
116	Kusatsu Yubatake (83_3)	36.623	138.597	Gas	7.01	±	0.21	1.3	n
117	Kusatsu Yubatake (83_3)	36.623	138.597	Gas	7.22	±	0.17	1.2	n
118	Kusatsu Yubatake (83_8)	36.623	138.597	Gas	6.8	±	0.13	2.3	n
119	Kusatsu Yubatake	36.623	138.597	Gas	7.12	±			p
120	Kusatsu Shirohata	36.623	138.597	Gas	7.66	±		1.28	o

Table 2. Continued

No.	Site name	Location		Sample	$^3He/^4He$ (x 10^{-6})			$^4He/^{20}Ne$	Ref.
		Latitude	Longitude	Type	[±1σ]				
121	Kusatsu Yubatake	36.623	138.597	Gas	6.98	±		1.02	o
122	Kazawa	36.436	138.414	Gas	4.36	±	0.07	1.9	q
123	Isobe	36.286	138.830	Gas	0.92	±	0.037	630	r
124	Arafune	36.220	138.730	Gas	1.29	±	0.13	9.1	r
125	Nasu Dake	37.153	139.957	Gas	6.66	±	0.29	97	s
126	Okunosawa fumarole	37.129	139.980	Gas	5.07	±	0.11	26	g
127	Chausudake fumarole 5	37.128	139.962	Gas	5.6	±	0.17	11	g
128	Chausudake fumarole 2	37.125	139.961	Gas	10.58	±	0.24	26	g
129	Chausudake fumarole 4	37.123	139.962	Gas	6.59	±	0.12	34	g
130	Arayu fumarole	36.956	139.780	Gas	8.36	±	0.31	120	g
131	Nikko Yumoto	36.803	139.413	Gas	4.09	±	0.11	7.1	r
132	Heiwajima (900m)	35.587	139.747	Gas	0.137	±	0.01	100	q
133	Mitsubisi	35.543	139.727	Gas	<0.3	±		23	s
134	Kawasahi (1016m)	35.533	139.717	Gas	0.86	±	0.32	0.72	s
135	Nakagawa	35.437	139.064	Water	1.92	±	0.02	0.72	t
136	Jakotsu-A(90_6)	35.242	139.055	Gas	8.65	±	0.08	2.3	t
137	Jakotsu-B(90_6)	35.242	139.055	Gas	7.93	±	0.08	2	t
138	Owakudani-A(90_6)	35.242	139.019	Gas	9.63	±	0.12	19	t
139	Owakudani-B1(90_3)	35.242	139.019	Gas	8.79	±	0.1	7.9	t
140	Owakudani-B2(90_6)	35.242	139.019	Gas	9.25	±	0.08	12	t
141	Owakudani-B3(90_9)	35.242	139.019	Gas	9.11	±	0.08	10	t
142	Owakudani-B4(91_3)	35.242	139.019	Gas	9.25	±	0.09	26	t
143	Sounjigoku-1	35.237	139.027	Gas	9.61	±	0.15	19	t
144	Sounjigoku-2	35.237	139.027	Gas	9.71	±	0.21	17	t
145	Owakudani	35.237	139.014	Gas	8.29	±	0.2	16	r
146	Hakone	35.233	139.017	Gas	8.51	±	0.4	23	s
147	Yumoto	35.232	139.095	Water	7.46	±	0.07	1.4	t
148	Iouyama-1	35.225	139.035	Gas	9.16	±	0.09	20	t
149	Iouyama-2	35.225	139.035	Gas	8.84	±	0.19	11	t
150	Yunohanazawa (CW-1)	35.220	139.030	Gas	8.6	±		40	u

Table 2. Continued

No.	Site name	Location		Sample Type	$^3He/^4He$ (x 10^{-6}) [±1σ]			$^4He/^{20}Ne$	Ref.
		Latitude	Longitude						
151	Yunohanazawa	35.220	139.030	Gas	7.62	±	0.07	7.1	r
152	Ashinoyu	35.218	139.039	Water	6.73	±	0.07	0.85	t
153	Kusashio	35.437	138.330	Gas	2.82	±	0.1	100	r
154	Shimobe	35.420	138.480	Gas	5.68	±	0.23	10	r
155	Shima	36.883	137.867	Gas	1.21	±	0.13	0.26	l
156	Yubara	36.870	137.880	Gas	8.27	±	0.11	4.7	m
157	Buno	36.823	137.927	Water	1.43	±	0.19	1.12	l
158	Kudarise	36.803	137.867	Water	5.63	±	0.14	2.91	l
159	Karamatuzawa	36.803	137.830	Gas	9.73	±	0.14	2.9	m
160	Karamatuzawa	36.803	137.830	Gas	10.61	±	0.15	36	m
161	Shiroumayari	36.753	137.767	Water	9.57	±	0.25	3.27	l
162	Kurashitanoyu	36.733	137.817	Water	2.11	±	0.1	2.41	l
163	Matsushiro (Helium spots)	36.570	138.214	Gas	9.12	±	0.58		v
164	Matsushiro (CO2-spring)	36.570	138.214	Gas	9.1	±	0.69	20	v
165	Matsushiro	36.570	138.197	Gas	6.41	±	0.1	29	r
166	Omachi onsen Go	36.533	137.817	Gas	0.7	±	0.07	0.55	l
167	Kuzu	36.523	137.747	Water	3.52	±	0.09	3.69	l
168	Nanakura	36.523	137.717	Water	1.26	±	0.01	0.41	l
169	Osa	36.486	138.380	Gas	4.62	±	0.09	5.3	r
170	Yumata	36.423	137.677	Gas	8.29	±	0.18	6.8	l
171	Ariake	36.403	137.767	Water	1.38	±	0.09	0.79	l
172	Nakabusa	36.403	137.747	Water	1.14	±	0.21	2.56	l
173	Nunobiki	36.336	138.380	Gas	5.71	±	0.04	9.1	r
174	Nunoshita	36.336	138.364	Gas	5.24	±	0.17		r
175	Kamikochi	36.253	137.617	Water	4.93	±	0.22	3.32	l
176	Nakanoyu	36.223	137.597	Gas	4.15	±	0.1	0.92	l
177	Myoko	36.183	137.797	Water	2.94	±	0.08	24.7	l
178	Sakamaki	36.183	137.617	Water	1.47	±	0.08	1.29	l
179	Shirahone	36.153	137.630	Gas	8.91	±	0.26	2.33	l
180	Shirahone	36.153	137.630	Gas	9.65	±	0.28	16	r

Table 2. Continued

No.	Site name	Location		Sample	$^3He/^4He$ (x 10^{-6})			$^4He/^{20}Ne$	Ref.
		Latitude	Longitude	Type	[±1σ]				
181	Midoriko	36.103	138.014	Gas	1.48	±	0.03	29	r
182	Nagawa	36.053	137.667	Gas	4.86	±	0.11	2.33	l
183	Meiji	36.036	138.297	Gas	9.24	±	0.07	12	r
184	Uminokuchi	36.020	138.480	Gas	7.22	±	0.15	63	r
185	Kiso (81_11)	35.903	137.580	Gas	6.8	±	0.07	150	w
186	Kiso (85_4)	35.903	137.580	Gas	7.25	±	0.11	43	x
187	Kiso (84_10)	35.903	137.580	Gas	7.06	±	0.29	49	x
188	Kiso (84_9)	35.903	137.580	Gas	7.83	±	0.06	120	x
189	Kiso (93_6)	35.903	137.580	Gas	3.06	±		0.42	y
190	Kiso (91_6)	35.903	137.580	Gas	8.49	±		2.3	y
191	Kiso (96_6)	35.903	137.580	Gas	7.66	±		2.45	z
192	Kanose (81_11)	35.903	137.564	Gas	8.13	±	0.27	83	w
193	Kanose (85_4)	35.903	137.564	Gas	8.33	±	0.35	91	x
194	Kanose (84_10)	35.903	137.564	Gas	8.69	±	0.18	92	x
195	Kanose (84_9)	35.903	137.564	Gas	8.44	±	0.2	92	x
196	Kanose (93_6)	35.903	137.564	Gas	8.62	±		8	y
197	Kanose (91_6)	35.903	137.564	Gas	9.84	±		51	y
198	Kanose (2000_6)	35.903	137.564	Gas	3.64	±		0.5	z
199	Kanose (98_6)	35.903	137.564	Gas	9.07	±		27	z
200	Kanose (96_6)	35.903	137.564	Gas	8.93	±		30.9	z
201	Yukawa	35.893	137.517	Gas	9.44	±		4.5	y
202	Ontake Kako A	35.893	137.477	Gas	9.95	±		360	p
203	Ontake Kako B	35.893	137.477	Gas	10.16	±		67	y
204	Shirakawa (84_10)	35.873	137.597	Gas	7.13	±	0.08	4.6	x
205	Shirakawa (85_4)	35.873	137.597	Gas	7.25	±		22	x
206	Shirakawa (93_6)	35.873	137.597	Gas	5.23	±		5.5	y
207	Shirakawa (91_6)	35.873	137.597	Gas	5.88	±		7.9	y
208	Shirakawa (2000_6)	35.873	137.597	Gas	7.35	±		1.4	z
209	Shirakawa (98_6)	35.873	137.597	Gas	7.13	±		1.57	z
210	Shirakawa (96_6)	35.873	137.597	Gas	4.48	±		0.58	z

Table 2. Continued

No.	Site name	Location		Sample	$^3He/^4He$ (x 10^{-6})			$^4He/^{20}Ne$	Ref.
		Latitude	Longitude	Type	[±1σ]				
211	Nigorigawa	35.837	137.464	Gas	7.63	±	0.18	4.3	w
212	Daikisen (Kamanuma)	35.823	137.627	Gas	5.59	±		3.67	y
213	Daikisen (Kamanuma)	35.823	137.627	Gas	5.43	±		2.29	z
214	Shojima (81_11)	35.803	137.714	Gas	2.83	±	0.1	17	w
215	Shojima (85_4)	35.803	137.714	Gas	2.87	±	0.17	3.2	x
216	Shojima (84_10)	35.803	137.714	Gas	2.93	±	0.08	3.4	x
217	Shojima (84_9)	35.803	137.714	Gas	2.9	±	0.15	9.3	x
218	Shojima (93_6)	35.803	137.714	Gas	3.53	±		7.9	y
219	Shojima (91_6)	35.803	137.714	Gas	3.5	±		61	y
220	Shojima (2000_6)	35.803	137.714	Gas	3.46	±		11.1	z
221	Shojima (98_6)	35.803	137.714	Gas	3.39	±		6.86	z
222	Shojima (96_6)	35.803	137.714	Gas	3.51	±		15.3	z
223	Kakehashi (2000_6)	35.803	137.680	Gas	5.73	±		44.3	z
224	Kakehashi (81_11)	35.803	137.680	Gas	4.17	±	0.06	23	w
225	Kakehashi (85_4)	35.803	137.680	Gas	4.8	±	0.07	96	x
226	Kakehashi (84_10)	35.803	137.680	Gas	4.7	±	0.1	47	x
227	Kakehashi (84_9)	35.803	137.680	Gas	4.26	±	0.06	41	x
228	Kakehashi (93_6)	35.803	137.680	Gas	5.23	±		82	y
229	Kakehashi (91_6)	35.803	137.680	Gas	5.41	±		150	y
230	Kakehashi (98_6)	35.803	137.680	Gas	5.59	±		107	z
231	Kakehashi (96_6)	35.803	137.680	Gas	3.88	±		0.89	z
232	Shikanoyu	35.720	137.680	Gas	2.39	±	0.04	71	w
233	Unazuki	36.817	137.584	Gas	5.21	±	0.07	1.1	c
234	Eiwa	36.872	136.983	Gas	0.42	±	0.02	640	c
235	Kanebo-cho	36.733	136.997	Water	4.77	±	0.14	0.23	d
236	Sano	36.733	136.997	Water	2.18	±	0.1	0.22	d
237	Beniya(29m)	36.723	136.997	Water	1.8	±	0.13	0.21	d
238	Wakamizu-cho (32m)	36.703	136.967	Water	1.47	±	0.11	0.22	d
239	Sanga (35m)	36.703	136.977	Water	1.41	±	0.11	0.22	d
240	Ayako(72m)	36.683	136.857	Water	1.21	±	0.12	0.22	d

Table 2. Continued

No.	Site name	Location		Sample	$^{3}He/^{4}He$ (x 10^{-6})			$^{4}He/^{20}Ne$	Ref.
		Latitude	Longitude	Type	[±1σ]				
241	Hiirata (54m)	36.633	136.847	Water	1.33	±	0.1	0.23	d
242	Tateyama	36.586	137.614	Gas	10.9	±	0.8	22	s
243	1Rindo	36.493	136.912	Water	6.048	±		3.89	aa
244	Nebuta	37.399	136.941	Gas	1.38	±	0.04	0.87	c
245	Jomon-mawaki	37.307	137.204	Gas	1.73	±	0.02	720	c
246	Wakura	37.090	136.915	Gas	0.96	±	0.02	5.4	c
247	Yugawa	37.083	137.042	Gas	0.47	±	0.02	1600	c
248	Chirihama	36.890	136.768	Gas	0.83	±	0.02	300	c
249	2Yuwaku	36.478	136.761	Water	7.168	±		93.9	aa
250	3Yuwaku soyu	36.478	136.761	Water	7.126	±		19.8	aa
251	4Tatsunokuchi	36.436	136.553	Water	5.334	±		2.18	aa
252	5Tatsunokuchi	36.436	136.553	Water	5.53	±		19	aa
253	6Yusenji	36.397	136.528	Water	4.396	±		14.1	aa
254	7Yusenji	36.397	136.528	Water	4.844	±		17.4	aa
255	12Kawati	36.342	136.686	Water	0.896	±		1.96	aa
256	8Awadu	36.328	136.453	Water	4.732	±		14.6	aa
257	9Awadu	36.328	136.453	Water	5.026	±		33.6	aa
258	11Suginoko	36.322	136.658	Water	5.866	±		147	aa
259	10Yamashiro	36.286	136.372	Water	4.858	±		6.24	aa
260	Iwama	36.236	136.764	Gas	11.4	±	1.1	8.1	s
261	Shirahone	36.170	136.630	Gas	5.65	±	0.27	470	s
262	Miyagawa	36.423	137.197	Gas	2.57	±	0.08	140	c
263	Wariishi	36.373	137.277	Gas	2.45	±	0.04	310	c
264	Nakao 6	36.233	137.527	Gas	3.28	±	0.08	3.4	l
265	Yakedake	36.223	137.567	Gas	11.18	±	0.15	242	l
266	Seizansou-6-gosei	36.223	137.517	Gas	6.14	±	0.28	50	l
267	Kyodai-Saboken	36.223	137.517	Gas	7.29	±	0.18	30.3	l
268	Akigami (81_11)	36.003	137.397	Gas	4.81	±	0.18	2.5	w
269	Akigami (84_9)	36.003	137.397	Gas	4.44	±	0.15	1.5	x
270	Akigami (84_10)	36.003	137.397	Gas	4.89	±	0.15	2.7	x

Table 2. Continued

No.	Site name	Location		Sample	$^{3}He/^{4}He$ (x 10^{-6})			$^{4}He/^{20}Ne$	Ref.
		Latitude	Longitude	Type	[±1σ]				
271	Akigami (85_4)	36.003	137.397	Gas	5.07	±	0.1	3.3	x
272	Akigami (91_6)	36.003	137.397	Gas	5.88	±		7.9	y
273	Akigami (93_6)	36.003	137.397	Gas	5.23	±		5.5	y
274	Akigami (96_6)	36.003	137.397	Gas	2.86	±			z
275	Akigami (2000_6)	36.003	137.397	Gas	2.17	±			z
276	Nigorigo (81_11)	35.920	137.447	Gas	8.61	±	0.11	32	w
277	Nigorigo (84_9)	35.920	137.447	Gas	8.4	±	0.13	79	x
278	Nigorigo (84_10)	35.920	137.447	Gas	8.79	±	0.15	55	x
279	Nigorigo (85_4)	35.920	137.447	Gas	8.48	±	0.24	29	x
280	Nigorigo (91_6)	35.920	137.447	Gas	8.88	±		39	y
281	Nigorigo (93_6)	35.920	137.447	Gas	8.23	±		15	y
282	Nigorigo (96_6)	35.920	137.447	Gas	6.12	±		0.99	z
283	Nigorigo (98_6)	35.920	137.447	Gas	7.29	±		2.22	z
284	Nigorigo (2000_6)	35.920	137.447	Gas	7.99	±		5.56	z
285	Shimojima (81_11)	35.920	137.330	Gas	2.6	±	0.16	0.77	w
286	Shimojima (84_9)	35.920	137.330	Gas	2.65	±	0.24	0.86	x
287	Shimojima (84_10)	35.920	137.330	Gas	1.44	±	0.15	0.36	x
288	Shimojima (85_4)	35.920	137.330	Gas	1.85	±	0.32	0.52	x
289	Shimojima (91_6)	35.920	137.330	Gas	2.96	±		1.8	y
290	Shimojima (93_6)	35.920	137.330	Gas	3.54	±		66	y
291	Yuya (81_11)	35.903	137.314	Gas	3	±	0.15	37	w
292	Yuya (84_9)	35.903	137.314	Gas	2.86	±	0.07	110	x
293	Yuya (84_10)	35.903	137.314	Gas	2.93	±	0.07	130	x
294	Yuya (85_4)	35.903	137.314	Gas	2.98	±	0.08	99	x
295	Yuya (91_6)	35.903	137.314	Gas	3.18	±		120	y
296	Yuya (93_6)	35.903	137.314	Gas	2.89	±		285	y
297	Yuya (96_6)	35.903	137.314	Gas	3.02	±			z
298	Yuya (98_6)	35.903	137.314	Gas	3	±			z
299	Yuya (2000_6)	35.903	137.314	Gas	3.04	±			z
300	Byakko	35.403	137.280	Gas	1.1	±	0.03	59	r

Table 2. Continued

No.	Site name	Location		Sample	$^3He/^4He$ (x 10^{-6})			$^4He/^{20}Ne$	Ref.
		Latitude	Longitude	Type	[±1σ]				
301	Takasago	35.373	137.247	Water	0.93	±		681	ab
302	Tono AN-6_14.0m	35.373	137.217	Water	1.26	±	0.04	0.5	ac
303	Tono AN-6_34.0m	35.373	137.217	Water	0.98	±	0.04	0.3	ac
304	Tono AN-6_50.0m	35.373	137.217	Water	0.341	±	0.012	0.93	ac
305	Tono AN-6_121.5m	35.373	137.217	Water	0.293	±	0.01	1.38	ac
306	Tono TH-3_45.5m	35.373	137.217	Water	1.48	±	0.05	0.45	ac
307	Tono TH-3_85.5m	35.373	137.217	Water	0.63	±	0.02	0.57	ac
308	Tono TH-3_124.0m	35.373	137.217	Water	0.179	±	0.006	9.38	ac
309	Tono TH-4_60m	35.373	137.217	Water	1.37	±	0.05	0.23	ac
310	Tono TH-4_83m	35.373	137.217	Water	1.28	±	0.05	0.28	ac
311	Tono TH-4_83.0m	35.373	137.217	Water	1.08	±	0.04	0.24	ac
312	Tono TH-4_99m	35.373	137.217	Water	1.26	±	0.05	0.22	ac
313	Tono TH-4_152m	35.373	137.217	Water	0.117	±	0.004	5.5	ac
314	Tono TH-4_152.5m	35.373	137.217	Water	0.126	±	0.005	2.5	ac
315	Tono TH-6_104m	35.373	137.217	Water	0.56	±	0.02	1	ac
316	Tono TH-6_104m	35.373	137.217	Water	0.493	±	0.017	1	ac
317	Tono TH-6_132m	35.373	137.217	Water	0.282	±	0.012	1.46	ac
318	Tono TH-6_153m	35.373	137.217	Water	0.309	±	0.011	1.31	ac
319	Tono DH-3_19m	35.373	137.217	Water	1.25	±	0.04	0.25	ac
320	Tono DH-3_130m	35.373	137.217	Water	0.451	±	0.017	0.71	ac
321	Tono DH-3_208m	35.373	137.217	Water	0.441	±	0.017	0.82	ac
322	Tono DH-3_330m	35.373	137.217	Water	0.347	±	0.015	2.5	ac
323	Tono DH-3_508m	35.373	137.217	Water	0.23	±	0.008	1.2	ac
324	Tono DH-3_646m	35.373	137.217	Water	0.236	±	0.01	2.75	ac
325	Tono DH-3_699.7m	35.373	137.217	Water	0.244	±	0.009	2.03	ac
326	Tono DH-3_984.4m	35.373	137.217	Water	0.167	±	0.006	2.82	ac
327	Tono DH-12_157.45-164.12m	35.373	137.217	Water	0.384	±	0.013	75.9	ac
328	Tono DH-12_157.45-164.12m	35.373	137.217	Water	0.42	±	0.015	97.9	ac
329	Tono DH-12_157.45-164.12m	35.373	137.217	Water	0.401	±	0.014	82.9	ac
330	Tono DH-12_178.00-280.17m	35.373	137.217	Water	0.326	±	0.012	117.4	ac

Table 2. Continued

No.	Site name	Location		Sample Type	$^3He/^4He$ (x 10^{-6}) [±1σ]			$^4He/^{20}Ne$	Ref.
		Latitude	Longitude						
331	Tono DH-12_178.00-280.17m	35.373	137.217	Water	0.333	±	0.012	64.3	ac
332	Tono DH-12_345.9-387.2m	35.373	137.217	Water	0.488	±	0.017	1555.6	ac
333	Tono DH-12_345.9-387.2m	35.373	137.217	Water	0.5	±	0.02	188.9	ac
334	Tono DH-12_388.9-429.7m	35.373	137.217	Water	0.468	±	0.017	131.3	ac
335	Tono DH-12_388.9-429.7m	35.373	137.217	Water	0.499	±	0.019	150	ac
336	Tono DH-12_388.9-429.7m	35.373	137.217	Water	0.486	±	0.017	136.5	ac
337	Tono DH-12_431.4-510.0m	35.373	137.217	Water	0.444	±	0.016	139.5	ac
338	Tono DH-12_431.4-510.0m	35.373	137.217	Water	0.435	±	0.017	78.3	ac
339	Tono DH-12_431.4-510.0m	35.373	137.217	Water	0.447	±	0.018	87	ac
340	Tono DH-13_60.50-74.15m	35.373	137.217	Water	0.448	±	0.018	0.42	ac
341	Tono DH-13_60.50-74.15m	35.373	137.217	Water	0.457	±	0.017	0.62	ac
342	Tono MIU-4_82.5-88.65m	35.373	137.217	Water	0.138	±	0.005	3.2	ac
343	Tono MIU-4_82.5-88.65m	35.373	137.217	Water	0.125	±	0.005	3.3	ac
344	Tono MIU-4_82.5-88.65m	35.373	137.217	Water	0.139	±	0.007	5.8	ac
345	Tono MIU-4_95.02-134.47m	35.373	137.217	Water	0.101	±	0.005	4.8	ac
346	Tono MIU-4_95.02-134.47m	35.373	137.217	Water	0.108	±	0.004	4.6	ac
347	Tono MIU-4_95.02-134.47m	35.373	137.217	Water	0.109	±	0.006	5.8	ac
348	Tono MIU-4_314.95-316.95m	35.373	137.217	Water	0.091	±	0.004	9.8	ac
349	Tono MIU-4_314.95-316.95m	35.373	137.217	Water	0.098	±	0.004	8.3	ac
350	Tono KNA-6_160m	35.373	137.217	Water	0.21	±		4.9	ab
351	Tono KNA-6_160m	35.373	137.217	Water	0.08	±		5.47	ab
352	Tono KNA-6_160m	35.373	137.217	Water	0.14	±		5.79	ab
353	Tono KNA-6_180m	35.373	137.217	Water	0.12	±		7.25	ab
354	Tono MIU-4_584.0-647.0m	35.373	137.217	Water	0.09	±		6.22	ab
355	Tono MIU-4_754.5-790.1m	35.373	137.217	Water	0.18	±		2.33	ab
356	Tono MIU-4_754.5-790.1m	35.373	137.217	Water	0.31	±		1.97	ab
357	Tono DH-3(Z-1)19.1m	35.373	137.217	Water	1.32	±		0.53	ab
358	Tono DH-3(Z-2)130.3-139.5m	35.373	137.217	Water	1.03	±		0.28	ab
359	Tono DH-3(Z-3)208.4-220.7m	35.373	137.217	Water	1.14	±		0.4	ab
360	Tono DH-3(Z-3)208.4-220.7m	35.373	137.217	Water	1.34	±		0.55	ab

Table 2. Continued

No.	Site name	Location		Sample	$^3He/^4He$ (x 10^{-6})			$^4He/^{20}Ne$	Ref.
		Latitude	Longitude	Type	[±1σ]				
361	Tono DH-3(Z-6)507.9-521.7m	35.373	137.217	Water	1.04	±		0.43	ab
362	Tono DH-3(Z-9)876.9-889.1m	35.373	137.217	Water	1.18	±		0.69	ab
363	Tono DH-7_482.4m	35.373	137.217	Water	1.27	±		0.33	ab
364	Tono DH-7_563.8m	35.373	137.217	Water	0.83	±		0.58	ab
365	Tono DH-7_837.6m	35.373	137.217	Water	0.92	±		0.3	ab
366	Tono DH-7_860.1m	35.373	137.217	Water	1.01	±		0.75	ab
367	Tono DH-7_967.5m	35.373	137.217	Water	0.47	±		0.74	ab
368	Tono DH-7_967.5m	35.373	137.217	Water	0.86	±		0.4	ab
369	Tono DH-13_182.0-252.0m	35.373	137.217	Water	0.59	±		0.69	ab
370	Tono DH-13_253.5-323.5m	35.373	137.217	Water	0.5	±		0.62	ab
371	Tono DH-13_474.5-544.5m	35.373	137.217	Water	0.62	±		0.59	ab
372	Tono DH-13_546.0-616.0m	35.373	137.217	Water	0.44	±		0.68	ab
373	Tono DH-13_713.0-783.0m	35.373	137.217	Water	0.3	±		1.08	ab
374	Tono DH-13_856.2-926.2m	35.373	137.217	Water	0.64	±		0.59	ab
375	Tono DH-13_995.0-1015.0m	35.373	137.217	Water	0.55	±		0.71	ab
376	Tono DH-15_84.5-97.4m	35.373	137.217	Water	0.72	±		0.682	ab
377	Tono DH-15_63.0-72.5m	35.373	137.217	Water	1	±		0.506	ab
378	Tono DH-15_152.1-216.3m	35.373	137.217	Water	0.276	±		36.9	ab
379	Tono DH-15_233.6-240.0m	35.373	137.217	Water	0.402	±		112	ab
380	Tono MSB-2_79.0-130.5m	35.373	137.217	Water	0.42	±		19.2	ab
381	Tono MSB-2_132.0-154.0m	35.373	137.217	Water	0.4	±		123	ab
382	Tono MSB-2_171.5-175.5m	35.373	137.217	Water	0.45	±		109	ab
383	Tono MSB-4_95.5-99.0m	35.373	137.217	Water	1.82	±		3.86	ab
384	Tono MIZ-1_215.0-225.7m	35.373	137.217	Water	0.43	±		20.6	ab
385	Yoro	35.301	136.555	Gas	0.66	±	0.04	53	ad
386	Kaidu	35.180	136.657	Gas	0.71	±	0.04	150	ad
387	Yubune	35.370	138.980	Water	6.94	±	0.07	5.1	t
388	Nakaizu	34.937	139.030	Gas	1.46	±	0.04	0.32	r
389	Atagawa	34.803	139.064	Water	1.7	±	0.09	0.67	r
390	Mine	34.753	138.964	Water	0.925	±	0.017	0.65	r

Table 2. Continued

No.	Site name	Location		Sample	$^{3}He/^{4}He$ (x 10^{-6}) [±1σ]			$^{4}He/^{20}Ne$	Ref.
		Latitude	Longitude	Type					
391	Ohsawa	34.753	138.830	Water	4.53	±	0.15	0.74	r
392	Omaezaki	34.603	138.230	Gas	0.102	±	0.004	6.7	q
393	Yuya	34.970	137.614	Gas	1.73	±	0.04	100	r
394	Kutsuki	35.342	135.921	Water	0.186	±	0.008	2.4	c
395	Biwako	35.120	135.947	Gas	1.82	±	0.1	45	r
396	Ogoto	35.097	135.899	Water	0.091	±	0.004	5.3	c
397	Junibo	35.013	136.105	Water	1.011	±	0.021	1.9	c
398	Shiono	34.937	136.147	Gas	3.63	±	0.07	177	ae
399	Miyano	34.920	136.164	Gas	3.15	±	0.06	170	ae
400	Ageki	35.153	136.517	Water	0.339	±	0.011	1.9	af
401	Nagashima R15	35.093	136.717	Gas	1.035	±	0.032	62	c
402	Nagashima R3	35.033	136.737	Gas	0.833	±	0.028	41	c
403	Yunoyama	35.020	136.507	Gas	0.54	±	0.03	29	ad
404	Sarubino	34.778	136.283	Water	1.157	±	0.021	23	c
405	Inokura	34.681	136.340	Water	1.01	±	0.05	0.305	ag
406	Omiya	35.603	135.081	Gas	0.55	±	0.01	7	ae
407	Iwataki	35.572	135.153	Gas	1.75	±	0.04	269	ae
408	Iwataki	35.572	135.153	Water	1.74	±	0.06	20	c
409	Momoyama	34.929	135.770	Water	1.14	±	0.05	6.08	ad
410	Joto (935m)	34.687	135.547	Water	0.564	±	0.011	317	ah
411	Joto (935m)	34.687	135.547	Water	0.509	±	0.023	260	ah
412	Naniwa (1000m)	34.670	135.481	Water	1.42	±	0.37	1.8	ah
413	Tondabayashi (1300m)	34.503	135.597	Water	2.72	±	0.04	472	ah
414	Tondabayashi (180m)	34.503	135.597	Water	1.71	±	0.16	0.45	ah
415	Sayama (1000m)	34.503	135.547	Water	4.6	±	0.06	3050	ah
416	Sayama (250m)	34.503	135.547	Water	4.43	±	0.49	1.9	ah
417	Tsukigase	34.714	136.049	Water	1.686	±	0.034	1.1	c
418	Heijougu	34.683	135.783	Gas	3.68	±	0.06	1020	af
419	Ikoma Oze	34.653	135.723	Water	0.35	±	0.01	1.4	af
420	Kusayama	35.168	135.275	Gas	0.45	±	0.03	4100	ad

Table 2. Continued

No.	Site name	Location		Sample	$^3He/^4He$ (x 10^{-6})			$^4He/^{20}Ne$	Ref.
		Latitude	Longitude	Type	[±1σ]				
421	Kagobo	35.037	135.331	Gas	0.888	±	0.041	8.3	r
422	Konda	35.003	135.097	Gas	0.24	±	0.01	202	ae
423	Shionoyu	34.953	135.397	Gas	0.87	±	0.07	86	ae
424	Kawanishi	34.903	135.417	Gas	1.53	±	0.06	110	s
425	Inagawa	34.895	135.375	Gas	1.16	±	0.02	36	ai
426	Ishimichi	34.883	135.384	Gas	1.53	±	0.03	251	ai
427	Ishimichi	34.882	135.383	Gas	1.71	±	0.06	45	af
428	Hirano	34.870	135.414	Gas	1.29	±	0.04	95	ae
429	Takaraduka	34.803	135.347	Gas	8.04	±	0.46	3	ae
430	Takaraduka	34.803	135.347	Gas	8.62	±		45	aj
431	Takaraduka	34.803	135.347	Gas	8.15	±	0.21	42	r
432	Arima	34.803	135.247	Gas	10.42	±	0.18	270	ae
433	Arima	34.803	135.247	Gas	11	±	0.4	210	s
434	Arima	34.803	135.247	Gas	9.22	±	0.11	37	r
435	Arima	34.803	135.247	Gas	9.69	±	0.12	530	ak
436	Kasagi	34.753	135.947	Gas	5.29	±	0.07	125	ae
437	Sumikawa-1 (93_4)	34.753	135.327	Water	1.44	±		0.253	al
438	Sumikawa-1 (95_1)	34.753	135.327	Water	1.2	±		0.308	al
439	Sumikawa-1 (95_8)	34.753	135.327	Water	1.36	±		0.267	al
440	Sumikawa-3 (93_4)	34.753	135.327	Water	3.5	±		0.262	al
441	Sumikawa-3 (95_1)	34.753	135.327	Water	2.2	±		0.338	al
442	Sumikawa-4 (93_4)	34.753	135.327	Water	2.55	±		0.246	al
443	Sumikawa-4 (95_1)	34.753	135.327	Water	1.39	±		0.341	al
444	Sumikawa-4 (95_8)	34.753	135.327	Water	2.74	±		0.246	al
445	Sumikawa-5 (93_9)	34.753	135.327	Water	1.41	±		0.26	al
446	Sumikawa-5 (95_1)	34.753	135.327	Water	1.11	±		0.32	al
447	Hiramatsu-2(93_9)	34.753	135.327	Water	2.28	±		0.285	al
448	Hiramatsu-2(95_8)	34.753	135.327	Water	1.51	±		0.357	al
449	Hiramatsu-1(93_9)	34.753	135.327	Water	1.47	±		0.274	al
450	Hiramatsu-1(95_8)	34.753	135.327	Water	1.17	±		0.325	al

Table 2. Continued

No.	Site name	Location		Sample	$^3He/^4He$ (x 10^{-6})			$^4He/^{20}Ne$	Ref.
		Latitude	Longitude	Type	[±1σ]				
451	Amagasaki (1370m)	34.720	135.447	Water	4.48	±	0.06	418	ah
452	Amagasahi (250m)	34.720	135.447	Water	4.08	±	0.19	1.29	ah
453	Kobe	34.703	135.197	Gas	2.88	±	0.1	0.8	ak
454	Kobe	34.703	135.197	Gas	2.74	±	0.07	0.9	ak
455	Minatoyama	34.687	135.164	Gas	3.77	±	0.06	43	r
456	Iwaya	34.591	135.020	Gas	0.67	±	0.04	1	ae
457	Iwaya	34.591	135.020	Gas	0.68	±	0.04	0.79	c

References: a, Wakita and Sano (1983); b, Wakita et al. (1990); c, this study; d, Takaoka and Mizutani (1987); e, Umeda et al. (2007a); f, Sano et al. (1985); g, NEDO (1990); h, Sakata et al. (1986); i, Sakata (1991); j, Sakata et al. (1989); k, Kita and Nagao (1996); l, Kusakabe et al. (2003); m, NEDO (1999), n, Marty et al. (1989); o, Sano et al. (1994); p, Sano and Marty (1995); q, Sano et al. (1982); r, Sano and Wakita (1985); s, Nagao et al. (1981), t, Sakamoto et al. (1992); u, Craig et al. (1978); v, Nagao et al. (1980); w, Sano et al. (1984); x, Sano et al. (1986); y, Sano et al. (1998b); z, Takahata et al. (2003); aa, Ohwada et al. (2007); ab, Nagao (2004); ac, Morikawa (2001); ad, Umeda et al. (2006b); ae, Matsumoto et al. (2003); af, Umeda et al. (2007b); ag, Umeda et al. (2006a); ah, Okada et al. (1994); ai, Sano et al. (2006), aj, Ono et al. (1993); ak, Wakita et al. (1987); al, Sano et al. (1998a).

Most hot springs in the southern part, except for Arima hot spring (Matsumoto et al., 2003), have lower $^3He/^4He$ ratios than the atmospheric value. In general, a crustal helium component is expected to be dominant in samples from non-volcanic regions due to the addition to groundwater of radiogenic helium produced by the decay of U and Th. The $^3He/^4He$ ratios observed on the southern NKTZ are similar to those in the forearc region of the Northeast Japan Arc (Sano and Wakita, 1985) and the Ryukyu Arc (Umeda et al., 2007c), indicating an insignificant contribution of mantle helium from the subcrustal lithosphere to the Earth's surface.

On the other hand, the $^3He/^4He$ ratios of volcanic gases or hot spring gases observed near active volcanoes in the central NKTZ are high and similar to the mean value of 5.4 ± 1.9 RA for arc-related volcanism worldwide (Hilton et al., 2002). Helium isotope ratios of gases in volcanic regions are usually higher than the atmospheric value because mantle melting is the most likely mechanism responsible for the transfer of mantle helium from the subcrustal lithosphere where the volatiles would be released directly from a magma body to the crustal fluids (e.g., Torgersen, 1993). However, the $^3He/^4He$ ratios in hot springs laterally away from active volcanoes are similar to those observed in the southern NKTZ, indicating that radiogenic helium dominates over any mantle helium input. Similar trends have been often observed around other active volcanoes in subduction zones (Sakamoto et al., 1992).

In the northern NKTZ, the $^3He/^4He$ ratios are significantly higher than those observed in the southern NKTZ, although both parts are located on non-volcanic regions. Given the presence of mantle helium beneath the northern NKTZ, the potential ^3He sources are expected to be the erupted and intruded Miocene basalts partially composing of Miocene sedimentary basin because of scavenging of mantle helium trapped in ancient magmas stored in the crust for millions of year (Torgersen and Jenkins, 1982). Umeda et al. (2007a) calculated the evolution in $^3He/^4He$ ratios of ancient magma that erupted in the Middle Miocene in Northeast Japan with time, and indicated that the expected $^3He/^4He$ ratio at the present time is 3.8 RA. Therefore, the $^3He/^4He$ ratios of most hot springs in the northern NKTZ can be interpreted as a three-component mixture of mantle helium associated with magmatism of Middle Miocene age, radiogenic crustal helium and atmospheric helium. However, it should be noted that there appear to be several areas with hot springs having the $^3He/^4He$ ratios higher than the value of 3.8 RA. These areas are close to SW-NE oriented active reverse faults causing large earthquakes, such as the 2004 mid-Niigata Prefecture earthquake (Mj 6.8), and the 2007 Niigataken Chuetsu-Oki earthquake (Mj 6.8).

4. Discussion

Seismic velocity and resistivity of the crust and the upper mantle provide important insight into the existence of mantle melting and/or aqueous fluid arising from the dehydration of the subducting slab probably controlling the regional variation in $^3He/^4He$ ratios of hot springs observed on the Earth's surface. Nakajima and Hasegawa (2007a), using data from the updated seismic network in Japan, imaged the three-dimensional seismic wave velocity structure along the NKTZ. Figure 4 shows the vertical-cross section of the P- and S-wave velocity structure and the Vp/Vs ratio together with variation in $^3He/^4He$ ratios along the NKTZ. The seismic velocity structure obtained indicates the existence of a highly heterogeneous structure beneath the NKTZ.

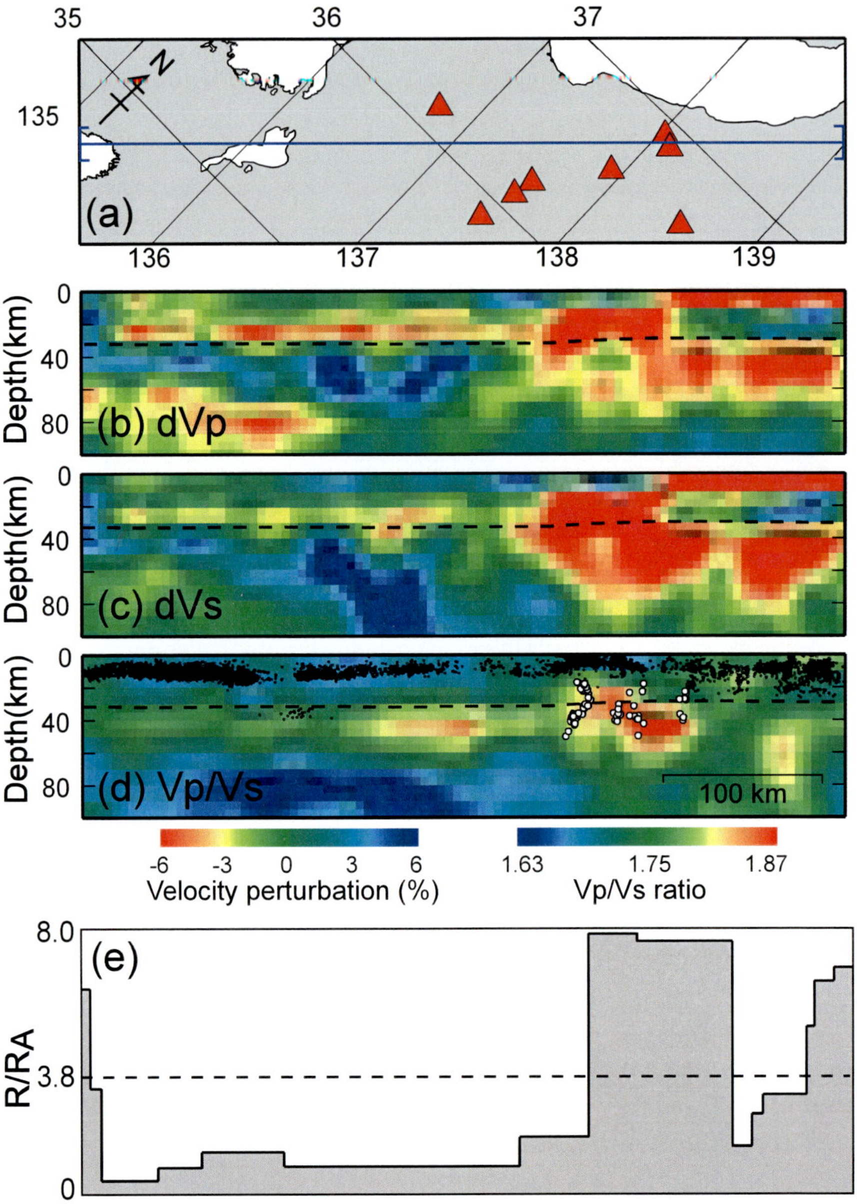

Figure 4. (a) Location of the profile along the NKTZ for which the vertical cross sections of velocity structure and distribution of helium isotopes are shown. The red triangles in (a) denote active volcanoes. Vertical cross sections of (b) P- and (c) S-wave velocity perturbations, and (d) Vp/Vs ratio along the profile in (a). Microearthquakes (dots) and low-frequency earthquakes (white circles) that occurred within a 10-km-wide zone along the profile are shown in (d). The broken line in each figure denotes the depth of the Moho adopted in the inversion (after Nakajima and Hasegawa, 2007a). (e) a profile of the upper limit of $^{3}He/^{4}He$ ratios within a 20-km-wide zone along the profile in (a).

In the southern NKTZ, the Philippine Sea slab subducting sub-horizontally around Lake Biwa can be imaged as a high-velocity anomaly at a depth of 40 km. In addition, a low-velocity zone exists above the Philippine Sea slab. Nakajima and Hasegawa (2007a) suggested that the low-velocity zone in the lower crust is attributed to aqueous fluids derived by dehydration of the subducting slab with the result that the existence of aqueous fluids reduces the strength of the lower crust and contributes to the large contraction in the southern NKTZ. Although aqueous fluids derived from the Philippine Sea slab may bring mantle helium from the mantle wedge to the Earth's surface so that high ^{3}He gas ratios would emanate from hot springs in the Kii Peninsula (e.g., Matsumoto et al., 2003), hot springs with high ^{3}He/^{4}He ratios are not be observed in the southern NKTZ.

To solve this problem, an important point to notice is the configuration of the subducting Philippine Sea slab. The geometry of the Philippine Sea slab is complicated and the depth of the slab varies along the arc (e.g., Nakamura et al., 1997). The dip of the slab is much gentler beneath Lake Biwa compared to surrounding regions so that the inland crust may come into direct contact with the subducted oceanic crust (Figure 1). There does not appear to be a mantle wedge, which is a possible source of mantle helium. In the absence of a mantle wedge, flux of mantle helium in aqueous fluids is insignificant for interpretation of ^{3}He/^{4}He ratios in the hot springs in the southern NKTZ.

Moreover, the leading edge of the aseismic Philippine Sea slab is not considered to have reached the most southern part of NKTZ (Nakajima and Hasegawa, 2007b) (Figure 1) where the exceptional Arima hot spring, with a significantly high ^{3}He/^{4}He ratio of 7.4 RA (Matsumoto et al., 2003) is located. Therefore, the potential ^{3}He source in the Arima hot spring can be interpreted to be upwelling of mantle helium in aqueous fluids derived from the Pacific slab through the mantle wedge, where the Philippine Sea slab does not act as a barrier to fluids rising from the Pacific slab.

Shallow subduction of the Philippine Sea slab beneath the southern NKTZ contrasts with steeper dips beneath the central and northern NKTZ (Nakajima and Hasegawa, 2007b). Thus, a mantle wedge as the plausible source of mantle helium exists beneath the central and northern NKTZ, in contrast to the southern NKTZ. The extremely low velocity anomaly extending from the upper mantle to the upper crust appears to occur beneath the central NKTZ, which is likely to be associated with volcanism in this region (Nakajima and Hasegawa, 2007a). A high Vp/Vs value of approximately 1.85 in the lower crust is similar to that observed in the volcanic region of northeast Japan. In general, it has been believed that observations of low Vp, low Vs and high Vp/Vs in the crust are related to the existence of melt inclusions (e.g., Nakajima et al., 2001). Ogawa and Honkura (2004) reported that wide-band magnetotelluric soundings also imaged an anomalously conductive body (<10 Ωm) beneath the same region. This observation suggests the existence of magmatic fluids in the lower crust and upper mantle under possibly high temperature conditions. Consequently, it is expected that ^{3}He emanation manifested in the central NKTZ results from the effective transfer of mantle helium by intrusion and degassing of mantle-derived magma in the crust.

In the northern NKTZ, low velocity anomalies can be observed in the upper crust and the uppermost mantle (Nakajima and Hasegawa, 2007a). The low velocity zone in the upper crust may reflect the Miocene sedimentary basin with the erupted and intruded basalts related to opening of the Sea of Japan, whose thickness is now more than 6 km (e.g., Sato et al., 1994). The low velocity zone is also clearly visible in the uppermost mantle. Nakajima and Hasegawa (2007a) indicated that the low velocity anomaly is caused by the upwelling of fluids derived from the Pacific slab, and the concentrated supply of fluids originating from the slab may weaken the uppermost mantle, causing contractive deformation to occur within the upper crust in the northern NKTZ.

As mentioned above, although most hot springs with $^{3}He/^{4}He$ ratios higher than the atmospheric value can be attributed to emanation of ^{3}He gases derived from the Miocene basalt, the $^{3}He/^{4}He$ ratios of hot springs close to active faults are higher than the expected value of 3.8 RA. Okada et al. (2005) and Wang and Zhao (2006) showed the detailed seismic velocity structure and the Vp/Vs ratio structure beneath the active fault system causing the 2004 mid-Niigata Prefecture earthquake. Their results demonstrate that a zone with low velocity and Poisson's ratio anomaly is clearly imaged in the lower crust and the uppermost mantle under the seismic source area. A conductive zone below the source area was also imaged in the lower crust and the uppermost mantle (Ogawa and Honkura, 2004; Uyeshima et al., 2005). It has been suggested that geophysical anomalies below the seismic source area related to the 2004 mid-Niigata Prefecture earthquake reflect the existence of aqueous fluids which are released due to dehydration of the subducting Pacific slab (Uyeshima et al., 2005; Wang and Zhao, 2006). Moreover, they proposed a model for large inland earthquake that, when aqueous fluid enters the active fault in the crust, fluids will reduce the strength of the fractured rock matrix, and thus fault ruptures occur and can trigger earthquakes.

It is known that hot springs with significantly high $^{3}He/^{4}He$ ratios are observed from seismically active zones rather than non-active zones (e.g., Güleç et al., 2002). Active faults associated with extensional tectonics play an important role in effective transfer of mantle fluid from the subcrustal lithosphere to the surface in non-volcanic regions (e.g., Kulongoski et al., 2005). On the other hand, Kennedy and van Soest (2006) reported that excess ^{3}He was found in groundwater from seismically active areas undergoing compression. Sibson (2007) suggested that a heterogonous distribution of anomalously low P and S-wave velocities and high electrical conductivity below the source area related to the 2004 mid-Niigata Prefecture earthquake point to strong fluid over-pressuring in the vicinity of the active fault system and in the lower crust. The increase of fluid pressure along the active fault makes the fluid pressure approach the critical pressure needed for opening the fracture as a permeable conduit (Xu and Oki, 2004), facilitating transfer of mantle helium from the lower crust to the Earth's surface. As a result, the emanations from hot springs with amorously high $^{3}He/^{4}He$ ratios observed around the active faults in the northern NKTZ could be interpreted as the contribution of mantle helium carried by aqueous fluids derived from the subducting Pacific slab.

Our study reinforces the clear relationship between helium isotope variations on the surface and the seismicity and resistivity structure of the crust and upper mantle along the Niigata-Kobe Tectonic Zone, Central Japan. In addition, the helium isotopes in hot spring gases may be recognized as a useful geochemical indicator of the presence of concealed active faults with compressional tectonics.

5. Conclusion

New helium isotope data from hot springs in and around the Niigata-Kobe Tectonic Zone, Central Japan, were obtained from 30 sampling points in order to elucidate the geographic distribution of $^{3}He/^{4}He$ ratios along the NKTZ. Additionally, we consider an interpretation of the helium isotope ratios and their relationship to seismotectonics along the NKTZ, especially with regards to seismic wave velocity of the crust and the upper mantle.

(1) In the southern NKTZ, the Philippine Sea slab subducting sub-horizontally may come into contact with the inland crust. As there does not appear to be a mantle wedge, which is a possible source of mantle helium, $^3He/^4He$ ratios lower than the atmospheric value observed in the southern NKTZ indicate that radiogenic helium dominates over any mantle helium input from the subcrustal lithosphere.

(2) The extremely low seismic wave velocity and high conductive anomalies in the crust and the upper mantle appears to occur beneath the central NKTZ, which can be attributed to Quaternary volcanism in this region. It is suggested that ^3He emanation manifested in the central NKTZ results from the effective transfer of mantle helium by intrusion and degassing of mantle-derived magma in the crust.

(3) The $^3He/^4He$ ratios of most hot springs in the northern NKTZ, a non-volcanic region, can be interpreted as a three-component mixture of mantle helium associated with magmatism of Middle Miocene age, radiogenic crustal helium and atmospheric helium. However, the $^3He/^4He$ ratios close to active faults causing recent large earthquakes in the northern NKTZ are higher than the calculated value for the erupted and intruded Miocene basalts. The geophysical anomalies beneath the seismic source area related to active faults suggest the existence of aqueous fluids which are released due to dehydration of the subducting Pacific slab. Therefore, active faults may play an important role in the effective transfer of mantle helium from the subcrustal lithosphere to the Earth's surface in non-volcanic regions.

Acknowledgements

We would like to thank Dr. K. Nagao, University of Tokyo, for helping with the helium isotope analyses. We also appreciate the constructive comments given by Dr. G. F. McCrank.

References

Aka, F. T.; Kusakabe, M.; Nagao, K.; Tanyileke, G. *Appl. Geochem.* 2001, 16, 323-338.

Craig, H.; Lupton, J. E.; Horibe, Y. *Terrestrial Rare Gases*; Alexander, Jr., E. C.; Ozima, M.; Ed.; Center for Academic Publications: Tokyo, 1978; pp. 3-16,

Güleç, N.; Hilton, D. R.; Mutlu, H. *Chem. Geol.* 2002, 187, 129-142.

Heki, K.; Miyazaki, S. *Geophys. Res. Lett.* 2001, 28, 2313-2316.

Hilton, D. R.; Fischer, T. P.; Marty, B. *Noble Gases in Cosmochemistry and Geochemistry*; Porcelli, D.; Ballentine, C. J.; Wieler, R.; Ed.; Reviews in Mineralogy and Geochemistry 47; Mineral. Soc. Am.: Washington D. C., 2002; pp. 319-370.

Iio, Y.; Sagiya, T.; Kobayashi, Y.; Shiozaki, I. *Earth Planet. Sci. Lett.* 2002, 203, 245-253.

Italiano, F.; Martelli, M.; Martinelli, G.; Nuccio, P. M. *J. Geophys. Res.* 2000, 105, 13569-13578.

Kennedy, B. M.; van Soest, M. C. *Geothermics* 2006, 35, 26-43.

Kita, I.; Nagao, K. *Chishitsu News* 1996, 506, 29-34.

Kulongoski, J. T.; Hilton, D. R.; Izbicki, J. A. *Geochimica et Cosmochimica Acta* 2005, 69, 3857-3872.

Kusakabe, M.; Ohwada, M.; Satake, H.; Nagao,K.; Kawasaki, I. In *Volcanic, Geothermal, and Ore-Forming Fluids: Rulers and Witnesses of Processes within the Earth*; Simmons,

S. F.; Graham, I.; Ed.; Society of Economic Geologists Special Publication 10; Economic Geol. Pub. Co.: Littleton, (CO), 2003; pp.75-89.

Marty, B.; Jambon, A.; Sano, Y. *Chem. Geol.* 1989, 76, 25-40.

Matsumoto, T.; Kawabata, T.; Matsuda, J.; Yamamoto, K.; Mimura, K. *Earth Planet. Sci. Lett.* 2003, 216, 221-230.

Morikawa, N. *Geochemical study on groundwater flow research: Groundwater survey by dissolved noble gases in groundwater from Tono area*; JNC Tech. Rep., TN7400 2001-003; Jpn. Nuclear Cycle Dev. Inst.: Tokai, 2001; pp.1-66.

Nagao, K. *Development of geochemical monitoring methods by suing Sr and noble gas isotopic compositions*; JNC Tech. Rep., TY7400 2004-001; Jpn. Nuclear Cycle Dev. Inst.: Tokai, 2004; pp.1-74.

Nagao, K.; Takaoka, N.; Matsubayashi, O. *Earth Planet. Sci. Lett.* 1981, 53, 175-188.

Nagao, K.; Takaoka, N.; Wakita, H.; Matsuo, S.; Fujii, N. *Geochem. J.* 1980, 14, 63-70.

Nakajima, J.; Hasegawa A. *Earth Planet. Space* 2007a, 59, e5-e8.

Nakajima, J.; Hasegawa, A. *J. Geophys. Res.* 2007b, 112, B08306, doi:10.1029/2006JB004770.

Nakajima, J.; Matsuzawa, T.; Hasegawa, A.; Zhao, D. *J. Geophys. Res.* 2001, 106, 21843-21857.

Nakamura, M.; Watanabe, H.; Konomi, T.; Kimura, S.; Miura, K. *Annu. Disaster Prevent. Res. Inst.* 1997, 40, 1-20.

NEDO *Regional exploration of geothermal fluid circulation system, Nasu Area, National Geothermal Resources Exploration Project: Project 3rd. phase*; New Energy and Industrial Technology Development Organization: Tokyo, 1990; pp.1-305.

NEDO *Report of Geothermal Development Promotion: Himekawa Area*; New Energy and Industrial Technology Development Organization: Tokyo, 1999; pp.1-885.

Ogawa, Y.; Honkura, Y. *Earth Plant. Space* 2004, 56, 1285-1291.

Okada, T.; Itaya, T.; Sato, M.; Nagao, K. *The Island Arc* 1994, 3, 221-231.

Okada, T.; Umino, N.; Matsuzawa, T.; Nakajima, J.; Uchida, N.; Nakayama, T.; Hirahara, S.; Sato, T.; Hori, S.; Kono, T.; Yabe, Y.; Ariyoshi, K.; Gamage, S.; Shimizu, J.; Suganomata, J.; Kita, S.; Yui, S.; Arao, M.; Hondo, S.; Mizukami, T.; Tsushima, H.; Yaginuma, T.; Hasegawa, A.; Asano, Y.; Zhang, H.; Thurber, C. *Earth Planet. Space* 2005, 57, 435-440.

Ono, A.; Sano, Y.; Wakita, H.; Giggenbach, W. F. *Geochem. J.* 1993, 27, 287-295.

Ozima M.; Podosek, F. A. Noble Gas Geochemistry (Second Edition); Cambridge University Press: Cambridge, 2002; pp.1-286.

Sagiya, T.; Miyazaki, S.; Tada, T. *Pure Appl. Geophys.* 2000, 157, 2303-2322.

Sakamoto, M.; Sano, Y.; Wakita, H. *Geochem. J.* 1992, 26, 189-195.

Sakata, S. *Geochimica et Cosmochimica Acta* 1991, 55, 1395-1405.

Sakata, S.; Takahashi, M.; Hoshino, K. *J. Jap. Assoc. Petrol. Technol.* 1986, 51, 228-237.

Sakata, S.; Takahashi, M.; Igari, S. I.; Suzuki, N. *Chem. Geol.* 1989, 74, 241-248.

Sano, Y.; Wakita, H. *J. Geophys. Res.* 1985, 90, 8729-8741.

Sano, Y.; Marty B. *Chem. Geol.* 1995, 119, 265-274.

Sano, Y.; Yakamura, Y.; Wakita, H. *Chem. Geol. Isot. Geosci. Sect.* 1985, 52, 1-8.

Sano, Y.; Takahata, N.; Seno, T. *Pure appl. geophys.* 2006, 163, doi:10.1007/S00024-006-0035-0.

Sano, Y.; Tominaga, T.; Nakamura, Y.; Wakita, H. *Geochem. J.* 1982, 16, 237-245.

Sano, Y.; Hirabayashi, J.; Oba, T.; Gamo, T. *Appl. Geochem.,* 1994, 9, 371-377.

Sano, Y.; Nakamura, Y.; Wakita, H.; Urabe, A.; Tominaga, T. *Science,* 1984, 224, 150-151.

Sano, Y.; Nishio, Y.; Sasaki, S.; Gamo T.; Nagao, K. *J. Geophys. Res.* 1998a, 103, 23863–23873.

Sano, Y.; Takahata, N.; Igarashi, G.; Koizumi, N.; Sturchio, N. *Chem. Geol.* 1998b, 150, 171-179.

Sano, Y.; Nakamura, Y.; Wakita, H.; Notsu, K.; Kobayashi, Y.; Sato, M.; Matsuo, S.; King C.-Y. *J. Geophys. Res.* 1986, 91, 12292-12295.

Sato, H. *J. Geophys. Res.* 1994, 99, 22261–22274.

Shimazaki, K.; Zhao, Y. *Earth Planet. Space* 2000, 52, 1091-1094.

Sibson R. H. *Earth Planet. Sci. Lett.* 2007, 257, 188-199.

Takahata N.; Yokochi, R.; Nishio, Y.; Sano, Y. *Geochem. J.* 2003, 37, 299-310.

Takaoka N.; Mizutani Y. *Earth Planet. Sci. Lett.* 1987, 85, 74-78.

Torgersen, T. *J. Geophys. Res.* 1993, 98, 16257-16269.

Torgersen, T.; Jenkins, W. J. *Geochimica et Cosmochimica Acta* 1982, 46, 739-748.

Umeda, K.; McCrank, G. F.; Ninomiya, A. *J. Geophys. Res.* 2007c, 112, B10206, doi:10.1029/2007JB005031.

Umeda, K.; Ogawa, Y.; Asamori, K.; Oikawa, T. *J. Volcanol. Geotherm. Res.* 2006a, 149, 47-61.

Umeda, K.; Sakagawa, Y.; Ninomiya, A.; Asamori, K. *Geophys Res. Lett.* 2007b, 34, L05310, doi:10.1029/2006GL028975.

Umeda, K.; Kanazawa, S.; Kakuta, C.; Asamori, K.; Oikawa, T. *Geochem. Geophys. Geosyst.* 2006b, 7, Q04009, doi:10.1029/2005GC001210.38.

Umeda, K.; Asamori, K.; Ninomiya, A.; Kanazawa, S.; Oikawa, T. *J. Geophys. Res.* 2007a, 112, B05207, doi:10.1029/2006JB004590.

Uyeshima, M.; Ogawa, Y.; Honkura, Y.; Koyama, S.; Ujihara, N.; Mogi, T.; Yamaya, Y.; Harada, M.; Yamaguchi, S.; Shiozaki, I.; Noguchi, T.; Kuwaba, Y.; Tanaka, Y.; Mochido, Y.; Manabe, N.; Nishihara, M.; Saka, M.; Serizawa, M. *Earth Planet. Space* 2005, 57, 441-446.

Wang, Z.; Zhao, D. *Earth Planet. Sci. Lett.* 2006, 244, 16-31.

Wakita H., and Sano, Y. *Nature* 1983, 305, 792-794.

Wakita, H.; Sano, Y.; Mizoue, M. *J. Geophys. Res.* 1987, 92, 12539-12546.

Wakita, H.; Sano, Y.; Urabe, A.; Nakamura, Y. *Appl. Geochem.* 1990, 5, 263-278.

Xu, H. L.; Oki, Y. *Island Arc* 2004, 13, 333-345.

Yamasaki, T.; Seno T. *Earth Planet. Sci. Lett.* 2005, 232, 13-27.

Zhao, D.; Hasegawa A. *J. Geophys. Res.* 1993, 98, 4333-4353.

In: The Lithosphere: Geochemistry, Geology and Geophysics ISBN: 978-1-60456-903-2
Editors: J.E. Anderson et al, pp. 171-212

Chapter 5

VOLATILES IN THE MANTLE LITHOSPHERE: MODES OF OCCURRENCE AND CHEMICAL COMPOSITIONS

Mingjie Zhang [1*], ***Yaoling Niu*** [2] ***and Peiqing Hu*** [1]

[1] School of Earth and Environment Sciences, Lanzhou University, Lanzhou 730000, China

[2] Department of Earth Sciences, Durham University, Durham DH1 3LE UK

Abstract

Volatiles play important roles in chemical differentiation of the Earth, in concentrating economic metals, and in regulating earth's surface environments by means of magmatism, metasomatism, degassing and recycling. Mantle rocks and rocks derived from the mantle such as basalts, mantle xenoliths and ophiolitic peridotites are materials available to investigate the ways in which volatiles may store in the mantle, their compositions, and probable histories. The laser Raman spectroscopy, Infrared spectrometry and ion microprobe in combination with micro-thermometry are non-destructive methods to analyze volatile compositions trapped in fluid inclusions. On the other hand, vacuum crushing and stepwise heating are methods employed to extract the volatiles and measure their chemical and isotopic compositions using mass spectrometry. An improved vacuum stepwise heating technique can effectively separate volatiles in different occurrence modes in mantle materials, which in combination with mass spectrometry can yield excellent and highly reproducible analytical data.

Volatiles in the mantle occur in various forms such as free element or molecular species along grain boundaries, carbonate, sulfide or hydrous minerals, fluid inclusions or charged species dispersed in mineral structures (e.g., OH^-), structural defects or vacancies. Volatiles trapped in structural defects and vacancies are volumetrically significant. Large amount of hydrogen occurs as free H_2 species, not OH^- as previously thought.

Volatiles in the mantle are mixtures of primordial volatiles and recycled volatiles with characteristic chemical compositions. Volatiles in the sub-continental lithospheric mantle (SCLM) vary with depth and mantle reservoirs. Deep portions of mantle lithosphere in the diamond stability field have higher contents of reduced volatile species such as H_2 and CO

[*] E-mail address: mjzhang@lzu.edu.cn. Phone: +86-931-891-2403; Fax: +86-931-891-2449, Corresponding author: Dr. Mingjie Zhang, Institute of Geological Science, Lanzhou University, P. R. China.

etc., whereas at shallow levels, the mantle lithosphere as reflected in mantle xenoliths displays varying volatile compositions; initial volatiles trapped during primary crystallization stage are dominated by reduced species like CO, H_2. In contrast, metasomatic volatiles are more oxidized such as CO_2 and SO_2 etc.

Volatiles in mantle source regions of oceanic basalts are all dominated by H_2O and CO_2 with minor CO, CH_4, N_2, and H_2; the abundances vary with tectonic settings. MORB are depleted in volatiles as a result of source depletion in its history, whereas abundant volatiles in IAB and BABB are probably originated from subduction devolatilization. OIB may have abundant volatiles inherited form the undegassed mantle with a recycled component. Volatiles in ancient oceanic lithosphere as recorded in ophiolites are all dominated by CO_2 with minor amounts of other volatile species.

Key words: volatiles, occurrence mode, chemical composition, mantle lithosphere.

1. Introduction

1.1. Implications of Mantle Volatiles

The Earth's mantle is an important reservoir for volatiles. Water of various forms, inorganic species (i.e. CO_2 and SO_2 etc.), hydrocarbon (i.e. CH_4 etc.) and noble gases have all been found in mantle xenoliths (Trull et al., 1993; Zhang et al., 1999a,b, 2007), diamond (Ivankin, 1988; Verchovsky et al., 1998; Ratan and Mohapatra, 2006; Frezzotti and Peccerillo,2007), basalts (Byers et al., 1983-1986; Javoy and Pineau, 1991), kimberlite (Canil and Bellis, 2008) and volcanic fumaroles (Yuri, 2000; Zimbelman et al., 2000), which provide convincing evidence for the presence of abundant volatiles in the Earth's mantle (Cartigny et al., 2001; Aubaud et al., 2005; Workman et al., 2006).

Degassing of the Earth's mantle has led to the formation of Earth's atmosphere and hydrosphere (Allegre et al., 1987), and continues to modify our global environment (Self et al., 2006; Moune et al., 2007). Volatiles play important roles in mantle differentiation and evolution by means of mantle metasomatism, magma generation and crust-mantle interactions (Porcelli and Wasserburg, 1995; O'Reilly and Griffin, 2000; Wyllie and Ryabchikov, 2000; Niu and O'Hara, 2003). Volatiles in the Earth's mantle are in the supercritical state under ambient conditions (Wyllie, 1987), and can transport quantities of economically valuable metals from the Earth's deep interiors (Olmez and Finnegan, 1986; Wyllie, 1987; Rubin, 1997; Yuri, 2000). Mantle volatiles continue presently to escape through Earth's crust (Weinlich, 1999; Hilton, 2007, Zhang et al., 2008). Thus, studies of mantle volatiles provide a means for understanding Earth's early history and its continued evolution.

Volatiles in the mantle are "mixtures" of volatiles with different origins. Some are primordial incorporated during Earth formation (Burnard et al., 1997; Hilton, 2007), some are likely recycled from the atmosphere, hydrosphere and crustal materials through lithosphere subduction (Trull et al., 1993; Fyfe, 1997; Zhang et al., 2004, 2007; Ratan and Mohapatra, 2006), and some may be added by nucleogenic or radiogenic ingrowths (Albarede, 2008). To characterize volatiles degassed from the mantle requires a good knowledge of the initial volatile abundances and isotopic compositions in the convective mantle. However, such mantle materials are rare, and the SCLM offers the best sample for such investigations. The types, abundances, compositions and characteristics of volatile species in mantle materials are likely to have diverse identities because of their different origins and complex histories, and can thus be used to decipher their ultimate origins and passageways in the context of crust-mantle recycling and mantle circulation.

1.2. Occurrence Modes of Mantle Volatiles

It has been recognized that volatile species of C-O-H-S composition must exist in the mantle in some physical state (free volatile phase, dissolved volatiles in melts or interstitial solid solutions, complex anions incorporated in minerals). Some volatiles are present as free species dispersed along grain boundaries of mantle minerals, but most mantle volatiles are stored in minerals. Sulfide, carbonates and hydrous phases, such as phlogopite and amphiboles, can hold abundant volatiles, but these phases are volumetrically insignificant and have limited stability fields. Hence, much of the mantle volatiles reside in nominally anhydrous minerals such as olivine, pyroxenes, garnet and other phases at great depths (Bell and Rossman, 1992; Beran and Libowitzky, 2006; Skogby, 2006). Volatiles in nominally anhydrous mantle minerals have long been thought to reside primarily in various types of fluid inclusions. However, the exact ways in which volatiles reside in the mantle minerals remain poorly understood, largely due to difficulties in obtaining representative mantle samples and inaccessibility for direct observations.

Mantle volatiles are conventionally considered to reside as free phases in different types of fluid inclusions in mantle minerals (Roedder, 1965; Anderson et al., 1984; Xia, 1984; Deines, 1992; Andersen and Neumann, 2001), to occur as anion complexes in hydrous minerals (e.g. OH^- in amphibole and mica, and Cl^- and F^- in apatite etc., Chazot et al., 1997; O'Reilly and Griffin, 2000; Aubaud et al., 2007) or to be trapped in structural defects and vacancies of crystals (Freund and Oberheuser, 1986; Bai and Kohlstedt, 1992). Their residence in mantle minerals is thought to reflect different origins by mantle events (Hansteen et al., 1991; Matsumoto et al., 2000; Landis and Rye, 2005). Hence, to identify the occurrence modes of volatiles in mantle minerals is key to understanding their origins.

1.3. Compositions of Mantle Volatiles

Volatiles are most active agents in Earth processes and have chemical and isotopic compositions that are characteristic of different major reservoirs such as the atmosphere, hydrosphere, crust and different mantle source components (e.g. DM, EMI-II and HIMU). Therefore, volatiles have been used as tracers for their origins (Norman and Musgrave, 1994; Eiler et al., 1997; Deines, 2002; Hansteen and Troll, 2003; Shaw et al., 2004; Zhang et al, 2004). Abundant literature exists on stable isotopes and noble gas isotopes of mantle-derived volatiles.

The $\delta^{13}C$ of ~ -5‰ has been identified as a characteristic value of mantle materials, e.g., kimberlite, mantle xenoliths and fresh glasses of mid-ocean ridge basalts (MORB) and ocean island basalts (OIB) (Exley et al., 1986). The $\delta^{13}C$ of diamonds (-27.9‰ to -3.2‰, Deines et al., 1997) and mantle xenoliths (-30‰ to 0 ‰) (Nadeau et al., 1990; Pineau and Methez, 1990; Trull et al., 1993; Deines, 2002) shows bimodal peaks at -5‰ and -25‰, respectively; whose origin remains poorly understood (Deines, 2002). The $\delta^{18}O$ values of "Normal" mantle is widely accepted as ~ +5.5‰ (Kyser, 1986; Mattey et al., 1994) and δD values as -40‰ to -

95‰ (a mean of -80‰, Kyser and O'Neil, 1984; Kyser, 1986), e.g. $\delta^{18}O$ values of oceanic gabbros (+3.3‰ to +5.1‰, Hansteen and Troll, 2003), fresh MORB or OIB glasses (+5.7‰ to +6.0‰, Eiler et al., 1997; Widom and Farquhar, 2003) and olivines in mantle xenoliths and of MORB phenocrysts (+5.2‰ on average, Mattey et al., 1994; Eiler et al., 1997).

Noble gases are inert and do not recycle into the mantle by subduction with the possible exception of Xe (Porcelli and Wasserburg, 1995). Because of the rarity, chemical inertness and significant isotopic fractionation noble gases are important tracers for volatile origins in the context of Earth evolution and specific Earth processes such as magmatism, metamorphism, fluid migration and melt segregation (Allegre et al., 1987; Norman and Musgrave, 1994; Burnard et al., 1998; Farley and Neroda, 1998; Ozima and Podosek, 2002). The shallow upper mantle contains a higher proportion of radiogenic noble gases sampled through mid-ocean ridge basalts (MORB) (e.g. ^{4}He, ^{21}Ne and ^{40}Ar; Allegre et al. 1987; Farley and Neroda, 1998). The depleted MORB mantle (DMM) shows a small range of $^{3}He/^{4}He$ (7-9Ra; O'Nions and Tolstikhin, 1994; Trieloff et al., 2000). Deep mantle plume sources as analyzed in OIB contain relatively high abundances of primordial noble gas isotopes, e.g. ^{3}He, ^{22}Ne, ^{36}Ar and ^{38}Ar (Trieloff et al., 2000).

The stable and noble gas isotopic compositions of volatiles are complex in mantle minerals and show differences between the sub-continental lithospheric mantle (SCLM) as revealed by mantle xenoliths and oceanic (asthenospheric) upper mantle as revealed by oceanic basalts (Inonov et al., 1994; Eiler et al., 1997; Valley and Cole, 2001; Deines, 2002; Ozima and Podosek, 2002; Hansteen and Troll, 2003; Shaw et al., 2004). However, whether there are systematic differences in terms of chemical compositions of the volatiles between the SCLM and upper mantle source regions of oceanic basalts remain poorly understood. Furthermore, theoretical modeling of mantle volatiles are hampered by lacking thermo-dynamic data at high temperature and pressure.

In this chapter we review our present knowledge about mantle volatiles and their redox state. We also review available mantle volatile data and analytical methods we use to acquire these data. We use a combined techniques of microthermometry, improved vacuum stepwise heating, grain-size reduction, mass spectrometry and infrared spectrum measurements (1) to assess analytical reproducibility of volatile compositions and the superimposed effects of volatiles in different occurrence modes, (2) to characterize occurrence modes of volatiles, and (3) to discuss in details the decrepitation mechanism of inclusions by stepwise heating, and (4) to elucidate chemical compositions of volatiles that vary between different tectonic settings and mantle reservoirs.

2. Samples and Analytical Methods

2.1. Representative Samples

Mantle volatiles can be rarely sampled directly. Hot-springs, summit fumaroles, diamonds, mantle derived melts (e.g., basalt, kimberlite and lamprophyre), mantle xenoliths, ultramafic intrusions and ophiolitic peridotites are all used to characterize mantle volatiles. Hot-springs, summit fumaroles, even natural gases can contain a large quantity of mantle volatiles, although they are always contaminated by crustal and atmospheric components (Kelley, 1996; Charlou et al, 1998; Welhan, 1988). Oceanic basalt glasses, including MORB, OIB, back-arc basin basalt (BABB), and island arc basalt (IAB) can provide volatile compositions of oceanic upper mantle source regions (Welhan, 1988; Kelley and Frueh, 1995; Burnard et al., 1997). Peridotite in the lowest part of ophiolite sequence may be unique samples for the volatiles of ancient oceanic lithosphere (Abrajano et al., 1988; Su et al., 1999; Hu et al., 2007).

Diamonds that act as an impermeable capsule for a variety of volatiles and fluid inclusions offer a unique insight into the physicochemical evolution of volatiles in deep portions of the SCLM, e.g. Africa, USA and Brazil (Deines, 1980; Eggler and Baker, 1982; Ivankin et al., 1988; Navon, 1991; Verchovsky et al., 1998; Frezzotti and Peccerillo, 2007).

Mantle xenoliths brought to the surface by alkali basalts are interpreted as direct samples of the SCLM to characterize volatiles in the mantle (Mathez, 1987; Litasov et al., 2000). Mantle xenoliths can preserve volatiles intact in primary fluid inclusions or crystal structural defects and vacancies in the interiors of constituent mineral crystals, thus prevent from subsequent contamination by crustal and atmospheric components. This is however not the case for volatiles from volcanoes, hot-springs, summit fumaroles, ultramafic intrusions, basalts or xenoliths hosted in kimberlite or lamproite (Andersen and Neumann, 2001; Deines, 2002). Therefore, mantle xenoliths hosted in alkali basalts are probably the best mantle material for reliable estimation of average compositions of mantle volatiles. Abundant spinel lherzolite, garnet lherzolite, harzburgite, and pyroxenite xenoliths and megacryst from eastern China (Figure 1 of Zhang et al., 2007), central Asia (Inono et al., 1992), Australia (Anderson et al., 1984), Siberia (Ionov et al., 1994), central Europe (Weinlich, 1999) and North America (Nadeau et al., 1990; Tingle et al., 1990) have been collected to analyze the chemical and isotopic compositions of volatiles by various methods.

2.2. Analytical Methods

Ideal sample sets and effective pretreatments are important for obtaining reproducible volatile concentration and isotope data. Less altered olivine (Olv), orthopyroxene (Opx) and clinopyroxene (Cpx) separates or megacrysts were hand-picked under an binocular. They were then ultrasonically cleaned with 0.3mol/L HCl and analytical grade CH_2Cl_2 and acetone, respectively, and rinsed with distilled water until pH value reaches around 7 at first HCl-cleaning step, finally dried up at 110°C prior to analysis. Such sample pretreatment procedure effectively removes carbonate and sulfide hosted in the sample as well as possible surface contaminants and volatiles absorbed on mineral surfaces and cracks (Miller and Pillinger, 1997).

Thin sections have been made for petrography and fluid inclusion observations and micro-thermometry studies. The petrographic microscope equipped with a heating-cooling stage or a high temperature stage is routinely employed to examine the type, size, distribution, and thermal data (e.g., melting and homogenization temperatures) of fluid inclusions (Roedder, 1990; Andersen and Neumann, 2001).

Laser Raman microprobe, or ion microprobe is non-destructive and easy to combine with microthermometry to directly observe and analyze the composition of volatiles trapped in fluid inclusions (Anderson et al., 1984; Bergman and Dubessy, 1984; Pasteris and Wanamaker, 1988; Roedder, 1990; Hansteen et al., 1991; Williams, 1996; Chazot et al., 1997; Workman et al., 2006; Frezzotti and Peccerillo, 2007), but has a relatively high detection limit on the order of 0.x mol% for most of potentially important volatile species of interest, such as water and noble gases (Andersen and Neumannr, 2001), and assume implicitly that mantle volatiles have the same equilibrium phases as those at room temperature. On the other hand, superimposed volatile fractionation during inclusion entrapment leads to volatile heterogeneity among inclusions (Burnard, 1999).

Infrared spectroscopy or Raman spectroscopy can be used directly to analyze volatiles incorporated as anion complexes in minerals (Dixon et al., 1988; Roedder, 1990). For example, the modes of occurrence of H and OH^- in nominally anhydrous mantle minerals have been widely studied using FTIR and ion microprobe (Skogby and Rossman, 1989; Smyth et al., 1991; Bell et al. 1995; Kohn, 1996; Zhang et al., 2005; Aubaud et al., 2007). However, it is not straightforward to determine the abundances and compositions of volatiles trapped in structural defects and vacancies of minerals.

Crushing, stepwise heating or laser extraction techniques can extract bulk volatiles trapped in fluid inclusions or structural defects and vacancies for chemical and isotopic analysis with both advantages and limitations (Anderson et al., 1984; Burnard et al., 1997; Marty and Zimmermann, 1999; Matsumoto et al, 2000; Cartigny et al., 2001; Aubaud et al., 2005). In most instances, crushing can extract volatiles in fluid inclusions of large size, thus yielding only a small fraction of volatiles (ca. 5-15% of heating, Landis and Rye, 2005). Stepwise heating or laser extraction could cause reactions among volatile species before or after released, meanwhile reactions between molecular carbonic species in the extracted volatiles and fragments of the host mineral is possible at high temperature (Andersen and Neumannr, 2001). The *pros* and *cons* of volatile extraction by stepwise heating in carbon isotopic measurements have been assessed by Miller and Pillinger (1997).

Vacuum stepwise heating is widely used to extract and analyze volatiles (e.g., species, concentrations and isotopic compositions etc.) in mantle derived melts and mantle xenoliths, which can distinguish volatiles stored in different occurrence modes by setting temperature increments according to incipient melting or homogenization temperatures for melt or fluid inclusions, and can extract volatiles trapped in structural defects and vacancies of crystals at high temperatures, even up to the melting point (Marty and Zimmermann, 1999; Matsumoto et al., 2000).

Volatiles released from mantle minerals at different temperature intervals during stepwise heating show distinct chemical, stable and noble gas isotopic compositions and have been interpreted as reflecting different modes of occurrence and different origins (Nadeau et al., 1990; Pineau and Methez, 1990; Javoy and Pineau, 1991; Trull et al., 1993; Miller and Pillinger, 1997; Deines, 2002; Shaw et al., 2004; Zhang et al., 2004, 2007). For example, volatiles released at high heating temperatures (>800°C) dominated by CO and H_2 with $\delta^{13}C$ value of ~ -5‰ vs. PDB and MORB-like noble gas isotopes are consistent with being fluid inclusions trapped at an early stage of crystal formation (Nadeau et al., 1990; Pineau and Methez, 1990; Trull et al., 1993; Marty and Zimmermann, 1999; Matsumoto et al, 2000; Zhang et al., 2004, 2007) or incorporated in crystal structures (Miller and Pillinger, 1997; Deines, 2002). On the other hand, volatiles released at low heating temperatures (<800°C) dominated by CO_2 with $\delta^{13}C$ values of ~ -25‰ are subject to a wide range of interpretations. These include surface contamination by ubiquitous airborne particulates of biological origin (Miller and Pillinger, 1997), graphitic and organic materials absorbed on sample cracks or surfaces (Pineau and Methez, 1990; Nadeau et al., 1990), mantle heterogeneities (Deines, 2002), and different types of fluid inclusions (Zhang et al., 2004, 2007).

Volatile extraction by vacuum stepwise heating has been widely used for stable and noble gas isotopic measurements, but shows difficulty in precise analysis of volatile concentrations, partly due to (1) reactions among volatile species released at high temperature (Nadeau et al., 1990; Roedder, 1990; Graney and Kesler, 1995; Zhang et al., 2000, 2005), and (2) superimposed effects of volatiles from different occurrence modes; all these often result in poor reproducibility of measured volatile concentrations. Zhang et al. (2000, 2004, 2007) have improved the procedure of stepwise heating to minimize these affects.

2.3. Improved Online Vacuum Stepwise Heating Method

2.3.1. Analytical Procedures

A well-established and improved online vacuum stepwise heating system connected to a MAT-271 mass spectrometer was used to extract volatiles from mineral separates (Zhang et al, 2000, 2007). The complete apparatus of online volatile extracting (Figure 1) are described in Zhang et al (2004, 2007). The efforts were made to use path length and diameter of the inlet ports as small as possible. The inlet ports were heated at a constant temperature of 120°C to minimize adsorption of volatiles onto surfaces. A cold trap held at the liquid nitrogen temperature was adopted during sample heating to condense and isolate H_2O, CO_2 and heavy hydrocarbon species released from the sample from the high temperature area, which can minimize or prevent contributions from reactions among released volatiles (Zhang et al., 2005a). Reactions among these volatiles in the interior of samples could be considered as the reconversion of equilibrated volatiles at mantle temperatures (Zhang et al., 2000).

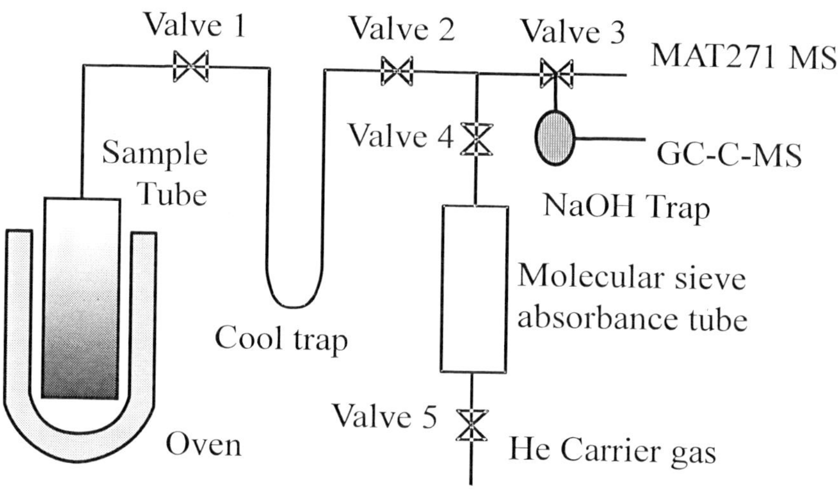

Figure 1. Apparatus of online stepwise heating system.

The improved analytical procedures of volatile concentration were previously described (Zhang et al., 2000, 2004, 2007) and are briefly summarized here. The sample was loaded into a quartz sample tube and combusted in high-purity O_2 gas for 4 hours in a furnace at 200°C to remove possible surface organic contaminant (Shaw et al., 2004), then preheated at 200°C and outgassed under ultrahigh vacuum for 4 hours prior to analysis. Volatiles in a given sample crushing size of mineral separates were extracted by stepwise heating from 200 to 1200°C with 100°C increment and 1 hour duration per step. A cold trap held at the liquid nitrogen temperature was adopted to prevent volatile reactions (see above). After heating for 1 hr at each step, volatiles have been released completely from samples. The condensed volatile species (except for H_2O) in cold trap are released by replacing cold trap into alcohol-liquid nitrogen mixture (ca. -60°C). All the volatiles released are then introduced to a MAT271 mass spectrometer for concentration measurement. The analytical errors are typically < 1 mol% for main volatile species such as CO_2, CO and H_2.

The cold trap at the liquid nitrogen temperature can rapidly trap the released volatiles and minimize the possibility for new species (e.g., H_2 and CO etc.) formation (Hu et al., 2007). The one hr duration of heating is adequate to release volatiles completely as demonstrated by negligible volatile yields for extra heating. The analysis suggests that the released volatiles are stored in the form of fluid inclusions as well as structural sites, defects and vacancies of the constituent minerals.

2.3.2. Analytical Reproducibility of Volatile Concentrations

Measurement of volatile concentrations in mantle minerals has always shown poor reproducibility due to heterogeneous distributions and asynchronous release of volatiles (Matson and Muenow, 1984). Magmatic pyroxenite xenoliths without granoblastic microstructures are often considered to be of magmatic origin under mantle conditions; they have least number of fluid-trapped stages with relatively homogeneous distribution of volatiles because of the simple history (Hansteen et al., 1991; Litasov et al., 2000). Hence, Cpx separates in the pyroxenite xenoliths are ideal samples for carrying out analytical reproducibility tests. On the other hand, the volatiles resided in different sites in host crystal may be released at different heating temperature intervals depending on experimental procedures (Nadeau et al., 1990; Pineau and Methez, 1990; Trull et al., 1993; Zhang et al., 2004, 2007). Therefore, one step heating can effectively eliminate superimposed effects due to asynchronous release of volatiles in different occurrence modes.

Volatiles in Cpx separates from pyroxenite xenoliths in Hanluoba and Dafangshan in eastern China were extracted by one step heating from 200 to 1200°C for 1 hour to test analytical reproducibility. Cpx separates from pyroxenite xenoliths show much higher volatile contents than Cpx from lherzolite minerals, which is consistent with magmatic petrogenesis of pyroxenite xenoliths, because volatiles are enriched as incompatible components in the late crystallized minerals, or are added as the allothigenetic volatiles into the parental melt of pyroxenite. The reproducibility for major volatile species CO_2, CO, H_2 and N_2 is excellent with relative standard deviation < 13.68%. The relative standard deviation is 2.82-12.35% for CO_2, 0.94-2.68% for CO, 4.4-13.68% for H_2, 2.54-13.63% for H_2S, 3.26-12.91% for N_2 (Zhang et al., 2009).

The excellent reproducibility of volatile abundances and compositions in our experiments indicates the effective analytical procedures we applied, including (1) sample pre-treatment to remove carbonate, sulfide and possible volatiles absorbed on mineral surfaces and cracks, (2) effective analytical procedures that rapidly trap released volatiles and prevent formation of new species, and (3) 1 hour duration of heating can release volatiles completely.

3. Occurrence Modes of Volatiles

Volatiles in the mantle occur in various forms, such as a free state of elements (e.g., graphite, diamond etc.; Melton et al., 1972; Melton and Giardini, 1974; Tingle et al., 1990; Deines, 1992) and molecules (e.g., CO_2, H_2O etc.), as structural components in carbonate, sulfide or hydrous minerals (Matson and Muenow, 1984), as fluid inclusions (Roedder, 1965; Andersen and Neumann, 2001) and as some unidentified "species" dispersed in structural defects and vacancies in mantle mineral crystals (Green and Radcliffe, 1975). These occurrence modes have been recognized by petrographic observations and analytical results using Laser Raman microprobe, Infrared spectroscopy and stepwise heating methods.

Mantle volatiles as free species dispersed along grain boundaries in the mantle can percolate through and escape into the crust or leak to the Earth's surface through hot-springs, summit fumaroles, volcano and faults, which are difficult to quantify due to surface contaminations. Graphite, diamond, carbonate, sulfide and hydrous minerals are volumetrically insignificant in the mantle. Therefore, fluid inclusions and structure defects and vacancies in mantle minerals are primary host of mantle volatiles.

Stepwise heating reveals that volatiles are released progressively from mineral separates and tend to concentrate at three temperature intervals of 300-600°C, 600-1000°C and 1000-1200°C in constituent minerals of mantle xenoliths in eastern China (Figure 2a, Zhang et al., 2004, 2007, 2009) and of 200-400°C, 400-800°C and 800-1200°C in harzburgite and dunite of the Yushigou ophiolite in western China (Figure 2b, Hu et al., 2007). The progressive release of volatiles with increasing temperature, the increase of volatile contents with decreasing sample crushing size (Figure 3, Zhang et al., 2009), and the chemical compositional differences of three volatile release peaks (Table 1) suggest that volatiles reside in more than one form in the constituent minerals of mantle xenoliths, e.g., different types of fluid or melt inclusions (Xia, 1984; Deines, 2002; Zhang et al., 2004), and crystal structural defects and vacancies (Freund and Oberheuser, 1986; Bai and Kohlstedt, 1992; Zhang et al., 2004, 2007). Volatiles in different occurrence mode may have complex origins and histories (Burnard et al., 1998).

3.1. Fluid Inclusions

3.1.1. Types of Fluid Inclusions

Fluid inclusions are traditionally thought to be the primary hosts of volatiles present in nominally anhydrous mantle minerals, which were trapped during mantle mineral crystallization (Andersen and Neumann, 2001) or exsolved from the decompressed solid (Green and Radcliffe, 1975). The most common fluid inclusion reported from mantle minerals is pure CO_2 (Murck et al., 1978; Bergman and Dubessy, 1984; Frezzotti and Peccerillo, 2007). The volatile phases in fluid inclusions are often associated with melt (Roedder, 1965; Murck et al., 1978; Andersen and Neumann, 2001).

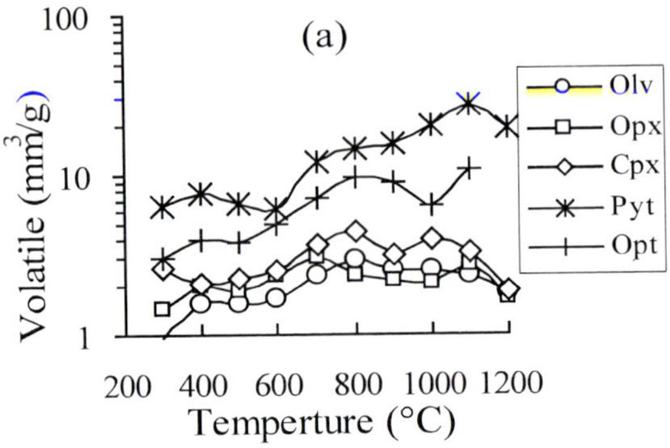

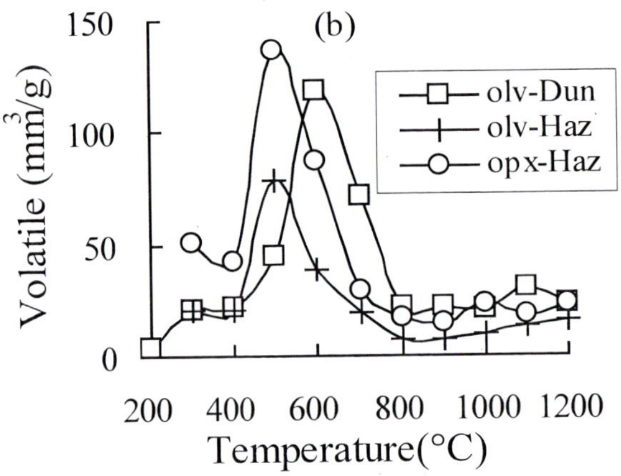

Figure 2. Average abundances of volatiles (mm^3.STP/g) released from mantle xenolith minerals (a) and the Yishugou ophiolitic peridotite (b) in China plotted as a function of heating temperature. Olv, Opx and Cpx are olivine, orthopyroxene and clinopyroxene in lherzolite xenoliths, respectively; Pyt is clinopyroxene in pyroxenite xenoliths, Opt is orthopyroxene megacryst, olv-Dun and olv-Haz are olivine in dunite and harzburgite respectively and opx-Haz is orthopyroxene in harzburgite.

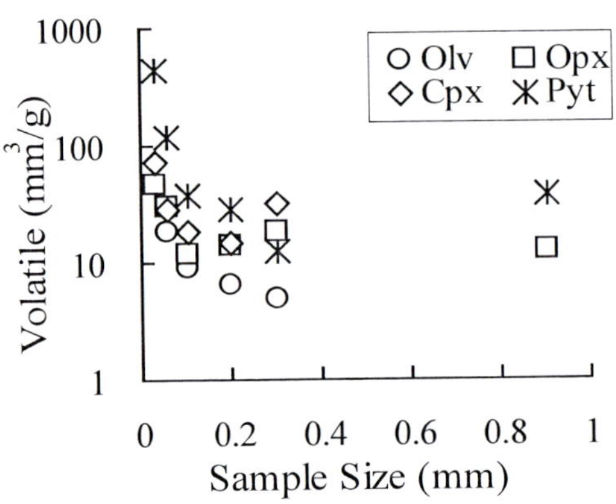

Figure 3. Total volatile contents (mm^3.STP/g) released at all steps plotted as a function of crushing size (mm) of olivine (Olv), orthopyroxene (Opx) and clinopyroxene (Cpx) in lherzolite and pyroxenite (Pyt) xenoliths.

Table 1. Chemical compositions (mm^3.STP/g) of volatiles released at different temperature intervals from constituent minerals of mantle xenoliths in Damaping, eastern China

Temperature	Mineral	Rock	H_2	CH_4	CO	N_2	CO_2	SO_2
300-600°C	Olv	lherzolite	0.90	0.33	1.70	0.08	1.81	0.00
	Opx	lherzolite	0.34	0.42	2.17	0.05	4.56	0.01
	Cpx	lherzolite	1.83	0.48	1.95	1.43	2.60	0.01
	Pyt	pyroxenite	1.29	0.57	6.79	0.23	12.15	0.05
600-1000°C	Olv	lherzolite	1.95	0.01	6.29	0.05	0.42	0.00
	Opx	lherzolite	0.82	0.09	5.00	0.16	3.44	0.12
	Cpx	lherzolite	4.11	0.96	10.21	1.24	2.72	0.79
	Pyt	pyroxenite	12.28	0.15	37.28	5.32	2.48	0.01
1000-1200°C	Olv	lherzolite	0.47	0.00	2.65	0.04	0.35	0.00
	Opx	lherzolite	0.42	0.00	1.99	0.15	1.91	0.11
	Cpx	lherzolite	0.64	0.20	4.81	0.20	0.95	0.17
	Pyt	pyroxenite	2.42	0.00	28.75	2.26	0.68	0.01

The early-stage and late-stage two type fluid/melt inclusions can be readily recognized in mantle xenolith minerals based on the shape, size, distribution pattern and homogenization temperature (Roedder, 1965; Xia, 1984; Zhang et al., 2007).

(1) Early-stage fluid inclusions: Primary fluid inclusions trapped during mantle mineral crystallization show high internal pressure (0.8~ 1.4GPa) (Andersen et al., 1984; Xia, 1984; Navon, 1991), which can result in fracturing of host crystal and leakage of volatiles by rapid decompression or even thermal shock prior to ascent and cooling (Roedder, 1965; Andersen et al., 1984; Tait, 1992; Rosenbaum et al., 1996; Ertan and Leemanr, 1999; Andersen and Neumann, 2001). Some decrepitated and leaked inclusions are recognized by·the star-shaped and ragged appearances, and often show signs of partial decrepitation in the form of planar arrays of tiny ellipsoidal inclusions radiating outward from a central spherical bubble (Roedder, 1984; Rosenbaum et al., 1996). Only crystal interiors preserve primary fluid inclusions intact, i.e., early-stage fluid inclusions, which usually are small in size (mostly 5 to 20 μ in diameter) with regular shapes (e.g., oval, egg or negative crystal shape), and are located in the crystal interiors, or distributed along crystal lattices of the host crystals, and have high homogenization temperatures (Xia, 1984; Andersen and Neumann, 2001).

(2) Late-stage fluid inclusions: Primary fluid inclusions distributed along edges of host crystals have been deformed, altered, leaked, or refilled during subsequent mantle metasomatism and other tectono-thermal events (Andersen et al., 1984; Ertan and Leemanr, 1999). The healed irregular or circle micro-fractures at edges of re-crystallized minerals trapped volatiles in subsequent mantle tectonic-thermal events as late-stage fluid inclusions (Roedder, 1965; Xia, 1984; Hansteen et al., 1991; Andersen and Neumann, 2001; Zhang et al., 2007), which are large in size (from 10 to 200μ in diameter or length) with irregular shapes (e.g., tree-branch, bead-string or tabular), and are located in healed micro-fractures or cracks at edges or oriented parallel to the cleavage traces of host minerals with low homogenization temperatures (Xia, 1984).

Volatiles in inclusions are compressible and condensed to the liquid state following an isochore, and form "shrinkage" bubble when mantle xenoliths ascend to surface and cool below entrapment temperature to room temperature (Roedder, 1965), which leads to reduced internal pressures within inclusions. Furthermore, silicate glasses or carbonaceous films deposited and encapsulated at inclusion lining are relatively resistant to volatile exchanges between inclusions and the host crystals (Murck et al., 1978; Xia, 1984; Mathez, 1987; Rosenbaum et al., 1996; Burnard et al., 1997). Depressurized inclusions with silicate glass or carbonaceous lining readily keep their volatiles intact during surface events after ascending and cooling; the volatiles stored can provide information on their origins.

3.1.2. Decrepitation of Fluid Inclusion

The decrepitation of fluid inclusion depends on physical strength of the host crystal as well as pressure difference between the internal fluid pressure in the inclusion and external pressure of the host crystal. The physical strength of the host crystal around the fluid inclusion depends on crystal type, density of cleavages/cracks, and wall thickness of inclusion (e.g., sample crushing size, distance to crystal edges) (Zhang et al., 2007). Mantle minerals such as Olv, Opx and Cpx are brashy (vs. quartz), which leads to reduced physical strength and readily fractured host crystals, thus reducing decrepitation temperatures of fluid inclusions. Small crushing size, presence of micro-cracks and thin walls also facilitate inclusion decrepitation.

During vacuum heating, liquid phases in inclusion are gasified, and the volatiles dissolved in glass/melt and daughter crystal in inclusion are separated into gas phase while glass and daughter crystal begin to melt progressively, which rapidly raises internal pressure in the inclusion as progressively heating due to the high thermal expansion coefficient of volatiles (Lowenstern, 1994), and internal pressure of inclusion will reach the entrapped pressure at homogenization temperature. On the other hand, mantle minerals release very low content of volatiles into the high vacuum line at each heating step, moreover most of the volatiles with low freezing-point are condensed into cold trap (Zhang et al., 2004, 2007), i.e., the pressure of host crystal exterior is very low (~10-200Torr). Therefore, a large pressure gradient exists between the inclusion interior (usually over-pressured) and the crystal exterior (nearly low vacuum) when sample has been heated upto homogenization temperature. The fluid inclusion will decrepitate if the host crystal is not strong enough to withstand large pressure differences at about the range of homogenization temperatures (Tait, 1992).

In principle, inclusions trapped at same stage of tectono-thermal event should simultaneously decrepitate for a given crushing size of sample and release volatiles during stepwise heating due to similar internal pressures, homogenization temperatures, residence sites, and ratios of vapor/liquid or glass (Hansteen et al., 1991; Andersen and Neumann, 2001). For a given crushing size of sample, small size of inclusions, such as early-stage fluid inclusions, are associated with thick walls of fluid inclusions (i.e. greater strength against decrepitation) and will decrepitate at higher temperatures than large size of inclusions, such as late-stage fluid inclusions.

(1) Early-stage fluid inclusions: The early-stage fluid inclusions trapped at ambient mantle P-T conditions of mineral crystallization are expected to decrepitate at a temperature range close to or slightly lower than homogenization temperatures due to anticipated high interior pressures of inclusions (Xia, 1984; Andersen and Neumann, 2001). The volatile release peak at 600-1000°C shows largest intensity for crushing size of 0.06mm, and declines

in intensity for samples of smaller or larger crushing size, indicating the degassing feature of fluid inclusion. The process of sample crushing size reduction is actually to reduce wall thicknesses (i.e. physical strength) of fluid inclusions with a given size range, which facilitates decrepitation of fluid inclusions. The volatile release peak at 600-1000°C are interpreted as decrepitation of early-stage fluid inclusions (Zhang et al., 2007), and are dominated by CO (7.17 mm^3.STP/g on average), which is supported by volatile compositions of same type fluid inclusions (Bergman and Dubcssy, 1984) and re-equilibrated experiment of pure CO_2 fluid inclusions (Pasteris and Wanamaker, 1988). Pasteris and Wanamaker (1988) reported that pure CO_2 fluid inclusions were re-equilibated to dominant CO at >1000°C based on their micro- thermometry and laser Raman microprobe study.

(2) Late-stage fluid inclusions: The late-stage fluid inclusions distributed along crystal edges have thin walls of fluid inclusions with weak physical strength, thus are readily decrepitated with reduced decrepitation temperatures. The volatile release peak at 300-600°C decreases in intensity with decreasing crushing size, and nearly disappears at crushing size of 0.03mm, which is about the smallest size of late-stage fluid inclusions (Zhang et al., 2009). This is caused by late-stage fluid inclusions to be broken and outgassed, and lost their volatiles during progressive crushing (Marty and Zimmermann, 1999; Matsumoto et al., 2000; Aubaud et al., 2005), which is supported by the similar compositions of volatiles (i.e., CO_2, ~ 2.99 mm^3.STP/g on average) to those of same type individual fluid inclusions analyzed by *in situ* Laser Raman Probe (Xia et al., 1984; Yang et al., 2001).

3.2. Volatiles in Mineral Structural Defects and Vacancies

Volatiles can also be trapped in structural defects and vacancies (Freund and Oberheuser, 1986; Bai and Kohlstedt, 1992), or dispersed in crystal structure (Green and Radcliffe, 1975) during crystallization, and possess similar chemical compositions to early stage inclusions. The former are similar to nm-sized inclusions below optical resolution, which are attached to crystal defects induced by deformation and exsolution (Green and Radcliffe, 1975; Koch-Müller et al., 2004) and are released at higher heating temperatures to overcome the larger physical strength around structural defects and vacancies than early stage inclusions (Matsumoto et al., 2000).

Grain size reduction experiment reveals that the total contents of volatiles as well as individual volatile species (CO_2, CO, H_2) increase with decreasing sample crushing size, especially increase rapidly towards smaller sample crushing size, which is consistent with increasing grain surface areas (Figure 3, Zhang et al., 2009). Volatile abundances released at the 1000-1200°C interval increase with decreasing sample crushing size and show additional concentration peak in samples of smaller crushing size. The release temperature of 1000-1200°C overlaps the final equilibration temperature ranges of lherzolite xenoliths (987°C to 1296°C, Fan and Hooper, 1989). Their chemical composition dominated by CO with minor H_2 and CO_2 are similar to volatiles released at the 600-1000°C interval for early-stage fluid inclusions. Hence, volatile release peak at the 1000-1200°C interval likely results from degassing of volatiles trapped in crystal defects and vacancies, and are volumetrically significant (Zhang et al., 2009).

3.3. Anion Complexes Incorporated in Crystal Structures

Anion complexes incorporated in crystal structures are considered as another primary occurrence mode of volatiles in hydrous mantle minerals, e.g., OH^- in amphibole and mica, Cl^- and F^- in apatite etc. (Chazot et al., 1997; O'Reilly and Griffin, 2000; Aubaud et al., 2007). Especially OH^- concentrations, incorporation and diffusion mechanism and their effect in olivine, pyroxene, and garnet etc. major mineral phases of upper mantle have been well studied by infrared spectroscopy (IR), NMR, X-ray and neutron diffractions, mass spectrometry (MS), laser Raman spectroscopy, hydrogen diffusion, hydrothermal and dehydrogenation experiments (Ackermann et al., 1983; Freund and Oberheuser, 1986; Kitamura et al., 1987; Smyth et al., 1991; Bai and Kohlstedt,1992; Bell et al,1995; Kohn,1996; Mierdel et al., 2007; Grant et al., 2007) since hydroxyl was first identified in nominally anhydrous minerals (Martin and Donnay, 1972).

The amount of structural OH^- is constrained by the replacement amount of Fe^{3+} by Fe^{2+} in mineral structure and hydrogen partial pressure (P_{H2}) of the formation environment (Skogby and Rossman, 1989; Ingrin et al,1995). OH^- in crystal structure is released when heating temperature is high enough to break OH bond in crystal lattices (Bell et al. 1995; Kohn, 1996). FTIR absorbance peak of OH^- at a wavelength of about 3450cm^{-1} in olivine and orthopyroxene shows that OH^- amounts are obviously reduced only when samples have been vacuum degassed at 1200°C (Zhang et al., 2005), which is supported by OH^- solubility and stability experiments (Bell et al. 1995; Kohn, 1996).

3.4. The Superimposed Effects of Volatiles in Different Occurrence Modes

Large size fluid inclusions can be broken and release their volatiles during progressive crushing (Marty and Zimmermann, 1999; Matsumoto et al., 2000; Aubaud et al., 2005). Late stage fluid inclusions with large size are always located at sample edges and can be readily decrepitated at lower temperatures, or broken in smaller crushing size of samples. So the volatiles in late stage fluid inclusions have little superimposed effects on volatile release peaks at high heating temperatures of 600-1000°C and 1000-1200°C.

The early stage fluid inclusions located in crystal edges are readily decrepitated with reduced decrepitation temperatures due to thin wall of fluid inclusions. Some early stage fluid inclusions in crystal interiors could be shifted to sample edges during sample crushing, so more early stage fluid inclusions in sample edges of small crushing size could be released at lower temperatures and have superimposed effects on volatiles at 300-600°C intervals. Early stage fluid inclusions in large crushing size samples are difficult to decrepitate due to thicker wall of fluid inclusions, a little of early stage fluid inclusions in crystal edges show little superimposed effect on volatiles at 300-600°C intervals.

Volatiles trapped in structural defects and vacancies of crystals are released at temperatures high enough to destruct the crystal lattices, so their superimposed effect on concentration peaks of 300-600°C and 600-1000°C is negligible.

4. The Redox State and Volatile Species in the Mantle

4.1. The Redox State in the Mantle

Oxygen fugacity (fO_2) is an important parameter in mantle processes. The redox state of lithospheric mantle is heterogeneous and varies over four log units, e.g., oxygen fugacity (f_{O2}) can vary from 3 log units below and 1 log unit above the fayalite-magnetite-quartz (FMQ) buffer using the olivine-orthopyroxene-spinel oxygen barometry (i.e., FMQ-3 to FMQ+1; Ballhaus, 1993; Wood and Virgo, 1989). The f_{O2} in the mantle may have changed to the more oxidized state over the Earth's history because of mantle degassing and introduction of recycled components (O'Neill and Wall, 1987), and also vary spatially from oceanic to continental mantle (Bernardj and Ltaras, 1992), from orogenic belts (Woodland et al, 1992) to craton (Ballhaus, 1993; Woodland and Koch, 2003), and from deep to shallow mantle.

The redox states in upper mantle become more oxidized from initial Earth to the present (Eggler and Baker, 1982; O'Neill and Wall, 1987). The f_{O2} in the upper mantle could be near nickel precipitation curve (iron-wustite buffer, IW) in olivine during late stage of Earth's formation, and then became more oxidized because oxygen-enriched components (e.g., H_2O, CO_2 and O_2) were added by plate subduction (O'Neill and Wall, 1987). For example, the f_{O2} of isolated samples from Archaean cratonic lithosphere is as low as FMQ-5, fertile spinel peridotite xenoliths (1.5GPa) is around FMQ-2 (Ballhaus, 1993). Heterogeneous redox state of Archean cratons may mostly result from heterogeneous nature of percolating fluids (Thomassota et al., 2007). The fO_2 in upper mantle show highest value nearby subduction zones due to large quantity of oxidized volatiles released (Wood et al., 1990).

The Earth's mantle is degassed along mid-ocean ridges, while is re-hydrated (and possibly re-carbonated) at subduction zones. These processes and the speciation of C-H-O fluids in the mantle are related to the redox state of mantle peridotite. Abyssal spinel peridotites and quenched MORB glasses demonstrated that the redox state in oceanic mantle is more reduced than in the SCLM (Bernardj and Ltaras, 1992; Ionov and Wood, 1992). The f_{O2} values of MORB and OIB range from FMQ-1 (N-MORB) to ~ FMQ +2 (OIB) (Ballhaus, 1993), the outer surface quench glasses of MORB lavas indicated the fO_2 of FMQ-1 to FMQ-2 in the mantle source regions of MORB (Christie et al.,1986). Peridotites from ocean ridges are reduced and have an average fO_2 of about -0.9 log unit relative to FMQ; whereas peridotite xenoliths from continental localities exhibit a fO_2 range from FMQ-1.5 to FMQ +1.5 (Wood et al., 1990). The Earth's interior is more reduced than the redox state reflected by the erupting tholeiitic magmas; basaltic melts are more oxidized than samples from lithospheric upper mantle. An initially highly reduced mafic magma may become oxidized by degassing of C-species under the shallow crust conditions (Mathez, 1984).

The fO_2 is sensitive to pressure, and falls with increasing pressure. The rate of reduction per unit GPa pressure increase may be of the order of 0.6 log unit (Ballhaus, 1995). For example, the fO_2 varies along CCO buffer ($C+O_2=CO_2$) at uppermost portions of upper mantle (Blundy et al., 1991). fO_2 may decrease up to ~3 log units with the upper mantle depth between about 80 and 150 km beneath the Kaapvaal craton in Southern Africa. The $\Delta\log fO_2$ lies just FMQ-4 at ~220 km depth (Woodland and Koch, 2003). Spinel lherzolites from the Baikal rift area have slightly lower fO_2 than depper garnet- and garnet-spinel lherzolites (Ionov and Wood, 1992). fO_2 is expected to decrease further with depth in the asthenospheric mantle, reaches conditions of metal saturation (Fe-Ni precipitation curve in olivine, IW) near the 410 km discontinuity, and meets the necessary condition for 'redox melting' in the deeper portions of the upper mantle (Wood et al., 1990; Woodland and Koch, 2003). The lowest fO_2 is consistent with dominant CH_4, H_2 and other reduced volatiles in fluid inclusions of diamond (Melton et al., 1972; Melton and Giardini, 1974; Eggler and Baker, 1982).

4.2. Volatile Species in the Mantle

Volatile species in the C-H-O system at mantle high pressure and temperature are mixtures of H_2O, CO_2, CO, CH_4, O_2 and H_2 (Bergman and Dubessy, 1984; Saxena and Fei, 1988; Ballhaus, 1993), which are potentially important redox-controlling reactants (Deines, 1992). In the f_{O2} range between IW and FMQ, upper mantle materials contain a number of elements that can attain more than one redox state (Ballhaus, 1995):

$$C+O_2=CO_2 \quad [1]$$

$$2C+O_2=2CO \quad [2]$$

$$CH_4+O_2=C+2H_2O \quad [3]$$

$$2H_2 + O_2 = 2H_2O \quad [4]$$

Spatial and temporal variations of oxygen fugacity in the mantle place important controls on volatile speciation in the mantle (Deines, 1980). Significant variability of oxygen fugacity in the upper mantle is consistent with the occurrence of both reduced (CH_4 and H_2, or CH_4 and H_2O bearing) and oxidized (CO_2 and H_2O bearing) fluids (Thomassota et al., 2007). The volatile accretion in the Earth are heterogeneous: the primary reduced materials, 99.8% of the solar planet, brought 6% of H, 48% of C and 70% of N of the upper Earth (upper mantle, crust, ocean-atmosphere), a late veneer of CI composition supplied the residual (Javoy, 1997). Mantle volatiles also vary in chemical compositions from continent to ocean, and from deep mantle to shallow mantle due to heterogeneous redox state in the mantle (O'Neill and Wall, 1987; Wood and Virgo, 1989; Bernardj and Ltaras, 1992; Ballhaus, 1993; Javoy, 1997; Woodland and Koch, 2003).

Volatiles identified in mantle samples are often oxidized and consist of CO_2 and H_2O (Roedder, 1984), such as high density CO_2 inclusions in worldwide mantle minerals (Roedder, 1984; Xia, 1984; Andersen and Neumann, 2001; Frezzotti and Peccerillo, 2007), which gives a false notion that CO_2 is a dominant volatile phase in the mantle. With increasing pressure or depth in the lithosphere, elemental carbon becomes progressively refractory and carbon-bearing equilibria become more important for f_{O2} control. Below a realistic lower f_{O2} limit given by the H_2O+ C= CH_4+ O_2 (C-H_2O) equilibrium, a fluid becomes CH_4 rather than CO_2-bearing, and carbonatitic melt is presumably unstable (Ballhaus, 1993). The volatile phase(s) would be dominantly composed of H_2O and CH_4 near the 410 km discontinuity (below FMQ-4) (Ballhaus, 1995; Woodland and Koch, 2003).

Different elements may occur as variably different volatile species with depths in the mantle, e.g., carbon solubility increases with pressure (Shcheka et al., 2006). Volatile speciation in the upper mantle has been examined by sufficient number of abyssal spinel peridotites and quenched MORB glasses as probes of the upper mantle (Bernardj and Ltaras, 1992). C-H-O system is likely to be dominated by CH_4, H_2 and H_2O in deeper upper mantle (>100 km), CH_4 and H_2 are dominant volatiles in the convecting upper mantle, while CO_2 only plays a minor role. CO_2 is unlikely a major contributor to the upper mantle's volatile budget at pressures exceeding 3GPa (Ballhaus, 1995). Primordial mantle volatiles in the asthenosphere were originally CH_4-N_2-rich corresponding to mantle redox melting (Liu and Fei, 2006). For example, diamonds were precipitated from reduced CH_4-bearing fluid(s) within the sub-continental mantle at depths > 150 km and T > 1200 °C (Thomassota et al., 2007). Conditions in the vicinity of mantle plumes are favorable for CH_4 dominant volatile phase in the upper mantle (Bernardj and Ltaras, 1992). But asthenospheric mantle at depths greater than 150 km in Hawaii could have high-density CO_2 (+H_2O+H_2S) volatiles (Frezzotti and Peccerillo, 2007).

5. Chemical Compositions of Volatiles in Sub-Continental Lithospheric Mantle

Volatiles in the sub-continental lithospheric mantle (SCLM) are still poorly understood, especially their chemical compositions. Mantle-derived minerals from different depths provide a rare opportunity to study a deep-seated compositional profile of volatiles in the SCLM. Continental basalts are always contaminated by crustal or atmosperic components during magma ascent, and contain more oxidized volatile species like CO_2 (58.4-79.78 mol%) and SO_2 (2.87-31.35 mol%) (Zhang et al., 1999a, b).

Among mantle minerals, only diamonds possess sufficient strength to encapsulate and transport volatiles trapped at depths in excess of 100 km to the surface (Navon, 1991). The fluid inclusions in different color/shape, corresponding to different origins, have distinct chemical compositions (Melton et al., 1972; Melton and Giardini, 1974; Deines, 1980; Eggler, and Baker,1982; Navon et al., 1988). Colorless diamonds have more reduced volatiles (e.g., higher contents of H_2 and CO etc.) than colored diamonds with recycled components (Table 2, Ivankin et al., 1988), which is similar to initial volatiles in the SCLM with primitive mantle-like reservoir (see below).

Table 2. Chemical compositions (mm^3.STP/g) of volatiles of fluid inclusions in different types of diamonds (summarized from Ivankin et al., 1988)

Colored	Location	Type	n	CO_2	N_2	H_2O	H_2	CH_4	CO
	Congo	Cubic	3	215.8	23.6	653.2	19.7	41.5	46.2
Colored	Brazil	Cubic	2	48.0	20.7	832.6	51.7	47.0	0.0
	Mean	Cubic	5	147.1	22.4	726.6	32.8	43.8	27.3
	Pamir	Octahedron	4	80.6	114.1	210.0	340.1	76.8	178.4
Colorless	USA	Dodecahedron	3	144.6	87.1	423.8	201.7	49.2	93.6
	Mean		7	108.0	60.6	264.5	386.7	81.5	98.8

Volatiles in mantle xenoliths record their nature and origin during mantle evolution. Alkali basalts always host mantle xenoliths with fresh Olv, Opx and Cpx minerals or megacrysts. Olv, Opx and Cpx behave in different way during mantle partial melting, magma crystallization and mantle metasomatism due to distinct chemical and physical properties (e.g., initial melting or crystallization temperatures), and hence preserved volatiles from different depths or origins (Albarede, 2008; Zhang et al., 2004, 2007). Volatiles in different occurrence modes of minerals in mantle xenoliths show various origins with information pertaining to the formation and evolution of the Earth's mantle. Based on the chemical and isotopic compositions of volatiles in mantle xenoliths, "initial" and "metasomatic" two types of volatiles in the SCLM can be distinguished.

5.1. The Initial Volatiles in the SCLM

The initial volatiles in the SCLM refer to volatiles trapped in crystals during initial stage of mantle mineral crystallization; they were trapped both in primary fluid inclusions and in structural defects and vacancies of the minerals. Olivine is the most refractory mineral of lherzolite xenoliths that can preserve the initial volatiles intact in crystal interiors, and survived subsequent mantle events. So initial volatiles in the SCLM can be characterized by the volatiles released at 600-1200°C from olivine of lherzolite xenoliths (Zhang et al., 2004).

Initial volatiles in the SCLM beneath eastern China vary in chemical compositions with mantle components defined by Sr-Nd-Pb isotopes (Zhu, 1995; Griffin et al., 1998). The chemical compositions of initial volatiles in different mantle reservoirs from eastern China and western USA are summarized in Table 3 (Zhang et al., 2004, 2005). Initial volatiles in the SCLM is dominated by CO with minor CO_2 and H_2 in primitive mantle-like reservoir (PM), such as the North China Craton; whereas they are dominated by CO_2 with minor SO_2 and H_2 in depleted mantle reservoir (DM), e.g., the South China Craton. Enriched mantle reservoir (EM), e.g., northern part of northeastern China fold belt and western America, display high contents of volatile and H_2 in initial volatiles in the SCLM, suggesting a mantle fluid with high H_2 contents. Initial volatiles in the SCLM with a "mixed" mantle reservoir are primarily composed of CO_2 and SO_2 which are related to recycled continental crustal fluids (Dabie ultra-high pressure metamorphic belt).

Table 3. Chemical compositions (mm^3.STP/g) of initial volatiles in the SCLM in different mantle reservoirs

Mantle	H_2	CH_4	H_2S	CO	N_2	CO_2	SO_2	Total
Primitive Mantle	2.76	0.12	0.01	7.77	0.41	3.04	0.02	14.13
Depleted Mantle	3.66	0.19	0.00	4.92	0.18	7.49	0.82	17.26
Enriched Mantle	18.73	0.20	0.00	4.60	0.38	6.26	0.18	30.35
Mixed Mantle	0.12	0.03	0.00	0.00	1.01	24.66	2.61	28.43

CO_2 contents in initial volatiles in the SCLM with various mantle reservoirs are all much lower than estimated values of primitive mantle (from 5.23-12.5 $\times 10^{-6}$ mol/g) (Zhang and Zindler, 1993; Jambon, 1994). The calculated H_2O contents in depleted and primitive mantle also show high values to be 13.89-64.44 $\times 10^{-6}$ mol/g (O'Neil and Palme, 1998). Low abundances of CO_2 and H_2O in olivine, i.e. initial volatiles in the SCLM, are supported by the fact that CO_2 and H_2O contents in whole rock samples of lherzolite are higher than in olivine, and that LREE-rich volatiles related to CO_2 were preserved mainly in Opx, Cpx or mineral interstices (Xu, 2000). Ultramafic xenoliths from alkali basalts or kimberlites have undergone voaltile fractionation etc. processes within the mantle, so that volatiles in Olv must be considered as the lowest values of initial volatiles in the SCLM. CO_2 and H_2O could be preserved in Opx, Cpx or mineral interstices.

5.2. The Metasomatic Volatiles in the SCLM

5.2.1. The Types of Metasomatic Volatiles

Mantle metasomatism is extensive throughout the SCLM, which may be macroscopic with conspicuous metasomatic minerals (e.g., amphiboles, phlogopite, apatite etc.) or cryptic with the enrichments as hydrous melts distributed along grain boundaries (e.g., O'Reilley et al., 1988; Hansteen et al., 1991; Wyllie, 1987). The metasomatic volatiles have been trapped as metasomatic agents in macroscopic metasomatic minerals or secondary inclusions in the edges and cracks of recrystallized mantle minerals.

The volatiles of macroscopic metasomatism can be infered from volatiles in different types of metasomatic minerals. The amphiboles from mantle xenoliths in Vulcan's Throne, Grand Canyon, Arizona, USA show volatiles of macroscopic metasomatism dominated by H_2O with high concentrations of O_2, CH_4, CO_2 and CO (Table 4, Matson and Muenow, 1984), differing from initial volatiles in the SCLM in the western USA (Zhang et al., 2005b) or volatiles in diamonds with mantle origin (Ivankin et al., 1988). The more oxidized volatile species could be originated from shallow mantle or recycled components.

Table 4. Chemical compositions (mm^3.STP /g) of macroscopic metasomatic volatiles from different types of amphiboles in Arizona, USA (summarized from Matson and Muenow, 1984)

Type	n	CO_2	H_2O	SO_2	H_2S	O_2	CH_4	CO
Hornblendite	1	356.4	11200.0	140.0	65.9	980.0	560.0	240.0
Selvage	1	152.7	14560.0	35.0	131.8	630.0	420.0	240.0
Oikocrysts	6	356.4	12091.9	46.7	175.7	1493.3	490.0	306.7
Megacrysts	2	305.5	10826.7	297.5	230.6	1260.0	350.0	160.0

Extensive cryptic mantle metasomatism in the SCLM has been widely observed in mantle xenoliths and is defined by the decoupling between enriched trace elements and depleted major elements and Sr-Nd-Pb isotopes (Griffin et al., 1998; Coltorti et al., 2000). Orthopyroxene and clinopyroxene with low melting point are readily altered by metasomatic fluids. The volatiles released at 600-1200°C from Opx and Cpx of mantle xenoliths are predominantly H_2 with minor amounts of CO_2 and CO in eastern China and western USA

(Zhang et al., 2004, 2005b), which can represent one type of volatiles in cryptic mantle metasomatism (Zhang et al., 2004, 2007), and are summarized as Type I in Table 5.

On the other hand, recrystallization of such metasomatized peridotites would incorporate the cryptic metasomatic volatile agents as fluid (or melt) inclusions at the edges and cracks of the recrystallized minerals (Peng et al., 1994). This type of metasomatic volatiles are released at 300-600°C from Olv, Opx and Cpx of mantle xenoliths in eastern China, and are primarily composed of CO_2 and SO_2 (Zhang et al., 2004), which differ from initial volatiles in the SCLM and type I cryptic metasomatic volatiles. So the volatiles released at low temperature interval represent another type of volatiles in cryptic mantle metasomatism, and are summarized as Type II in Table 5.

Table 5. Chemical compositions (mm^3.STP/g) of cryptic metasomatic volatiles in the SCLM in different mantle reservoirs

Type	Description	Location	Mantle reservoirs	H_2	CH_4	H_2S	CO	N_2	CO_2	SO_2
		NCC	PM	83.65	0.77	0.80	31.05	1.09	8.91	0.05
I	Opx, Cpx	SCC	DM	138.91	0.92	0.08	32.31	1.01	11.32	0.00
	600-1200°C	W-USA	EM	13.08	0.29	0.01	7.25	0.23	8.05	1.18
		NCC	PM	1.47	0.68	0.07	4.63	0.16	7.81	0.23
II	Olv,Opx,Cpx	SCC	DM	1.07	0.66	0.00	3.11	0.22	8.64	0.42
	300-600°C	W-USA	EM	0.21	0.17	0.00	1.64	0.04	5.26	0.00

NCC- northern China craton, SCC- southern China craton, W-USA- western margin of USA.

5.2.2. The Residence of H_2 in Metasomatic Volatiles

H_2 is one of the most important volatile components in the solar system. A certain amount of H_2 as one of the primordial components remains in the Earth's interior. H can be incorporated as OH^- in crystal structures of mantle-derived olivine, pyroxene and garnet etc. minerals (Skogby and Rossman, 1989; Smyth et al, 1991; Bai and Kohlstedt, 1992; Ingrin et al, 1995; Bell et al, 1995; Kohn, 1996). Structural OH^- has been considered as the main occurrence mode of H in anhydrous mantle minerals (Martin and Donnay, 1972).

H can enter into or degas from Si-deficient and Fe^{3+}-rich pyroxene structure by the following reaction, which controls the amount of OH^- in pyroxene (Skogby et al., 1989).

$$2Fe^{3+}+2O^{2-}+H_2 = 2Fe^{2+}+2OH^- \quad [5]$$

Olivine is volumetrically the most important mineral in the upper mantle. The incorporating way of H in olivine could be distinctive from pyroxene, yet little is known about the mechanisms of OH^- incorporation in olivine structure (Grant et al., 2007). OH^- could take octahedral or tetrahedral vacancies in olivine (Freund and Oberheuser, 1986). H may enter into olivine structure by placing a neutral site and incorporating OH^- with vacancy O^{2-}, or placing a metal position with 2 negative charges and forming H (Bai and Kohlsted, 1992). Hydroxyl solubility in olivine depends on fO_2, but current data from relatively high fO_2 experiments overestimate the OH^- solubility in mantle olivine (Grant et al., 2007).

Olv, Opx and Cpx show varying H_2 contents and infrared spectra with grain-size and heating temperature during stepwise heating, (1) H_2 mainly release at two temperature intervals of 600-1000 and 1000-1200°C. (2) H_2 contents decrease from Cpx (av. 3.667mm^3.STP/g), to Olv (av. 3.183 mm^3.STP/g) and to Opx (av. 1.381 mm^3.STP/g). H_2 contents released from olivine increase rapidly with decreasing sample crushing size. H_2 in orthopyroxene and clinopyroxene has the highest content at crushing size of 0.2 and 0.1mm, respectively, and then decreases with decreasing sample crushing size. (3) Infrared absorbance peaks of OH^- in Olv and Opx at a wavelength of about 3450cm^{-1} are not reduced while degassed at 600°C and 900°C compared with samples dried at 120°C, but are obviously reduced after degassed at 1200°C.

Because organic contaminants and gases adsorbed on the sample surface and in cracks and secondary inclusions have been removed completely by our efficient sample pretreatment and analytical procedures, H_2 released from mantle minerals could be derived from three sources: (1) reactions [6-9] among volatiles at high-temperature:

$$C + H_2O = CO + H_2 \quad [6]$$

$$CO_2 + CH_4 = 2CO + 2H_2 \quad [7]$$

$$CH_4 + 2H_2O = CO_2 + 4H_2 \quad [8]$$

$$2Fe_3O_4 + H_2O = 3Fe_2O_3 + H_2 \quad [9]$$

(2) OH^- in mineral structure and H^0 in structural vacancies, and (3) fluid inclusions.

The efficient analytical procedures, i.e., adopting a liquid-N_2 cold trap etc., can eliminate the contribution of H_2 derived from reactions [6-9] of released volatiles, which are supported by negligible contents of H_2 and CO in analyses of other type samples using the same experimental procedure (Zhang et al., 1999a, b; Hu et al., 2007). New volatile phases formed by possible reactions among volatiles inside fluid inclusions or samples are considered as re-equilibrated volatile compositions at mantle high temperature conditions (Anderson et al., 1984; Pasteris and Wanamaker, 1988; Zhang et al., 2000).

Following lines of evidence indicate that little H_2 released at 600-1000°C has been derived from degassing of structural OH^-:

1) Infrared spectra of Olv and Opx separates from lherzolite xenoliths indicate that OH^- contents are not reduced in samples degassed at 600°C and 900°C compared with samples dried at 120°C, but are obviously reduced when samples are degassed at 1200°C (Zhang et al., 2005).
2) H_2 contents decrease with decreasing sample crushing size in Opx and Cpx, which is inconsistent with increasing sample surface area. H_2 from either structural OH^- or H^0 in structural defects and vacancies should increase by sample crushing (Skogby and Rossman, 1989).
3) H_2 in metasomatic volatiles has heavier δD ranging from -31.1 to -96.4‰ (Zhang et al., 2007) than OH^- in mantle-derived garnet, enstatite and augite (-92 to -113‰, Bell and Ihinger, 2000).

4) H_2 releasing peaks at 600-1000°C show synchronous release with volatiles derived from early stage fluid inclusions.

All these suggest that high concentrations of H_2 released at 600-1000°C may be primarily stored in fluid inclusions; H_2 released at 1000-1200°C could be from the loss of structural OH^-, which is also supported by OH^- solubility and stability experiments (Bell et al., 1995; Kohn, 1996; Grant et al., 2007). Hydroxyl solubility in olivine at low fO_2 (below Fe-FeO) is substantially less than that solubility laws might suggest (Grant et al., 2007). Structural OH^- in pyroxene is more difficult to lose, and could be released when pyroxene has been heated at 1000°C for 24 hrs in air (Bell et al.,1995).

OH^- can be incorporated efficiently in pyroxene structures when pyroxene was heated only in H_2 atmosphere (Skogby and Rossman, 1989), so high concentrations of H_2 from either early stage fluid inclusions or the loss of structural OH^- suggested that an H_2-rich (in free state) volatile may have been involved in the metasomatism of the SCLM in eastern China and western USA.

5.3. The Origins of Volatiles in the SCLM

5.3.1. The Volatiles of Mantle Origin

C, O and H isotopes and concentrations of volatiles in mantle xenoliths should allow faithful identification of various source components in the volatiles. Initial volatiles in the SCLM beneath eastern China are characterized by $\delta D_{H2,\ H2O}$, $\delta^{13}C_{CH4}$ and $\delta^{18}O_{CO2}$ of mantle origin. δD values of H_2O and H_2 (from -33.8‰ to -88.2‰) are restricted to the δD range of "normal" upper mantle, $\delta^{13}C$ values of CH_4 towards the methane $\delta^{13}C$ value from Murchison meteorite and olivine in oceanic gabbros (Yuen et al., 1984; Abrajano et al, 1988; Kelley and Frueh, 1995), suggesting that a large proportion of initial volatiles in the SCLM could be upper mantle origin.

5.3.2. The Origin of H_2 in Metasomatic Volatiles

H_2 in metasomatic volatiles has δD ranging from -31.1 to -96.4‰. Garnet lherzolite from deeper mantle shows higher content of H_2 (145.10 mm^3.STP/g) than spinel lherzolite (Zhang et al., 2007); H_2 contents in Cpx (av. 3.667mm^3.STP/g) are higher than Olv (av. 3.183) and Opx (av. 1.381) etc. crystalized early minerals. All thses prove that H_2 could come from deep mantle at late stage of the SCLM formation and evolution beneath eastern China.

5.3.3. Recycled Crustal Volatiles

The recycled crustal volatiles (RCVs) are identified in both initial volatiles and metasomatic volatiles in the SCLM beneath eastern China based on C-O-H isotopic compositions of volatiles:

1) RCVs in initial volatiles: Initial volatiles released at 600-1200°C from Olv show difference in volatile compositions between North China Craton (NCC, dominated

by CO) and South China Craton (SCC, dominated by CO_2), suggesting different formation and development condition between NCC and SCC. They are characterized by $\delta D_{H2, H2O}$, $\delta^{13}C_{CH4}$ and $\delta^{18}O_{CO2}$ of mantle origin with lighter $\delta^{13}C_{CO2, CO}$, and suggested that a little RCVs were mixed into volatiles of old Cratonic lithosphere during a Proterozoic continental collision (1.9Ga) (Zhang et al., 2008).

2) RCVs in type I cryptic metasomatic volatiles, which are released at 600-1200°C from Opx and Cpx, are dominated by H_2 both in NCC and SCC, suggesting their similar development condition. δD_{H2} and $\delta^{13}C_{CH4}$ with mantle characteristics implied their mantle origin. Lighter $\delta^{13}C$ of CO_2 and CO, and heavier $\delta^{18}O$ of CO_2 suggested that a recycled crustal volatiles with terrigenous features have been involved possibly during a common crust-mantle interaction process during Mesozoic collision of NCC and SCC (Zhang et al., 2008).

3) RCVs in type II cryptic metasomatic volatiles, which were released at 300-600°C from Olv, Opx and Cpx of mantle xenoliths, and composed mainly of CO_2 both in NCC and SCC. Their crustal $\delta^{13}C$ of CO_2 and CO, biogenic $\delta^{13}C$ of CH_4, lighter and variable $\delta^{18}O$ of CO_2, and lighter δD of H_2 suggested the presence of a recycled oceanic crustal volatile during the SCLM thinning, which could be derived from dehydration of the subducted paleo- Pacific lithosphere (or predecessor) that lies horizontally in the transition zone beneath eastern China (Zhang et al., 2007).

6. Chemical Compositions of Volatiles in Oceanic Upper Mantle

The chemical composition of volatiles in oceanic upper mantle (OUM) can be deduced from volatiles of fresh oceanic basaltic glasses because of the inaccessibility of fertile mantle source regions. MORB, OIB, BABB and much of IAB are originated from OUM. Volatiles in modern OUM can be investigated by means of oceanic basalts and mantle xenoliths, mantle xenoliths are reliable samples for mantle volatiles relative to oceanic basalts that are always contaminated by air and seawater during magma generation and ascent, and are in a more oxidized state than OUM (Mathez et al., 1984). The volatiles in ancient OUM can be investigated by means of ophiolitic peridotite or ancient ocean basalts which always altered.

6.1. Modern Oceanic Upper Mantle

Volatiles in modern (refer to currently extant) OUM have been the subject of a small number of broad and global studies. Available volatile investigations in modern OUM are conducted by oceanic basalt glasses and mantle xenoliths, Volatiles within the asthenospheric mantle at depths greater than 150 km in Hawaii could be high-density CO_2 ($+H_2O+H_2S$) in the diamond stability field (Frezzotti and Peccerillo, 2007). Volatiles in oceanic basalts are higher in abundance than in SCLM xenoliths. MORB, OIB and IAB formed under distinct conditions and by different mechanisms show difference in redox state (Ballhaus, 1993), REE, Pb-Sr-Nd isotopes (Hofmann, 1997; Niu and Batiza, 1997; Niu et al., 2003, 2005), and display distinct volatile compositions, C-O-H and noble gas isotopes (Eiler et al., 1995, 1997; Ozima and Podosek, 2002; Hansteen and Troll, 2003). The chemical compositions of volatiles released from different types of oceanic basalt are summarized in Table 6.

Table 6. Chemical compositions (mm^3.STP/g) of volatiles released from different type of oceanic basalts

Type	Ocean	Location*	Rock	n	CO_2	N_2	H_2O	CH_4	CO	Method#	Ref.+
MORB	Indian	IRT,CIR,SI	Glass	52	172	0.19				Crushing	[a]
		IRT	Basalt	6	28.1	0.18	9.7067			Crushing	[b]
	Pacific	GSC, EPR	Glass	33	99.1	0.02	2940.8	140	320	H-T MS, IR	[c]
		GSC, EPR	Basalt	15	716		2204.4	97	211	H-T MS	[d]
	Atlantic	13.5°N	Popping	5	464		77.907			Crushing	[e]
		FAMOUS	Glass	9	672		2613.3			H-T MS	[f]
OIB	Hawaii	Loihi	Basalt	5	468		7740.4			IR, H-T MS	[g]
		Loihi, Kilauea	Glass	29	295		7831.1			H-T MS	[h]
	Samoa		Glass	18	30.4		11455			IMP	[i]
IAB	Mariana	Arc	Basalt	4	1246		13263			H-T MS	[j]
BABB	Mariana	Trough	Basalt	4	1060		12731			H-T MS	

* IRT- Indian Ridge Triple Junction, CIR-Central Indian Ridge, SI- Southeast Indian, GSC-Galapagos Spreading Center, EPR-East Pacific Rise.

IR- infrared spectroscopy, H-T MS- high temperature mass spectrometry, IMP-Ion microprobe.

+ Ref. [a] Marty and Zimmermann, 1999; Cartigny et al., 2001; Nishio Y, 2007; [b] Marty and Zimmermann, 1999; [c] Byers et al., 1984; Michael, 1995; Marty and Zimmermann, 1999; Roux et al., 2006; [d] Byers et al., 1983, 1986; [e] Javoy and Pineau, 1991; [f] Delaney et al., 1978; [g] Dixon and Clague, 2001; [h] Muenow et al., 1979; Byers et al., 1985; [i] Workman et al., 2006; [j] Garcia et al., 1979.

6.1.1. Mid-Ocean Ridge Basalts (MORB)

Volatiles in MORB are dominated by H_2O (1284 mm^3.STP/g) and CO_2 (263 mm^3/g) with minor CO, CH_4, N_2, and H_2 components. CO_2 is richer in the crystalline basalts (av. 509 mm^3/g) than in glasses (av. 194.4mm^3/g), whereas H_2O is richer in glasses (av. 2870.6mm^3/g) than in the more crystalline basalts (av. 1289 mm^3/g). The MORB volatiles vary with region (Table 6). Atlantic MORB have higher CO_2 (av. 598 mm^3/g) than Pacific and Indian (av. 292 and 157 mm^3/g, respectively). Pacific MORB show higher content of H_2O (av. 2271 mm^3/g) than Atlantic (av. 1708 mm^3/g). N-MORB and E-MORB also are uniform in H_2O/Ce ratios (av. 155-213) within a region but show significant regional variations. N-MORB and E-MORB sources may share a common heritage (Michael, 1995).

Volatiles in MORB usually are contaminated by recycled atmospheric component and seawater during magmatism. Estimating the volatile species degassed along mid ocean ridges (MORs) requires a good knowledge of the initial volatile characteristics in terms of abundance and isotopic composition of MOR magmas (Pineau and Javoy, 1983; Cartigny et al., 2001). Defining the initial volatile composition of MORB is a key to understanding volatile cycling and mixing in the upper mantle.

The popping rock on Mid Atlantic Ridge (MAR) with typical MORB noble gas isotopic composition ($^3He/^4He$=8Ra, $^{40}Ar/^{36}Ar$= 28000) provides probably so far the best sample of upper mantle volatiles with negligible (3-7%) atmospheric and seawater contamination, and places a constraint on initial volatiles in undegassed MORB at depth of MORB formation. Their vesicles have volatile content of 7.56 cm^3 STP/g, and is mainly composed of CO_2 (94.95%) with minor H_2O (4.93%) and N_2 (0.12%) as well as trace He, Ar, H_2, CO and CH_4 (Javoy and Pineau, 1991). Meanwhile CO_2 is also a dominant volatile in East Pacific Rise (EPR) with an average content of 831.5 mm^3/g (Byers et al., 1986). CO_2 abundances decrease from fractionated ferrobasalt, to andesite and to rhyodacite glasses in Galapagos Spreading Center (GSC) (Byers et al., 1984). These suggested that CO_2 should be a dominant volatile species in initial volatiles of undegassed MORB.

H_2O is the most abundant volatile in MORB, and increases from fractionated ferrobasalt, to andesite and to rhyodacite glasses in GSC (Byers et al., 1984). Average content of H_2O in Pacific MORB glasses increase from 1203 mm^3/g in GSC 95.5°W propagating rift (Byers et al., 1984), to 1493.5 mm^3/g in the EPR at 21°N (Byers et al., 1986), and to 2613 mm^3/g in Woodlark Spreading Center, Southwest Pacific (Muneow et al., 1991). H_2O content should be low in initial volatiles of undegassed MORB. Michael (1995) argued that H_2O in MORB is derived from recycled, subducted, altered oceanic crust. The poor correlation of H_2O/Ce with $^3He/^4He$ imply that H_2O/Ce variations are not related to variations in juvenile H_2O in the source. Higher H_2O/Ce ratios in MORB (e.g., in north MAR, av. 240-280) may be a regional characteristic of the mantle, which might be related to a period of rapid subduction in the past that resulted in depressed isotherms and less dehydration in the slab.

Indian MORB differ from the EPR and MAR MORB by having higher $^{87}Sr/^{86}Sr$ and lower $^{206}Pb/^{204}Pb$ (Schiano et al., 1997; Rehkamper and Hofmann, 1997) because of the effect of Dupal anomaly, which could be due to large-scale mantle contamination either by ancient recycled sediments or by delaminated subcontinental lithosphere (Castillo, 1988; Schiano et al., 1997; Rehkamper and Hofmann, 1997; Regelous et al., 2009). Indian MORB volatiles are mainly composed of CO_2 (9.63-52.4, av. 28.1 mm^3/g) and H_2O (0.005-0.022, av. 0.01mm^3/g) with minor N_2, Ar, CO, CH_4 and H_2, and show the lowest contents of CO_2 and H_2O with unusual $N_2/^{40}Ar$, C/N, and $CO_2/^3He$ ratios in vesicles (120, 26190, and 0.3×10^9, respectively), whereas $N_2/^{40}Ar$, C/N, and $CO_2/^3He$ in the MAR and EPR are 91, 2154600, and 1.5×10^9, respectively (Nishio et al, 1999). Major volatile signatures of Indian MORB could be produced more likely by mantle heterogeneity (Nishio et al, 1999), degassing-induced fractionation or late atmospheric contamination (Cartigny et al, 2001).

EPR and GSC glasses contain trace amounts of reduced carbonaceous species, such as methane (CH_4 <0.01 wt.%) and carbon monoxide (CO, 0.04 wt.%). The abundance of reduced carbon species in GSC glass decrease with increasing magma differentiation (Byers et al., 1983, 1984). Indian MORB also contains minor CO, CH_4 and H_2 (Nishio et al, 1999). These suggested that MORB sources are probably dominantly more reduced state of the upper mantle (below FMQ buffer).

6.1.2. Ocean Island Basalts

Volatiles in ocean island basalts (OIB) are characterized by the lowest CO_2 content (av. 210mm^3 /g) among oceanic basalts and higher H_2O content (9063.3 mm^3 /g) than MORB, but H_2O content in OIB lavas is generally lower than in arc lavas (up to ~6 wt%, Workman et al., 2006). CO_2 in OIB glass (av. 192 mm^3/g) is lower than OIB basalt (468 mm^3/g), whereas H_2O

in OIB glass (9100 mm^3/g) is much higher than OIB basalt (7740 mm^3/g). Volatiles in OIB also vary with region.

Volatile contents of submarine basalt glasses from Loihi and Kilauea, Hawaii range from 0.73 to 1.40 wt%. H_2O contents range from 4355.6 to 12568.9 mm^3/g and CO_2 from 14.8 to 967.3 mm^3/g (Muenow et al., 1979; Byers et al., 1985; Dixon and Clague, 2001). Minor reduced carbonaceous species (CO and CH_4) in most of Loihi alkali samples, combined with lower H_2O contents (av. 0.54 wt.%) of tholeiites in Loihi than in Kilauea indicate a reduced primitive source for Loihi magmas with higher $^3He/^4He$ (Byers et al., 1985).

The extent of OIB lava degassing is depth dependent (Heide and Schmidt, 2003). CO_2 may have been degassed from the tholeiitic and transitional magmas in the storage of a shallow magma chamber prior to eruption. For example, CO_2 content in glass decreases with increasing vesicle abundance in the alkali lavas from Loihi (Byers et al., 1985). Alkalic basalt glasses from the North Arch Volcanic Field, Hawaii have varying H_2O contents from 8586.7 to 17671.1 mm^3/g and CO_2 from 132.4 to 407.3 mm^3/g, which could be related to extensive degassing prior to deep submarine eruption. Estimated bulk volatile contents of the vesicular vent samples are up to 23644 mm^3/g for H_2O and 27490 mm^3/g for CO_2, which are interpreted to have formed by closed system degassing (Dixon et al., 1997).

OIBs from Samoan islands have higher H_2O content (11455 mm^3/g) and lower CO_2 (30.4 mm^3/g) than Hawaii. The Samoan lavas show a large range in H_2O (7840-15928 mm^3/g) and CO_2 (3.1-91.1 mm^3/g) contents (Workman et al., 2006). Volatiles in glasses from seamounts flanking the East Pacific Rise (EPR) are commonly enriched in H_2O, Cl, F and S; H_2O and Cl abundances range from 0.111 to 1.021 wt.% and 0.003 to 0.226 wt.%, respectively (Aggrey et al., 1988a).

6.1.3. Island Arc and Back-arc Basin Basalts

Basalts from back-arc basin (BABB) or island arc (IAB) are similar in chemical composition of volatiles to OIB (Garcia et al, 1979). The volatiles in BABB are slightly lower in H_2O and CO_2 contents than in IAB.

Volatiles in IAB are dominated by H_2O (13263 mm^3/g) with minor CO_2 (1246 mm^3/g), such as Mariana Island arc (Garcia et al, 1979). H_2O contents in Sumisu Rift basalts are at least four times greater than in MORB (Hochstaedter et al., 1990). H_2O is the principal volatile species in Solomon Island arc (av. 336×10^{-6} mol/g) and in Grand Canyon, where total volatiles range from 1.27 to 1.75 wt.% (Dean, 1984). Higher volatile contents in IAB imply partial melting and degassing of the subducted slab in production of island arc magmas.

Volatiles in BABB are also dominated by H_2O (av. 568.33×10^{-6} mol/g) with minor CO_2 (47.33×10^{-6} mol/g). H_2O and CO_2 abundances in glasses display a large range from depleted, primitive MORB to enriched, transitional tholeiites in North Fiji and Lau back-arc basins, and show distinct volatile characteristics from lavas from the Mariana Trough and Scotia Sea (Aggrey et al., 1988b).

6.1.4. The Origins of Volatiles in Modern OUM

The MAR popping rock constrains initial volatiles of undegassed MORB which are mainly composed of CO_2 (94.95%) with minor H_2O. CO_2 contents in oceanic basalts increase from OIB (av. 210 mm^3.STP /g), to MORB (263), to BABB (1060) and to IAB (1246, see Table 6 for details and references). MORB are depleted in volatiles relative to OIB, IAB and BABB; low CO_2 contents in MORB could be caused by magma degassing, whereas high CO_2 contents in BABB and IAB could be responsible by degassing of subducted slabs.

H_2O is the main volatile component of most oceanic basalts. H_2O is most depleted in MORB (1284 mm^3.STP/g), and increases from OIB (9063 mm^3/g), to BABB (12731) and to IAB (13263). So much higher H_2O contents in IAB and BABB could be near surface addition of volatiles (Garcia et al, 1979), but most likely caused by subduction-zone slab dehydration (e.g., Stolper and Newman, 1994) including water released from seafloor serpentinized peridotites (Niu, 2004).

H_2O must be a more important component in the generation and evolution of back-arc basin lavas than in the genesis of mid-ocean ridge basalts. Assimilation- fractional crystallization- degassing allows efficient degassing of mantle-derived volatiles and contribution of crust- and atmosphere- derived volatiles in Manus Basin. Basaltic andesite, dacite glasses with high vesicles (up to 18%) show saturated H_2O (av. 826×10^{-6} mol/g) in the glasses. The volatiles in vesicles consist predominantly of H_2O (average 179×10^{-6} mol/g) with minor amounts of CO_2 (0.18×10^{-6} mol/g) (Marty et al., 2001).

The chemical compositions of volatiles in modern OUM still need more data due to following causes:

(1) Wholly or partially crystalline basaltic lavas but quench glasses from the outer surfaces of lavas show more oxidized state than mantle source regions (Christie et al.,1986), an initial highly reduced mafic magma may be oxidize by degassing of C-species in the conditions of the shallow crust, the volatiles in modern OUM are more reduced phases than volatiles in basalts (Mathez et al., 1984), similar to alkali basalts in eastern China, which composed of more oxidized CO_2 and SO_2 species in volatiles (Zhang et al., 1999) than hosted mantle xenoliths which are dominated by H_2 and CO (Zhang et al., 1999, 2000).

(2) The chemical compositions of volatiles recorded in oceanic basalts depend on the volatile fractionation, which result in heterogeneous volatiles in vesicles, matrix glasses and crystalline phases. Basalt glasses are quenched melts whose volatile contents represent those in the melt. Non-class basalts are partly solidified with significant volatile loss or gain depending on erupting environments. For example, glass-vapor inclusions within phenocrysts in pillow tholeiitic basalts from the Marianas Interarc Basin contain more water than in the adjacent matrix glasses, whereas inclusions in phenocrysts from MORB and Hawaii pillows contain more CO_2 than adjacent matrix glasses (Delaney et al., 1978). The volatile compositions of He, Ar and CO_2 in individual vesicles in basaltic glass from MAR vary from vesicle to vesicle due to Rayleigh-type fractional degassing. Volatile fractionation during vesicle formation may be superimposed on the overall Rayleigh degassing trend in one of the samples (Burnard, 1999).

(3) Different analytical methods may yield different results due to various distributions and occurrence modes of volatiles, e.g., H_2O in MORB from the FAMOUS in the Atlantic, which could be dissolved in melts, trapped in vesicles, or incorporated in structural site of crystalline phases, shows much lower contents by secondary ion mass spectrometry (SIMS) (2.9×10^{-6}mol/g, Roux et al., 2006) than by high temperature mass spectrometry (116.7×10^{-6}mol/g, Delaney et al., 1978). SIMS may detect small proportion of H_2O on spots.

6.2. Ancient Oceanic Upper Mantle

Most of ancient oceanic lithosphere has been subducted into the convective mantle, which result in mantle chemical and isotopic heterogeneity (Fyfe, 1997), thus no fresh samples of ancient OUM are available. Ancient oceanic basalts have been altered or weathered for volatile investigating of ancient OUM. The peridotite in the mantle section of ophiolites may be representative materials of ancient OUM for volatile analysis (Hu et al., 2007; Sachan et al., 2007).

Chemical and stable isotopic compositions of volatiles in ophiolitic assemblages can trace the nature and evolution of volatiles in ancient OUM, e.g., pillow basalts and sheet dykes in the upper crustal section of ophiolites show higher $\delta^{18}O$ values in bulk rock samples (6.8-12.7‰) than the mantle value, whereas gabbro, layer cumulate and peridotite etc. are depleted in ^{18}O, and have lower $\delta^{18}O$ values in bulk rock samples (3.7- 6.2‰). Such variations of oxygen isotopic composition with rock types are caused by water-rock isotopic exchanges at different temperatures (Agrinier et al., 1988; Cartwright and Barnicoa, 1999). The fluid inclusions with high concentration of CH_4 found in some ophiolitic peridotites (Su et al., 1999; Liu and Fei, 2006; Sachan et al., 2007) could trace primary volatile components in ancient OUM, including reduced fluids produced by serpentinization (Rona et al., 1992; Sachan et al., 2007) or volatile fractionation during inclusion entrapment (Burnard, 1999).

Fresh olivine and orthopyroxene in harzburgite and dunite of the Yushigou ophiolite in northwestern China with a $^{206}Pb/^{238}U$ age of 550±17 Ma using zircon of cumulate gabbro (Shi et al., 2004) have been measured for volatile compositions by stepwise heating to reveal the abundances and compositiosn of volatiles in ancient OUM (Hu et al., 2007; Zhang et al., 2008). Three concentration peaks of volatiles at 200-400°C, 400-800°C and 800-1200°C (Figure 2b) appear to correspond to the decrepitation degassing of secondary, metasomatic and primary three types of fluid inclusions, respectively (Su et al., 1999; Hu et al., 2007). The release temperature (400-800°C) of main volatile peak is lower than mantle xenoliths in eastern China (Zhang et al., 2004, 2007).

Table 7. Chemical compositions (mm^3.STP/g) of different type of volatiles in Qilian ancient OUM

Type		H_2	CH_4	C_2H_6	C_3H_8	CO	H_2S	N_2	Ar	O_2	CO_2	SO_2	H_2O
Initial	OUM	1.68	0.26	0.23	0.07	2.29	3.48	0.05	0.02	0.18	39.99	13.33	6.44
Volatiles	Magmatism	19.28	0.07	0.00	0.00	13.53	2.02	0.33	0.02	0.08	59.23	0.61	9.23
Metasomatic Volatiles		3.19	0.61	0.26	0.11	1.70	2.32	0.36	0.01	0.52	201.37	5.60	67.60
Alteration Volatiles		1.16	0.27	0.22	0.17	0.47	1.06	0.00	0.01	0.23	58.65	0.16	7.47

The volatiles in Yushigou ophiolitic peridotite are composed mainly of CO_2 with trace amounts of H_2, N_2, H_2O and SO_2, and vary with types of mineral and rock although Su et al (1999) reported more reduced compositions. Olv in dunite has higher CO_2 (338.42 mm^3.STP/g) with minor H_2O, H_2 and CO (94.52, 27.32 and 15.79mm^3.STP/g, respectively) than Olv in harzburgite (181.96mm^3.STP/g) with trace SO_2, H_2O and CO (32.23, 15.84 and 5.07mm^3.STP/g, respectively). Opx in harzburgite has highest volatile contents and is characterized by highest CO_2 (408.51mm^3.STP/g) with trace H_2O, SO_2, H_2 and CO (16.33, 8.13, 6.51 and 6.04mm^3.STP/g, respectively). The following 3 types of volatiles in the Qilian ancient OUM are identified based on the petrogensis, chemical compositions and C-O-H isotopes of volatiles, combined with the types, size, shapes and homogenization temperature of fluid inclusions (Table 7).

6.2.1. The Initial Volatiles in Qilian Ancient OUM

The volatiles released at 800-1200°C intervals are likely derived from volatiles trapped in primary fluid inclusions and structural defects and vacancies of the minerals, which could represent the initial volatiles in Qilian ancient OUM. The different chemical compositions of volatiles at 800-1200°C intervals between dunite and harzburgite with distinct petrogenesis indicate the presence of following two type of initial volatiles.

(1) The Initial Volatiles in Qilian Ancient OUM

Yushigou ophiolitic harzburgite is interpreted as melting residue (Feng and He, 1995; Song and Su, 1998; Shi et al., 2004). The volatiles released from Olv and Opx of harzburgite at the 800-1200°C interval are characterized by dominant CO_2 (av. 57.95 mm^3.STP/g) and high concentrations of SO_2 (av. 13.33 mm^3.STP/g), which represent the volatiles during initial stage of Qilian ancient OUM formation, and semble volatile compositions of mantle xenoliths from lithospheric mantle in Su-Lu continental subduction belt (Zhang et al., 2004, 2007).

(2) The Initial Volatiles in Magmatism

Yushigou ophiolitic dunite was considered as cumulate from very primitive melt (Feng and He, 1995; Niu, 2005). The volatiles released from olivine in dunite at 800-1200°C interval are characterized by dominant CO_2 (59.23 mm^3.STP/g) with trace H_2 (19.28) and CO (13.53), which may represent initial volatiles of a magmatic event in Qilian ancient oceanic mantle.

The variable $\delta^{18}O$ of CO_2 (4.5-8.8‰) and δD of H_2O (-78.3‰), lighter $\delta^{13}C$ (-21.2 to -23.0 ‰) in Olv of dunite relative to the values of upper mantle suggested that the volatiles may have originated in early SCLM extension stage of Qilian initial ocean (Hu et al., 2007). Heavier δD of H_2O (av. -27.2‰), lighter $\delta^{13}C_{CO2}$ (-29.1~ -19.5‰) and $\delta^{18}O_{CO2}$ (1.0‰) in Olv and Opx of harzburgite than the values of upper mantle indicated possible recycled seawater and crustal components.

The higher volatile contents in dunite Olv than in harzburgite Olv imply that magmatism is consistent with the harzburgite being melting residue and the dunite being cumulate from the primitive basaltic melt. It is also possible that the magmatism may have taken place under volatile-rich conditions. The fluid inclusions with high concentration of CH_4 found in some ophiolitic peridotites (Su et al., 1999) suggest that the CH_4 could come from deeper mantle (Ballhaus, 1995; Su et al., 1999; Woodland and Koch, 2003) and undergone fractionation of volatile species during fluid inclusion entrapment (Burnard, 1999), or perhaps from inorganic synthesizing during serpentinization (Rona et al., 1992; Sachan et al., 2007).

6.2.2. The Metasomatic Volatiles in Qilian Ancient OUM

Mantle metasomatism was involved in Qilian ancient OUM, which is indicated by light REE enriched in harzburgite (Shi et al., 2004). The volatiles as metasomatic agents had been preserved as metasomatic fluid inclusions in healed fractures of constitute minerals in ophiolitic peridotite, which are decrepitated at similar temperature range (400-800°C) to homogenization temperature of metasomatic fluid inclusion (Hu et al., 2007). The metasomatic volatiles in Qilian ancient OUM are characterized by much high volatile concentration and dominant CO_2 (201.37mm^3.STP/g on average) with trace H_2O (67.60), SO_2 (5.60) and H_2 (3.19). Lighter δD_{H2O} (-306.3‰), $\delta^{13}C_{CO2}$ (-15.5~-12.5‰) and $\delta^{18}O_{CO2}$ (1.4~1.9‰, av. 1.6‰) suggest the processes of mantle degassing and addition of seawater.

6.2.3. The Alteration Volatiles in Qilian Ancient OUM

Ophiolitic peridotites have always underdone some low-temperature alteration, i.e. serpentinization and carbonatization with accompanied deformation. Alteration volatiles had been trapped in mineral cleavages or micro-fractures along mineral margins as secondary fluid inclusions that decrepitated at a temperature interval of 200-400°C. The alteration volatiles are mainly composed of CO_2 (58.65 mm^3.STP/g) and relatively high concentration of hydrocarbons, which appear to have been generated as a by-product of Fischer-Tropsch synthesis during a subsequent serpentinization of ultramafic rocks (Abrajano et al., 1988; Charlou et al., 1998; Sachan et al., 2007). Their low contents of volatiles could be caused by the outgassing of secondary fluid inclusions during sample pretreatment before analysis.

7. Summary

Volatiles in lithospheric mantle minerals occur in varied modes. The volatile speciation varies with oxygen fugacity and tectonic settings. The improved vacuum stepwise heating is an effective technique to extract volatiles resided in different sites in mantle minerals. The chemical compositions of mantle volatiles recorded in basalts, diamonds and mantle xenoliths differ from continental to oceanic settings. The main conclusions are as follows.

(1) The occurrence modes of mantle volatiles

1) Volatiles in the mantle occur as free states of elements (e.g., graphite, diamond etc.) and molecules (CO_2, H_2O etc.), as crystal structural components in hydrous and nominally anhydrous minerals or carbonate, or as fluid inclusions or as dispersed forms in structure defects and vacancies in mantle minerals.
2) Volatiles trapped in constituent minerals of mantle xenoliths occur as both fluid inclusions and structural sites, defects and vacancies. Hydrogen mainly occurs as H_2 in fluid inclusions rather than structural OH^- as previously thought. Much of the mantle volatiles may in fact reside in structural defects and vacancies, probably more so than in mineral fluid inclusions.
3) Oxygen fugacity in the mantle is spatially and temporally variable, and controls mantle volatile speciation.

(2) The improved vacuum stepwise heating technique

1) The improved vacuum stepwise heating can effectively separate volatiles in fluid inclusions from crystal structural defects and vacancies for analysis, and shows excellent analytical reproducibility of volatile compositions. Volatiles derived

from fluid inclusions of different stages have little superimposed effects on each other.

2) Fluid inclusions can release volatiles by progressive crushing and stepwise heating. Late-stage fluid inclusions with large size can be broken completely by crushing to small grain size or released at lower temperatures by stepwise heating.
3) The decrepitation temperature of fluid/melt inclusion depends on physical strength of crystals and pressure difference between the interior and exterior of the inclusions. The early-stage fluid inclusions are preserved intact in mineral interiors of mantle xenoliths, and are expected to be decrepitated at temperatures similar to or slightly lower than the homogenization temperature due to high anticipated internal pressures of inclusions and brickle constituent minerals.
4) Volatiles trapped in structural defects and vacancies are released at higher heating temperatures close to the formation temperatures of mantle minerals.

(3) Chemical compositions of volatiles in the SCLM

1) The volatiles in the SCLM are a mixture of volatiles of various origins, and vary with depth and mantle reservoirs. Volatiles in the deeper mantle corresponding to diamond stability field have higher contents of reduced volatile species such as H_2 and CO, whereas volatiles in the SCLM recorded in mantle xenoliths vary with mineral species and mantle reservoirs.
2) Mantle xenoliths show volatiles of different origin as revealed by the varying chemical and isotopic compositions of volatiles. Some volatiles are primordial and some are of metasomatic origin. Initial volatiles in the SCLM, i.e., volatiles at a primary stage of upper mantle crystallization, are associated with mantle reservoirs. Initial volatiles in the SCLM are dominated by CO with minor CO_2 and H_2 in primitive mantle-like reservoir indicted by Nd-Sr-Pb isotopes, and have high contents of volatile and H_2 in enriched mantle reservoir; whereas initial volatiles in the SCLM are dominated by CO_2 with minor SO_2 and H_2 in depleted mantle reservoir, and are primarily composed of CO_2 and SO_2 in a "mixed" mantle reservoir related to recycled crustal fluids. Most of the initial volatiles are of mantle origin.
3) The metasomatic volatiles in the SCLM result from subsequent mantle events, either exhibiting high contents H_2 from depths or containing oxidized species like CO_2 and SO_2 resulting from recycled crustal volatiles.

(4) Chemical compositions of volatiles in OUM

1) Volatiles in modern OUM could be more in reduced species than volatiles in modern oceanic basalts, which are dominated by H_2O and CO_2 with minor CO, CH_4, N_2 and H_2. H_2O is most depleted (57.32×10^{-6}mol/g) in MORB, and increases from OIB (404.61×10^{-6} mol/g), to BABB (568.33) and to IAB (592.08), whereas CO_2 content increases from OIB (9.36×10^{-6} mol/g) to MORB (11.72), to BABB (47.33) and to IAB (55.63). MORB are overall depleted in volatiles. Enriched volatiles in OIB, IAB and BABB are probably related to recycled volatiles by subduction.

2) Very few data of volatiles in ancient OUM are available. The volatiles released from several olivine and orthopyroxene samples of peridotites in the Yushigou ophiolite, western China reveal that CO_2 is a major constituent in the volatiles in ancient OUM. The volatiles vary in chemical composition, depending on mineral species and rock types. Initial volatiles of Qilian ancient OUM are characterized by dominant CO_2 (mean 57.95mm^3.STP/g) and high concentrations of SO_2 (mean 13.33 mm^3.STP/g). Initial volatiles of magmatism are dominated by CO_2 (59.23 mm^3.STP/g) and characterized by high concentrations of H_2 (19.28) and CO (13.53), volatiles in a metasomatism in Qilian ancient OUM are mainly composite of CO_2 (201.37 mm^3.STP/g on average) and trace H_2O (67.60), SO_2 (5.60) and H_2 (3.19).

Acknowledgements

This study was supported by NSFC (Grants 40273009, 40772058, 40534020, 40873005), the Program for New Century Excellent Talents (NCET-04-0980) by the Chinese Ministry of Education and State Key Lab of Gas-Geochemistry, CAS. We thank the Royal Society for support through China-UK Science Network program. We are very grateful to Prof. X. B. Wang, Q.B. Wen and L.W. Li for their helps with analytical work.

References

Abrajano, T. A.; Sturchio, N. C.; Bohlke, J. K.; Lyon, G. L.; Poreda, R. J.; Stevens, C. M. Methane- hydrogen gas seeps, Zambales Ophiolite, Philippines: Deep or shallow origin? *Chemical Geology* 1988, 71, 211-222.

Ackermann, L.; Cemič, L.; Langer, K. Hydrogarnet substitution in pyrope: a possible location for "water" in the mantle. *Earth and Planetary Science Letters* 1983, 62, 208-214.

Aggrey, K. E.; Muenow, D. W.; Batiza, R. Volatile abundances in basaltic glasses from seamounts flanking the East Pacific Rise at 21°N and 12-14°N. *Geochimica et Cosmochimica Acta* 1988a, 52, 2115-2119.

Aggrey, K. W.; Muenow, D. W.; Sinton, J. M. Volatile abundances in submarine glasses from the North Fiji and Lau back-arc basins. *Geochimica et Cosmochimica Acta* 1988b, 52, 2501-2506.

Agrinier, P.; Javoy, M.; Girardeau, J. Hydrothermal activity in a peculiar oceanic ridge: oxygen and hydrogen isotope evidence in the Xigaza ophiolite (Tibet, China). *Chemical Geology* 1988, 71, 313-335

Albarede, F. Rogue mantle helium and neon. *Science* 2008, 319, 943-945.

Allegre, C. J.; Staudacher, T.; Sarda, P. Rare gas systematics: formation of the atmosphere, evolution and structure of Earth's mantle. *Earth and Planetary Science Letters* 1987, 81, 127-150.

Andersen, T.; Neumann, E. R. Fluid inclusions in mantle xenoliths. *Lithos* 2001, 55, 301-320.

Anderson, T.; O'Reilly, S. Y.; Griffin, W. L. The trapped fluid phase in upper mantle xenoliths from Victoria, Australia: implication for mantle metasomatism. *Contributions to Mineralogy and Petrology* 1984, 88, 72-85.

Aubaud, C.; Pineau, F.; Hekinian, R.; Javoy, M. Degassing of CO_2 and H_2O in submarine lavas from the Society hotspot. *Earth and Planetary Science Letters* 2005, 235, 511- 527.
Aubaud, C.; Withers, A. C.; Hirschmann, M. M.; Guan, Y.; Leshin, L. A.; Mackwell, S. J.; Bell, D. R. Intercalibration of FTIR and SIMS for hydrogen measurements in glasses and nominally anhydrous minerals. *American Mineralogist* 2007, 92, 811- 828.
Bai, Q. and Kohlstedt, D. L. Substantial hydrogen solubility in olivine and implications for water storage in the mantle. *Nature* 1992, 357, 672-674.
Ballhaus, C. Redox states of lithospheric and asthenospheric upper mantle. *Contributions to Mineralogy and Petrology* 1993, 114, 331-348
Ballhaus, C. Is the upper mantle metal-saturated? *Earth and Planetary Science Letters* 1995, 132, 75-86.
Bell, D. R. and Rossman, G. R. Water in Earth's mantle. *Science* 1992, 255, 1391-1397.
Bell, D.R., and Ihinger, P.D. The isotopic composition of hydrogen in nominally anhydrous mantle minerals. *Geochimica et Cosmochimica Acta* 2000, 64, 2109-2118.
Bell, D. R.; Ihinger, P. D.; Rossman, G. R. Quantitative analysis of trace OH in garnet and pyroxenes. *American Mineralogist* 1995, 80, 465-474.
Beran, A. and Libowitzky, E. Water in natural mantle minerals II: Olivine, garnet and accessory minerals. *Reviews in Mineralogy and Geochemistry* 2006, 62, 169-191.
Bergman, S. C.; Dubessy, J. CO_2-CO fluid inclusions in a composite peridotite xenolith: implications for upper mantle oxygen fugacity. *Contributions to Mineralogy and Petrology* 1984, 85, 1-13.
Bernardj, W. and Ltaras, B. Determination of methane stability in the upper mantle and lower crust. *American Journal of Science* 1992, 290, 1-24.
Blundy, J. D.; Brodholt, J. P.; Wood, B. J. Carbon-fluid equilibria and the oxidation state of the mantle. *Nature* 1991, 349, 321-324.
Burnard, P. The bubble-by-bubble volatile evolution of two mid-ocean ridge basalts. *Earth and Planetary Science Letters* 1999, 174, 199-211.
Burnard, P. G.; Farley, K. A.; Turner, G. Multiple fluid pulses in a Samon harzburgite. *Chemical Geology* 1998, 147, 99-114
Burnard, P. G.; Graham, D.; Turner, G. Vesicle-specific noble gas analysis of "popping rock": implications for primordial gases in earth. *Science* 1997, 276, 568-571.
Byers, C. D.; Garcia, M. D.; Muenow, D. W. Volatile in pillow rim glasses from Loihi and Kilauea Volcanoes, Hawaii. *Geochimica et Cosmochimica Acta* 1985, 49, 1887-1896.
Byers, C. D.; Garcia, M. O.; Muenow, D. W. Volatile in basaltic glasses from the East Pacific Rise at 21°N. *Earth and Planetary Science Letters* 1986, 79, 9-20.
Byers, C. D.; Muenow, D. W.; Garcia, M. O. Volatile in basalts and andesites from the Galapagos spreading center, 85° to 86°W. *Geochimica et Cosmochimica Acta* 1983, 47, 1151-1558.
Byers, C. D.; Christie, D. M.; Muenow, D. W.; Sinton, J. M. Volatile contents and ferric-ferrous ratios of basalt, ferro-basalt, andesite, and thyodacite glasses from Galepagos, 95.5°W propagating rift. *Geochimica et Cosmochimica Acta* 1984, 48, 2239-2245.
Canil, D. and Bellis, A. J. Phase equilibria in a volatile-free kimberlite at 0.1 MPa and the search for primary kimberlite magma. *Lithos* 2008, doi:10.1016/j.lithos.2008.02.011.
Castillo, P. R. The Dupal anomaly as a trace of the upwelling lower mantle. *Nature* 1988, 336, 667-670.

Cartigny, P.; Jendrzewjewski, N.; Pineau, F.; Petit, E.; Javoy, M. Volatile (C, N, Ar) variability in MORB and the respective roles of mantle source heterogeneity and degassing. *Earth and Planetary Science Letters* 2001, 194, 241-257.

Cartwright, I. and Barnicoat, A. C. Stable isotope geochemistry of Alpine ophiolites: a window to ocean-floor hydrothermal alteration and constraints on fluid-rock interaction during high-pressure metamorphism. *Int J. Earth Sciences* 1999, 88, 219-235.

Charlou, J. L.; Fouquet, Y.; Bougault, H.; Donval, J. P.; Etoubleau, J.; Jean-Baptiste, P.; Dapoigny, A.; Appriou, P.; Rona, P. A. Intense CH_4 generated by serpentinization of ultramafic rocks at the intersection of the 15° 20'N fracture zone and the Mid-Atlantic Ridge. *Geochimica et Cosmochimica Acta* 1998, 62, 2323-2333.

Chazot, G.; Lowry, D.; Menzies, M. Oxygen isotopic composition of hydrous and anhydrous mantle peridotites. *Geochimica et Cosmochimica Acta* 1997, 61,161-169.

Christie, D. M.; Carmichael, I. S. E; Langmuir, C. H. Oxidation states of mid-ocean ridge basalt glasses. *Earth and Planetary Science Letters* 1986, 79, 397-411.

Coltorti, M.; Beccaluva, L.; Bonadiman, C. Glasses in mantle xenoliths as geochemical indicators of metasomatic agents. *Earth and planetary science letters* 2000, 183, 303-320.

Deines, P. The carbon isotopic composition of diamonds relationship to diamond shape, color, occurrence and vapour composition. *Geochimica et Cosmochimica Acta* 1980, 44, 943-961.

Deines, P. Mantle carbon: concentration, mode of occurrence and isotopic composition. In *Early organic evolution: Implications for mineral and energy resources*. Schidlowski, M. et al. eds.; Springer-Verlag: Berlin, 1992, pp133-146.

Deines, P. The carbon isotope geochemistry of mantle xenoliths. *Earth Science Reviews* 2002, 58, 247-278.

Deines, P.; Harris, J. W.; Gurney, J. J. Carbon isotope ratios, nitrogen content and aggregation state, and inclusion chemistry of diamonds from Jwaneng, Botswana. *Geochimica et Cosmochimica Acta* 1997, 61, 3993-4005.

Delaney, J. R.; Muenow, D. W.; Graham, D. G. Abundance and distribution of water, carbon and sulfur in the glassy rims of submarine pillow basalts, *Geochimica et Cosmochimica Acta* 1978, 42, 581-594.

Dixon, J. E. and Clague, D. A. Volatiles in Basaltic Glasses from Loihi Seamount, Hawaii: Evidence for a relatively dry plume component. *Journal of Petrology* 2001, 42, 627-654

Dixon, J. E.; Clague, D. A.; Wallace, P.; Poreda, R. Volatiles in alkalic basalts from the North Arch volcanic field, Hawaii. *Journal of Petrology* 1997, 38, 911-939.

Dixon, J. E.; Stolper, E.; Delaney, J. R. Infrared spectroscopic measurements of CO_2 and H_2O in Juan de Fuca Ridge basaltic glasses. *Earth Planetary Science Letters* 1988, 90, 81-107.

Eggler, D. H. and Baker, D. R. Reduced volatiles in the system C-O-H: Implications to mantle meting, fluid formation and diamond genesis, In *Advances in Earth and Planetary Sciences, High pressure research in geophysics,* Akimoto, S. and Manghnani, M.H., Eds.; Ceter for Academic Pub.: Reidel, Dordrecht, Holland, 1982; 12, pp237-250.

Eiler, J. M.; Farley, K. A.; Valley, J. W.; Stolper, E. M.; Hauri, E. H.; Craig, H. Oxygen isotope evidence against bulk recycled sediment in the mantle sources of Pitcairn Island lavas. *Nature* 1995, 377, 138-141.

Eiler, J. M.; Farley, K. A.; Valley, J. W.; Hauri, E.; Craig, H.; Hart, S. R.; Stolper, E. Oxygen isotope variations in ocean island basalt phenocrysts. *Geochimica et Cosmochimica Acta* 1997, 61, 2281-2293.

Ertan, I. E. and Leeman, W. P. Fluid inclusions in mantle and lower crustal xenoliths from the Simcoe volcanic field, Washington. *Chemical Geology* 1999, 154, 83-95.

Exley, R. A.; Mattey, K. P., Clague, K. A., Pillinger, C. T. Carbon isotope systematics of a mantle "hot spot": a comparison of Loihi Seamount and MORB glasses. *Earth and Planetary Science Letters* 1986, 78, 189-199.

Fan, Q.-C. and Hooper, P. R. The mineral chemistry of ultramafic xenoliths of eastern China: Implications for upper mantle composition and the palaeogeotherms. *Journal of Petrology* 1989, 30, 1117-1158.

Farley, K. A. and Neroda, E. Noble gases in the Earth's mantle. *Annu. Rev. Earth Planetary Science Letters* 1998, 26, 189-218.

Feng, Y.-M. and He, S.-P. Research for geology and geochemistry of several ophiolites in the North Qilian Mountains,China. *Acta Petrologica Sinica* 1995, 11, 125-146 (in Chinese)

Freund, F. and Oberheuser, G. Water dissolved in olivine: a single-crystal infrared study. *Journal of Geophysical Research* 1986, 91, 745-761.

Frezzotti, M.-L. and Peccerillo, A. Diamond-bearing COHS fluids in the mantle beneath Hawaii. *Earth and Planetary Science Letters* 2007, 262, 273-283.

Fyfe, W. S. Deep fluids and volatile recycling: crust to mantle. *Tectonophysics* 1997, 275, 243-251.

Garcia, M. O.; Liu, N. W. K.; Muenow, D. W. Volatiles in submarine volcanic rocks from the mariana Island Arc and Trough. *Geochimica et Cosmochimica Acta* 1979, 43, 305-312.

Graney, J. and Kesler, S. E. Factors affecting gas analysis of inclusion fluid by quadrupole mass spectrometry. *Geochimica et Cosmochimica Acta* 1995, 59, 3977-3986.

Grant, K. J.; Brooker, R. A.; Kohn, S. C.; Wood, B. J. The effect of oxygen fugacity on hydroxyl concentrations and speciation in olivine: Implications for water solubility in the upper mantle. *Earth and Planetary Science Letters* 2007, 261, 217-229

Green, H. W. and Radcliffe, S. V. Fluid precipitates in rocks from the Earth's mantle. *Geol. Soc. Am, Bul.*1975, 86, 846-852.

Griffin, W. L.; Zhang, A.; O'Reilly, S. Y.; Ryan, C. G. Phanerozoic evolution of the lithosphere beneath the Sino-Korean craton. In: Flower, M., Chung, S. L., Lo, C. H., Lee, Y. Y. (eds.), Mantle dynamics and plate interactions in east Asia. *Amer. Geodynamics Series* 1998, 27, 107-126.

Hansteen, T. H. and Troll, V. R. Oxygen isotope composition of xenoliths from the oceanic crust and volcanic edifice beneath Gran Canaria (Canary Islands). *Chemical Geology* 2003, 193, 181-193.

Hansteen, T. H.; Andersen, T.; Neumann, E. R.; Jelsma, H. Fluid and silicate glass inclusion in ultramafic and mafic xenoliths from Hierro, Canary Islands. *Contributions to Mineralogy and Petrology* 1991, 107, 242-254.

Heide, K. and Schmidt, C. M. Volatiles in vitreous basaltic rims, HSDP2, big island, Hawaii. *Journal of Non-Crystalline Solids* 2003, 323, 97-103.

Hilton, D. R. The leaking mantle. *Science* 2007, 318, 1389-1390.

Hochstaedter, A. G.; Gill, J. B.; Kusakabe, M.; Newman, S.; Pringle, M.; Taylor, B.; Fryer, P. Volcanism in the Sumisu Rift, I. Major element, volatile, and stable isotope geochemistry. *Earth and Planetary Science Letters* 1990, 100, 179-194.

Hofmann, A W. Mantle geochemistry: the message from oceanic volcanism. *Nature* 1997, 385, 219-229.

Hu, P.-Q.; Zhang, M.-J.; Tang, Z. L. Chemical and stable isotopic compositions of volatiles in mantle peridotites of the Yushigou Ophiolite in Qilian orogenic belt, western China. *Acta Petrologica Sinica* 2007, 23, 169-176.

Ingrin, J.; Hercule, S.; Charton, T. Diffusion of hydrogen in diopside: results of dehydration experiments. *Journal of Geophysical Research* 1995, 100, 15489-15499.

Inono, D. A.; Hoefs, J.; Wedepohi, K. H.; Wiechert, U. Contents and isotopic composition of sulfur in ultramafic xenoliths from Central Asia. *Earth and Planetary Science Letters* 1992, 111, 269-286.

Ionov, D. A. and Wood, B. J. The oxidation state of subcontinental mantle: oxygen thermobarometry of mantle xenoliths from central Asia. *Contributions to Mineralogy and Petrology* 1992, 111, 179-193.

Ionov, D. A.; Harmon, R. S.; France-Lanord, C.; Greenwood, P. B.; Ashchepkov, I. V. Oxygen isotope composition of garnet and spinel peridotites in the continental mantle: Evidence from the Vitim xenolith suite, southern Siberia. *Geochimica et Cosmochimica Acta* 1994, 58, 1463-1470.

Ivankin, P. F.; Argunov, K. P.; Boris, Ye. I. Stages of kimberlite development and evolving conditions of diamond formation. *International Geology Review* 1988, 30, 268-274.

Jambon, A. Earth degassing and large-scal geochemical cycling of volatile elements In: Carroll, M. R. and Holloway, J. R. (eds.) Volatiles in magmas. Mineralogical Scociety of America, *Reviews in Mineralogy* 1994, 30, 479-517.

Javoy, M. Major volatile elements of the earth: their origin, behavior and fate. *Geophysical Research Letters* 1997, 24, 177-180.

Javoy, M. and Pineau, F. The volatiles record of a "popping" rock from the Mid-Atlantic Ridge at 14°N: chemical and composition of gas trapped in the vesicles. *Earth and Planetary Science Letters* 1991, 107, 598-611.

Kelley, D. S. Methane-rich fluids in the oceanic crust. *Journal of Geophysical Research* 1996, 101, 2943-2962.

Kelley, D. S. and Frueh-Green, G. Methane concentrations and isotopic compositions in Layer 3 of the Oceanic crust. *EOS* 1995, 76, 675.

Kitamura, M.; Kondoh, S.; Morimoto, N.; Miller, G. H.; Rossman, G. R.; Putnis, A. Planar OH-bearing defects in mantle olivine. *Nature* 1987, 328, 143-145.

Kohn, S. C. Solubility of H_2O in nominally anhydrous mantle minerals using ^1H MAS NMR. *American Mineralogist* 1996, 81, 1523-1526

Koch-Müller, M.; Matsyuk, S. S.; Wirth, R. Hydroxyl in omphacites and omphacitic clinopyroxenes of upper mantle to lower crustal origin beneath the Siberian platform. *American Mineralogist* 2004, 89, 921-931.

Kyser, T. K. Stable isotope variations in the mantle. In: Valley, J. W., Taylor, Jr., H.P., O'Neil, J.R. (Eds.), Stable isotopes in high temperature geological processes. *Reviews in Mineralogy* 1986, 16, 141-164.

Kyser, T. K. and O'Neil, J. R. Hydrogen isotope systematics of submarine basalts. *Geochimica et Cosmochimica Acta* 1984, 48, 2123-2133.

Landis, G. P. and Rye, R. O. Characterization of gas chemistry and noble-gas isotope ratios of inclusion fluids in magmatic-hydrothermal and magmatic-steam alunite. *Chemical Geology* 2005, 215, 155-184

Litasov, K. D.; Foley, S. F.; Litasov, Y. D. Magmatic modification and metasomatism of the subcontinental mantle beneath the Vitim volcanic field East Siberia: evidence from trace element data on pyroxenite and peridotite xenoliths from Miocene picrobasalt. *Lithos* 2000, 54, 83-114.

Liu, W.; Fei, P. -X. Methane-rich fluid inclusions from ophiolitic dunite and post-collisional mafic- ultramafic intrusion. *Earth and Planetary Science Letters* 2006, 242, 286-301

Lowenstern, J. B. Dissolved volatile concentrations in an oreforming magma. *Geology* 1994, 22, 893-896.

Marty, B. and Zimmermann, L. Volatiles (He, C, N, Ar) in mid-ocean ridge basalts: Assment of shallow-level fractionation and characterization of source composition. *Geochimica et Cosmochimica Acta* 1999, 63, 3619-3633.

Marty, B.; Sano, Y.; France-Lanord, C. Water-saturated oceanic lavas from the Manus Basin. *Journal of Volcanology and Geothermal Research* 2001, 108, 1-10.

Mathez, E. A. Carbonaceous matter in mantle xenoliths: composition and relevance to the isotopes. *Geochimica et Cosmochimica Acta* 1987, 51, 2339-2347.

Mathez, E. A.; Dietrich, V. J.; Irving, A. J. The geochemistry of carbon in mantle peridotites. *Geochimica et Cosmochimica Acta* 1984, 48, 1849-1859.

Matson, D. W. and Muenow, D. W. Volatiles in amphiboles from xenoliths, Vulcan's Throne, Grand Canyon, Arizona, USA. *Geochimica et Cosmochimica Acta* 1984, 48, 1629-1636.

Matsumoto, T.; Honda, M.; McDougall, I.; O'Reilly, S. Y.; Norman, M.; Yaxley, G. Noble gases in pyroxenites and metasomatised peridotites from the Newer Volcanics, southeastern Australia. *Chemical Geology* 2000, 168, 49-73.

Mattey, D.; Lowry, D.; Macpheson, C. Oxygen isotope composition of mantle peridotite. *Earth and Planetary Science Letters* 1994, 128, 231-241.

Martin, R.F.; Donnay, G. Hydroxyl in the mantle. *American Mineralogist* 1972, 57, 554-570.

Melton, C. E. and Giardini, A. A. The composition and significance of gas released from natural diamonds from Africa and Brazil. *American Mineralogist* 1974, 59, 775-782.

Melton, C. E.; Salotti C.A.; Giardini, A. A. The observation of nitrogen, water, carbon dioxide, methane and argon as impurities in natural diamonds. *American Mineralogist* 1972, 57, 1518-1523.

Michael, P. Regionally distinctive sources of depleted MORB: evidence from trace elements and H_2O. *Earth and Planetary Science Letters* 1995, 131, 301-320.

Mierdel, K.; Keppler, H.; Smyth, J. R.; Langenhorst, F. Water solubility in aluminous orthopyroxene and the origin of Earth's asthenosphere. *Science* 2007, 315, 364-368.

Miller, M. F. and Pillinger, C. T. An appraisal of stepwise heating release of fluid inclusion CO_2 for isotopic analysis. *Geochimica et Cosmochimica Acta* 1997, 61, 193-205.

Moune, S.; Sigmarsson, O.; Thordarson, T.; Gauthier, P.-J. Recent volatile evolution in the magmatic system of Hekla volcano, Iceland. *Earth and Planetary Science Letters* 2007, 255, 373-389

Muenow, D.W.; Graham, D.G.; Liu,N.; Delaney, J.R. The abundance of volatiles in Hawaiian tholeiitic submarine basalts, *Earth and Planetary Science Letters* 1979, 42, 71-76.

Muneow, D. W.; Perfit, M. R.; Aggrey K. E. Abundances of volatiles and genetic relationships among submarine basalts from the Woodlark Basin, southwest Pacific. *Geochimica et Cosmochimica Acta* 1991, 55, 2231-2239.

Murck, B. W.; Burruss, R. C.; Hollister, L. S. Phase equilibria in fluid inclusions in ultramafic xenoltihs. *American Mineralogist* 1978, 63, 40-46.

Nadeau, S.; Pineau, F.; Javoy, M.; Francis, D. Carbon concentrations and isotopic ratios in fluid inclusion-bearing upper mantle xenoliths along the northwestern margin of North America. *Chemical Geology* 1990, 81, 271-297.

Navon, O. High internal pressures in diamond fluid inclusions determined by infrared absorption. *Nature* 1991, 353, 746-748

Navon, O.; Hutcheon, I. D.; Rossman, G. R.; Wasserburg, G. J. Mantle-derived fluid in diamond micro-inclusions. *Nature* 1988, 335, 784-789.

Nishio, Y.; Ishii, T.; Toshitaka, G.; Sano, Y. Volatile element isotopic systematics of the Rodrigues Triple Junction Indian Ocean MORB. *Earth and Planetary Science Letters* 1999, 170, 241-253.

Niu, Y. L., Bulk-rock major and trace element compositions of abyssal peridotites: Implications for mantle melting, melt extraction and post-melting processes beneath ocean ridges, *Journal of Petrology* 2004, 45, 2423-2458.

Niu, Y. L. and Batiza, R. Trace element evidence from seamounts for recycled oceanic crust in the eastern equatorial Pacific mantle. *Earth and Planetary Science Letters* 1997, 148, 471-484.

Niu, Y. L. Generation and evolution of basaltic magmas: Some basic concepts and a new view on the origin of Mesozoic-Cenozoic basaltic volcanism in eastern China. *Geological Journal of China Universities* 2005, 11, 9-46.

Niu, Y. L.; O'Hara, M. J.; Pearce, J. A. Origin of ocean island basalts: A new perspective from petrology, geochemistry and mineral physics considerations. *Journal of Geophysical Research* 2003, 108, 10.1029/ 2002JB002048.

Norman, D. I. and Musgrave, J. A. N_2-Ar-He compositions in fluid inclusions: indicator of fluid source. *Geochimica et Cosmochimica Acta* 1994, 58, 1119-1131.

O'Neill, H. S. C. and Wall, V. J. The olivine-orthopyroxene-spinel oxygen geobarometer 'the nickel precipitation curve' and the oxygen fugacity of the earth's upper mantle. *Journal of Petrology* 1987, 28, 1169-1191.

O'Reilly, S. Y. and Griffin, W. L. Apatite in the mantle: implications for metasomatic processes and high heat production in Phanerozoic mantle. *Lithos* 2000, 53, 217-232.

O'Reilly, S.Y. and Griffin, W. L. Mantle metasomatism beneath western Victoria, Australia: I, Metasomatic processes in Cr-diopside lherzolites. *Geochimica et Cosmochimica Acta* 1988, 52, 433-447.

Ozima, M. and Podosek, F. A. *Noble gas geochemistry*. Cambridge University Press: Cambridge, 2001; pp1-280.

Olmez, I. and Finnegan, D. L. Iridium emission from Kilauea volcano. *Journal of Geophysical Research* 1986, 91, 653-663.

O'Nions, R. K. and Tolstikhin, I. N. Behaviour and residence time of lithophile and rare gas tracers in the upper mantle. *Earth and Planetary Science Letters* 1994, 124, 131-138.

Pasteris, J. D. and Wanamaker, B. J. Laser Raman microprobe analysis of experimentally re-equilibrated fluid inclusions in olivine: Some implications for mantle fluids. *American Mineralogist* 1988, 73, 1074-1088.

Peng, L.-G., Liu, R.-X., Fan, Q.-C., Fluid and melt inclusion releases of pyrolite xenoliths from the shore of southern China. *Acta Petrologica Sinica* 1994, 10,440-448. (in Chinese with English abstract).

Pineau, F. and Mathez, E. A. Carbon isotopes in xenoliths from the Hualalai volcano, Hawaii, and the generation of isotopic variability. *Geochimica et Cosmochimica Acta* 1990, 54, 217-227.

Pineau, F. and Javoy M., Carbon isotopes and concentrations in mid-oceanic ridge basalts, *Earth and Planetary Science Letters* 1983, 62, 239-257.

Porcelli, D. and Wasserburg, G. J. Mass transfer of helium, neon, argon and xenon through a steady-state upper mantle. *Geochimica et Cosmochimica Acta* 1995, 59, 4921-4937.

Ratan, K., and Mohapatra, M. H. "Recycled" volatiles in mantle-derived diamonds. *Earth and Planetary Science Letters* 2006, 252, 215-219.

Regelous, M.; Niu, Y. L.; Abouchami, W.; Castillo, P. R. Shallow origin for South Atlantic Dupal Anomaly from lower continental crust. *Lithos* doi:10.1016/j.lithos.2008.10. 012.

Rehkamper, M. and Hofmann, A. W. Recycled ocean crust and sediment in Indian Ocean MORB. *Earth and Planetary Science Letters* 1997, 147, 93-106.

Roedder, E. Liquid CO_2 inclusions in olivine-bearing nodules and phenocrysts from basalts. *American Mineralogist* 1965, 50, 1746-1782.

Roedder, E. Fluid inclusions. *Reviews in Mineralogy* 1984, 12, 1-164.

Roedder, E. Fluid inclusion analysis-prologue and epilogue. *Geochimica et Cosmochimica Acta* 1990, 54, 495-507.

Rosenbaum, J. M.; Zindler, A.; Rubenstone, J. L. Mantle fluids: evidence from fluid inclusions. *Geochimica et Cosmochimica Acta*, 1996, 60, 3229-3252.

Roux, P. J.; Shirey, S. B.; Hauri E. H.; Perfit M. R.; Bender, J. F. The effects of variable sources, processes and contaminants on the composition of northern EPR MORB (8-10°N and 12-14°N): Evidence from volatiles (H_2O, CO_2, S) and halogens (F, Cl). *Earth and Planetary Science Letters* 2006, 251, 209-231.

Rubin, K. H. Degassing of metals and metalloids from erupting seamount and mid-ocean ridge volcanoes. *Geochimica et Cosmochimica Acta* 1997, 61, 3525-3542.

Sachan, H. K.; Mukherjee, B. K.; Bodnar, R. J. Preservation of methane generated during serpentinization of upper mantle rocks. *Earth and Planetary Science Letters* 2007, 257, 47-59.

Saxena, S. K. and Fei, Y.-W. Fluid mixtures in the C-H-O system at high pressure and temperature. *Geochimica et Cosmochimica Acta* 1988, 52, 505-512.

Schiano, P.; Birck, J.L.; Allegre, C.A. Osmium- strontium- neodymium- lead isotopic covariations in mid-ocean ridge basalt glasses and the heterogeneity of the upper mantle. *Earth and Planetary Science Letters* 1997, 150, 363-379.

Self, S.; Widdowson, M.; Thordarson, T.; Jay, A. E. Volatile fluxes during flood basalt eruptions and potential effects on the global environment: A Deccan perspective. *Earth and Planetary Science Letters* 2006, 248, 518-532

Shaw, A.M.; Hilton, D.R.; Macpherson, C.G.; Sinton, J.M. The CO_2-He-Ar-H_2O systematics of the Manus back-arc basin. *Geochimica et Cosmochimica Acta* 2004, 68, 1837-1856.

Shcheka, S. S.; Wiedenbeck, M.; Frost, D. J.; Keppler, H. Carbon solubility in mantle minerals. *Earth and Planetary Science Letters* 2006, 245, 730-742.

Shi, R.-D.; Yang, J.-S.; Wu, C.-L. The evidence of SHRIMP age formed Late Sinina of Yushigou ophiolite in North Qilian mountains, China. *Acta Geol. Sin.* 2004, 78, 649-657.

Skogby, H. Water in natural mantle minerals I: Pyroxenes. *Reviews in Mineralogy and Geochemistry* 2006, 62, 155-167.

Skogby, H. and Rossman, G. R. OH^- in pyroxene: an experimental study of incorporation mechanisms and stability. *American Mineralogist* 1989, 74, 1059-1069.

Smyth, J. R.; Bell, D. R.; Rossman, G. R. Incorporation of hydroxyl in upper mantle clinopyroxene. *Nature* 1991, 351, 723-735.

Song, S.-G. and Su, L. Plastic rheology of the Yushigou mantle peridotite and implications for dynamics of Paleao-Plate movement in the North Qilian Mountains. *Acta geologica Sinica* 1998, 72, 131-141

Su, L.; Song, S.-G.; Wang, Z.-H. CH_4-rich fluid inclusions in the Yushigou mantle peridotite and their implications, North Qilian Mountains, China. *Chinese Bulletin Science* 1999, 44, 1992-1995.

Stolper, E. and Newman, S. The role of water in the petrogenesis of Mariana trough magmas. *Earth and Planetary Science Letters* 1994, 121, 293-325.

Tait, S. Selective preservation of melt inclusions in igneous phenocrysts. *American Mineralogist* 1992, 77, 146-155.

Thomassota, E.; Cartignya, P.; Harrisc, J. W.; Viljoend, K.S. Methane-related diamond crystallization in the Earth's mantle. *Earth and Planetary Science Letters* 2007, 257, 362-371.

Tingle, T. N.; Hochella, M. F.; Becket, C. H.; Malhotra, R. Organic compounds on crack surfaces in olivine from San Carlos, Arizona, and Hualalai Volcano, Hawaii. *Geochimica et Cosmochimica Acta* 1990, 54, 477-485.

Trieloff, M.; Kunz, J.; Clague, D. A.; Harrison, D.; Allegre, C. J. The nature of pristine noble gases in mantle plumes. *Science* 2000, 288, 1036-1038.

Trull, T.; Nadean, S.; Pineau, F.; Polve, M.; Javoy, M. C-He systematics in hotspot xenoliths: Implications for mantle carbon contents and carbon recycling. *Earth and Planetary Science Letters* 1993, 118, 43-64.

Valley, J. W. and Cole, D. R. *Stable isotope geochemistry*. Mineralogical Society of America, 2001; pp1-662.

Verchovsky, A. B.; Fisenko, A. V.; Semjonova, L. F. C, N and noble gas isotopes in grain size separates of presolar diamonds from Efremovka. *Science* 1998, 281, 1165-1167.

Weinlich, F. H. An active subcontinental mantle volatile system in the western Eger rift, central Europe. *Geochimica et Cosmochimica Acta* 1999, 63, 3653-3671

Welhan, J. A. Methane and hydrogen in mid-ocean-ridge basalt glasses: analysis by vacuum crushing. *Can. J. Earth Science* 1988, 25, 38-48.

Widom, E. and Farquhar, J. Oxygen isotope signatures in olivines from Sao Miguel (Azores) basalts. *Chemical Geology* 2003, 193, 237-255.

Williams, A. E. Mass spectrometric analysis of volatiles in fluid inclusions: aliquot calibration valve to simulate inclusion rupture. *Chemical Geology* 1996, 131, 155-165

Wood, B. J. and Virgo, D. Upper mantle oxidation state: Ferric iron contents of lherzolite spinels by 57Fe Mössbauer spectroscopy and resultant oxygen fugacities. *Geochimica et Cosmochimica Acta* 1989, 53, 1277-1291

Wood, B. J.; Bryndzia, L. T.; Johnson, K. E. Mantle oxidation state and its relationship to tectonic environment and fluid speciation. *Science* 1990, 248, 337-345.

Woodland, A. B.; Kornprobst, J.; Wood, B. J. Oxygen thermobarometry of orogenic lherzolite massifs. *Journal of Petrology* 1992, 33, 203-230.

Woodland, A. B. and Koch, M. Variation in oxygen fugacity with depth in the upper mantle beneath the Kaapvaal craton, Southern Africa. *Earth and Planetary Science Letters* 2003, 214, 295-310.

Workman, R. K.; Hauri, E.; Hart, S. R.; Wang, J. H.; Blusztajn, J. Volatile and trace elements in basaltic glasses from Samoa. *Earth and Planetary Science Letters* 2006, 241, 932-951.

Wyllie, P.J.; Ryabchikov, I. D. Volatile components, magmas, and critical fluids in upwelling mantle. *Journal of petrology* 2000, 41, 1195-1206.

Wyllei, P. J. Metasomatism and fluid generation in mantle xenoliths, In *Mantle xenoliths*. Nixon, P. H., ed.; John Wiley and Sons Ltd, 1987; pp599-621.

Xia, L.-Q. Highly dense CO_2 fluid inclusions in peridotitic inclusions in some alkali-basalts from east China. *Acta Mineral. Sin.* 1984, 4, 133-142 (in Chinese with English abstract).

Xu, Y.-G. The distributions of trace elements in spinel- and garnet- peridotite. *Science in China (D)* 2000, 30, 307-314.

Yang, X.-Y.; Zheng, Y.-F.; Liu, D.; Dai, J. Study on chemical compositions of fluid inclusions from peridotite and eclogites in Eastern China: The geodynamic implications.. *Phys. Chem. Earth* 2001, 26, 705-718

Yuen, G.; Blair, N.; Des Marais, D. J.; Chang, S. Carbon isotope composition of low molecular weight hydrocarbons and monocarboxylic acids from Murchison meteorite. *Nature* 1984, 307, 252-254.

Yuri, A. T.; Alain, B.; Juan, G. Native gold in mineral precipitates from high-temperature volcanic gases of Colimma volcano, Mexico. *Applied Geochemistry* 2000, 15, 337-346.

Zhang, M. -J.; Hu, P. -Q.; Niu, Y. L., Su, S. -G. Chemical and stable isotopic constraints on the origin and nature of volatiles in sub-continental lithospheric mantle beneath eastern China. *Lithos* 2007, 96, 55-66.

Zhang, M. -J.; Hu, P. -Q.; Zheng, P. The occurrence modes of H_2 in mantle-derived rocks, In *Mineral Deposit Research,* Mao, J.; Bierlein, F. P. eds., Springer: New York, 2005a; *Chapter*19, pp73-76.

Zhang, M. -J.; Hu, P. -Q.; Wang, X.-B.; Liu, G. The fluid compositions of lherzolite xenoliths in Eastern China and Western American. *Geochim. Cosmochim. Acta* 2005b, 69*A*, 146.

Zhang, M.-J.; Wang, X. -B.; Liu, G.; Zhang, T. -W.; Bo, W. -R. The compositions of upper mantle fluids beneath Eastern China. *Acta Geologica Sinica* 2004, 78, 125-130.

Zhang, M. -J.; Wang, X. -B.; Li, L. -W. An appraisal of different experimental methods in the determination of fluid composition in mantle-derived rocks. *Geological Review* 2000, 46, 160-166. (in Chinese with English abstract)

Zhang, M. -J.; Wang, X. -B.; Liu, G.; Wen, Q. -B. Fluid compositions and carbon and oxygen isotope geochemistry of Cenozoic alkali basalts in Eastern China. *Chinese Journal of Geochemistry* 1999a, 18, 276-282.

Zhang, M.-J.; Wang, X.-B.; Liu, G.; Li, L.-W. The compositions of the fluids in alkali basalts and mantle-derived xenoliths in eastern China. *Acta Geologica Sinica* 1999b, 73, 162-166. (in Chinese with English abstract).

Zhang, M. -J.; Niu, Y. L.; Hu, P. -Q.; Li, L. -W. Huge amounts of volatiles in lithospheric mantle. *Chemical Geology* 2009, *(in submitted).*

Zhang, M. -J.; Meng G. -L.; Hu, P. -Q.; Li, L. -W.; Li, G. -Z. Chemical composition of mantle volatiles in ancient Qilian ocean -Evidence from Yushigou ophiolitic peridotites. *Journal of Lanzhou University* 2008, 44, 1-8. (in Chinese with English abstract)

Zhang, T. -W.; Zhang, M. -J.; Bai, B. -J.; Wang, X. -B.; Li, L. -W. Origin and accumulation of carbon dioxide in the Huanghua depression, Bohai Bay basin, China. *AAPG Bulletin* 2008, 92, 341-358.

Zhang, Y. and Zindler, A. Distribution and evolution of carbon and nitrogen in Earth. *Earth and Planetary Science Letters* 1993, 117, 331-345.

Zhu, B. -Q. The mapping of geochemical provinces in China based on Pb isotopes. *Journal of Geochemical Exploration* 1995, 55, 171-181.

In: The Lithosphere: Geochemistry, Geology and Geophysics ISBN: 978-1-60456-903-2
Editors: J.E. Anderson et al, pp. 213-265

Chapter 6

DEFORMABLE LITHOSPHERIC PLATES: CONTROLLING ACTION OF NETLIKE PLASTIC-FLOW

Sheng-zu Wang*

State Key Laboratory of Earthquake Dynamics, Institute of Geology, China Earthquake Administration, Beijing 100029, China

Abstract

Lithospheric plate assumed originally to be "rigid" in the plate tectonics theory is in fact deformable. To explain intraplate deformation geoscientists have proposed a variety of mechanical models, in which the "slip-line field" and "viscous flow" models, as two influential ones, emphasize the deformation to be controlled mainly by shear localization and viscous flow, respectively. The "netlike plastic-flow (NPF)" model for continental dynamics argues that shear localization can occur in viscous flow field, forming NPF, which includes a plastic-flow network in the flow field, and it is the NPF in the lower lithosphere that plays a major role for the long-range transmission of driving forces and therefore controls the intraplate deformation, resulting in the netlike distributions of various geological-geophysical phenomena. Plastic-flow network is similar to but different from the slip-line network which is regarded, in the classical plasticity mechanics, only as a critical yield state. Based on the geometrical and mechanical features of plastic-flow network the "conjugate-angle-bisector" and "conjugate-angle-increment" methods have been suggested for estimating the stress directions and strains in the lower lithosphere, respectively. The results of the studies dealing with NPF are stated briefly in this chapter, involving the mechanism of NPF, its physical and numerical simulations, and the NPF-controlled network patterns, multi-layered tectonic deformation and stress/strain fields in the central-eastern Asian continent. It is mentioned that NPF and its controlling action are also exist in other regions of the Eurasian and other plates which subject to indentation at their boundaries. The NPF model provides an effective approach to understanding the deformable lithospheric plates.

Keywords: lithosphere; netlike plastic-flow; plastic-flow network; multi-layered tectonic deformation; Stress field; Strain field.

* E-mail address: shengzuwang@263.net

1. Introduction

As considered in the plate tectonics theory the lithosphere is divided into "rigid" plates, which drift upon the asthenosphere and converge to, diverge from, or slide along one another. It seems now to be a consensus that the "rigid" plates are really deformable since a large number of strong tectonic and seismic activities have occurred within them. An appropriate approach to understanding the deformable lithospheric plates is to reveal its mechanical mechanism as done by more and more geoscientists.

Generally speaking, the mechanical mechanisms for intraplate deformation are associated with the following three sorts of driving forces:

(1) Plate-boundary driving force, which is induced by the interaction between adjacent plates (Molnar and Tapponnier, 1975; Vilotte et al., 1982; England and Houseman, 1986; Wang, 1993a; Wang et al., 2001);

(2) Plate-bottom driving force, which is induced by the local mantle convection, mainly involving various mantle plumes (Rousseau, 1992; Maruyama, 1994; Wang, 1998; Wang et al., 2000);

(3) Body force, which is associated with the rotation of the Earth and its variation (Wegener, 1924; Lee, 1984) and the attractions between celestial bodies (Du and Xu, 1989).

In addition, the gravitational forces related to the differences in surficial elevation, lithospheric thickness and rock density are also the body forces, but classified as those corresponding to the above-mentioned three sorts in terms of their origins. For instance, the gravitational force of the Qinghai-Tibet plateau derived from the indentation of the Indian plate is regarded as belonging to the first sort.

The contributions of these driving forces are various in different regions on the Earth. In the central-eastern Asian continent the plate-boundary driving force plays a dominant role accompanied with the subsidiary contributions of the plate-bottom and body driving forces and the intraplate deformation related to the plate-boundary driving force is stated in this chapter. As is known, there are two kinds of mechanical models, the "slip-line flied" and "viscous flow" models, suggested to explain the tectonic deformation in this region, which have been quite influential in the field of continental dynamics since 1970s.

The "slip-line field" model has applied the theory of slip-line field in the plasticity mechanics to explaining the large-scale strike-slip shear zones in the central-eastern Asian continent under the indentation of the Indian plate (Molnar and Tapponnier, 1975, 1978; Tapponnier and Molnar, 1976, 1977) and, as an extension of this idea, the "strike-slip extrusion" or "lateral escape of continental blocks" has been emphasized (Tapponnier et al., 1982, 1990; Peltzer and Tapponnier, 1988; Tapponnier et al., 2001). Note that the concept of "slip-line field" in the model was originally regarded as the shear localization in plastic medium but, in its practical application, it involves the fracturing along "slip-lines" and the formation of isolated blocks between "slip-lines" and is therefore changed to brittle shear localization, while the contribution of viscous (plastic) flow to tectonic deformation has been neglected.

On the contrary, the various "viscous flow" models emphasize the contribution of viscous (plastic) flow in the lithosphere to intraplate deformation (Evison, 1960; McKenzie, 1972;

England and McKenzie, 1982; Vilotte et al., 1982, 1986; England and Houseman, 1986; Houseman and England, 1993; England and Molnar, 1997). Some researchers (Vilotte et al., 1986) have considered the heterogeneity of media, taking into account the influence of "stable" blocks in the viscous flow field. However, the possibility and significance of shear strain localization for intraplate deformation are neglected at all in these models.

It should be mentioned that some ideas have involved both of these two aspects, shear localization and viscous flow. For instance, in addition to the application of the slip-line field theory, the action of viscous (plastic) flow in the lower crust and uppermost mantle was also presented by Molnar and Tapponnier (1978) for explaining the long-range transmission of plate-boundary driving force, but both the shear localization and viscous flow have not been combined into a unified mechanical system. Furthermore, a multi-layered structure of the lithosphere with the combined effect of shear localization in the brittle upper layer and viscous flow in the ductile lower layer has been suggested by some researchers (Davy and Cobbold, 1988, 1991; Cobbold and Davy, 1988; Sornette et al., 1993; Davy et al., 1995) and the relative contributions of the two layers to intraplate deformation have been estimated quantitatively (Davy et al., 1995). However, the possibility of the occurrence of shear localization in the ductile lower layer has been ruled out from their analogue models owing to the materials with lower viscosities to be used in their experiments (Wang et al., 2007).

The studies of netlike plastic-flow (NPF) (Wang, 1985, 1993a,b, 1997, 2006a; Wang et al., 2001, 2007, 2008; Li et al., 1997, 1998) indicate that ductile shear localization can occur in viscous flow field, resulting in NPF, which can be regarded as a combination of viscous (plastic) flow and "slip-line" network, forming the so-called "plastic-flow network"; the conditions of temperatures and confining pressures in the lower lithosphere are suitable for NPF; and it is the NPF in the lower lithosphere that transmits the driving forces over long distance from plate boundaries and, as one of the principal mechanisms, controls intraplate deformation, tectonic stress field and the netlike distributions of earthquakes, magmatic rocks and other geological-geophysical anomalies. Such kind of NPF-controlled netlike patterns have been recognized from the central-eastern Asian continent and other regions, e.g. those in western Asia and southern Europe (Wang, 1993c) and North America (Wang, 1995).

The appearance of shear localization in viscous flow implies that the yield can occur not only in deformed elastic medium, which is named conventionally as Bingham body, but also in viscous medium, which can be regarded as an independent kind of viscoplastic body and expressed with a "power/linear-binomial" combined flow law (Wang, 1997; Wang et al., 2001) instead of the conventional power flow law. The concept of NPF provides a theoretical basis for explaining the netlike distributions of a variety of geological-geophysical phenomena and estimating the distributions of stress directions and strains in the lower lithosphere. As a summary of the studies of NPF, the theoretical basis and physical/numerical simulations of NPF, the recognitions of various NPF-controlled phenomena in the central-eastern Asian continent and their applications will be stated briefly in this chapter.

2. The Layered Lithosphere and Netlike Plastic-Flow

2.1. The Lithosphere Layered in Terms of Deformation Regime

As is well known, deformation and failure of rocks transit from brittle to ductile regime with increasing temperature and confining pressure (Heard, 1960, 1976; Wang and Zhang, 1982; Kirby and McCormick, 1984; Shimamoto, 1986; Evans et al., 1990; Wang, 1993d). Based on a number of experimental results of rocks, including quartzite, granite, granodiorite, gabbro and basalt for the crust (Nicolas et al.,1976; Wang and Zhang, 1982; Wang, 1985, 1993d; Wang, Z.-c., 1988; Wang, Z.-c. et al., 1992) and peridotite for the mantle, (Nicolas et al.,1976; Chopra and Paterson, 1981; Turcotte et al.,1982; Chopra,1986), the variation of shear failure regime is summarized in Table 1, in which the shear angle θ (i.e. the angle between the directions of maximum compressive stress and normal to fracture plane) and the conjugate angle ψ (i.e. the angle facing or opposing compression direction) are shown in the schemes, n is the stress exponent in power flow law, and silicate rocks and peridotite represent those in the crust and mantle, respectively.

Table 1. Transition of shear failure from brittle to ductile regime in rocks. (modified after Wang, 1992, 1993a)

Deformation Regime	Brittle	Semi-Brittle	Semi-Ductile	Ductile	Ductile
Mechanism	Shear or tension-shear fracturing	Fracturing accompanied locally by plastic flow	Plastic flow accompanied locally by fracturing	Plastic flow (mainly, dislocation movement)	Plastic flow (mainly, diffusion or super-plasticity
Shear structure	Monoclinal	Sparse-conjugate	Dense-conjugate	Dense-conjugate	No
Shear localization	Extreme localization	Localization	Fine localization	Fine localization	No
Homogeneity	Extreme inhomogeneous	Inhomogeneous	Pseudo-homogeneous	Pseudo-homogeneous	Homogeneous
Shear angle θ(°)	75 - 55	65 - 50	50 - 45	~45	—
Conjugate angle ψ(°)	—	50 - 80	80 - 90	~90	—
Barrel-like shape	No	Yes	Obvious	Obvious	Obvious
Stress exponent n	—	—	2.4 - 6.2		~1.0
Instability mode	Gradual or sudden	Gradual or sudden	Gradual	Gradual	Gradual
Temperature(□): Silicate rocks Peridotite	≤500	≤700	500 - 1000	600 - 1100 1000 - 1400	≥800 ≥1100
Section of deformed specimen	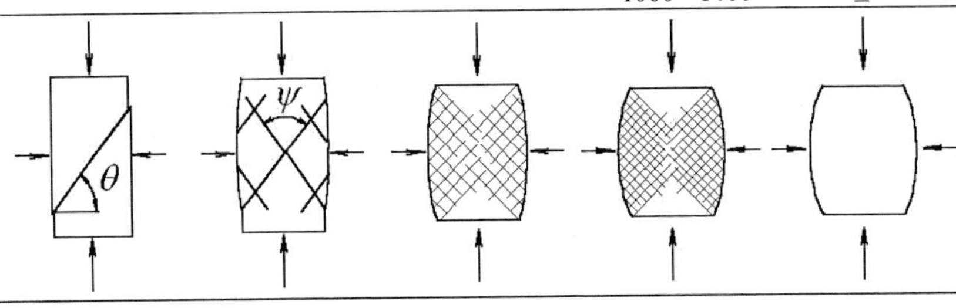				

Note that the semi-ductile and ductile failure regimes appear at temperatures ranging from 500℃ to 1400℃, which correspond roughly to the depths of the lower lithosphere, including the lower crust and lithospheric mantle. It is known that the depths of the top and bottom of the lower lithosphere in the central-eastern Asia continent are about 25–35 km and 100–200 km (Ma, 1987, 1989; Yuan, 1996) with temperatures of 350–700℃ and 1000–1700℃, respectively (Jeffreys, 1976; Pollack and Chapman, 1977), which are essentially identical with the temperatures suitable for the "fine shear localization" and the formation of "dense conjugate shear structure" as shown in Table 1. Consequently, plastic-flow network with its initial conjugate angles approximate or equal to 90° is formed mainly in the lower lithosphere.

In addition, the results of experimental deformation also indicates that under the condition of triaxial compression the transverse-tension-induced longitudinal fracturing occurs only at confining pressures lower than 150–300 MPa for rocks in the upper crust (Wang and Zhang, 1982; Wang, Z.-c., 1988), corresponding to the depth up to about 10–20 km. As a matter of fact, the depths of tension fractures in the China continent are generally not beyond about 10 km (Wang, 1993b) probably owing to the heterogeneous distribution of rocks and the dependence of transverse-tension-induced fracturing upon the weakest rock mass.

Based on the geological-geophysical data (Ma, 1987, 1989; Wang, 1993b) and considering the transition of deformation and failure from brittle to ductile regime with the increase of depth, the continental lithosphere can be divided generally into the following four tectonic deformation layers from bottom to top (Wang, 1993b; Wang et al., 2001) as shown in Fig.1:

(1) Netlike-flow tectonic layer (NFTL), including the lower crust and lithospheric mantle, is characterized by the existence of plastic-flow network as a manifestation of dense-conjugate shear localization in the semi-ductile and ductile layer, which is composed of two families of ductile shear belts, called plastic-flow belts;

(2) Low-velocity / high-conductivity tectonic layer (LVTL), or called the middle-crustal weakness layer, with thickness 0–5 km and lower viscosity is distributed discontinuously between its overlying layer and NFTL;

(3) Shear-fracture tectonic layer (SFTL), i.e. the seismogenic layer, in the depth range of 5 – 25 km in the eastern part of the China continent and of 10–35 km in the western part, is dominated by strike-slip or thrust faulting and includes most of intraplate shallow earthquakes with $M \geq 5$ occurring in it, showing netlike distribution of epicenters controlled by NPF in NFTL;

(4) Tension-fracture tectonic layer (TFTL), or called the shallow fracture tectonic layer, is limited generally to the depth less than 10 km and characterized or accompanied by extensional tectonics, including normal faults, extensional basins and grabens.

TFTL may be absent from some regions, mainly from those in the western part of central-eastern Asia, where SFTL tends to outcrop directly at the surface, while under the strong action of the Indian plate's indentation a series of near-south-north-trend extensional basins have been developed in the shallow crust owing to the transverse-tension-induced fracturing in the region near the Himalayan collision boundary (Cui et al., 1992). The seismic activities concentrating within SFTL implies that this layer is brittle or semi-brittle while the underlying layers are semi-ductile or ductile. Moreover, it is inferred that the difference in horizontal flow velocity between SFTL and NFTL and its shear-heating effect result probably in the weakening of media between them, being favorable for the formation of LVLT.

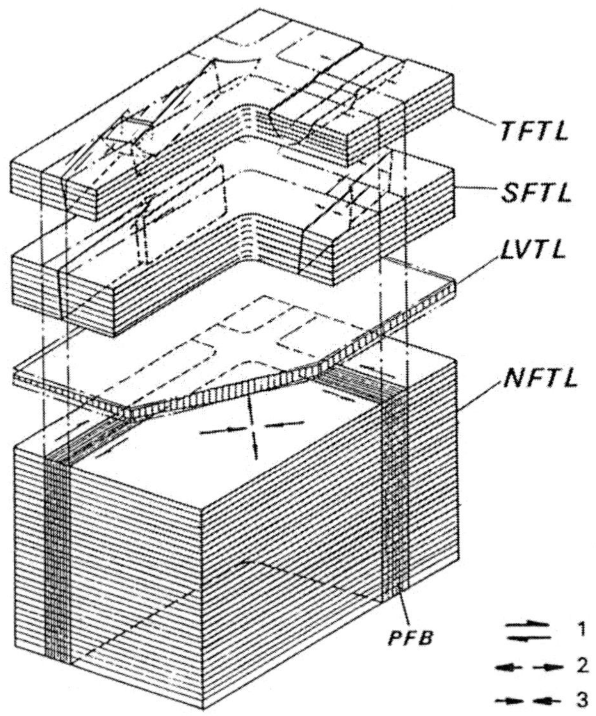

Figure 1. Scheme of multi-layered tectonic deformation in continent (Wang, 1993b; Wang et al., 2001). TFTL: Tension-fracture tectonic layer; SFTL: Shear-fracture tectonic layer; LVTL: Low-velocity / high-conductivity tectonic layer; NFTL: Netlike-flow tectonic layer; PFB: Plastic-flow belt; 1, 2and 3: Shearing, tension and compression, respectively.

2.2. A Combined Flow Law for NPF

Instead of the conventional power flow law, which is suitable only for the "homogeneous" flow, i.e. the flow without shear localization, a combined flow law has been suggested for describing the rheological behavior of NPF (Wang, 1997). It is a "power / linear-binomial" combined flow law expressed as follows (Fig.2):

$$\dot{\varepsilon} = (1-\beta)\,\dot{\varepsilon}_{\mathrm{v}} + \beta\,\dot{\varepsilon}_{\mathrm{vp}} \tag{1}$$

$$\dot{\varepsilon}_{\mathrm{v}} = C_{\mathrm{v}.i}\,\sigma^{n_i} \tag{2}$$

$$\dot{\varepsilon}_{\mathrm{vp}} = C_{\mathrm{vp}.i}\,(\sigma - \sigma_{\mathrm{y}.i}) \tag{3}$$

where $\dot{\varepsilon}$ is the total strain rate; $\dot{\varepsilon}_v$ and $\dot{\varepsilon}_{vp}$ are the viscous and viscoplastic strain rates as the power and linear-binomial functions of stress, respectively; β is the netlike-flow coefficient ranging from 0 to 1 for measuring the development level of shear localization; $C_{v.i}$ and $C_{vp.i}$ are the fluidities of viscous and viscoplastic flows, respectively; n_i is the stress exponent; $\sigma_{y.i}$ the yield limit, i.e. a critical stress beyond which shear localization occurs or its mechanism changes; i indicates the parameters belonging to the ith segment of the curve. We have the expressions of the parameters related to strains $\dot{\varepsilon}, \dot{\varepsilon}_v$ and $\dot{\varepsilon}_{vp}$ as shown in Table 2.

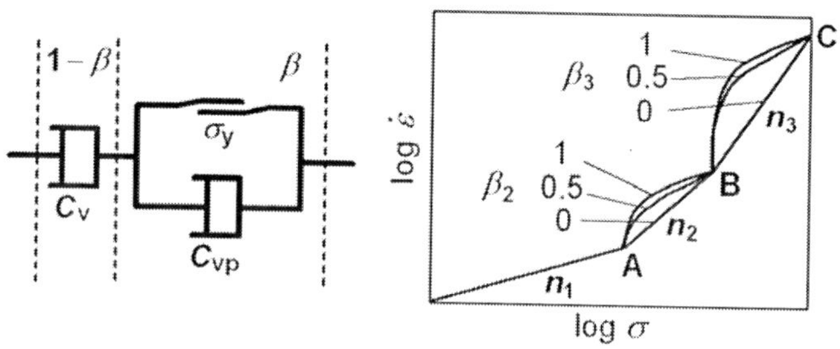

Figure 2. "Power / binomial-linear" combined flow law for NPF (Wang, 1997). Left: A rheological model for NPF. Right: A scheme of $log\ \dot{\varepsilon}$ versus $log\ \sigma$. $\dot{\varepsilon}$: strain rate; σ : differential stress; β, β_2 and β_3: netlike-flow coefficients; C_v and C_{vp}: fluidities of viscous and viscoplastic flows, respectively; σ_y: yield limit; n_1, n_2 and n_3: stress exponents.

Table 2. Expressions of the parameters related to strains $\dot{\varepsilon}$, $\dot{\varepsilon}_v$ and $\dot{\varepsilon}_{vp}$.

i	β	n	C_v	C_{vp}	$\sigma_{y.}$	Condition
1	0	n_1=1	$C_{v.1}=\dot{\varepsilon}_A/\sigma_A$	—	—	as $\sigma \leq \sigma_A$;
2	β_2	n_2>1	$C_{v.2}=\dot{\varepsilon}_A/\sigma_A^{n_2}$	$C_{vp.2}=(\dot{\varepsilon}_B-\dot{\varepsilon}_A)/(\sigma_B-\sigma_A)$	$\sigma_{y.2}=\sigma_A-\dot{\varepsilon}_A/C_{vp.2}$	as $\sigma_A \leq \sigma \leq \sigma_B$;
3	β_3	n_3>n_2	$C_{v.3}=\dot{\varepsilon}_B/\sigma_B^{n_3}$	$C_{vp.3}=(\dot{\varepsilon}_C-\dot{\varepsilon}_B)/(\sigma_C-\sigma_B)$	$\sigma_{y.3}=\sigma_B-\dot{\varepsilon}_B/C_{vp.3}$	as $\sigma_B \leq \sigma \leq \sigma_C$

(σ_A, $\dot{\varepsilon}_A$), (σ_B, $\dot{\varepsilon}_B$) and (σ_C, $\dot{\varepsilon}_C$) are the stresses and strain rates at the points A, B and C shown in Fig.2, respectively.

It is shown in Fig.2 that, in addition to the segmentation of the curve of $log\ \dot{\varepsilon}$ versus $log\sigma$ as mentioned by some researchers (Nicolas and Poirier, 1976; Turcotte and Schubert, 1882; Walker et al., 1990), implying the changes in rheological mechanism from one to next segment, the combined flow law emphasizes further the convexity of the curve within some segments, showing the development of shear localization in viscous flow field, and the coefficient β is introduced into it for measuring the development level of NPF. The combined flow law has been examined by the experimental results of some materials (Wang, 1997), including the wax/rosin/cement-powder mixture (WRC), limestone and aggregate of olivine.

The result of uniaxial compression creep tests of WRC at room temperature is given in Fig.3 and Table 3. WRC is a mixture of wax (W), rosin (R) and cement powder (C) with the weight ratio W: R: C = 1 : (R/W) : 2.4, in which wax and rosin (colophony) are mixed in their molten state at elevated temperature with cement powder as filling material. The size of cylindrical specimens is 38.5 mm in diameter and 61–79 mm in height. The fluidity of the mixture increases with decreasing the content of rosin which is brittle at room temperature. The experimental data (Li et al., 1993) are fitted using the combined flow law, in which n_1, n_2 and n_3 are equal to 1, 1.98–1.91 and 3.90–3.96, respectively, showing the ith segments of the curves with different *R/W* values to be essentially parallel to each other. All the first segments "—A" for the three curves correspond to Newtonian flow with n_1 =1 and β_1=0, while the subsequent segments are changed to non-Newtonian with β_2 =1 for "A—B" as *R/W* = 0.3 and 0.6, and with β_3 =1 for "B—C" as *R/W* = 0.6, showing the full development of plastic-flow network in the specimens.

It has been determined in terms of the values of β for the deformed specimens that NPF takes place fully at *R/W* = 0.2–0.6 and partially at *R/W* = 0.1, while disappears at *R/W*=0. As shown in Fig.3 (right), the conjugate netlike pattern as a manifestation of plastic-flow network can be seen from the surface of the specimen with *R/W*=0.4.

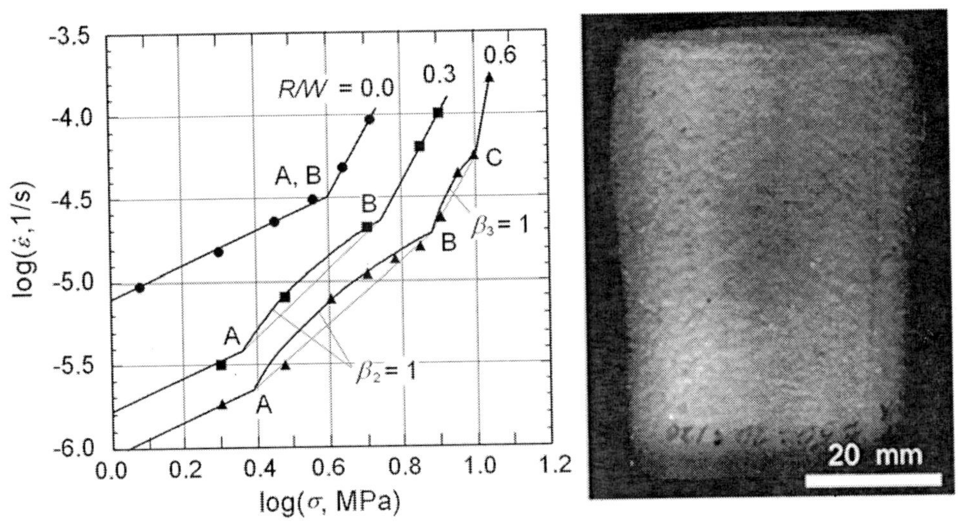

Figure 3. NPF in WRC specimens at room temperature. Left: The curves of *log* $\dot{\varepsilon}$ versus *log*σ (Wang, 1997); Right: Deformed specimen with *R/W* =0.4. Symbols are the same as in Fig.2.

Also, the experimental data of Solenhofen limestone (Walker et al., 1990) and aggregate of olivine (Carter and Ave'Lallemant, 1970; Kirby and Raleigh, 1973; Kohlstedt and Goetze, 1974) have been fitted well with the combined flow low, showing the segmentation of the *log*$\dot{\varepsilon}$ versus *log*σ curves and the convexity of some segments, which imply the existence of NPF in the specimens.

Note that, in addition to the yield limit being a critical point between elastic and plastic deformation states as expressed by Bingham's model, according to the "power / linear binomial" combined flow law, there is also a yield limit to distinguish netlike flow from homogeneous one as shown in Fig.3 and Table 2. That is to say, shear localization is possible to occur not only in elastic deformation but in viscous flow field.

Table 3. Rheological parameters of WRC (Wang, 1997).

Segment	Parameter	Value of parameter		
		R/W =0.0	*R/W* =0.3	*R/W* =0.6
— A	n_1	1	1	1
	η_1 (Pa.s)	6.20x10^{10}	2.99x10^{11}	5.56x10^{11}
A — B	n_2		1.98	1.91
	$\dot{\varepsilon}_A$ (s^{-1})	3.18x10^{-5}	3.55x10^{-6}	2.19x10^{-6}
	σ_A (MPa)	3.95	2.13	2.43
	σ_{y2} (MPa)		1.51	1.77
	C_{v2} (MPa$^{-n_2}$ s^{-1})		8.01x10^{-7}	4.00x10^{-7}
	C_{vp2} (MPa^{-1} s^{-1})		5.81x10^{-6}	3.33x10^{-6}
	β_2		1	1
B — C	n_3	3.95	3.90	3.96
	$\dot{\varepsilon}_B$ (s^{-1})	3.18x10^{-5}	2.29x10^{-5}	1.95x10^{-5}
	σ_B (MPa)	3.95	5.46	7.64
	σ_{y3} (MPa)			6.40
	C_{v3} (MPa$^{-n_3}$ s^{-1})			6.24x10^{-9}
	C_{vp3} (MPa^{-1} s^{-1})			1.58x10^{-5}
	β_3	0	0	1

2.3. Physical Simulation of NPF

NPF in ductile one-layer model

As mentioned above, the difference between NPF and "homogeneous" flow is whether plastic-flow network as a manifestation of shear strain localization is formed in flow field or not. Note that deformation in "homogeneous" flow process is not always exactly homogeneous. Compressive folding can, for example, be regarded as "compressive strain localization", which tends to be attributed to viscous buckling or boudinaging (Ramberg, 1963, 1981; Smith, 1975, 1979; Cobbold, 1975; Dubey and Cobbold, 1977; Zuber, 1987; Davy and Cobbold, 1991; Liu and Dixon, 1991; Bull et al., 1992).

The physical simulations have verified that the shear strain localization in viscous flow field is an important kind of intraplate tectonic deformation (Li et al., 1997; Wang et al., 2001). The ductile layer in physical model is made of plasticized rosin, i.e. a mixture of gum rosin (*R*, called "rosin" for short) and plasticizer (*P*, i.e. turpentine oil), and its viscosity increases with increasing the weight ratio *R/P* and decreases with increasing temperature. As a result of the experiments, the plastic-flow network is formed in the ductile one-layer model as shown in Fig.4 (Top view), where the indenter is located at the "lower" boundary of the model and displaced "upward"; T and η are the temperature and viscosity of the model, respectively; D is the inner diameter of the circular frame; b the width of the indenter; h_d the thickness of the ductile layer; ΔL the indentation; t the duration of modeling; and V the average velocity of the indenter. As shown in the figure, there are two families of curved plastic-flow belts intersecting each other with the conjugate angles greater than 90° and each of the belts is composed of a series of small-scale secondary folds arranged en echelon.

Figure 4. Plastic-flow network in the ductile one-layer model M272 (modified after: Li et al., 1997; Wang et al., 2001). R/P=6, T=20℃, η = 2.80×10^5 Pa s, D=160 mm, b=20 mm, h_d=12 mm, ΔL=11.5 mm, t=31 hr, V=0.37 mm/hr.

It is also indicated by the experiments of the ductile one-layer models that there is a tendency in deformation structure with the increase of viscosity, namely, the transition from compressive fold zones occurring in the medium with lower viscosity to conjugate shear network with higher viscosity. The transitional states between both of them are the fold zones with kinks, which show shear displacements across the zones. Connecting the corresponding kinking points of the fold zones the shear belts appear. As shown in Fig.5, for example, there are seemingly the compressive fold zones distributed in the ductile one-layer model, but, looking at it more carefully, a series of shear belts can be recognized in terms of the connection of kinks or the en echelon distribution of secondary folds.

Figure 5. Compressive fold zones and shear belts in the ductile one-layer model M10 (modified after Li et al., 1997; Wang et al., 2001). R/P=6, T=28℃, η = 3.78×10^4 Pa s, b=22 mm, h_d=10 mm, ΔL=20 mm. Solid and dotted lines denote the secondary folds and kinks, respectively. Others as in Fig.4.

It is therefore confirmed further by the physical simulations that shear localization and the formation of plastic-flow network occur in ductile one-layer model made of viscous (plastic) medium with viscosity higher than some critical value. The shear belts of plastic-flow network can be the en-echelon-distributed secondary compression-shear folds or shown by those connecting the kinking points of compressive fold zones.

Moreover, NPF can also occur under the action of gravity potential caused by high elevation. As shown in Fig.6, the "plateau" in the model is getting lower under the action of its self-gravity and disappears finally. Correspondingly, the conjugate angles between two families of plastic-flow belts are getting larger with the development of plastic-flow network and, finally, the belts become a cluster of folding traces distributed compactly within a narrow zone.

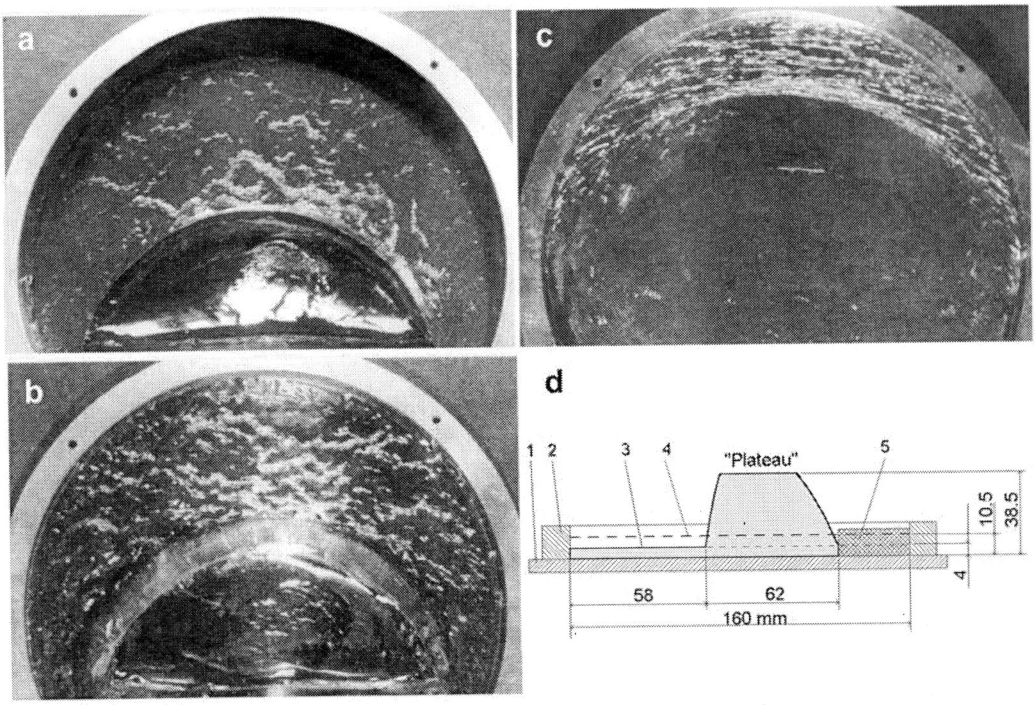

Figure 6. Gravity-driven NPF in the ductile one-layer model M26 (modified after Li et al., 1997). R/P=10, T=18-20℃, η = 6.55×10^7 Pa s, D=160 mm, h_d=4 mm. The semi-circle-shaped "plateau" is 34.5 mm high above the layer's surface, forming a pressure difference of about 367 Pa at the root of the "plateau". (a) The duration t = 8 days. (b) t =18 days. (c) t =119 days. (d) Longitudinal vertical profile of the model. 1: glass plate; 2: metallic ring-shaped frame; 3: initial state of the model, being a layer with a "plateau" upon it; 4: final state of the model, being a 10.5mm-thick layer; 5: retaining block.

NPF in brittle/ductile two-layer model

To model the multi-layered tectonic deformation, mainly, to understand the relation between the ductile lower and brittle upper layers, the brittle/ductile two-layer model has a thin brittle upper layer with thickness h_b formed by the consolidation of talc-powder slurry for modeling the upper crust and a ductile lower layer made of plasticized rosin for modeling the lower lithosphere. A typical pattern of NPF-controlled deformation (Fig.7) shows a conjugate tectonic network on the surface of the model. A series of secondary folds or thrust faults are arranged en echelon or intermittently along the tectonic belts, forming "mountain ranges", while the "mesh" areas between the belts are relatively flat, forming "intermountain basins". The anticlinal fold zone close to the indenter was formed in the early stage of deformation and then has been broken up into several segments by radial-trend tensile fractures, showing the strong action of transverse tension in that area. This arcuate fold zone and the fracture zone close to it separate the indenter from the broad area of the upper layer, indicating the impossibility of the "long-range" transmission of the indenter's driving force directly along the upper layer. Moreover, it is indicated further by the conjugate angles equal to or greater than 90° that the development of the tectonic network in the brittle upper layer is controlled mainly by NPF in the ductile lower layer. In contrast, the conjugate angle between brittle faults is only possible to be obtuse at large deformation, since their initial conjugate angles are only about 30-70° (see Table 1).

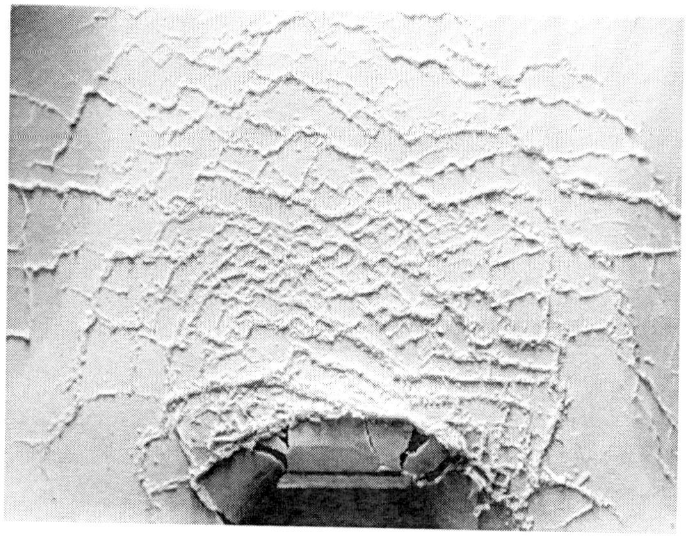

Figure 7. NPF-controlled tectonic network in the brittle/ductile two-layer model M225 (Li et al., 1997; Wang et al., 2001). R/P=16, T=20℃, η = 1.10×10^9 Pa s (ductile lower layer), D=160 mm, b=20 mm, h_d=7 mm, h_b=0.4 – 0.5 mm, ΔL =18.8 mm , t=49 h, V=0.38 mm/h.

We can also use the so-called strength ratio between the brittle and ductile layers, R_s, suggested by Davy et al. (1995) for judging the possibility of shear localization to occur independently in the brittle upper layer by its lower limit R_s = 5–10. The expression of R_s is simplified for the case of brittle/ductile two-layer model as:

$$R_s = \rho_b\, g\, h_b^{\,2}\, L \,/\, (2\, h_d\, \eta_d\, V) \tag{4}$$

where g is the gravity acceleration (9.81 m/s^2); ρ_b the density the brittle upper layer (1720 kg/m^3); η_d the viscosity of the ductile lower layer; V the converging velocity, i.e. the velocity of indentation; and L the horizontal dimension of model, which can be approximately taken as half of the influence range of the indentation, i.e. half of the diameter of the circular frame, D. As a result for the model M225, R_s = 0.00021 is obtained, which is considerably less than 5–10, showing the formation of the tectonic network in the brittle upper layer to be controlled by NPF in the ductile lower layer.

In addition, the influence of discontinuously-distributed weak intercalation on the NPF-controlled deformation were also tested in some models, in which some "patches" made of plasticine with viscosity considerably lower than that of the lower layer are buried between the upper and lower layers. As shown in Fig.8, three gaps corresponding to the soft "patches" appear in the tectonic network that some strongly-developed fracturing belts originating from the indenter end up at the sides of the "patches" and then reappear from their other sides, indicating that the existence of the weak intercalation influences or even restricts the lower layer to drag the upper one and verifying further that the extending of the fracturing belts in the brittle upper layer is not related directly to the force transmission along itself, but controlled by NPF in the lower layer.

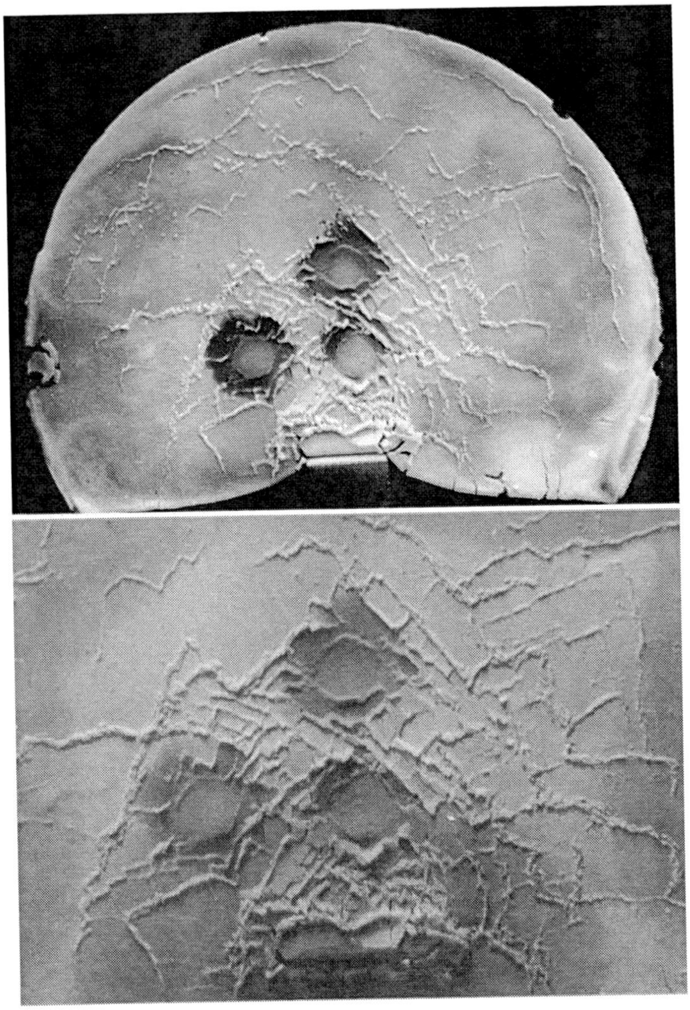

Figure 8. Influence of discontinuously-distributed weak intercalation on deformation in the brittle/ductile two-layer model M226 (Li et al., 1997; Wang et al., 2001). Top: Top view of the whole model. Bottom: Its magnified main part. R/P=16, T=20℃, η = 1.10×10^9 Pa s (ductile lower layer), D=160 mm, b=20 mm, h_d=7 mm, h_b=0.3-0.5 mm, ΔL =14.5 mm , t=23 h, V=0.63 mm/h; Weak intercalation is made of plasticine with about 10 mm in diameter, 1.5 mm thick and the viscosity considerably lower than that of the ductile lower layer.

2.4. Numerical Simulation of NPF

The numerical simulation methods, including the finite difference method (Höeg et al., 1968; Christian, 1977) and the finite element (FE) method (Meguid and Klair, 1984; Kotake et al., 2001; Li et al., 2002; Alsaleh, 2004; Siribumrungwong et al., 2004), have been used extensively for studying the bearing capacity of indented system and the generation of plastic zones or shear bands in it since 1960's. However, these studies have not involved the whole process of the development of shear strain localization and the formation of conjugate shear network.

As a preliminary trial of the numerical simulations of NPF the study is focused firstly on the development of plastic-flow network as a consequence of shear localization in ductile layer and some relevant problems (Wang et al., 2008). A plane-strain rectangular model is used for modeling the deformation of semi-space elastoplastic medium indented locally, which can be regarded as a viscoplastic medium deformed for a very short duration. The model is 400 mm wide and 160 mm high and the indenter with the width of 38 mm. is placed at the central part of its boundary (see Fig.12). Considering the symmetry of the model, the finite element (FE) analysis is carried out only concerning its left half portion. The central line and left boundary of the model are subjected to lateral constraint (U_X =0), the lower boundary to longitudinal constraint (U_Y=0) and the upper boundary is free except its central part being indented. The simulations were performed using the FE software ANSYS University Research 10.0 (Operations Guide, ANSYS Inc., 2005). The Plane42-type 4-node isoparametric square element with 2x2 integration points and side length of 2 mm is adopted to construct the model automatically in ANSYS system. There are 8000 elements and 8181 nodes in it. The sparse-matrix direct-elimination method and the Newton-Raphson method are used for the solutions of simultaneous linear equations and the iteration of non-linear equations in the simulations, respectively.

The mechanical parameters taken for the model are as follows: the elastic modulus $E = 2\times10^{10}$ Pa, the Poisson ratio ν= 0.495, and the yield limit $\sigma_{\text{yld}} = 2\times10^{8}$ Pa. The constitutive relation is given in Fig.9, along which element yields once its von Mises equivalent stress σ_{eqv} to be equal to the yield limit σ_{yld}, and then the stress will drop to a lower level, $R\sigma_{\text{yld}}$, and keep constant in consideration of the weakening effect of plastic-flow belt (Wang, 2004a), where R is the weakening coefficient, i.e. the yield-limit ratio of original medium to intra-belt medium. The von Mises equivalent stress of element is expressed as:

$$\sigma_{\text{eqv}} = \{(1/2)\,[(\sigma_x - \sigma_y)^2 + (\sigma_y - \sigma_z)^2 + (\sigma_z - \sigma_x)^2 + 6\,(\sigma_{xy}{}^2 + \sigma_{yz}{}^2 + \sigma_{zx}{}^2)]\}^{1/2} \tag{5}$$

where σ_x, σ_y, σ_z, σ_{xy}, σ_{yz} and σ_{zx} are the stress components.

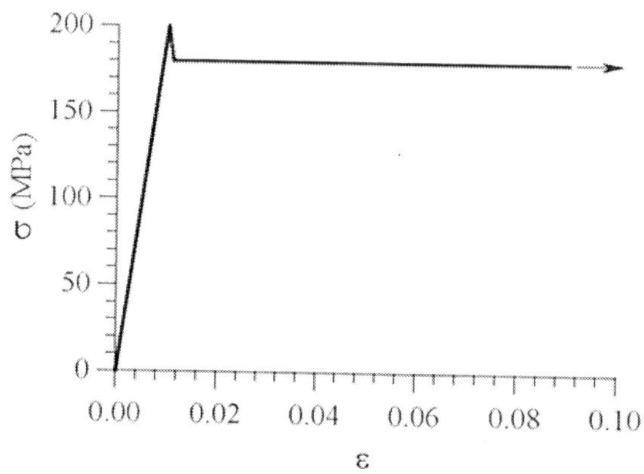

Figure 9. Constitutive relation (Wang et al., 2008). σ: Stress; ε: Strain.

The model is loaded by displacing the indenter step by step till the system to be unstable, i.e. its solution to be not converged. As shown in Figs.10 to 12, the model P14 has undergone 27 steps, which can be divided into the following three stages:

Stage I, Locally-Compacting, during which the compaction of medium in front of the indenter and the propagation of plastic-flow belts (i.e. ductile shear bands) from both the corners of the indenter (Fig.11, step 1) result in the formation of a triangular compacted area (Fig.11, step 2);

Stage II, Network-Spreading, during which the divergence and propagation of plastic-flow belts result in two families of belts to intersect normally each other, forming a conjugate network, and the triangular compacted area is disintegrated by some belts (Fig.11, steps 3 and 6) and, meanwhile, once tensile fractures occur in a number of elements beside the indenter, the pressure-displacement curve shows a small drop between the points 6 and 7 (Fig.10), by which the stage II is divided into two substages, IIa and IIb, and, as another sign of the substage IIb, there are some new belts appearing within the triangular area, resulting in its further "disintegration" (Fig.11, step 7);

Stage III, Surface-Bulging, during which the major plastic-flow belts, originating from the corners of the indenter and propagating obliquely upward, run through the free surface (Fig.11, step 22), and, consequently, resulting in the extrusion of the blocks confined by the belts and the slow descending of the curve (Fig.10) until the instability of the indented system occurs.

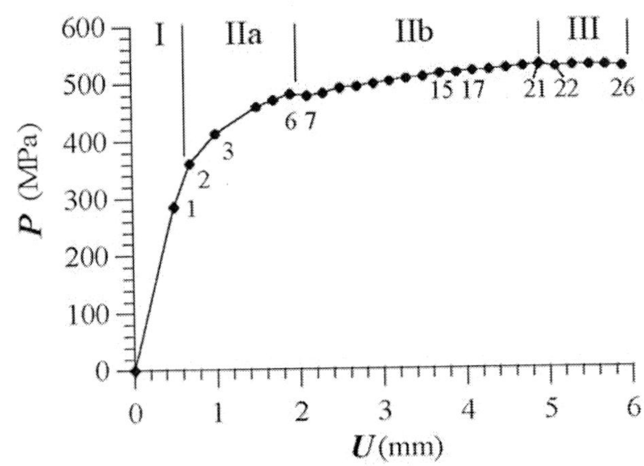

Figure 10. Pressure-displacement curves of the indenter in the FE model P14 (Wang et al., 2008). P and U are the average pressure and indentation (i.e. longitudinal displacement of the indenter), respectively. The deformation stages: I, locally-compacting; IIa and IIb, network- spreading; III, surface-bulging. The numbers marked under the curve denote the loading step.

As shown in Fig.12, the area confined by the contours of $\varepsilon_{eqv} = 0.012$ (dotted lines), corresponding to the yield limit σ_{yld}, is larger than that covered with the plastic-flow network, showing that the stresses of all the plastic-flow belts and the blocks between them are beyond the yield point, that is to say, plastic flow occurs not only inside but also outside the belts, implying the strain localization to occur against a plastic-flow background.

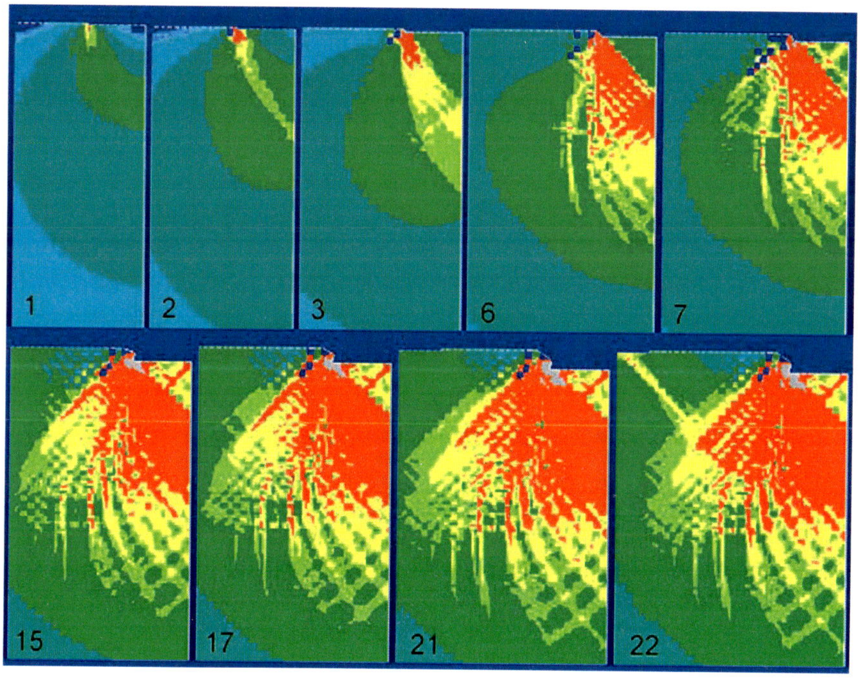

Figure 11. The development of plastic-flow network: Distribution of von Mises equivalent strains (main part of the model P14) (Wang et al., 2008). The numbers show the sequence of loading steps and the indentations are as follows: Step 1, 0.5 mm; Step 2, 0.7 mm; Step 3, 1.0 mm; Step 6, 1.9 mm; Step 7, 2.1 mm; Step 15, 3.7 mm; Step 17, 4.1 mm; Step 21, 4.9 mm; Step 22, 5.1 mm. Color scale is the same as in Fig.12. The small blue squares beside the indenter denote the fractured elements.

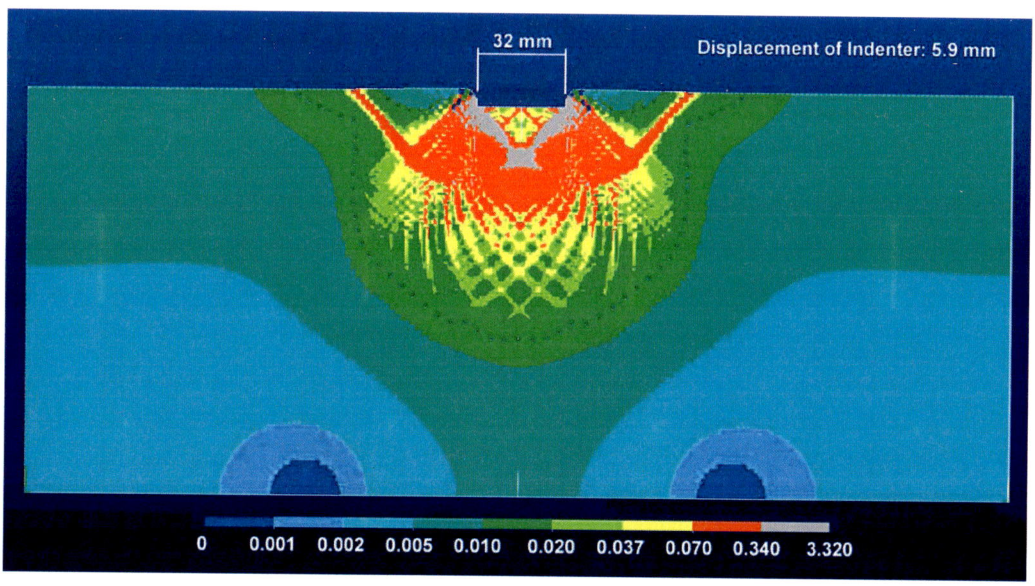

Figure 12. The final distribution of von Mises equivalent strains in the model P14 (full view) (Wang et al., 2008). This is the pattern for the loading step 26 with the indentation of 5.9 mm. Dotted lines are the contours of $\varepsilon_{eqv} = 0.012$, corresponding to the yield limit. Others as in Fig.11.

It can be seen from Figs.11 to 13 that the conjugate angles are equal to or greater than 90°, corresponding to ductile shear. Moreover, it is shown in Fig.13 that the general tendency of the maximum compressive stress directions is sectorially divergent from the vicinity of the indenter and identical essentially with the directions of the bisectors of conjugate angles, while the anomalies appear locally near the boundary of the triangular compacted area. In the stage IIa the development of plastic-flow belts outside the triangular compacted area contrasts badly with the intact state inside it (see Fig.11, steps 3 and 6), forming a discontinuity between them, across which the stress directions deflect abruptly. After that, the triangular compacted area is disintegrated and then some plastic-flow belts run through the boundary AC, indicating the degeneration of the discontinuity, which is divided into two segments, the discontinuity AB and the continuity BC, as shown in Fig.13.

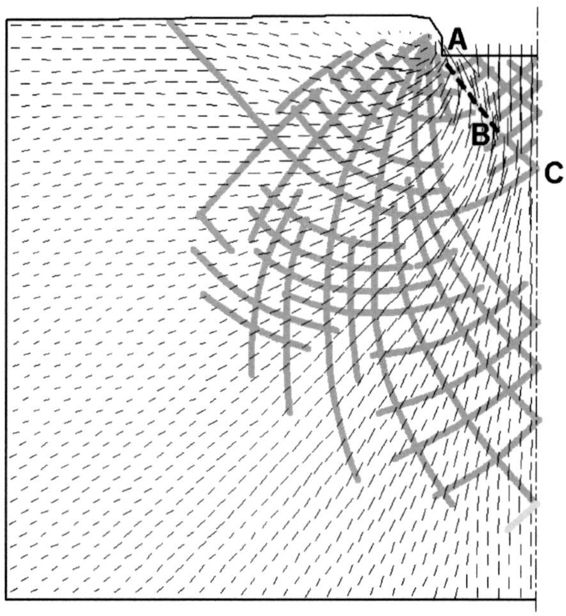

Figure 13. The distribution of the maximum compressive stress directions in the model P14 (its main part) (Wang et al., 2008). The plastic-flow network is drawn in terms of the equivalent strain map (Fig.12). AB and BC are the segments of discontinuity and continuity, respectively.

As a preliminary result of the numerical simulation, there are three points to be indicated as follows:

(1) Ductile shear strain localization occurs not only as a critical state, at which stress is equal to the yield limit and the ductile shear zone appears in the original elastic medium, but is also possible to occur against a plastic-flow background, showing strain crest in the yielded medium. As a matter of fact, the ductile shear localization in elastoplastic medium stated here can be regarded as a special case of viscoplastic one with high viscosity or high strain rate, or saying, with the duration considerably shorter than the relaxation time of the medium.

(2) Plastic-flow network is similar to but different from the so-called "slip-line network" involved in the classical plasticity mechanics. "Slip-line network" is determined in terms of yield limit only as a critical failure state of medium and is in fact a pattern of conjugate

network composed of virtual "slip-lines" without actual widths, locations and evolution processes. A variety of traditional solutions of "slip-line network" suggested by different researchers (Prandtl, 1920; Hill, 1950; Sokolovskii, 1950; Prager and Hodge, 1951; Bishop, 1953) can be considered as some critical states in the development process of plastic-flow network.

(3) The results of the numerical simulation is similar in the structural framework, geometric feature and some of mechanical behaviors to the plastic-flow networks in the lower lithosphere (see Section 3 below), involving the initial conjugate angle of 90°, the approximation between the directions of maximum compressive stress and bisector of conjugate angle, the "disintegration" of triangular compacted area, the degeneration of discontinuity, the shear-weakening effect of plastic-flow belts, the plastic-flow background of shear localization, and therefore verify theoretically the possibility of NPF in the lower lithosphere and the reasonability of the conjugate-angle-bisector and conjugate-angle-increment methods for estimating stress directions and strains, respectively.

3. NPF-Controlled Tectonic and Seismic Activities in the Central-Eastern Asian Continent

3.1. "Plastic-Flow / Seismic" Networks

As mentioned above, plastic-flow network in the lower lithosphere controls intraplate tectonic and seismic activities, resulting in the relatively strong tectonic activity along/upon the belts, the netlike distribution of seismic activity in the brittle upper crust and therefore the formation of "plastic-flow / seismic" network (Fig.14). Note that earthquakes occur along seismic faults, which are located in SFTL (i.e. the seismogenic layer) and run over plastic-flow belts with different angles, while the energy necessary for earthquakes is transmitted mainly from the underlying plastic-flow belts, so that the epicenters are distributed along the belts.

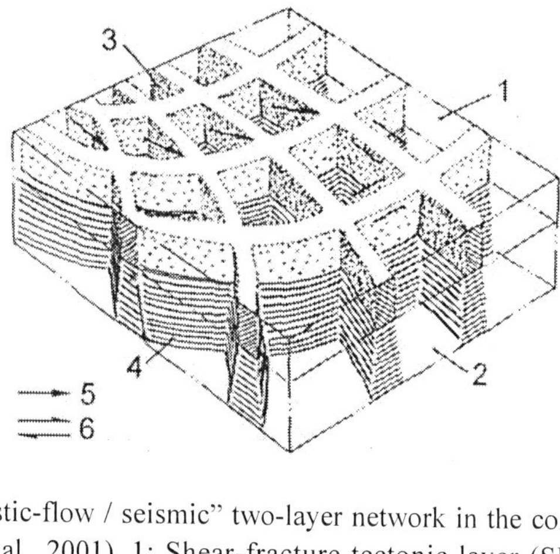

Figure 14. Scheme of "Plastic-flow / seismic" two-layer network in the continental lithosphere (Wang and Zhang, 1995; Wang et al., 2001). 1: Shear-fracture tectonic layer (SFTL), i.e. the seismogenic layer in the upper crust; 2: Netlike-flow tectonic layer (NFTL), i.e. the lower lithosphere; 3: Large-scale seismic belt; 4: Plastic-flow belt; 5: Direction of maximum compressive stress; 6: Shear deformation.

The NPF-controlled earthquake networks have been recognized from the distribution of epicenters in the central-eastern Asian continent as shown in Fig.15. There are three independent plastic-flow network systems distributed in this area, i.e. those of central-eastern Asia (CEAs), southeastern China (SECh) and Burma (Burm), which have the Himalayan arc (H), Taiwan arc (T) and Burmese arc (B) as their driving boundaries, respectively, showing the compressive actions of the Indian plate applying to both CEAs and Burm and that of the Philippine Sea plates to SECh. Each of the networks is composed of two families of plastic-flow belts, the left- and right-extended belts. The existence of the relatively-stable blocks beneath the large-scale compressive basins, TR, AX, OR and SC, form the gaps of the network, showing the obstruction and force-transmission effects of the relatively-stable blocks on NPF.

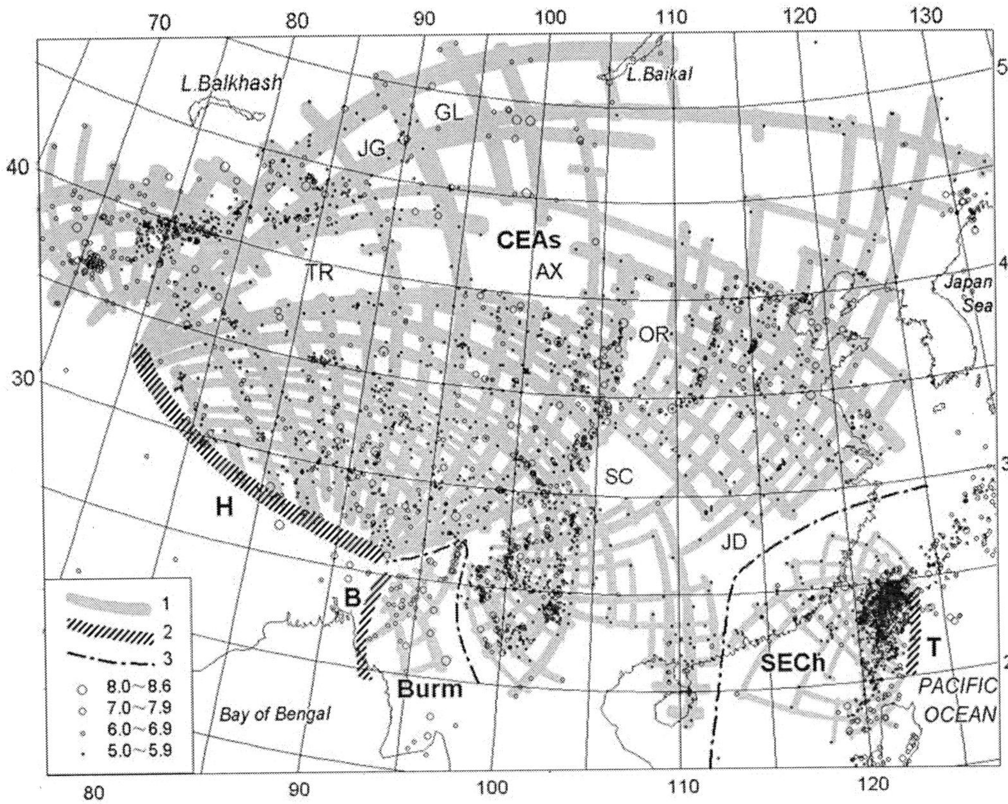

Figure 15. Earthquake epicenter distribution and "plastic-flow / seismic" network systems in the central-eastern Asian continent (Wang et al., 2001; Wang, 2006a). 1: "Plastic-flow / seismic" belt; 2: Driving boundary marked with H, T or B, indicating the Himalayan, Taiwan or Burma arc, respectively; 3. Boundary line between network systems, involving the central-eastern Asia (CEAs), southeastern China (SECh) or Burma (Burm) network systems. Circles with various sizes denote epicenters with different magnitudes. TR, AX, OR, SC, JD, JG and GL show the Tarim, Alxa, Ordos, Sichuan, Jianghan-Dongting, Junggar and Great-Lake basins, respectively.

3.2. Distributions of Stress Directions and Strains in the Lithosphere

Distribution of stress directions in the lithosphere

Using the conjugate-angle-bisector method (Wang, 1993a; Wang et al., 2001), in which the bisector of conjugate angle in plastic-flow network is regarded approximately as the direction of the maximum compressive stress, the distribution of stress directions in the lower lithosphere has been estimated as shown in Fig.16. Meanwhile, the stress directions estimated in terms of the focal mechanism solutions of earthquakes (Xu et al., 1989, 1999; Zoback et al., 1989; Du and Shao, 1999; Chandra, 1981; Mikumo and Ishikawa, 1987; Burtman and Molnar, 1993) are also shown in the figure, manifesting the stress state of the seismogenic layer in the upper crust.

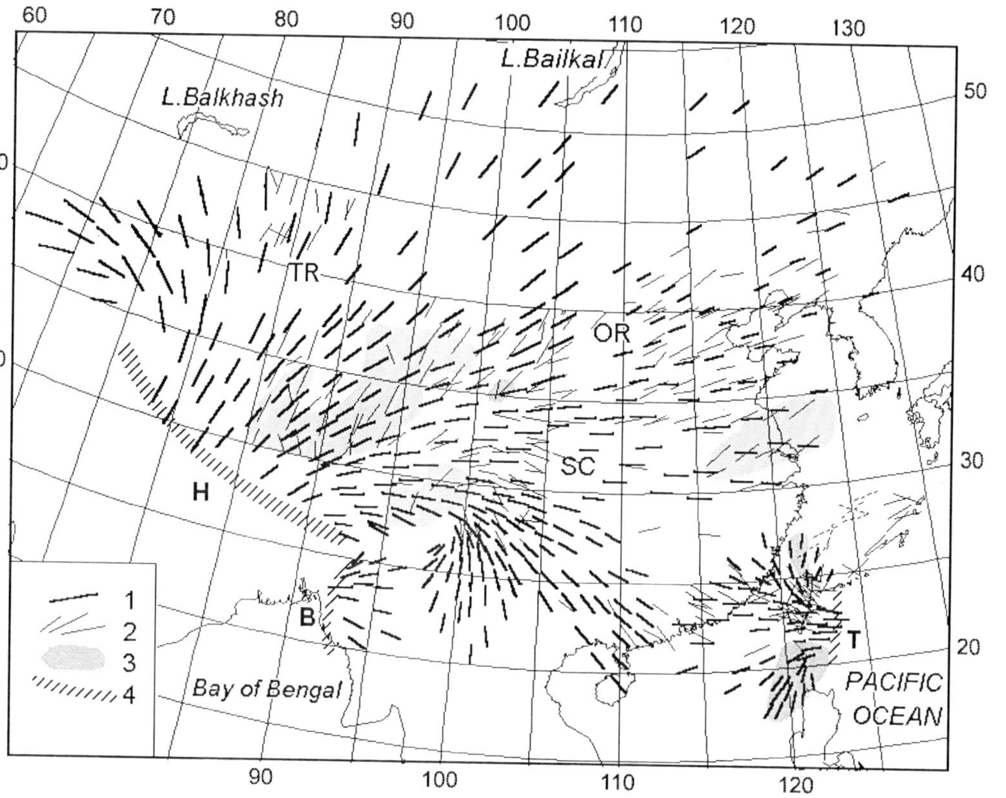

Figure 16. The distribution of maximum compressive stress directions in the central-eastern Asian continent (Wang et al., 2001). 1 and 2: Directions of maximum compressive stresses in the lower lithosphere and the upper crust (seismogenic layer), respectively; 3: The area where the differences of stress directions between the two layers are greater than 30°; 4: Driving boundary. Others as in Fig.15.

It can be seen from the figure that the stress directions in both the layers are identical with each other in general tendency for most areas, where the deviations between them are not larger than 30° and can be neglected in comparison with the large errors possible for the focal-mechanism-solution method itself. It is therefore confirmed further that the stress state in the upper layer is controlled in general tendency by that in the lower layer, while the significant differences are shown mainly in some areas close to the driving boundaries, being associated probably with the action transmitted from the driving boundary directly along the upper layer.

Generally speaking, this is a multi-layered stress field in the continental lithosphere, including those in the lower lithosphere, seismogenic layer and shallow crust, which can be estimated using the conjugate-angle-bisector method, the focal-mechanism-solution method and the methods of in-situ measuring (such as hydraulic-fracturing and borehole-breakout methods), respectively.

Distribution of strains in the lower lithosphere

The strain state in the ductile lower layer can be estimated using the conjugate-angle-increment method, which considers the initial conjugate angle of plastic-flow belts to be 90° and the horizontal maximum compressive strain to be related to the increment of conjugate angle (Wang, 2001). As is well known, the volume of a plastically deforming body is kept constant during its deformation process and, meanwhile, the vertical strain ε_2 is approximate to the horizontal lateral strain ε_3 in a ductile layer. Thus, the expressions of strains are given approximately as follows:

$$\varepsilon_1 = (2/3) \ln [\tan (\psi/2)] \tag{6}$$

$$\varepsilon_3 = -(1/3) \ln [\tan (\psi/2)] \tag{7}$$

$$\varepsilon_2 = \varepsilon_3 = -\varepsilon_1/2 \tag{8}$$

where ψ is the conjugate angle; ε_1, ε_2 and ε_3 are the maximum (horizontal longitudinal), intermediate (vertical) and minimum (horizontal lateral) principal strains, respectively, in which compression is positive and tension negative. Correspondingly, we have the longitudinal contract ΔL and vertical elongation ΔH as follows:

$$\Delta L = L\,[\exp(\varepsilon_1) - 1] \tag{9}$$

$$\Delta H = H\,[1 - \exp(\varepsilon_2)] \tag{10}$$

where L and H are the longitudinal length in deformed plastic-flow network and the thickness of deformed ductile layer, respectively.

In terms of the strains calculated using eq.(6) the contour map of the maximum compressive strain ε_1 is given in Fig.17, which can be regarded relatively as the map of ε_2 or ε_3 in terms of eq.(8), and, furthermore, considered as the contribution of ε_2 to the thickening of the lower lithosphere, ΔH, as shown in eq.(10).

It is shown in Figs.17 and 18 that the strain for each network system is highest at the area close to the driving boundary and descends fluctuantly along with the distance apart from the boundary, in which the "stable" blocks play an important role in constraining strains.

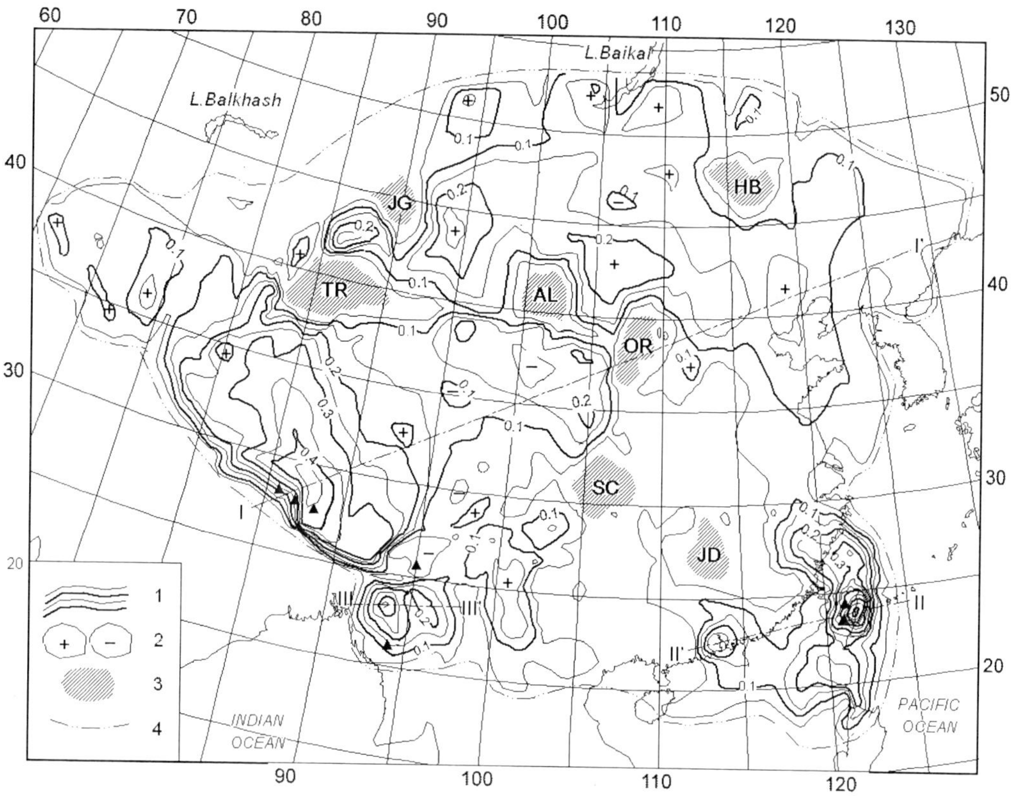

Figure 17. The distribution of strains (ε_1) in the lower lithosphere in the central-eastern Asian continent (Wang, 2001). 1: contours of maximum compressive strains; 2: strain "convex" (with positive sign) and strain "concave" (with negative sign); 3: "stable" block, in which HB indicates that beneath the Hulun-Buir basin and others are the same as in Fig.15; 4: Boundary of the researched area. Triangles denote the topographic peaks (see Table 4). The strain distribution along the lines I - I', II - II' and III - III' are shown in Fig.18.

Comparing the distribution of strains in the lower lithosphere with the three-step staircase-like topography in the China continent indicates the coincidence between them in general framework. As is known, The so-called three-step staircase-like topography includes the first "step" being the Qinghai-Tibet plateau, which is obstructed by the "stable" blocks TR, AL, OR and SC in its northern and eastern margins, the third "step" being the plains and hilly regions in the eastern part of the continent, and the second "step" located between them. Being similar to the three-step staircase, the first "step" for the strain distribution is the area with the maximum compressive strains ranging from 0.10 up to about 0.50 corresponding to the Qinghai-Tibet plateau, the third "step" with strains lower than 0.10 in the eastern part of the continent and the second "step" with strains in the range of about 0.10 to 0.25. It is interesting that both the strain "peaks" and the topographic peaks in CEAs and SECh correspond roughly to each other, which are located near the driving boundary as shown in Fig.17 and Table 4.

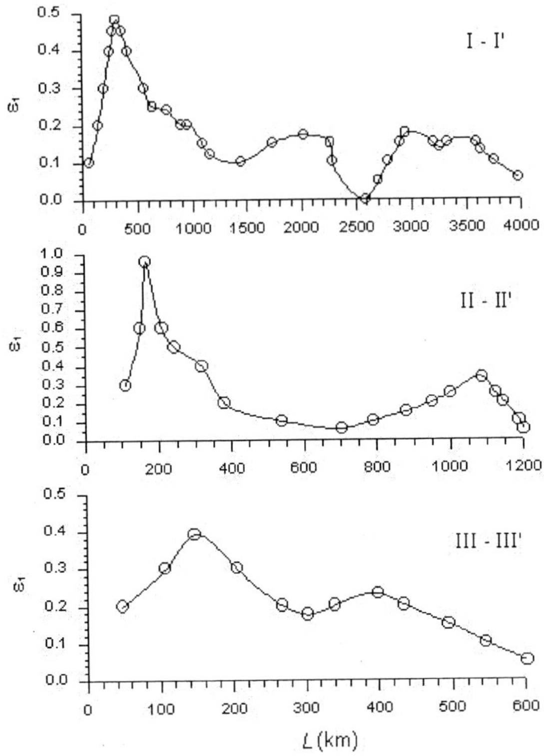

Figure 18. The maximum compressive strain in the lower lithosphere, ε_1, as a function of the distance from driving boundary, *L*, in the central-eastern Asian continent (Wang, 2001). Top, middle and bottom plots show the strain distributions along the lines I - I', II - II' and III - III' (see Fig.17), respectively.

Table 4. Comparison between strain "peaks" in the lower lithosphere and topographic peaks in central-eastern Asia.

Plastic-flow network	Strain "peak" ε_1	$\lvert\varepsilon_2\rvert$	Location	Topographic peak Peak	Elevation(m)	Location
CEAs	0.500	0.250	87.7°E / 29.0°N	Qomolangma	8848	86.9°E / 27.9°N
				Kanchenjanga	8585	88.2°E / 27.7°N
				Xixiabangma	8012	85.8°E / 28.3°N
SECh	0.960	0.480	121.8°E / 23.8°N	Yu Shan	3952	121.0°E / 23.5°N
				Xue Shan	3884	121.2°E / 24.4°N
				Hsiukuluan Shan	3833	121.1°E / 23.5°N
Burm	0.390	0.195	93.5°E / 23.3°N	Saramati Peak	3840	95°E / 25.7°N
				Mt. Victoria	3658	94°E / 21.2°N

The coincidence in relative fluctuation between the distribution of strains in the lower lithosphere and the topographic feature in the central-eastern Asian continent indicates that the contribution of strains in the lower lithosphere plays an important role in the topographic variation, although it is influenced by other factors, such as the deformation of the brittle upper lithosphere and the erosion of the Earth's surface.

3.3. Multi-layered Tectonic Deformation

As has been stated in Section 2.1, lithospheric plate generally includes four tectonic deformation layers from bottom to top, i.e. NFTL, LVTL, SFTL and TFTL. In order to show the relation between them more clearly, the situation in North China is described as an example in this section. In addition to the discontinuously-distributed LVTL, which has not been marked, there are three kinds of tectonic deformation shown in Fig.19: the plastic-flow network in the lower lithosphere, the seismic faults in the seismogenic layer and the faults with outcrops in the shallow crust. Note that all the seismic faults shown by seismic lineaments, focal mechanism solutions and major axis of meizoseismal area are distributed upon the plastic-flow belts, while there are almost no earthquakes with $M \geq 5.0$ occurring along the faults outside plastic-flow belts, showing the close relation between seismic activities and plastic-flow belts.

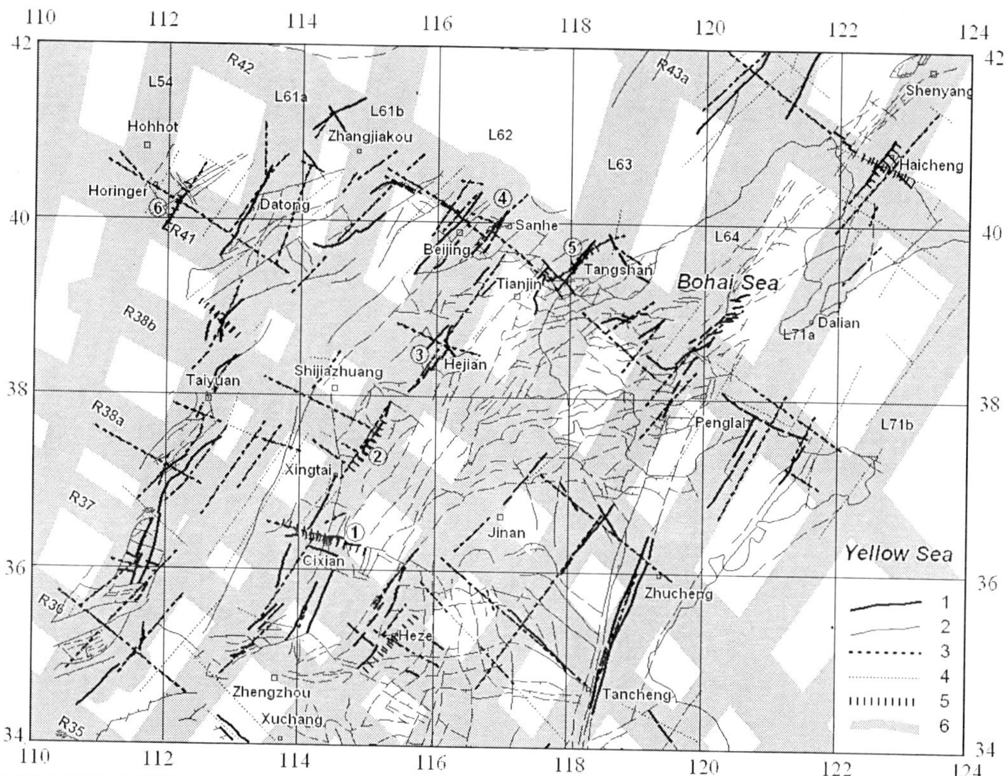

Figure 19. Multi-layered tectonic deformation in North China (Wang and Zhang, 2002). 1 and 2: Shallow active faults (mainly after: Deng, 1986 and Ma, 1989), in which the thick line is related to the earthquakes with $M \geq 5.0$; 3 and 4: Seismic faults inferred from seismic lineaments and focal mechanism solutions, in which the thick line is related to the earthquakes with $M \geq 5.0$ (Zhang et al., 1995); 5: Fault trend inferred from the major axis of isoseismic-line loop (Compilation Group, SSB, 1979); 6: Plastic-flow belt in the lower lithosphere. Faults related to the multi-layered tectonic composite (MLTC): ① Western segment of Linzhang–Daming fault; ② Xinhe fault; ③ (Unnamed, near Hejian); ④ Xiadian fault; ⑤ Tangshan fault; ⑥ Jiucaizhuang– Haolaigou fault. L54, L61a, ... and R35, R36, ...are the codes of left- and right-extended plastic-flow belts, respectively.

The multi-layered tectonic deformation forms tectonic composite for each group of seismic events, called multi-layered tectonic composite (MLTC). There are six typical examples of MLTCs shown in Fig.19 and their relevant parameters are listed in Table 5. It is indicated in the table that the angles between shallow faults and corresponding seismic faults, $\delta_{1\text{-}2}$, are in the range of 0.4° to 8.5°, showing the approximation in strike between them and therefore implying that the strikes of seismic faults can be inferred roughly by those of the shallow faults. The seismic faults can be classified as the following three sorts in terms of the value of $\delta_{2\text{-}3}$, i.e. the angles between seismic fault and its underlying plastic-flow belt: Longitudinal seismic fault with $\delta_{2\text{-}3}$ = 0–30°, oblique seismic fault with $\delta_{2\text{-}3}$ = 30–60° and lateral seismic fault with $\delta_{2\text{-}3}$ = 60–90°. Most of the seismic faults listed in Table 5 have the values of $\delta_{2\text{-}3}$ in the range of 0.2° to 27.3°, being "longitudinal", and only the seismic fault in the composite No.6 has $\delta_{2\text{-}3}$ = 83.8°, being "lateral".

Table 5. Several multi-layer tectonic composites (MLTC) in North China (Wang and Zhang, 2002).

No.	β_1 (°)	β_2 (°)	β_3 (°)	α (°)	$\delta_{1\text{-}2}$ (°)	$\delta_{2\text{-}3}$ (°)	θ_1 (°)	θ_2 (°)	Seismic fault type and largest earthquake: Type	Year / Site / Magnitude
1	94.7	102.5	126.3	75.6	7.8	23.8	70.9	63.1	L	1830 / Cixian, HB / 7.5
2	32.0	34.2	34.4	77.6	2.2	0.2	44.4	46.6	R	1966 / Xingtai, HB / 7.2
3	43.1	34.6	29.8	71.6	8.5	4.8	61.5	53.0	R	1068 / Hejian, HB / 6.5
4	41.8	47.9	20.6	71.5	6.1	27.3	60.3	66.4	R	1679 / Sanhe-Pinggu, HB / 8.0
5	40.3	46.5	20.5	73.7	6.2	26.0	56.6	62.8	R	1976 / Tangshan, HB / 7.8
6	30.0	30.4	114.2	59.8	0.4	83.8	60.2	60.6	R	1976 / Horinger, IM / 6.2

"No." shows the number of MLTC (see Fig.19); α is the direction of the maximum compressive stress; β, δ and θ are the trend azimuth, the angle between trends and the horizontally-sliding angle of tectonic trace(s), respectively; Subscript 1, 2 and 3 denote fault in the shallow crust, seismic fault in the seismogenic layer and plastic-flow belt in the lower lithosphere, respectively; 1-2 or 2-3 the relationship between two layers; L and R the left- and right-lateral strike-slip faults, respectively; HB and IM represent Hebei province and Inner Mongol autonomous region.

A stratigraphic-tectonic profile for the seismic region of the 1966 *M*7.2 Xingtai earthquake (No.2 in Table 5) is given in Fig.20. The Xinhe normal fault as the boundary of the Sulu extensional basin extends down to the depth less than 9 km, while the 1966 *M*7.2 earthquake occurred along a fault plane coincident with the strike-slip fault F3-F2 and its hypocentral depth is about 9 km. The hypocenters with magnitudes of M = 5.0–7.2 are limited to the depth range of about 8–20 km, which, together with the layer between the surface C and LVL (i.e. LVTL), corresponds roughly to the seismogenic layer (SFTL). The layer overlying the SFTL is the tension-fracture tectonic layer (TFTL), in which no earthquake or only those with M<5.0 occurred, while the lower crust beneath LVTL, as a semi-ductile layer with a few of hypocenters (M<5.0), is favorable for the development of plastic-flow belts. It is therefore considered that the Xinhe normal fault in TFTL, the strike-slip seismic fault F3-F2 in SFTL and the plastic-flow belt in NFTL with LVTL between SFTL and NFTL form a MLTC.

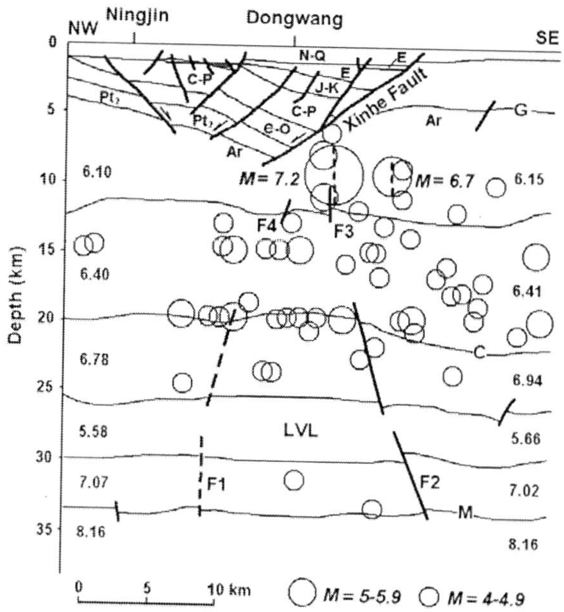

Figure 20. Stratigraphic-tectonic profile in the Xingtai seismic area, Hebei province, China (modified after: Xu et al., 1988; Wang et al., 2001). Thin and thick lines denote the bedding surface and fault, respectively; M: Moho surface; C: Conrad surface; G: Crystalline basement surface; 6.10, 6.40, 6.78, ...: Longitudinal wave velocities (km/s) for the different layers; LVL: Low-velocity layer, i.e. the weakness layer in the middle crust; Circles with different scales denote the hypocenters occurring in the period of 1966–1968; Dashed lines in circles are the seismic fault planes of 1966 *M*7.2 and *M*6.7 earthquakes. Strata: Ar, Archean; Pt2, Mid-Proterozoic; Є, Cambrian; O, Ordovician; C, Carboniferous; P, Permian; J, Jurassic; K, Cretaceous; E, Paleogene; N, Neogene; Q, Quaternary.

4. Analogue Modeling of NPF-Controlled Tectonic Deformation

4.1. Model As146

The analogue model As146 is composed of a ductile lower layer made of plasticized rosin and a thin brittle upper layer of consolidated talc-powder slurry just as those stated in Section 2.3. The model is placed on a glass plate, which has been smeared with silicon oil as lubricant on its surface. To model the situation existing in the earlier stage of Quaternary (Wang et al., 2001), the indentations are applied to the driving boundaries corresponding to the Himalayan arc (H), Taiwan arc (T) and that near the Xiao Hingganling Mountains (X), as shown in Fig.21. The indenter is pushed using a shim-spring-rod unit, which applies compressive force on the boundary by increasing the shim(s) step by step.

The similarity between the natural prototype and the models is expressed using a simplified approach (Tapponnier et al., 1982; Peltzer and Tapponnier, 1988), in which K_L, K_t, K_V, K_η, K_ρ, K_g and K_p denote the scale ratios for length, time, velocity, viscosity, density, gravity acceleration and pressure (stress, strength), respectively, where by K we refer to the scale ratio of prototype to model. The length ratio $K_L = 3.16\times10^7$ is taken corresponding to the ratio of the length of the Himalayan arc (about 1930 km) to the width of the indenter H in the model; the velocity ratio K_V is the ratio of the average velocity of the Himalayan collision boundary (about 50 mm/a) to that of the driving boundary H in the model; and, therefore, we have the time ratio $K_t = K_L / K_V$. In addition, the pressure ratio for the ductile lower layer is calculated approximately as $K_p = K_\eta / K_t$ in terms of the relation between pressure (stress, strength) and viscous force.

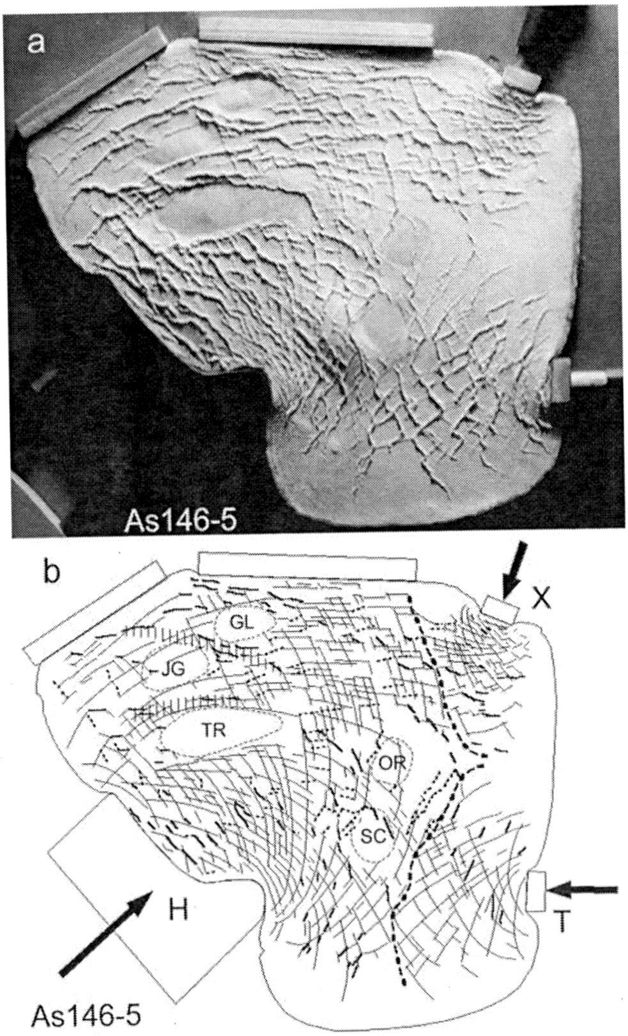

Figure 21. NPF-controlled tectonic deformation in the model As146 (top view) (Wang et al., 2007). (a) Photo of the deformed model. (b) Line drawing of the tectonic deformation pattern. Thin solid lines: NPF-controlled tectonic belts, forming tectonic networks. Short thick lines: Compressive folds or thrust faults. Thick dot lines: Traces deviating from plastic-flow belts. Hatched zones: Shear zones occurring in the earlier stage, which are rebuilt later by the plastic-flow network. Thick dot-dash lines: Boundaries between adjacent plastic-flow networks. Thick arrows show the compressive actions at the driving boundaries. H, T and X denote the driving boundaries corresponding to the Himalayan arc, Taiwan arc and that near the Xiao Hingganling Mts., respectively. SC, OR, TR, JG and GL indicate the Sichuan, Ordos, Tarim, Junggar and Great-Lake basins, respectively, beneath which "stable" blocks with higher viscosity have been buried in the lower layer.

The relevant parameters of the model As146 and the extrapolated values for the prototype are given in Table 6, where p_H is the average horizontal pressure at the driving boundary H and other symbols are the same as in Section 2.3. The data following the symbols H/T/X in Table 6 (or H/T/R/J/X in Table 7) indicate the values of parameters for different driving boundaries.

Table 6. Parameters of the model As146 (Step 5) and their extrapolated values for the natural prototype.

Parameters	Model	Prototype
h_b	0.2 (mm)	6.3 (km)
h_d	6.0 (mm)	190 (km)
η	2.03×10^8 (Pa s)	10^{21} (Pa s)
B (H/T/X)	60 / 15 / 11 (mm)	1896 / 474 / 348 (km)
p_H	4790 (Pa)	9.72 (MPa)
ΔL (H/T/X)	17.5 / 5.0 / 3.0 (mm)	553 / 158 / 95 (km)
t (H/T/X)	40 / 21 / 21 (h)	11.10 / 5.83 / 5.83 (Ma)
V (H/T/X)	0.44 / 0.24 / 0.14 (mm/h)	50.0 / 27.3 / 15.9 (mm/a)

R/P=12, T=20□. Scale ratio: $K_L = 3.16\times10^7$, $K_\eta = 4.93\times10^{12}$, $K_t = 2.43\times10^9$, $K_V = 1.30\times10^{-2}$, $K_p = 2.03\times10^3$.

The upper layer made of the weakly-consolidated talc powder with the compressive strength of about 0.05 MPa is originally stronger than the ductile lower layer, but gets weaker once faults and folds occur in it, and then the lower layer will subject to the major fraction of driving forces in the subsequent deformation process.

It is shown in Fig.21 that there are three tectonic network systems driven by the indenters H, T and X, respectively, and the H- and T-driven networks are quite similar in general framework to the "plastic-flow / seismic" networks CEAs and SECh in the central-eastern Asian continent shown in Fig.15. The belts of the tectonic networks in the model As146 are composed of a series of secondary folds and thrust faults and form "mountain ranges" with "intermountain basins" between them just as in the model M225 (Fig.7). In addition, a number of compressive folds or thrust faults shown as short thick lines in Fig.21 are roughly normal to the directions of maximum compressive stresses (see Fig.23b below) and some scattered traces shown as dot lines deviate from belts of the tectonic networks.

The large-scale compressive basins, SC, OR, TR, JG and GL, are developed upon the "stable" blocks, which have been buried in advance in the ductile lower layer. There are two tectonic zones shown hatched in Fig.21, one distributed between the basins TR and JG and another between TR and GL, which contain a series of secondary compressive folds or thrust faults arranged en echelon and intersect obliquely the belts of the tectonic network. These two zones are comparable in locations roughly to the Tianshan and Altay mountain ranges (tectonic zones) in the prototype.

The compressive folds developed in front of the indenters, the conjugate angles of the tectonic networks being equal to or greater than 90° and the extremely low strength ratio (R_s = 0.00041) of the brittle upper layer to the ductile lower layer indicate that the development of the tectonic networks in the model are controlled by NPF in the lower layer rather than by the action of driving force transmitted directly along the brittle upper layer.

4.2. Model As153

Most of the tectonic phenomena in the model As153 (Fig.22 and Table 7) are similar to those in As146, such as the development of tectonic networks, large-scale compressive basins, "mountain ranges" and "intermountain basins". It is more obvious in this model than in As146 that an embryonic form of the "Qinghai-Tibet plateau" appears between the driving boundary H and the "stable" blocks TR, OR and SC, probably owing to the indentation of H (ΔL=632 km) to be larger than that in As146 (ΔL=553 km). Moreover, some differences between them in tectonic deformation result mainly from the additional driving boundaries J and R at the eastern side and the absence of the basins JG and GL in the model As153.

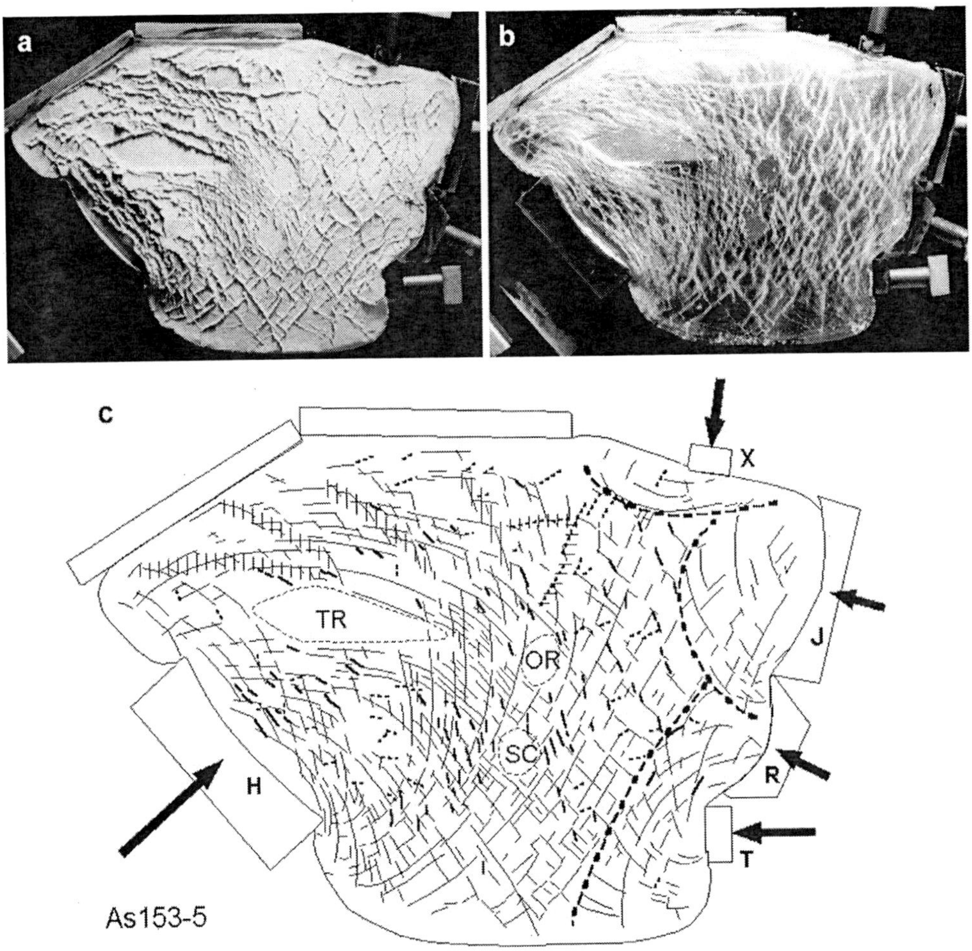

Figure 22. NPF-controlled tectonic deformation in the model As153 (top view) (Wang et al., 2007). (a) Photo of the model. (b) Photo of the model's lower layer (the upper layer has been stripped), in which no fracture occurs and the white traces are the remains of the upper layer's talc powder stuck on the plastic-flow belts and other structures in the lower layer. (c) Line drawing of the tectonic deformation in the upper layer of the model. R and J denote the Ryukyu and Japan arcs. Others as in Fig.21.

Table 7. Parameters of the model As153 (Step 5) and their extrapolated values for the natural prototype.

Parameters	Model	Prototype
h_b	0.2 (mm)	6.3 (km)
h_d	7.4 (mm)	234 (km)
η	9.06×10^{7} (Pa s)	10^{21} (Pa s)
B (H/T/R/J/X)	58.2 / 15 / 33 / 48 / 11 (mm)	1839 / 474 / 1043 / 1417 / 348 (km)
p_H	6720 (Pa)	19.08 (MPa)
ΔL (H/T/R/J/X)	20.0 / 5.4 / 4.7 / 4.5 / 0 (mm)	632 / 171 / 149 / 142 / 0 (km)
t (H/T/R/J/X)	28.5 / 8.0 / 8.0 / 8.0 / – (h)	12.62 / 3.54 / 3.54 / 3.54 / – (Ma)
V (H/T/R/J/X)	0.70/ 0.68/ 0.59/ 0.56/ 0(mm/h)	50.0 / 48.5 / 42.1 / 40.0 / 0 (mm/a)

R/P=10, T=18□. Scale ratio: $K_L = 3.16\times10^{7}$, $K_\eta = 1.10\times10^{13}$, $K_t = 3.88\times10^{9}$, $K_V = 8.15\times10^{-3}$, $K_p = 2.84\times10^{3}$.

The existence of the driving boundary J together with a part of the boundary R, which subjects to the westward driving force corresponding to the horizontal compressive component of the Pacific plate subduction, results in the development of the J-driven tectonic network, which obstructs the eastward spread of the H-driven network and gives rise to the reduction of the X-driven network. The eastern boundary between T and X in the model As146 displaces eastward up to 6.2 mm, corresponding to 196 km in the prototype, while once the model is constrained by J and R at its eastern side in As153, the eastward displacement is replaced by the increase of intraplate strains, which results in the development of NNE-trend tectonic belts in the areas on the east and northeast of OR and SC (Fig.22), corresponding to North China and its northern and southern adjacent areas in the prototype. These NNE-trend tectonic belts can be regarded as the left-extended belts of the H-driven network.

It is seen from Fig.22b that the plastic-flow belts in the lower layer are considerably narrower than those in the one-layer models (Figs.4 and 6) and the surface bulges upon the belts in the one-layer models are replaced by the thin wrinkles which are filled with the remains of talc powder of the stripped upper layer, appearing as white traces. The formation of the narrower belts with flatter top surface in the lower layer can be attributed mainly to the vertical constraint of the upper layer.

In addition to the facts similar to those shown in the model As146, including the development of folds and faults in front of the indenter, the conjugate angles of tectonic networks being equal to or greater than 90° and the extremely low strength ratio (R_s = 0.00021 for As153), the coincidence between the tectonic pattern seen from the upper layer (Fig.22a) and the plastic-flow belts shown by the white traces on the surface of the ductile lower layer (Fig. 22b) in the model As153 indicates further the close relation between both of them and the controlling action of NPF in the ductile lower layer.

4.3. The Distribution of Stress Directions and Strains in the Ductile Lower Layer

Using the "conjugate-angle-bisector" method the distributions of maximum compressive stress directions in the lower layer of the models As146 and As153 are estimated as shown in

Figs.23b and 23c. Comparing them with Fig.23a indicates that the distributions of stress directions in most areas of the H- and T-driven tectonic networks in the models, especially in As153, are similar in general tendency to those in the prototype, while the differences between them appear mainly in the wing areas of the driving boundaries H and T, where the clockwise or counter-clockwise deflections of stress directions in the models are considerably smaller than those in the prototype. These differences are attributed mainly to the underachievement of gravity-driven action of the "plateau" owing to the durations extrapolated from the models to be some few orders of magnitude smaller than the actual duration in the prototype if the time scale ratio K_t was determined to follow the similarity law exactly.

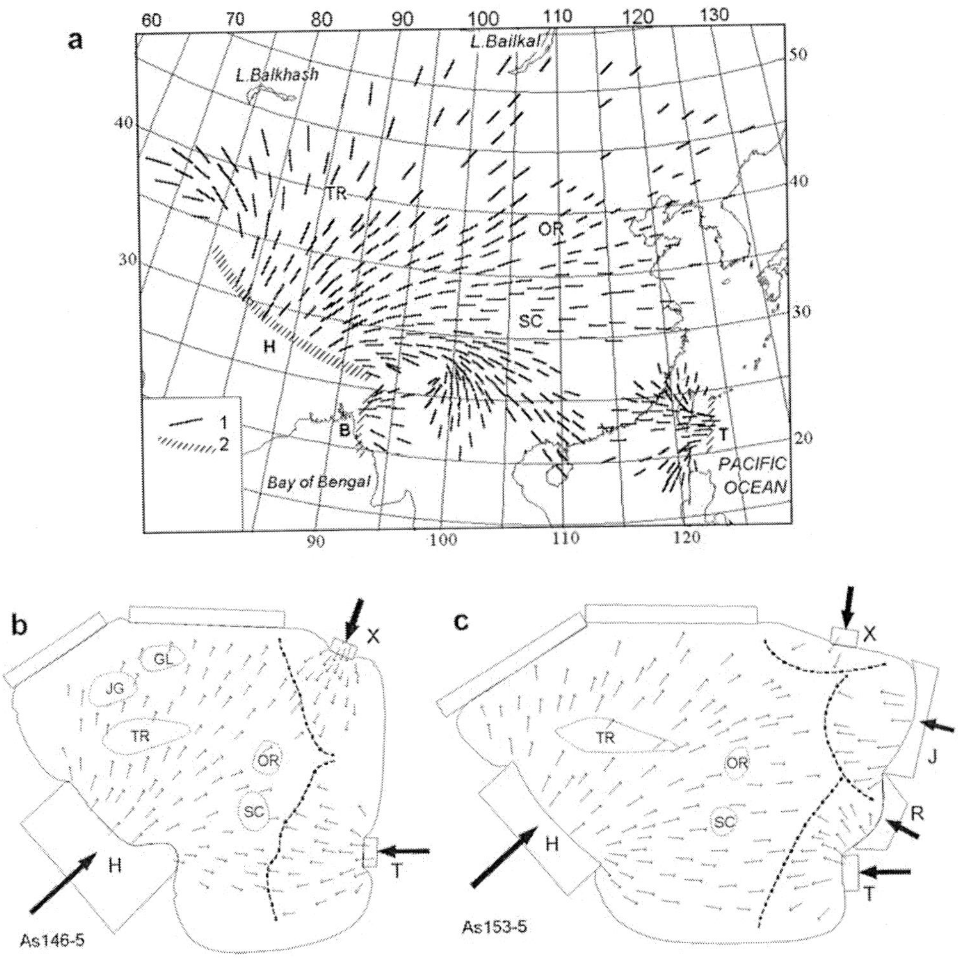

Figure 23. The distribution of maximum compressive stress directions in the ductile lower layer. (a) Distribution of stress directions in the central-eastern Asian continent (see Fig.16). 1: Direction of maximum compressive stress; 2: Driving boundary. (b) and (c) Distributions of stress directions in the ductile lower layer of the models As146 and As153 (Wang et al., 2007). Small arrows denote the directions of maximum compressive stresses and large ones the driving forces at boundaries. Others as in Fig.21.

Moreover, using the conjugate-angle-increment method, the strains in the lower layer of the models are estimated in terms of the conjugate angles of the NPF-controlled tectonic networks as shown in Fig.24, which also show relatively the distribution of the thickening of the lower lithosphere as shown in eqs.(8) and (10). The strain distribution in the lower lithosphere shown in Fig.24 (Top) is equivalent to that in Fig.17, which shows roughly the three-step staircase-like topography in the China continent as mentioned in Section 3.2. Comparing the situations shown in the models (Fig.24, middle and bottom) with it (Fig.24, top) indicates that they are similar to each other in general framework, although the models

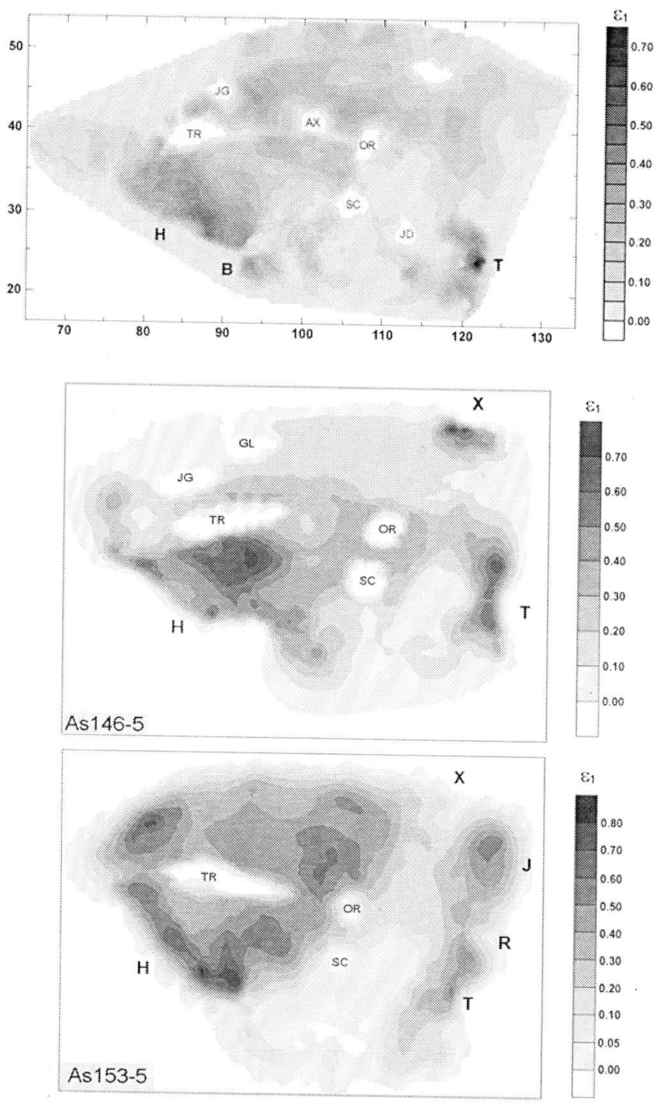

Figure 24. Distribution of maximum compressive strains in the ductile lower layer of the models. Top: Strain distribution in the lower lithosphere in the central-eastern Asian continent, being equivalent to Fig.17. Middle and bottom: Strain distributions in the ductile lower layer of the models As146-5 and As153-5 (Wang et al., 2007). ε_1: Maximum compressive strain. Others as in Figs.15 and 23.

are imperfect in time scaling, boundary condition and setting up of "stable" blocks. The strain distributions in the models are also characterized roughly by the three-step staircase, including the first "step", i.e. an embryonic "plateau" near the driving boundary H with vertical strains greater than about 0.15-0.2 (As146) or 0.25-0.30 (As153), the third "step" in the eastern part with vertical strains less than about 0.10-0.15, and the second "step" between them.

Note that an independent high-strain region with strains up to 0.73 has been developed under the indentation at the driving boundary X in the northeastern margin of the model As146 (Fig.24, middle), which simulates the situation in an earlier stage of Quaternary, while there is no independent high-strain region near the driving boundary X in the model As153 (Fig.24, bottom) owing to the strong westward indentation at the driving boundary J. It is inferred that the vanishing of driving boundary X in the recent prototype (Fig.15) may be related to the westward action of the horizontal component of the Pacific plate subduction. Furthermore, the lack of the Burmese driving boundary (B) in the models is responsible for the absence of the independent high-strain region near the boundary B.

5. Other NPF-Controlled Geological-Geophysical Phenomena

5.1. "Plastic-Flow / Magmatic Rock" Networks

In addition to the NPF-controlled netlike distribution of earthquakes stated above, shear deformation and its viscous-friction heating along plastic-flow belts give rise to weakening and lightening of the media within and upon the belts. The so-called "lightening" refers to the decrease of density of the media owing to the rise of temperature and the fracturing in the crust, and the "weakening" to the decrease of viscosity and strength of the media and the development of fractures in the crust. Both of them provide the condition favorable for local mantle uplifting and magma upwelling by means of the gravity balance and the formation of transport paths. Furthermore, friction heating is also possible to promote the partial melting of rocks along the belts. Thus, upwelling and solidifying of magmas, including those derived from the deeper mantle and produced in the plastic-flow belts, form magmatic rocks in the lithosphere as shown in Fig.25, resulting in the belt-like and netlike distributions of them. The NPF-controlled magmatic-rock networks seen from geological map is in fact the netlike distribution of magmatic rocks, including extrusive and uplifted intrusive rocks, which were formed in some geological period.

According to the geological maps and relevant references (China Geological Academy, 1975; Li et al., 1982; Ma, 1989), the netlike distribution patterns of magmatic rocks for the different geological periods in the central-eastern Asian continent have been recognized, including those for Late Paleozoic, Mesozoic and Cenozoic eras. For instance, there were two stages, PzA and PzB, in the Late-Paleozoic era as shown in Fig.26 and the NPF-controlled magmatic-rock networks for both of them were superposed each other in the northern region, including the North network system (NN) in the PzA stage and two systems, the Northwest (NW) and Northeast (NE) ones, in the PzB stage. In addition, there was another network system, the Indo-China system (IC), existing in the southern region.

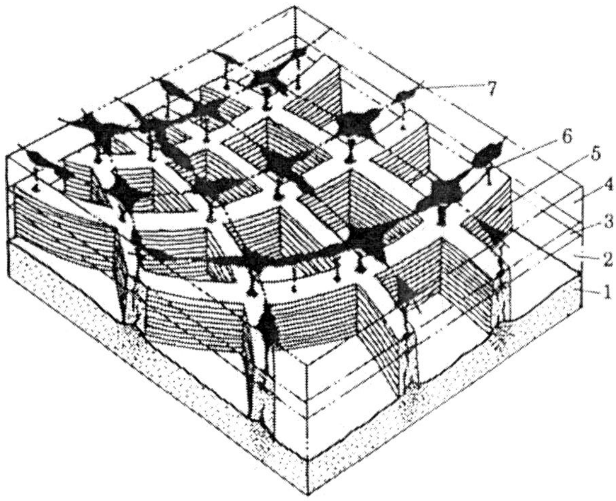

Figure 25. Scheme of "plastic-flow / magmatite" network (Wang and Zhang, 1997; Wang et al., 2001). 1: Asthenospheric mantle; 2: Lithospheric mantle; 3: Lower crust; 4: Upper crust; 5: Plastic-flow network; 6: Magmas or magmatic rocks shown by black blocks, columns and points; 7: Magmatic-rock network shown by dot-dash lines.

The distributions of maximum compressive stress directions in the lower lithosphere estimated using the conjugate-angle-bisector method for the network systems in the stages PzA and PzB are shown in Fig.27. It seems that the major driving forces was applied southward to the boundaries located at 55°-57°N (present latitudes), involving one driving boundary for the NN network system in the Stage PzA and two for the NW and NE systems in the Stage PzB. Meanwhile, the network system IC subjected to the eastward driving force.

Moreover, the strain and displacement fields can be estimated using the conjugate-angle-increment method. In the NN network system for the PzA stage, for instance, the strains between Lake Baikal and Ordos (OR) near the line of longitude 110°E are estimated by means of eq.(6) and the result is given in Table 8. Substituting the average strain ε_1 and the current distance between Lake Baikal and OR, L = 1230 km, into eq.(9), the contraction is obtained as ΔL = 562 km and it is therefore inferred roughly that the distance between Lake Baikal and Ordos (OR) was about 1790 km in the Late Paleozoic era.

Table 8. Conjugate angles and strains distributed between Lake Baikal and Ordos (OR) in the period PzA.

No.	1	2	3	4	5	6	7	8	9	Average
ψ(°)	116	119	112	111	117	121	125	128	135	
ε_1	0.314	0.353	0.2625	0.250	0.327	0.380	0.435	0.479	0.588	0.376

The NPF-controlled magmatic-rock network as a "fossil" of ancient tectonic deformation opens up possibility for understanding the states of stresses and strains in the lower lithosphere in different geological times.

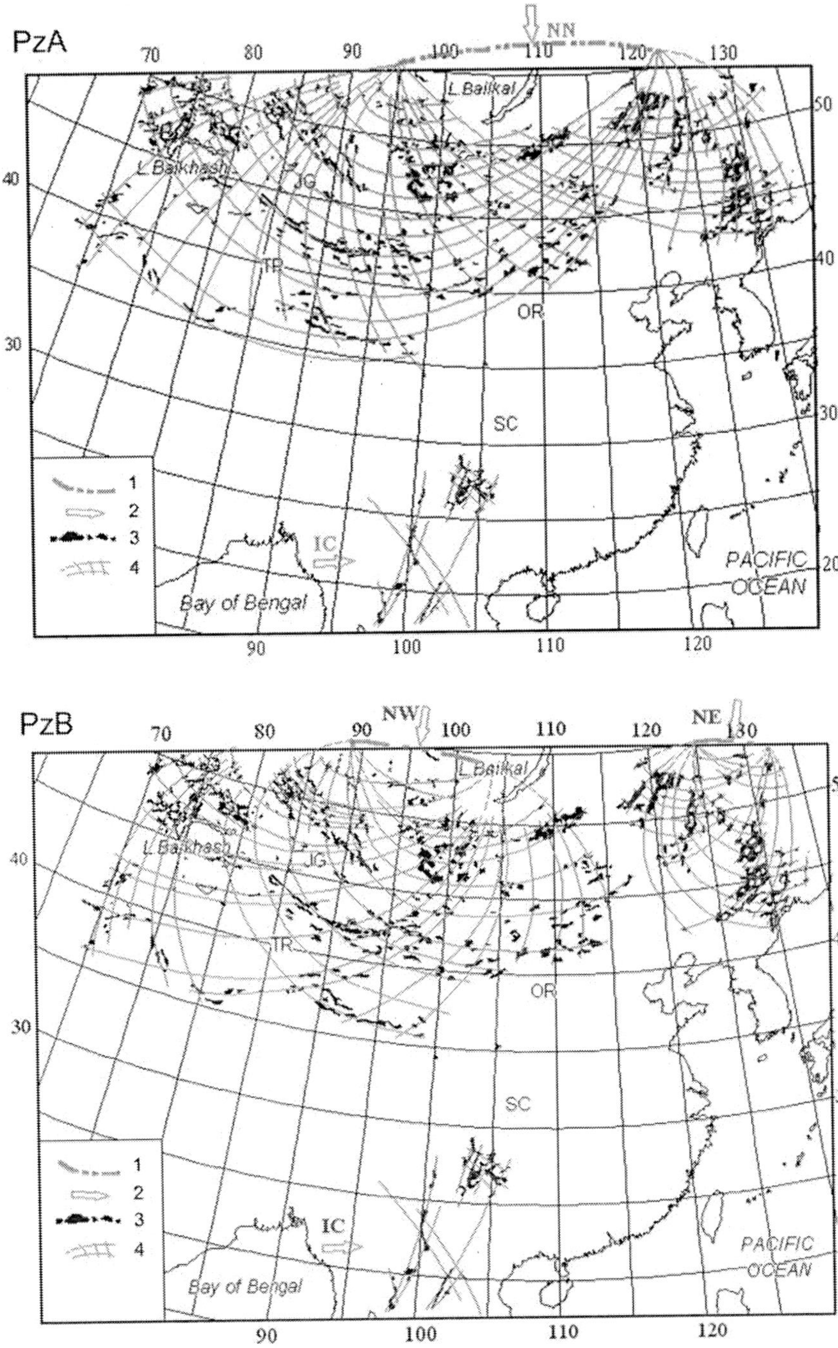

Figure 26. NPF-controlled late-Paleozoic magmatic-rock networks in the central and eastern Asian continent (Wang and Zhang, 1997; Wang et al., 2001a). Top: Stage PzA, involving the North (NN) and Indo-China (IC) network systems; Bottom: Stage PzB, involving the Northwest (NW), Northeast (NE) and Indo-China (IC) network systems. 1: Driving boundary; 2: Direction of driving force; 3: Magmatic rocks; 4: "Plastic-flow / magmatic-rock" network; TR, JG, OR and SC show the present locations of the Tarim, Junggar, Ordos and Sichuan basins.

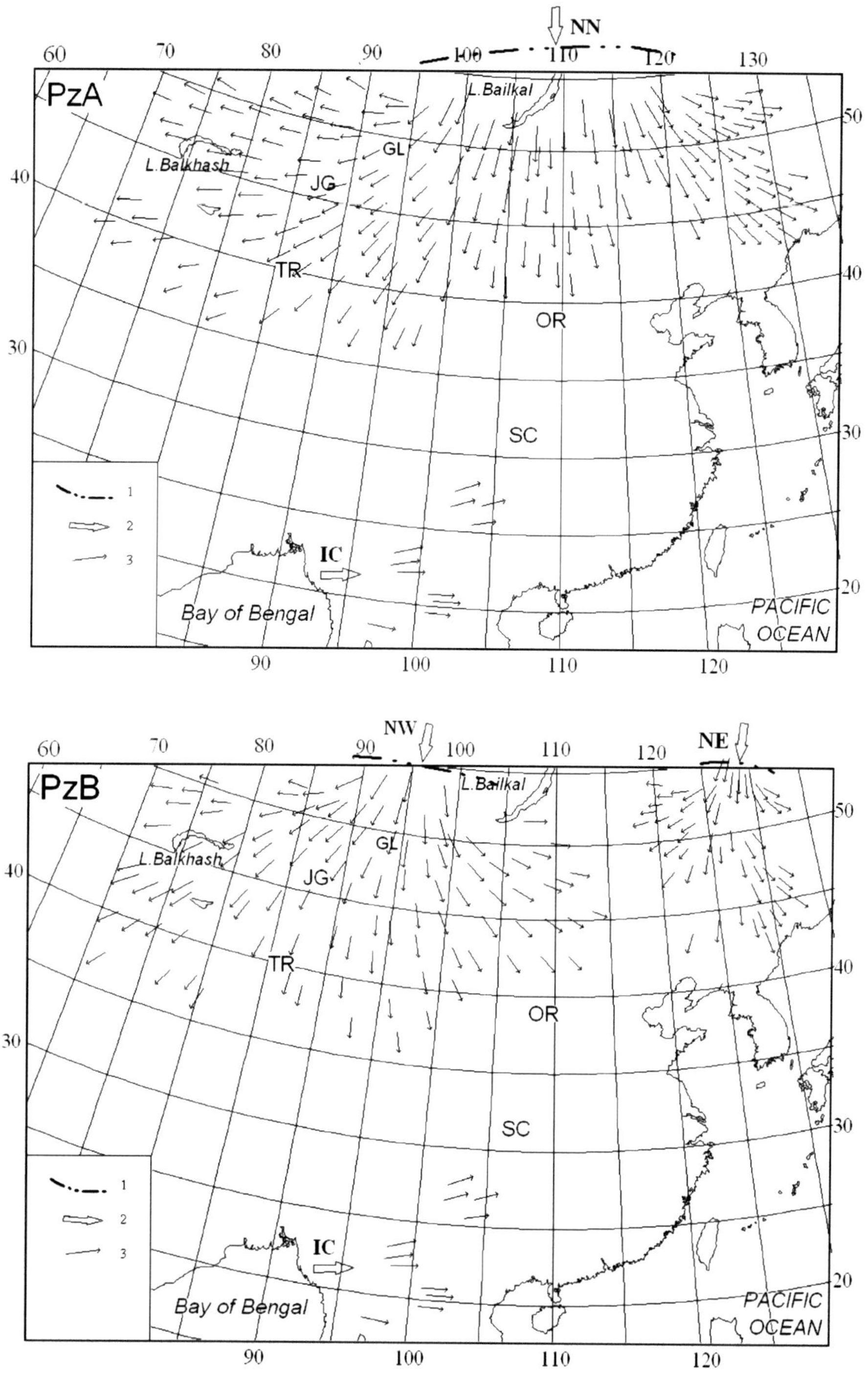

Figure 27. Distribution of maximum compressive stress directions estimated in terms of the late-Paleozoic magmatic-rock networks in the central and eastern Asian continent (Wang and Zhang, 1997; Wang et al., 2001). Top: Stage PzA; Bottom: Stage PzB. 1: Driving boundary; 2: Direction of driving force; 3: Direction of maximum compressive stress. Others as in Fig.26.

5.2. "Plastic-Flow / Velocity Anomaly" Networks

The NPF-controlled magmatic activities result not only in the development of magmatic-rock networks as described above but in the netlike distributions of geophysical anomalies, such as those of seismic-wave-velocity, gravity and magnetism. For instance, the netlike distribution of velocity anomaly in Xinjiang and adjacent regions is stated in this section. Being different from the magmatic-rock network composed of extrusive and uplifted intrusive rocks, which are distributed on surface or in the shallowest crust, showing a planimetrical pattern of network, the velocity-anomaly images at different depths show a stereo pattern of network.

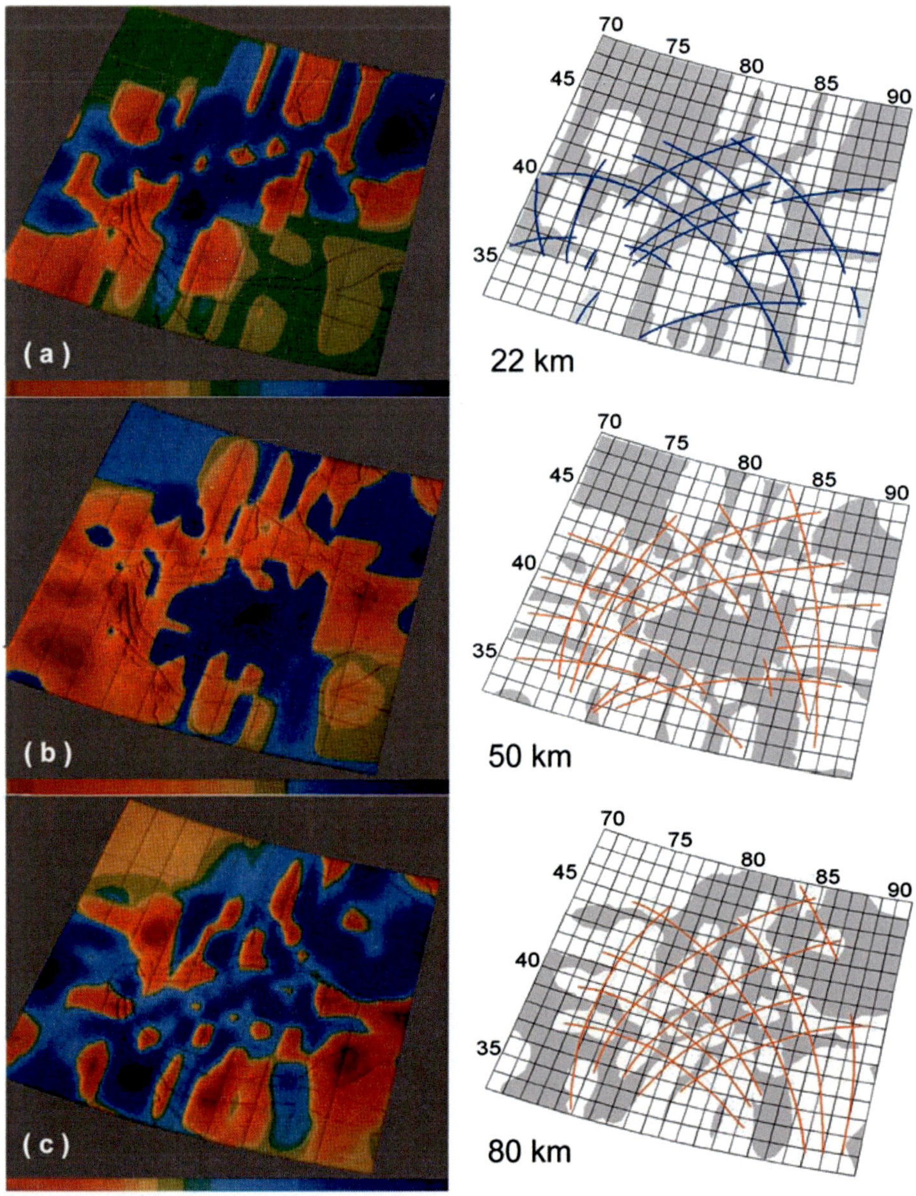

Figure 28. Continued on next page.

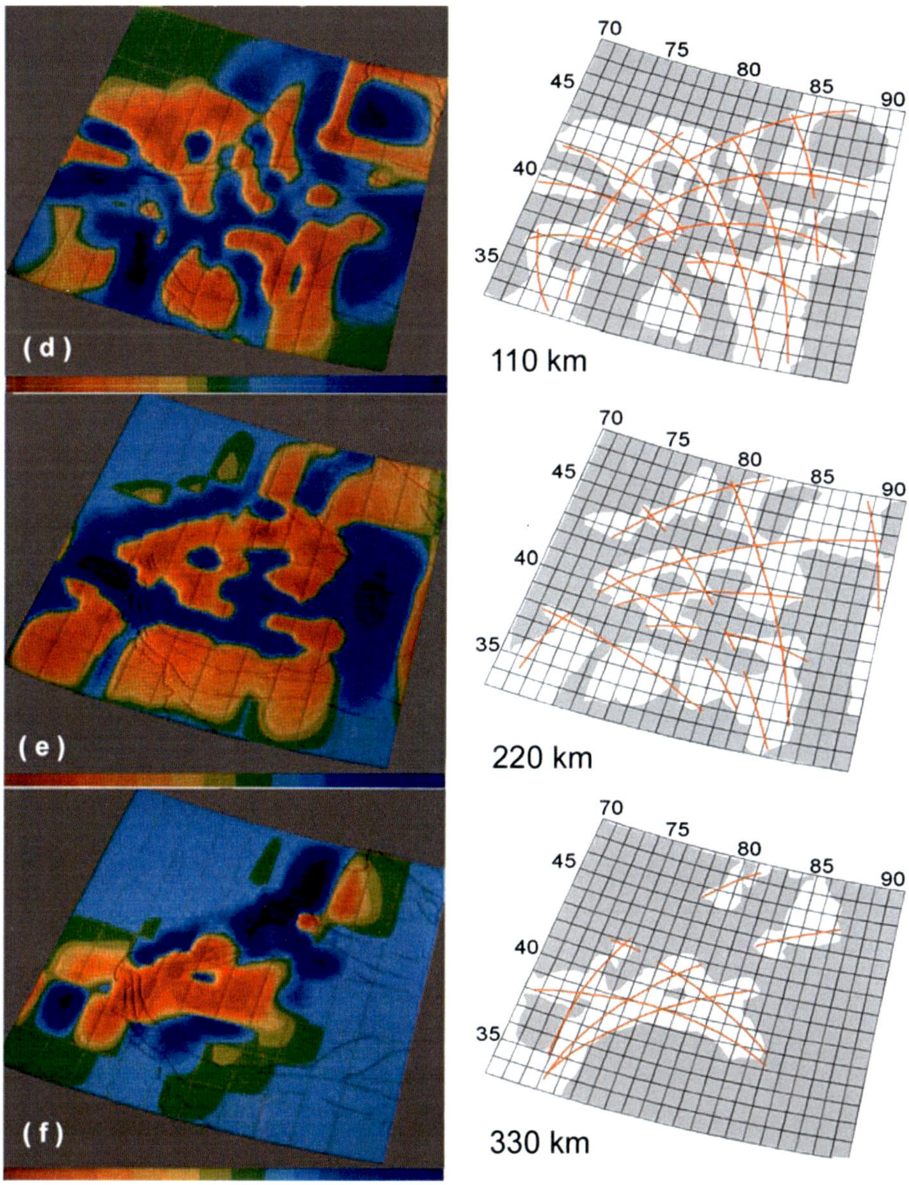

Figure 28. The seismic tomographic images and NPF-controlled network patterns for the different depths in Xinjiang and adjacent regions (Wang et al., 2001). Left: Seismic-wave-velocity images at different depths, in which red and blue denote the lower and higher velocities, respectively (Li et al., 1994). Right: Velocity-anomaly network patterns, in which grey area denotes the relatively-high velocity and red and blue lines show the plastic-flow belts corresponding to lower and higher velocities, respectively. Depths: (a) 22 km; (b) 50 km; (c) 80 km; (d) 110 km; (e) 220 km; (f) 330 km.

The tomographic images of seismic velocities shown in Fig.28 (left) are obtained by the inversion of arrival-times of initial P-waves (Li et al., 1994) and the network patterns recognized from them are shown in Fig.28 (right) (Wang et al., 2001), in which the plastic-flow belts are drawn up in terms of velocity highs at the depth of 22 km in consideration of the solidification of magma, while in terms of velocity lows at other deeper levels where the densities of the upwelling mantle materials along the belts are lower than those of the rocks outside the belts.

It is known that the thickness of the crust in this researched region is in the range of 44–60 km (Ma, 1987) and the depth of 22 km lies in the upper crust. As shown in Fig.29 (left), the comparison of the network pattern at 22 km depth with the "plastic-flow / seismic" network in the same region shown in Fig.15 indicates that both of them are essentially identical with each other, implying that they belong to the same present network system. It can be seen further from Fig.29 (right) that the velocity-anomaly network patterns at different depths are identical in general tendency but somewhat different from one another in the trends of the belts, showing the vertical variations of the "plastic-flow/ velocity-anomaly" network.

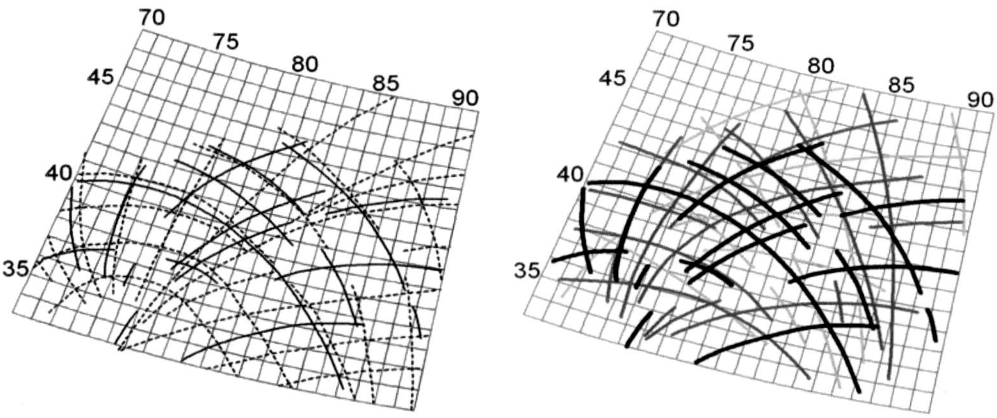

Figure 29. "Plastic-flow / velocity-anomaly" network patterns at different depths and "plastic-flow / seismic" network in Xinjiang and adjacent regions (Wang et al., 2001). Left: Comparison between the velocity-anomaly network pattern at 22 km depth (solid line) and the "plastic-flow / seismic" network (dotted line, see Fig.15). Right: Comparison of velocity-anomaly network patterns at depths of 22, 50 and 220 km shown with black, dark grey and light grey lines, respectively.

The depth range of the "plastic-flow / velocity-anomaly" network may be helpful to examine the thickness of the lithosphere, which is possible to be larger than 200 km in some regions of the China continent as shown in the literature (Yuan, 1996). The obvious network pattern at 220 km depth shown in Fig.28 indicates that the thickness of the lithosphere is not less than this depth, while the netlike pattern appearing locally at the depth of 330 km seems to be a problem: Does the bottom of the lithosphere reach so deep level or is this netlike pattern only some traces in the asthenosphere resulting from the influence of NPF in the lithosphere? It remains to be answered in the future.

5.3. NPF-Controlled Distribution of Positive Heat Flow Anomalies

Based on the facts of the shear strain localization and the viscous-friction-heating effect along plastic-flow belts, it is inferred that the positive heat flow anomaly would appear upon the belts. The published heat flow data of the China continent open up possibility to a certain extent for investigating the relation of the distribution of heat flow to plastic-flow network. The compilation of heat flow data of the China continent (Wang and Huang, 1990; Hu et al., 2001) has collected 816 heat flow data with different qualities published by some research groups in China in the period from 1978 to 2001. In order to ensure the reliability of the study, 639 heat flow values with "high quality" or "second-high quality" are adopted. The distribution of the measuring points of these heat flow data in the "plastic-flow / seismic" network systems is shown in Fig.30. Suppose d is the distance from measuring point to network line, i.e. the midline of plastic-flow belt, and its positive value is taken for the point situated on the line's side facing the driving boundary while negative on its opposite. The correlation between the heat flow q and the distance d for the CEAs network system is shown in Fig.31, in which the apparent width of plastic-flow belt, b, is equal to the average width of NPF-controlled seismic belts shown in Fig.15.

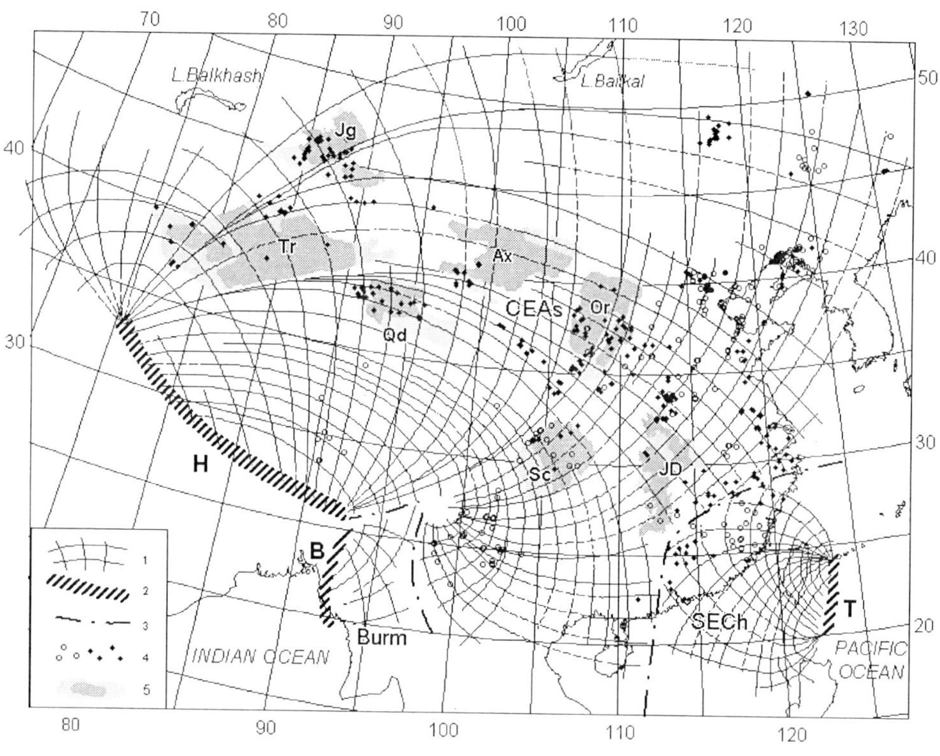

Figure 30. "Plastic-flow / seismic" networks, relatively-stable blocks and distribution of heat-flow measuring points in the central-eastern Asian continent (Wang, 2004b, 2006a). 1. Network lines, which are approximate to the mid-lines of plastic-flow belts; 2. Driving boundary; 3. Boundary line between network systems; 4. Heat flow measuring points shown by empty and black points, corresponding to the second (Wang and Huang, 1990) and third (Hu et al., 2001) edition data, respectively; 5. Relatively-stable block, shown by large-scale compressive basin, in which the darker part is the area without earthquake of $M \geq 5$; Others as in Fig.15.

The confidence interval of the normal heat flow is expressed as:

$$q = q_{av} \pm t_{0.05}\, S \qquad (11)$$

where q_{av} is the average heat flow, S the standard deviation and $t_{0.05}$ the probability coefficient with 95% confidence for the number of accepted data, n. Using the t-test method (Statistical Research Group, 1973), it is obtained finally through several rounds of rejecting operations for the CEAs system that n = 525, S = 12.1 mW/m^2, $t_{0.05}$ = 1.97 and q = 60.5±23.8 mW/m^2 (Wang, 2004b, 2006a). The data points higher than the upper limit and lower than the lower limit of the confidence interval show the positive and negative anomalies of heat flow, respectively.

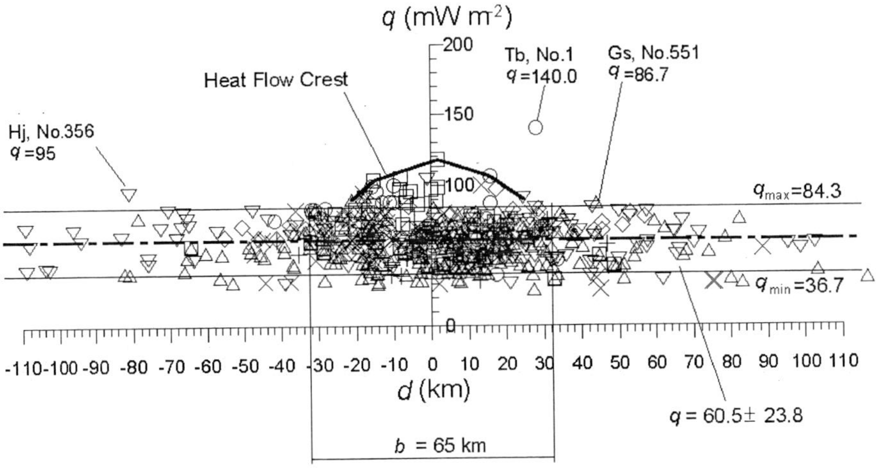

Figure 31. Distribution of heat flow relative to "plastic-flow / seismic" belts in the CEAs plastic-flow network system (Wang, 2004b, 2006a). Data points for following regions: Circles: Tibet, Qinghai (excluding the Qaidam basin) and the portions of Gansu and Sichuan inside the Qinghai-Tibet plateau; Squares: Yunnan and Sichuan (excluding the Sichuan basin); Triangles: Tarim, Junggar, Qaidam, Ordos, the Sichuan basin, a portion of Gansu associated with the Alxa relatively-stable block, and the JD basin involving the portions of Hunan, Hubei and Henan; Diamonds: Hainan, Anhui, Jiangsu, Shandong and the portions of Guangdong, Jiangxi and Zhejiang; Crosses: Hebei, Beijing, Tianjin and Bohai sea; Inverted triangles: Inner Mongolia, Liaoning and Heilongjiang. No.1, No.356 and No.551: Numbers of measuring points; q: the terrestrial heat flow value with "high" or "second high" quality (Wang and Huang, 1990; Hu et al., 2001); d: The distance from measuring point to network line, which has positive or negative value for the point located at the line sides near or away from the driving boundary, respectively; q_{max} and q_{min}: The upper and lower limits of the normal heat flow; b: The average apparent width of plastic-flow belts.

As shown in Fig.31, there are 34 data points of positively-anomalous heat flow in the CEAs network system, in which, except two points in Gansu and Heilongjiang provinces (Nos.356 and 551), 32 of them (amount to 94%) in Xizang, Qinghai, Yunnan, Sichuan, Guangdong, Henan, Hebei, Shandong and Liaoning are distributed within the average width of plastic-flow belts (65 km), showing a heat flow "crest" higher than the upper limit of the normal heat flow (84.3 mW/m^2) in the chart of q versus d and the maximum value near the mid-line of the "crest" to be 118.1 mW/m^2 (the point No.1 with q=140.0 mW/m^2 in Tibet as a particular case is beyond the envelope of the "crest"). On the other hand, 36 data points of negatively-anomalous heat flow in the system are scattered along the d–axis, showing no considerable correlation with the plastic-flow belts, while 20 of them (amount to 55%) are related to the relatively-stable blocks, involving the Tarim, Junggar, Qaidam, Alxa, Ordos, Sichuan and Jianghan-Dongting basins. The manifestation of positively-anomalous heat flow "crest" and the correlation of negatively-anomalous heat flow with relatively-stable blocks indicate the effects of plastic-flow belts and relatively-stable blocks on heat flow distribution.

Although the heat flow data points are not enough to form a complete pattern of NPF-controlled heat-flow network, the existence of the heat-flow-anomaly "crest" indicates to a certain extent the distribution of positive anomalies of heat flow to be controlled by the plastic-flow network. In addition, the heat-flow-anomaly "crest" can be verified further through the derivation and solution of the energy equation of thermal dynamics, in which the shear heating effect of plastic-flow belt and the heat conduction normal to the belt have been considered (Wang, 2006c).

The width of positive-thermal-anomaly belt, called "thermal belt" for short, is larger than that of shear-localization belt, called "shear belt" for short. Suppose y_b and y_1 are the half widths of shear and thermal belts, respectively, T_b is the temperature at boundary of shear belt and T_0 at that of thermal belts, i.e. the original temperature without the influence of shear heating, the distribution of temperature along the profile normal to the belt as a result of the derivation of the energy equation can be expressed as follows (Wang, 2006c):

$$(T - T_0) = (T_b - T_0) + \{[(T_c - T_0) - (T_b - T_0)] / y_b + (1/2)\,(\zeta\, \eta_{blt}\, \dot{\gamma}_{blt}^{\,2} / k_{blt})\, y\}\,(y_b - y), \quad 0 \leq y \leq y_b \tag{12}$$

$$(T - T_0) = (T_b - T_0)\,(y_1 - y)^2 / (y_1 - y_b)^2, \quad y_b \leq y \leq y_1 \tag{13}$$

where T and y denote the temperature and the distance apart from the belt's midline, respectively; T_c is the highest temperature at the midline; $\dot{\gamma}_{blt}$ and η_{blt} are the shear strain rate and initial viscosity of shear belt, respectively; k_{blt} is the thermal conductivity of thermal belt; and ζ the viscosity-correction coefficient. The distribution of temperatures and the relevant parameters are shown in Fig.32 and Table 9, in which the temperature anomaly ($T - T_0$) is normalized as Ω:

$$\Omega = (T - T_0) / (T_c - T_0) \tag{14}$$

where ($T_c - T_0$) is the largest positive temperature anomaly being estimated roughly as

$$(T_c - T_0) = H_{uc}\, \Delta q_{t.C} / k_{uc} \tag{15}$$

H_{uc} and k_{uc} are the average thickness and the vertical thermal conductivity of the upper crust and $\Delta q_{t.C}$ is the largest positive heat-flow anomaly near the midline of the belt.

Table 9. Parameters related to the shear-heating effect of plastic-flow belt (Wang, 2006c).

Parameter	Value	Notation
y_b (m)	22000	Fitting it to the data points
y_l (m)	32500	$2y_l \approx b$, i.e. the apparent width of plastic-flow belt in CEAs
$(T_c - T_0)$ (°K)	210	Calculated in terms of Eq.(15)
$(T_b - T_0)$ (°K)	67	Fitting it to the data points
$\dot{\gamma}(s^{-1})$	1.950×10^{-14}	Average strain rate of the lower lithosphere in the CEAs network system[1]
K	1.30[1]	Strain-rate ratio $K = \dot{\gamma}_{blt} / \dot{\gamma}$
R	0.82	Weakening level of plastic-flow belt[2]
$\dot{\gamma}_{blt}$ (s^{-1})	2.535×10^{-14}	$\dot{\gamma}_{blt} = K\dot{\gamma}$
η (Pa s)	8.579×10^{20}	Weighted average viscosity of the lower crust and lithospheric mantle
η_{blt} (Pa s)	5.412×10^{20}	$\eta_{blt} = (R/K)\eta$
ζ	5.574	Fitting it to the data points
$\zeta\eta_{blt}$ (Pa s)	2.922×10^{21}	Fitting it to the data points
k_{blt} (W m^{-1} $°K^{-1}$)	3.3[3, 4]	$k_{blt} \approx k$, i.e. the horizontal thermal conductivity in the lower lithosphere
k_{uc} (W m^{-1} $°K^{-1}$)	4.0[3, 4]	$k_{uc} > k_{blt}$ in consideration of the development of fractures in the upper crust
H_{uc} (m)	25000	Being less than the average thickness of the upper crust in the CEAs system
$\Delta q_{t.C}$ (mW m^{-2})	33.8	The largest positive heat-flow anomaly at the midline (see Fig.31)

References: [1] Wang, 2006b; [2] Wang, 2004a; [3] Yuen et al., 1978; [4] Turcotte and Schubert, 1982.

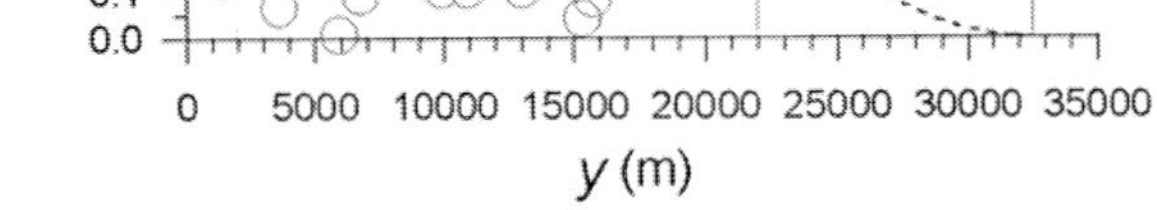

Figure 32. Relative temperature anomaly, Ω, versus distance from the midline of plastic-flow belt, y, in the lower lithosphere (Wang, 2006c). $\Omega = (T - T_0) / (T_c - T_0)$, where T denotes temperature, T_c is the temperature at the midline of plastic-flow belt, and T_0 the original temperature in the lithosphere without the influence of shear heating. y_b and y_l are the half widths of shear and thermal belts, respectively.

It is shown in Fig.32 that, on the average, the shear-belt width for the CEAs plastic-flow network system, $2y_b$, is equal to 44 km, and its thermal-belt width, i.e. its influence range depending on heat conduction, $2y_l$, is approximate to 65 km being identical with the average width of seismic belts shown in Fig.15. Correspondingly, the positive temperature-anomaly is about 210 °K (i.e. the normalized temperature-anomaly $\Omega = 1$) at the midline of shear belt, about 67 °K ($\Omega = 0.32$) at its boundary, and decreases down to zero at the boundary of the thermal belt.

As mentioned by a number of researchers, shear heating is one of the mechanisms for flow-softening (Yuen et al., 1978; Poirier J P, 1980; Kameyama et al., 1999). It is the shear heating of plastic-flow belt that, as a dominant mechanism, results in the positive thermal-anomaly and therefore the shear weakening of the belt. It should be mentioned that a similar idea has been proposed by Hochstein and Regenauer-Lieb (1998), who argued the "heat bands" with width of 30–50 km to be induced by the major "slip-lines" in the "ductile crust" in the Qinghai-Tibet plateau resulting from the plate collision at Himalayas. According to the viewpoint of NPF, some segments of these locally-distributed "slip-lines" correspond roughly to some of plastic-flow belts as a part of the CEAs network system.

It is expected that the NPF-controlled netlike distribution pattern of heat flow will be shown more and more obviously with the accumulation of heat flow data and, finally, an essentially complete "plastic-flow / thermal-anomaly" network will be recognized in the central-eastern Asian continent.

6. Conclusion

To sum up, there are two parts of basic viewpoints to be the core idea of the NPF model for continental dynamics. The first part involves the following aspects, which have been accepted or partially accepted by most of geoscientists:

— Deformable lithospheric plate;

— Multi-layered tectonic deformation resulting from the transition of deformation regime from brittle to ductile with the increase of depth in the lithosphere;

— Viscous (plastic) flow or shear strain localization playing an important role in intraplate deformation processes;

—Intraplate tectonic deformation to be driven by various forces, mainly including plate-boundary-driving force, plate-bottom-driving force and body force, which are associated directly with plate collision or subduction, mantle convection and rotation of the Earth or interaction between celestial bodies, respectively.

The second part has been suggested by the author on the basis of the first part viewpoints and the studies of netlike plastic-flow (NPF), mainly involving:

— NPF and its "power / linear-binomial" combined flow law instead of the conventional power flow law, implying that shear strain localization can occur in a viscous (plastic) flow field, forming plastic-flow network similar to but different from the "slip-line network" in the classical plasticity mechanics;

— The lower lithosphere, including the lower crust and lithospheric mantle, being suitable to NPF, which controls intraplate deformation, resulting in the netlike distributions of earthquakes, magmatic activities, and wave-velocity, geothermal and other geological-geophysical anomalies;

— Plate-boundary driving force together with the action of intraplate gravity potential playing the major role for the NPF-controlled intraplate deformation;

— Multi-layered tectonic deformation, generally including 4 layers from bottom to top of the lithosphere, i.e. the netlike flow (NFTL), low-velocity (LVTL), shear fracture (SFTL) and tensile fracture (TFTL) tectonic layers, in which TFTL may be absent in some regions and LVTL as a transitional weakness layer is distributed discontinuously between NFTL and SFTL;

— The estimations of stress directions and strains in the lower lithosphere using the conjugate-angle-bisector and conjugate-angle-increment methods, respectively.

As is well known, the theory of plate tectonics has drawn up a general structural framework of the global lithosphere, which is composed of a series of relatively-rigid lithospheric plates with various boundaries between them, forming a "global network" controlled mainly by mantle convection, while the NPF model brings the complicated, seemingly-irregular geological-geophysical phenomena within plate into a unified, rational "intraplate network" controlled by NPF in the lower lithosphere.

The NPF model has been being verified through the experimental, numerical and theoretical studies, the recognitions of a variety of NPF-controlled network patterns and the examinations in its applications since 1980s and lays a foundation for further understanding intraplate tectonic deformation and dynamic processes.

Acknowledgements

Thanks are extended to Jian-guo Li, Zong-chun Zhang, Yong-sheng Zhou, Zi-chao Wang, Liu Zhang, Jun-Xiu Sun and my other colleagues in the Institute of Geology for their participating in and supporting the studies of NPF, to Qiang Li for providing the tomographic images and to Hao Wang for helping in computing. This work has been performed under the supports of the Institute of Geology, China Earthquake Administration and the grants from the Natural Science Foundation of China (Nos.49070196 and 40472111), the Joint Earthquake Science Foundation of China (Nos.85080103, 91088, 199061 and 196065) and the National Key Scientific and Technological Project (No.9691309020302) since 1980s.

References

Alsaleh, M.I., 2004. Numerical modeling of strain localization in granular materials using Cosserat theory enhanced with microfabric properties. Ph D Dissertation, Dept of Civil and Environmental Engineering, Louisiana State Univ.

Bishop, J.F.W., 1953. On the complete solution to problems of deformation of a plastic-rigid material. *Journal of Mechanics and Physics of Solids*, **2**: 43-53.

Bull, J.M., Martinod, J. and Davy, Ph., 1992. Buckling of the oceanic lithosphere from geophysical data and experiments. *Tectonics*, **11**(3): 537-548.

Burtman,V.S. and Molnar,P.,1993. Geological and geophysical evidence for deep subduction of continental crust beneath the Pamir. *Special Paper* No.281, The Geological Society of America, Inc., Boulder, Colorado.

Carter, N.L. and Ave'Lallemant, H.G., 1970. High temperature flow of dunite and peridotite. *Geol. Soc. Amer. Bull.*, **81**: 2181-2202.

Chandra, U., 1981. Focal mechanism solutions and their tectonic implications for the eastern Alpine-Himalayan region. In: Gupta, H.K., Delany, F.M. (Eds.), *Zagros, Hindu Kush, Himalaya, Geodynamic Evolution*. Am. Geophys. Union, Washington, D. C. pp. 243-271.

China Geological Academy, 1975. Geological Map of Asia (1:5000000). China Cartographic Publishing House, Beijing.

Chopra,P.N. and Paterson,M.S.,1981. The experimental deformation of dunite. *Tectonophysics*, **78**: 453-473.

Chopra, P.N., 1986. The plasticity of some fine-grained aggregates of olivine at high pressure and temperature. In: Hobbs, B.E. and Heard, H.C. (eds.), *Mineral and Rock Deformation: Laboratory Studies (The Paterson Volume)*, Monogr.36, Am. Geophys. Union, Washington, D.C., 25-33.

Christian, J.T., 1977. Shallow foundations. In: Desai, C.S. and Christian, J.T. (eds), *Numerical Methods in Geotechnical Engineering*, McGraw-Hil, New York, pp.211-234.

Cobbold, P.R., 1975. Fold propagation in single embedded layers. *Tectonophysics*, **27**: 333-351.

Cobbold, P.R. and Davy, P., 1988. Indentation tectonics in nature and experiment: 2. Central Asia. *Bull. Geol. Inst. Univ. Uppsala, NS*, **14**: 143-162.

Compilation Group of Seismic Intensity Regionalization in China, State Seismological Bureau, 1979. Atlas of Curves of Equal intensity in China. Seismological Press, Beijing. (in Chinese)

Cui, Z.-z., Yin, Z.-x., Gao, E.-y., Lu, D.-y. and Fu, W.-z., 1992. Velocity Structure and Deep-Seated Structure of the Qinghai-Tibet Plateau. Geological Publishing House, Beijing. (in Chinese)

Davy, P. and Cobbold, P.R., 1988. Indentation tectonics in nature and experiment: 1. Experiments scaled for gravity. *Bull. Geol. Inst. Univ. Uppsala, NS*, **14**: 129-141.

Davy, Ph. and Cobbold, P.R., 1991. Experiments on shortening of 4-layer model of the continental lithosphere. *Tectonophysics*, **188**: 1-25.

Davy, P., Hansen, A., Bonnet, E. and Zhang, S.-Z., 1995. Localization and fault growth in brittle-ductile systems: Implications to deformations of the continental lithosphere. *J. Geoph. Res.*, **100**: 6281-6294.

Deng, Q.-d., 1986. Active tectonics in North China Region (Scale 1: 3 000 000). In: Center for Earthquake Analysis and Prediction and Institute of Geology, State Seismological Bureau (eds.), *Atlas for Assessments of Earthquake Risk Areas (1987-2000) in North China Region*. Xi'An Cartographic Publishing House, Xi'An, Map No. 5. (in Chinese)

Du, P.-r. and Xu, D.-y., 1989. An Introduction to Astroseismology. Seismological Press, Beijing. (in Chinese with English summary)

Du, X.-x. and Shao, H.-c., 1999. Modern tectonic stress field in the Chinese mainland inversed from focal-mechanism solutions. *Acta Geophysica Sinica*, **21**(4): 354-360. (in Chinese)

Dubey, A.K. and Cobbold, P.R., 1977. Noncylindrical flexural slip folds in nature and experiment. *Tectonophysics*, **38**: 223-239.

England, P. and McKenzie, D., 1982. A thin viscous sheet model for continental deformation. *Geophys. J. R. astr. Soc.*, **70**: 295-321.

England, P. and Houseman, G., 1986. Finite strain calculations of continental deformation, 2. Comparison with the India-Asia collision zone. *J. Geophys. Res.*, **91**(B3): 3664-3676.

England, P. and Molnar, P., 1997. Active deformation of Asia: From kinematics to dynamics. *Science*, **278**: 647-650.

Evans, B., Fredrich, J.T. and Wong, T.-f., 1990. The brittle-ductile transition of rocks: Recent experimental and theoretical progress. In: Duba, A.G. et al.(eds.), *The Brittle-Ductile Transition in Rocks (The Heard Volume)*, Geophys. Monogr.56, Am.Geophys.Union, Washington, D.C., 1-20.

Evison, F.F., 1960. On the growth of continents by plastic flow under gravity. *Geophys. J.*, **3**: 155-190.

Heard, H.C., 1960. Transition from brittle fracture to ductile flow in Solenhofen limestone as a function of temperature, confining pressure, and interstitial fluid pressure. In: Griggs, D.T. and Handin, J. (eds.), *Rock Deformation*, Memoir **79**, Geol. Soc. Am., New York, 193-226.

Heard, H.C., 1976. Comparison of the flow properties of rocks at crustal conditions. *Philos.Trans.R.Soc.London*, **A 283**: 173-186.

Hill, R., 1950. The Mathematical Theory of Plasticity. Clarendon Press, Oxford.

Hochstein, M.P. and Regenauer-Lieb, K., 1998. Heat generation associated with collision of two plates: the Himalayan Geothermal Belt. *Journal of Volcanological and Geothermal Research* , **83**: 75-92.

Houseman, G. and England, P., 1993. Crustal thickening versus lateral expulsion in the Indian-Asian continental collision. *J. Geophys. Res.*, **98**(B7): 12233-12249.

Höeg, K., Christian J.T. and Whitman R.V., 1968. Settlement of strip load on elastic plastic soil. *J. Soil Mech. Found. Div. ASCE*, **94**(SM2): 431-445.

Hu, S.-B., He, L.-J. and Wang, J.-Y., 2001. Compilation of heat flow data in the China continental area (3rd edition). *Chinese J. Geophys.*, **44**(5): 611–626. (in Chinese with English abstract)

Jeffreys, H., 1976. The Earth: Its Origin History and Physical Constitution. Cambridge Univ. Press, Cambridge London, New York, Melbourne.

Kameyama, M., Yuen, D.A and Karato, S.-I., 1999. Thermal-mechanical effects of low-temperature plasticity (the Peierls mechanism) on the deformation of a viscoelastic shear zone. *Earth Planet. Sci. Lett.*, **168**(1-2): 159-172.

Kirby, S.H. and Raleigh, C.B., 1973. Mechanisms of high-temperature, solid state flow in minerals and ceramics and their bearing on creep behavior of the mantle. *Tectonophysics*, **19**: 165-194.

Kirby, S.H. and McCormick, J.W., 1984. Inelastic properties of rocks and minerals: Strength and rheology. In: Carmichael, R.S. (ed.), *Handbook of Physical Properties of Rocks*, Volume III, CRC Press, Boca Raton, Florida, 139-280.

Kohlstedt, D.L. and Goetze, C., 1974. Low-stress high-temperature creep in olivine single crystals. *J.Geophys.Res.*, **79**(14): 2045-2051.

Kotake, N., Tatsuoka, F., Tanaka, T., Siddiquee, M.S.A. and Huang, C.C., 2001. FEM simulation of the bearing capacity of level reinforced sand ground subjected to footing load. *Geosynthetics International*, **8**(6): 501-549.

Lee, J.S., 1984. Introduction to Geomechanics. Science Press, Beijing and Gordon and Breach, Science Publishers, Inc., New York.

Li, C.-y., Wang, Q., Liu, X.-y. et al., 1982. Geotectonic Map of Asia (1:8000000). China Cartographic Publishing House, Beijing.

Li, J., Van Vilet, K.J., Zhu, T., Yip, S. and Suresh, S., 2002. Atomistic mechanisms governing elastic limit and incipient plasticity in crystals. *Nature*, **418**: 301-310.

Li, J.-g. Song, R.-q. and Wang, S.-z., 1993. Experimental study of brittle-ductile transition and macroscopic structures of some solid materials, *Progress in Geophysics*, **8**(4): 70-80. (in Chinese with English abstract).

Li, J.-g., Zhou, Y.-s., Song, R.-q. and Wang, S.-z., 1997. Physical modeling of plastic-flow network and multi-layer tectonic deformation in thc lithosphere. *Seismology & Geology*. **19**(3): 248-258. (in Chinese with English abstract)

Li, J.-g., Zhou, Y.-s. and Wang, S.-z., 1998. Physical modeling of plastic-flow-network-controlled tectonic deformation in the central-eastern Asian continent. *Seismology & Geology*. **20**(1): 63-72. (in Chinese with English abstract)

Li, Q., Liu, R.-f., Du, A.-l. and Yang, M.-j., 1994. Seismic tomography of Xinjiang and adjacent region. *Acta Geophysica Sinica*, **37**(3): 311-320, Plate 1. (in Chinese with English abstract)

Liu, S. and Dixon, J.M., 1991. Centrifuge modeling of thrust faulting: structural variation along strike in fold-thrust belts. *Tectonophysics*, **188**: 39-62.

Ma, X.-y. (chief compiler), 1987. *Lithospheric Dynamics Map of China and Adjacent Seas (Scale 1:4000000) and Its Explanatory Notes*. Geological Publishing House, Beijing.

Ma, X.-y. (chief compiler), 1989. Lithospheric Dynamic Atlas of China. China Cartographic Publishing House, Beijing.

Maruyama, S., 1994. Plume tectonics. *J. Geol. Soc. Japan*, **100**(1): 24-49.

McKenzie, D.P., 1972. Active tectonics of the Mediterranean region. *Geophys. J. R. astr. Soc.*, **30**: 109-185.

Meguid, S.A. and Klair, M.S., 1984. Finite element studies into incomplete coverage in shot-peening. In: Fuchs, H.O. (ed), *Proc. 2nd. Int. Conf. Shot Peening*, American Shot Peening Society, Paramus, Chicago, 306-315.

Mikumo, T. and Ishikawa, D., 1987. Major earthquakes along the eastern to southern margin of the Japan Sea and temporal variations of their activity, in relation to regional tectonics. In: National Committee for Seismology, Science Council of Japan and Seismological Society of Japan (Eds.), *Proceedings of Earthquake Prediction Research Symposium, 1987*. Seismological Society of Japan, Tokyo, pp.259-269. (in Japanese)

Molnar, P. and Tapponnier, P., 1975. Cenozoic tectonics of Asia: Effects of a continental collision. *Science*, **189**(4201): 419-426.

Molnar, P. and Tapponnier, P., 1978. Active tectonics of Tibet. *J.Geophys.Res.*, **83**(B11): 5361-5375.

Nicolas, A. and Poirier, J.P., 1976. *Crystalline Plasticity and Solid State Flow in Metamorphic Rocks*. John Wiley & Sons, New York.

Peltzer, G. and Tapponnier, P., 1988. Formation and evolution of strike-slip faults, rifts, and basins during the India-Asia collision: an experimental approach. *J. Geophys. Res.*, **93**(B12): 15085-15117.

Poirier, J.P, 1980. Shear localization and shear instability in materials in ductile field. *J. Struct. Geol.*, **2**(1/2): 135-142.

Pollack, H.N. and Chapman, D.S., 1977. On the regional variation of heat flow, geotherms and lithospheric thickness. *Tectonophysics*, **38**: 279-296.

Prager, W. and Hodge, P.G., 1951. *Theory of Perfectly Plastic Solids*. John Wiley, New York.

Prandtl, L., 1920. Über die Härte plastischer Körper. *Nachr. Ges. Wissensch, Göttingen, Math.-phys. Klasse*, 74-85.

Ramberg, H., 1963. Fluid dynamics of viscous buckling applicable to folding of layered rocks. *Bull. Am. Assoc. Petrol. Geologists*, **47**: 484-505.

Ramberg, H., 1981. Gravity, Deformation and the Earth's Crust (2nd edition). Academic Press, London.

Rousseau, A., 1992. A new geodynamic model for the seismicity and in situ stresses of the Mediterranean region. *Tectonophysics*, **202**: 183-201.

Shimamoto, T., 1986. Transition between frictional slip and ductile flow for halite shear zone at room temperature. *Science*, **231**: 711-714.

Siribumrungwong, B., Oka, F., Kodaka, T., Kimoto, S. and Higo, Y., 2004. An elasto-viscoplastic finite element study of the effect of microstructure change on bearing capacity of clay. In *Proc. of the International Symposium on Lowland Technology, Bangkok, Thailand*, Sep.1-3, pp.177-184.

Smith, R.B., 1975. Unified theory of the onset of folding, boudinage and mullion structure. *Geol. Soc. Am. Bull.*, **86**: 1601~1609.

Smith, R.B., 1979. The folding of a strongly non-Newtonian layer. *Am. J. Sci.*, **279**: 272~287.

Sokolovskii, V.V., 1950. Theory of Plasticity. Gostechizdat, Moscow. (in Russian)

Sornette, A., Davy, P. and Sornette, D., 1993. Fault growth in brittle-ductile experiments and the mechanics of continental collisions. *J. Geophys. Res.*, **98**: 12111-12139.

Statistical Research Group, Institute of Mathematics, China Academy of Sciences, 1973. Commonly-Used Methods of Mathematical Statistics. Science Press, Beijing. (in Chinese).

Tapponnier, P. and Molnar, P., 1976. Slip-line field theory and large-scale continental tectonics. *Nature*, **264**: 319-324.

Tapponnier, P. and Molnar, P., 1977. Active faulting and tectonics in China. *J. Geophys. Res.*, **82**(20): 2905-2930.

Tapponnier, P., Peltzer, G., Le Dain, A.Y., Armijo, R. and Cobbold, P., 1982. Propagating extrusion tectonics in Asia: New insights from simple experiments with plasticine. *Geology*, **10**: 611-616.

Tapponnier, P., Lacassin, R., Leloup, P.H., Scharer, U., Zhong, D., Wu, H., Liu, X., Ji, S., Zhang, L. and Zhong, J., 1990. The Ailao Shan/Red River metamorphic belt: Tertiary left-lateral shear between Indochina and South China. *Nature*, **343**: 431-437.

Tapponnier, P., Xu, Z., Roger, F., Meyer, B., Arnaud, N., Wittlinger, G. and Yang, J., 2001. Oblique stepwise rise and growth of the Tibet plateau. *Science*, **294**: 1671-1677.

Turcotte, D.L. and Schubert, G., 1982. Geodynamics: Applications of Continuum Physics to Geological Problems. Wiley & Sons, New York.

Vilotte, J.P., Daignières, M. and Madariaga, R., 1982. Numerical modeling of intraplate deformation: Simple mechanical models of continental collision. *J. Geophys. Res.*, **87**(B13): 10709-10728.

Vilotte, J.P., Madariaga, R., Daignières, M. and Zienkiewicz, O., 1986. Numerical study of continental collision: influence of buoyancy forces and an initial stiff inclusion. *Geophys. J. R. astr. Soc.*, **84**: 279-310.

Walker, A.N., Rutter, E.H. and Brodie, K.H., 1990. Experimental study of grain-size sensitive flow of synthetic, hot-pressed calcite rocks. In: Knipe, R.J. (ed.), *Deformation*

Mechanisms, Rheology and Tectonics, Geological Society Special Publications, **54**, pp.259-284.

Wang, J.-y. and Huang, S.-p., 1990. Compilation of heat flow data in the China continental area (2nd edition). *Seismology and Geology*, **12**(4): 351–366. (in both Chinese and English)

Wang, S.-z., 1985. Experimental investigation of conjugate shear networks in rocks. In: Institute of Geology, China Scismological Bureau (ed.), *Research on Recent Crustal Movement* (1). Seismological Press, Beijing. pp.171-178, 190-192. (in Chinese)

Wang, S.-z., 1992. A multi-layer tectonic model for interior deformation of continental plate. In: Ma, Z. (chief editor), *Research on Continental Seismogenic Layer: Selected Papers of Int. Symp.on Continental Seismogenic Layer*, Seismological Press, Beijing, pp.191-203. (in Chinese)

Wang, S.-z., 1993a. Net-like earthquake distribution and plastic-flow network in central and eastern Asia. *Phys. Earth Planet. Inter.*, **77**(3-4): 177-188.

Wang, S.-z., 1993b. Multi-layer Tectonic model for intraplate deformation and plastic-flow network in the Asian continental lithosphere. *Acta Geologica Sinica*, **6**(3): 247-271.

Wang, S.-z.,1993c. Earthquake distribution and plastic-flow networks in the southern Eurasia continent. In: Ding, G. et al. (eds.), *Continental Earthquakes, Selected Papers of the 2nd Int.Conf. on Continental Earthquakes, IASPEI Publication Series for the IDNDR*, Volume No.3, Seismological Press, Beijing, 142-147.

Wang, S.-z., 1993d. Brittle-ductile transition and plastic-flow networks in rocks. *Progress in Geophysics*, **8**(4): 25-37. (in Chinese with English abstract)

Wang, S.-z., 1995. Netlike magmatic distribution controlled by plastic-flow network tectonics in Asia and North America. In: *Abstracts, IUGG 21st General Assembly 1995, Boulder, Colorado,* **B368** (SB31A-06).

Wang, S.-z., 1997. Netlike and homogeneous plastic flows in the crust and mantle and the flow law. In: Zheng, Y., Davis, G. A. and Yin, A. (eds.), *Structural Geology and Geomechanics, Proc.30th Int. Geol. Congr., 1996, Beijing*, Vol.14, VSP (International Science Publishers), Utrecht, The Netherlands. pp.3-15.

Wang, S.-z., 1998. Mantle vortex convection in central-eastern Asia and the "Vortex/Network" continental dynamics model. *Seismology and Geology*, **20**(4): 379-390. (in Chinese with English abstract)

Wang, S.-z., 2001. Netlike Plastic-flow and strain field in the lower lithosphere in central-eastern Asia. *Geological Review*, **47**(5): 459-466. (in Chinese with English abstract)

Wang, S.-z., 2004a. Ductile weakness-plane effect of plastic-flow networks in the continental lithosphere. *Seismology and Geology*, **26**(3): 405-415. (in Chinese with English abstract)

Wang, S.-z., 2004b. The effects of plastic-flow networks and relative stable blocks on the distribution of heat flow in the China continent. In: Zhang, Z., Gao, R., Lü, Q. and Liu, Z. (eds.), *Studies on Structure and Dynamics of Deep Earth in the China Continent*, The Teng Jiwen Volume. Beijing: Science Press, 434–447. (in Chinese with English abstract)

Wang, S.-z., 2006a. Quantitative Expression of Heat Flow versus Tectonic Deformation in the China Continent: The Effects of Plastic-Flow Network and Stable Block. *Acta Geologica Sinica*, **80**(1): 97-109.

Wang, S.-z., 2006b. Multi-Layer Strain Rate Field Controlled by Netlike Plastic-Flow in the Lithosphere in Central-Eastern Asia. *Earthquake Research in China*, **20**(3): 326-338.

Wang, S.-z., 2006c. Shear-Heating effect of Plastic-Flow Belts in the Lower Lithosphere. *Seismology and Geology*, **28**(3): 336-350. (in Chinese with English abstract)

Wang, S.-z., Li, J.-g. and Zhang, Z.-c.., 2000. Experimental study of mantle convection: Non-columnar plume and mantle vortex. *Seismology and Geology*, **22**(2): 155.166. (in Chinese with English abstract)

Wang, S.-z., Li, J.-g. and Zhou, Y.-s., 2007. Experimental evidence for intraplate deformation controlled by netlike plastic-flow in central-eastern Asia. *Phys. Earth Planet. Inter.*, **165**(3-4): 208-228. doi: 10.1016/j.pepi. 2007.10.001.

Wang, S.-z. Sun, J.-x. and Ning, J.-y., 2008. Numerical simulation of ductile shear localization: Plastic-flow network and slip-line field. *Int. J. Num. Anal. Meth. Geomech.*, **32**(3): 307-324. doi:10.1002/nag.633 (www.interscience. wiley.com).

Wang, S.-z. and Zhang, L., 1982. Deformation and failure of Zhoukoudian granodiorite at crustal temperatures and pressures. *Seismology and Geology*, **4**(4): 68, plates 3-4. (in Chinese)

Wang, S.-z. and Zhang, L., 2002. Tectonic stress field and seismic tectonics controlled by plastic-flow network in North China. *Seismology and Geology*, **24**(1): 69-80. (in Chinese with English abstract)

Wang, S.-z., Zhang, S.-c., Tian, Q.-j., Li, J.-g., Shi, L.-b. and Zhang, Z.-c., 2001. Continental Dynamics—Netlike Plastic-Flow and Hierarchical Tectonic Deformation. Seismological Press, Beijing. (in Chinese with English Summary)

Wang, S.-z. and Zhang, Z.-c.., 1995. "plastic-flow/seismic" network systems and tectonic units in central-eastern Asia. *Earthquake Research in China*, **9**(3): 321-333.

Wang, S.-z. and Zhang, Z.-c., 1997. Netlike distribution of magmatite and plastic-flow networks in central and eastern Asia. *Seismology and Geology*, **19**(3): 235-247. (in Chinese with English abstract)

Wang, Z.-c., 1988. Study of Semi-Brittle, Non-homogeneous Creep of Crustal Rocks. Ph D Dissertation, Institute of Geology, State Seismological Bureau, China.

Wang, Z.-c., Wang, S.-z. and Wang, W., 1992. An experimental study on deformation networks of rocks at crustal pressures and temperatures. *Acta Geophysica Sinica*, **35** (suppl.): 114-124. (in Chinese with English abstract)

Wegener, A., 1924. The Origin of Continents and Oceans. Methuen & Co., Ltd.

Xu, j., Fang, Z.-j. and Yang, L.-h., 1988. Tectonic background and causative fault of1966 Xintai M_s 7.2 earthquake. *Seismology and Geology*, **10**(4): 51–59. (in Chinese with English abstract)

Xu, Z.-h., Wang, S.-y., Huang, Y.-r. and Gao, A-j., 1989. China's tectonic stress field deduced from earthquakes. *Acta Geophysica Sinica*, **32**(4): 509-520.

Xu, Z.-h., Xu, G.-q. and Wu, S.-w., 1999. Present-day tectonic stress in the East China Sea region and its possible origin. *Acta Seismologica Sinica*, **21**(5): 495-501. (in Chinese)

Yuan, X.-c. (chief compiler), 1996. Geophysics Atlas of China. Geological Publishing House, Beijing.

Yuen, D.A, Fleitout, L., Schubert, G. and Froidevaux, C., 1978. Shear deformation zones along major transform faults and subducting slabs. *Geophys. J R. astr. Soc.*, **54**: 93-119.

Zhang, S.-c., Zhao, J. and Diao, G.-l., 1995. A preliminary study of the relation between focal faults and deep and shallow structures in North China area. *North China Earthquake Sciences*, **13**(2): 1-10. (in Chinese)

Zoback, M.L., Zoback, M.D., Adams, J. et al., 1989. Global patterns of tectonic stress. *Nature*, **341**: 291-298.

Zuber, M.T., 1987. Compression of an oceanic lithosphere: An analysis of intraplate deformation in the Central Indian Basin. *J.Geophys. Res.*, **92**: 4817-4825.

In: The Lithosphere: Geochemistry, Geology and Geophysics ISBN: 978-1-60456-903-2
Editors: J.E. Anderson et al, pp. 267-294

Chapter 7

ELASTIC-ANELASTIC PROPERTIES OF THE AEGEAN LITHOSPHERE- ASTHENOSPHERE INFERRED FROM LONG PERIOD RAYLEIGH WAVES

I. Kassaras[1], F. Louis[1], K. Makropoulos[1], A. Magganas[2] and D. Hatzfeld[3]

[1] National and Kapodistrian University of Athens, Faculty of Geology and Geoenvironment, Department of Geophysics-Geothermics,, Panepistimioupolis, Zografou, Athens 15784 Greece

[2] National and Kapodistrian University of Athens, Faculty of Geology and Geoenvironment, Department of Mineralogy & Petrology, Panepistimioupolis, Zografou, Athens 15784 Greece

[3] Laboratoire de Géophysique Interne et Tectonophysique, Université Joseph Fourier, Grenoble, France.

Abstract

The highest deformation rate along the Africa/Eurasia convergence zone is well documented in the Aegean area, being >4 cm/yr. However, it is still under question whether continental deformation is distributed along major faults which extend through the whole lithosphere or over large areas. Furthermore, our knowledge concerning the implication of lithosphere-asthenosphere coupling in lithospheric plates driving forces is poor. These questions can not easily be answered as most of the available information is mainly located at or close to the surface (geodesy, tectonics, seismicity). The high rates and type of surface continental deformation within the Aegean constitute this region particularly interesting in this perspective.

This work is towards contributing to the better knowledge of the physical properties of the Aegean lithosphere by introducing experimental elastic and anelastic parameters inferred from long period Rayleigh wave. For this scope path-average phase velocities and attenuation coefficients of fundamental Rayleigh wave crossing the Aegean were extracted over the period range 10-100 s. The wavetrains were recorded at the temporary broadband stations installed some years ago in the Aegean region within the framework of a large scale experiment (SEISFAULTGREECE project).

The stochastic inversion algorithm has been used to derive 36 path-average models of shear velocity and 17 path-average models of inverse shear Q down to 200 km depth. Furthermore, the elastic and anelastic 1-D path-average models were combined in a continuous regionalization tomographic scheme to obtain a 3-D model of shear velocity variation and a 3-D model of Q_b^{-1} variation down to 120 km.

The most prominent features in the tomograms are: a) A low shear velocity zone in the back-arc region, particularly in the central and north Aegean. This region is located south of the North Aegean Trough (the western edge of the North Anatolian Fault) and correlates well with the derived anelastic tomograms which present high attenuation in this area. b) A high velocity/low attenuation zone in South Aegean indicating the subducted African lithosphere beneath the Aegean.

The low velocities/high attenuation zone in central and north Aegean is compatible with a region of high extensional strain rates, recent volcanism and high heat flow. These observations suggest a hot or perhaps partially molten upper lithospheric-asthenospheric mantle and/or distributed deformation beneath the study region, probably related with the slab roll-back that has accompanied back-arc extension.

Keywords: Rayleigh wave; Phase velocity; Attenuation; Aegean Sea

1. Introduction

As part of the Arabia-Africa-Eurasia collision zone, the Aegean region has been shaped by the compression, extension and rotation of different lithospheric blocks. The Arabia/Eurasia collision in eastern Turkey has caused westward extrusion of the Anatolian block and southwestward motion and counter clockwise rotation of Aegean (e.g. [4, 38]) which compounded the collision with the African plate in producing the subduction of oceanic lithosphere along the Hellenic arc [65].

This chapter presents a new shear velocity and attenuation model for the lower crust and lithosphere of the Aegean region, obtained by inversion of broadband surface-wave phase velocities and attenuation coefficients. We have two main motivations for conducting this study. First, knowledge of the regional structure of the Aegean lithosphere is fundamental for understanding the tectonic framework and mantle dynamics, posing constraints on possible models of deformation and evolution. Second, knowledge of attenuation can provide information on temperature, fluid content, phase change, and density of solid-state defects in the lithosphere, phenomena that are not easily studied using only seismic wave velocities. The exponential dependence of Q with temperature implies that attenuation tomography could explain better hot regions (high attenuation) than elastic tomography [53]. Furthermore, elastic velocity is highly influenced by the constitution of the medium. In case of strong lateral Q variations, elastic parameters also vary. Hence, it is important to combine results of elastic and anelastic tomography. Apparently, the high rates of deformation in the Aegean [38] constitute this region particularly interesting in this perspective.

An overall description of the velocity structure of the deep lithosphere and upper mantle in the Aegean area is given by several researchers. The majority of those studies are based on P wave propagation (e.g. [55, 47, 59]). Seismic surface waves have been also used to study the shear-wave velocity distribution in the Aegean (most recently [37, 8, 28, 27]). On the other hand, several Q studies have been carried out in Greece [46] were the first to point out the high attenuation of seismic rays that travel through the southern Aegean Sea. Shear and Coda waves for high frequencies have been also used to study Qb and Qc in Greece inferring a strong frequency dependence (e.g. [30, 60, 20, 50]). However, evident is the lack of surface wave attenuation studies, which, in contradiction to body or coda waves can provide information on the depth dependence of Q, inferring resources to understand the material and physical conditions in the lithosphere.

Aiming to supplement previous tomography studies in the area, we investigate lateral variations of shear velocity and attenuation. For this scope, we analyzed about 1100 fundamental mode Rayleigh wave seismograms generated by teleseismic earthquakes in the period range 10-100 sec, to extract interstation phase velocity and attenuation coefficient dispersion curves. Those were inverted to obtain 36 path-specific S-wave models with depth and 17 path-specific Q_b^{-1} models with depth. The resulting elastic and anelastic models were employed in a continuous regionalization tomographic scheme, in order to determine regional variations of shear velocity and Q_b^{-1}.

2. Data Treatment

In the present study, phase velocities and attenuation coefficients of the fundamental mode of Rayleigh wave propagated across the Aegean were used. Approximately 430 teleseismic

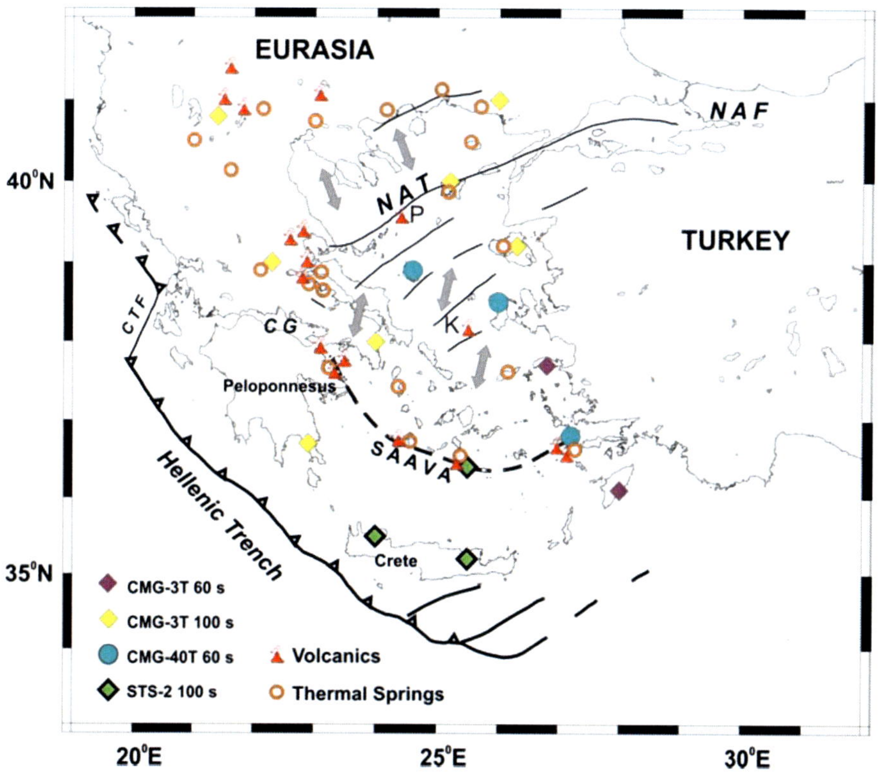

Figure 1. Map showing the location of the stations used in the present study, main active tectonic features of the Aegean after [3], Pliocene to Recent volcanic centers and main thermal springs. Abbreviations for geographical names: CG = Corinth Gulf; NAT = North Aegean Trough; NAF = North Anatolian Fault; SAAVA = South Aegean Active Volcanic Arc; CTF = Cephalonia Transform Fault; P = Psathoura; K = Kalogeri.

earthquakes with M>5, recorded during a 6 month experiment by 12 broadband stations installed in the Aegean were considered. Data from 3 GEOFON stations in south Aegean were also used. Figure 1 presents the locations of the 15 stations used in this study.

Out of all the recorded earthquakes of epicentral distance larger than 30^0, waveforms generated by 386 events showing a good-quality signal at most of the stations were analyzed. Given the different sensors and recorders, the data were first corrected for the instrument response. Phase velocities and attenuation coefficients for Rayleigh wave were determined between selected pairs of stations located on the same great circle as the epicenter, within $2\text{-}3^{\circ}$.

2.1. Phase Velocities Measurements

Phase velocities have been determined for periods 10-100 sec by the two-station method [54] along 36 interstation pathways covering the Aegean region. In order to select only the fundamental mode Rayleigh wave and eliminate heavily scattered energy (coda) and higher modes, a phase-matched filter which uses the group velocity information was applied [21, 31]. The obtained clean seismograms were used to determine interstation phase velocities. Phase velocities were estimated by wavefield transformation to the slowness-frequency domain, slant-stacking of the input signals and 1-D Fourier transformation [39]. A code written by [24] and modified by [5] was used. Given the small interstation distances, multipathing effect and contamination of Rayleigh waves from other seismic waves propagating with similar group velocities, several measurements presented strong biases, and were consequently rejected. Finally, 255 dispersion curves concerning 176 events, meeting several quality criteria, were adopted [28]. The geographic distribution of the 176 teleseismic events is shown in Figure 2.

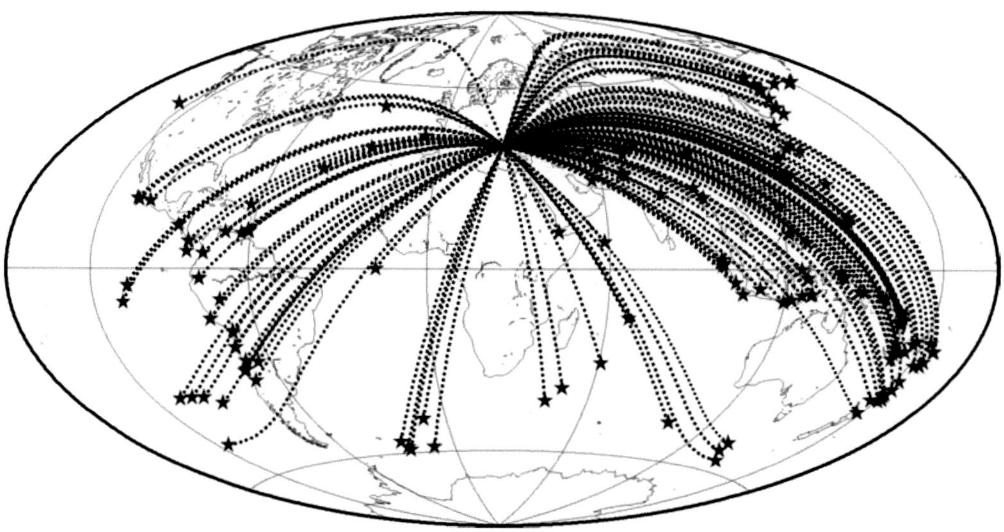

Figure 2. 176 used teleseismic events and propagation great circles.

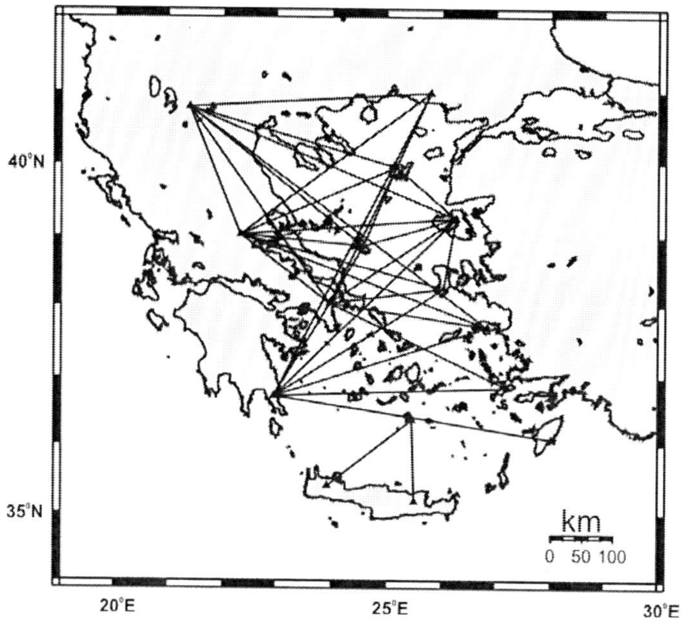

Figure 3. Paths for which phase velocity measurements were obtained.

The period range our measurements is not the same for all paths. This firstly arises by the fact that the instruments used were not of the same response. Six of the instruments were equipped with 60 sec sensors and 10 with 100 sec sensors. Thus, several paths were not sampled beyond 60 sec. The second reason was that not all paths were sampled below ~20 sec, due to multiple arrivals and multipathing effects, which could not be excluded through the filtering procedure. In order to encompass homogeneous sampling of the under study area, we only consider waves with periods 20-60 sec. The geographic distribution of the 36 paths for which phase velocities were extracted is shown in Figure 3. For each path the average phase velocities and the respective standard errors (σ) were estimated. Individual standard deviation may reach values as high as 0.09 km/s at low periods (<25 s) and 0.07 km/s at periods >25 s, corresponding to a realistic peak velocity error estimate ±2.6% for the upper layers and ±1.7% for the lower lithosphere and upper mantle.

2.2. Attenuation Coefficient Measurements

The decay of surface wave amplitudes, in excess of that produced by geometrical spreading, can be described by $e^{-\gamma x}$, where γ is termed the attenuation coefficient. Attenuation coefficients have been determined for periods 10-100 sec by the two-station method along 17 interstation pathways across the Aegean. Advantages of the two-station method are that it does not require knowledge of the earthquake source and that in the absence of effects produced by lateral refraction, it should produce average values for the attenuation coefficients along the path between the two stations. Average values should be obtained, even if the anelastic properties along that path vary with position.

The Multiple Filtering Technique (MFT) [23] was utilized for the identification of the fundamental Rayleigh waves and to extract group velocities and spectral amplitudes. About

1000 seismograms were analyzed and spectral amplitudes of the fundamental mode vertical components of Rayleigh waves were obtained for 510 pairs of stations. After rejecting largely erroneous data, the two-station method was applied for 330 pairs of stations to obtain the attenuation coefficients in the period range 10-100 sec. According to the two-station method, the averaged coefficient for the k_{th} path and period T between stations s_i and s_j is given by:

$$\gamma_k(T) = D_k^{-1} \ln \frac{A_{si}(T)}{A_{sj}(T)} \tag{1}$$

where D_k is the distance between stations and $A_{si}(T)$ and $A_{sj}(T)$ are spectral amplitudes of the fundamental mode, recorded at stations s_i and s_j. The spectral amplitudes were corrected for the geometrical spreading effect.

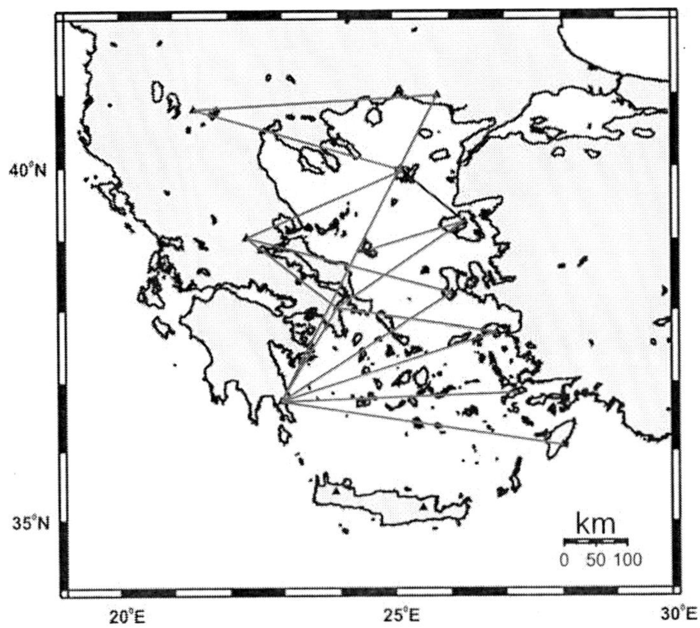

Figure 4. Paths for which attenuation coefficient measurements were obtained.

Attenuation data have to be carefully verified especially when the area of interest is subject to lateral heterogeneities due to effects of focusing, defocusing, departures from the great-circle path and multipathing, which impact the quality of the empirical attenuation coefficient. Other relevant error aspects are the site effect and the instrumental response of the seismometers if they are not correctly calibrated. To minimize those effects two tasks were performed. First, we performed a strict seismogram selection procedure. For those selected seismograms we computed the quotient of Equation 1. Secondly, we conducted a synthetic experiment adequately parameterized for our study area. By utilizing an average 1-D shear velocity model derived for the region [28] and perturbing Q_b between 10-100 we calculated the theoretical quotients of Equation 1 for various interstation distances, relevant to the lengths of our paths. Considering the results of the two procedures we set the tolerance range for our measurements. Furthermore, to control the errors for each individual path, we tried to collect as many events for each station pair as possible, adopting only similar observations, reducing thus the standard deviations of the mean values. Given that short interstation distances can cause problems when using the two-station method, results for the station pairs with shorter interstation distances were compared with those with longer interstation distances, and were also compared with the average values to examine the usability of the data.

As expected, given the strong impact of the above described effects, the largest part of the analyzed dataset exceeded the tolerance range and was consequently rejected. In conclusion, we adopted 75 interstation attenuation curves as representative of 17 paths across the Aegean (Figure 4) and calculated the average value and the corresponding standard deviation of γ_R for each period and each path. For the single period measurements, a value of 0.5×10^{-3} km^{-1} was chosen as a conservative error estimate.

2.3. Inversion for Shear Velocities

For each profile the average experimental dispersion curve has been inverted to obtain 1-D horizontal average shear wave velocity models. The method of damped least-squares solution (stochastic inverse solution) was applied and an algorithm written by [24] was used. Although error in the phase velocities propagates into the final model, the main source of error in the model is due to velocities in the starting model hence a realistic earth model is perquisite. In order to obtain an initial model for the inversion of our dispersion curves we inverted the average dispersion curve for the whole modeled area adopting the results of [48] for the average deep structure of the area.

Despite the small length of the paths in the study area, the influence of the crustal model may be sufficient to affect the fundamental Rayleigh wave mode at periods <40 s and to impose structure in the uppermost mantle [44, 57]. Crustal thickness varies from about 35-45 km in continental Greece, 25-35 km in North Aegean and 22-26 km in South Aegean [27]. To constrain our dispersion curve inversions, the depth of the Moho discontinuity was set for every path following a 2-D Moho depth model proposed by [27]. The damping parameter was estimated by the compromise of considering the "knee" of the empiric trade-off curve between the remaining residual variance after inversion and the squared model length. Considering one standard deviation to be a too optimistic error estimate, an error twice the standard deviation (2σ) was attached to each phase velocity. The inversion process was performed through iterations, where the model was perturbed and the resulting fit was observed. The number of iterations was monitored to minimize extraneous artifacts being introduced into the model (e.g. low-velocity zones). The inversion algorithm used allowed the computation of resolution of the solution, which is controlled by the density and uncertainties of phase velocity observations. For each path we analyzed the resolution matrix to define the number of model parameters which could be resolved by the experimental data. Following a manual procedure, consisting of averaging similar adjacent blocks, 13 homogeneous horizontal layers overlying a half-space at 200 km depth were adopted, as the most appropriate number of parameters. Via output resolving kernels, it was defined that for depths below 30 km the solution is robust for more than 80% of the profiles.

As mentioned in section 2.1, although the analysis was performed in the range 10-100 sec, here we consider only waves with periods 20-60 sec. This is done for two reasons; firstly, to achieve homogeneous sampling of phase velocities for all the paths, and secondly, because the structural estimate is likely to suffer from lack of resolution if all data are taken into account, as phase velocity measurements for periods 60-100 sec are available for only 18 paths. Thus, the 36 horizontal average shear velocity models determined by the phase velocities inversion process were restricted between 30-120 km depth, sampling the lower crust and the upper mantle of the region.

2.4. Inversion for Q_b^{-1}

Anelastic structures in terms of Q_b^{-1} distribution with depth were derived by stochastic inversion of attenuation coefficients [24]. At first, an average over the entire region anelastic model with depth was obtained by inverting the average attenuation coefficient dispersion curve. As a starting shear velocity model we used the one derived by [28], slightly modified. Given the lack on Q_b information for the area, after some tests with synthetics, starting Q_b was set to 40. The inversion process was repeated until a minimum misfit between empirical and theoretical values was achieved.

Additionally, average over the entire region phase velocities were employed together with the obtained average attenuation coefficients into a stochastic uncoupled causal inversion scheme [22], in order to obtain joint average Q_b^{-1} and shear velocity model with depth, representative for the Aegean region. In the uncoupled-causal procedure, in order to obtain the best possible structural model, each of the iterations was performed in two steps. Firstly, leaving unchanged the elastic part of the model and improving the anelastic part. Secondly, using the improved anelastic part, a new approach to a better elastic model was obtained. The average anelastic model for the region deduced for both uncaual and uncoupled-causal inversion is listed in Table 1. Figure 5 depicts the Q_b^{-1} and shear velocity models derived by ignoring and including causality effects.

As expected, elastic models derived without considering causal effects, especially for layers deeper than 20 km, are characterized by shear velocities lower than those obtained taking into account causal effects. The anelastic structures manifest a quite similar pattern, with the misfit between the two procedures been larger deeper than 60 km. Average Q_b values at depths from 0 to 200 km range between 29±13 and 23±10 for the non-causal and uncoupled-causal inversion respectively. Maximum attenuation is observed between 45-120 km depth. The observed low Q_b values contrast sharply with results from stable continental regions where Q_b may reach 200 or more. These low values may indicate that fluids reside in faults, cracks, and permeable rock at lower crustal depth due to intense deformation in the area. In the upper mantle it is possible to explain low Q_b by temperature increase. High temperatures produce a greater degree of unpinning of dislocations and eventually to a transition to viscous behavior [42].

Furthermore, we calculated the average value of attenuation coefficient for each period and each path and its standard deviation. The average attenuation data were then simplified to make a roughly even distribution over a linear period range. For the periods during which only a single measurement is available, a value of 0.5×10^{-3} km^{-1} was chosen as a conservative error estimate. After applying stochastic inversion to the 17 path-average attenuation coefficient curves, we obtained 17 path-average Q_b^{-1} with depth. In the inversion scheme, as starting V_s and Q_b^{-1} we used those listed in Table 1. As attenuation coefficient measurements for periods <20 and >60 sec were possible for only a few paths, we restrict those models between 30-120 km depth.

Table 1. Summary of the results for the average anelastic structure of the Aegean region.

Depth (km)	Starting Vs (km/s)	Uncausal Inversion			Uncoupled-Causal Inversion		
		Vs (km/s)	Q_b^{-1} ($\times10^{-2}$)	Q_b	Vs (km/s)	Q_b^{-1} ($\times10^{-2}$)	Q_b
10	3.43	3.37	2.47	40	3.43	2.75	36
20	3.57	3.51	2.65	38	3.58	2.88	35
30	3.97	3.93	3.34	30	4.05	3.50	29
45	4.05	4.02	4.54	22	4.21	4.71	21
60	4.23	4.22	5.72	17	4.48	6.15	16
75	4.46	4.46	6.37	16	4.78	7.19	14
90	4.43	4.44	6.35	16	4.80	7.71	13
105	4.44	4.44	5.82	17	4.83	7.64	13
120	4.40	4.40	5.02	20	4.79	7.12	14
140	4.41	4.41	4.03	25	4.78	6.25	16
160	4.40	4.40	3.07	33	4.73	5.10	20
180	4.45	4.44	2.26	44	4.71	3.80	26
200	4.45	4.44	1.64	61	4.64	2.43	41

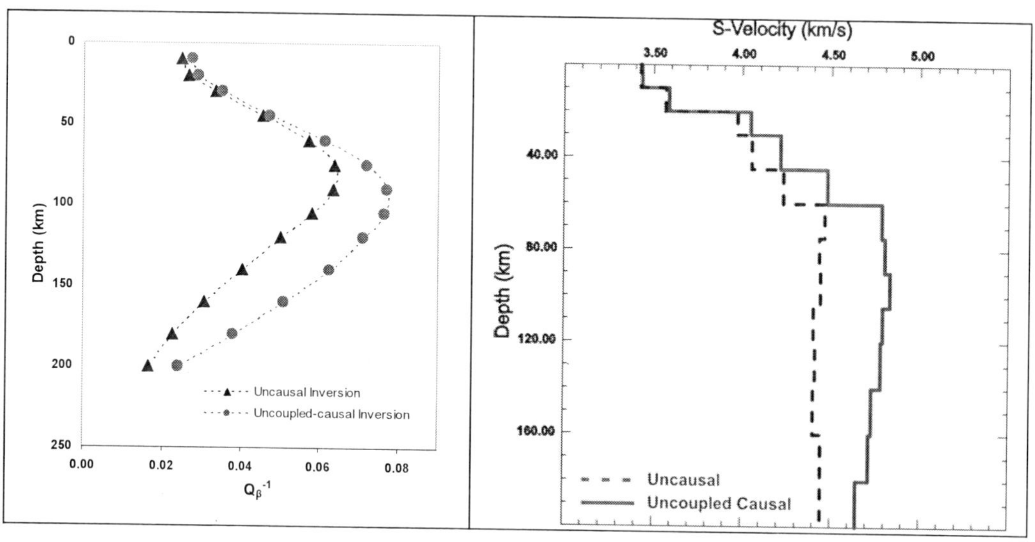

Figure 5. Average Q_b^{-1} (left) and shear velocity (right) models deduced from uncausal and uncoupled-causal stochastic inversion.

3. Tomography of Elastic and Anelastic Parameters

The effect of the finite frequency of surface waves is expected to be significant in the reconstruction of small scale heterogeneities in a region of small dimensions as the under study region. For the scale of velocity features we aim to image, off-great-circle propagation and scattering might be important. The tomographic scheme used in the study is based on the assumptions that the observed surface waveform can be represented by multimode surface waves propagating independently and along the great circle. These assumptions are valid for a smoothly varying medium without strong lateral velocity gradients [64]. In the velocity and energy domain a direct allowance for the influence zone surrounding a wave path should be made, approximately one-third of the first Fresnel zone, which arises from the finite frequency of surface waves. This zone has a typical half-width, transverse to the path, of about 50 km for the fundamental mode at 20 s period, and increases to close to 200 km at 100 s period. The effect of the influence zone is a natural smoothing out of short-wavelengths from the tomographic images due to the healing of small perturbations in the wavefronts by diffraction processes. Significant deviations from great-circle propagation have been observed for short periods surface waves (<40 s) [33, 2, 10], but surface wave ray tracing in Earth models confirms that off-great circle propagation can reasonably be neglected for the fundamental and first few higher modes at periods greater than ~40 sec [66]. [29] examined the validity of the path average approximation for surface wave propagation at regional continental scale and concluded that it should be suitable for periods between 30 and 100 sec. Furthermore, the influence of scattering effects in regional tomography was studied by [16] by applying ray theory and scattering approximation to a regional dataset, defining that ray theory approximation is valid for surface wave periods up to 100 sec. Hence, the great circle approximation inversion scheme adopted in the framework of the present study is reasonably valid in our case.

3.1. Continuous Regionalization

A classical approach in surface wave tomography is to create a 3-D model in two stages. The first step is the calculation of 1-D path average models and the second step is the regionalization of these models to obtain local models. The tomographic scheme used consists of a continuous formulation of the inverse problem and the least-square criterion. The algorithm used in this study was developed by [11] and it is an optimized version of the [43] approach for continuous regionalization of surface wave path-average measurements. In the next section the [43] continuous regionalization approach and the optimized scheme is briefly described.

3.1.1. Forward Problem

If we assume wave propagation along a source-receiver great circle *i*, regionalization consists in retrieving local measurements $C(T,\theta,\varphi)$ from path average measurements $Ci(T)$:

$$1/C_i(T) = 1/L_i \int 1/C(T,\theta,\phi)ds_i \qquad (2)$$

where $C_i(T)$ is the parameter (phase/shear velocity/attenuation) measured along i at period T, θ and φ are the coordinates of the geographical points along i, and L_i is the source-receiver distance. $C(T,\theta,\varphi)$ is the local parameter (phase/shear velocity/attenuation at point (θ,φ) and period/depth T.

The inverse problem can be expressed as a linear relationship between the data vector d which contains the average $C_i(T)^{-1}$ along each path and a parameter vector m which contains the local $C(T,\theta,\varphi)^{-1}$ at each geographical point:

$$d=Gm \tag{3}$$

where the matrix G contains the partial derivatives $\delta s/L_i$. δs are segments of path i.

3.1.2. Inversion

If we consider a large number of regular blocks, a least squares solution for discrete linear inverse problem is [58]:

$$m = m_0 + C_{m0} G' S^{-1}(d - Gm_0) \tag{4}$$

with

$$S = (GC_{m0}G^t + C_{d0}) \tag{5}$$

where t denotes the transpose, m is the inverted model, m_0 is the a priori model, C_{m0} and C_{d0} the a priori covariance matrices on the model and data, respectively.

Continuous regionalization consists of regionalization in blocks of decreasing size while their number increases towards infinity. The problem is undetermined. Following Montagner (1986) the solution is stabilized by using a continuous formulation of the inverse problem proposed by [58], through the definition of an a priori model m_0 and an a priori covariance function $C_{m0}(r,r')$:

$$C_{m0}(r,r') = \sigma(r)\sigma(r')\exp\left(\frac{-\Delta^2_{r,r'}}{2L^2_{corr}}\right) \tag{6}$$

where r and r' are two geographical points separated by a distance Δ. The exponential term has a standard deviation given by L_{corr}. This functions as a spatial filter, imposing correlation between points separated by distances of order L_{corr}. The function $\sigma(r)$ controls the amplitude of the model perturbation allowed at r. The solution of the inverse problem is:

$$m(r) = m_0 + \frac{1}{L_i}\sum_i \left[\int ds_i \sigma(r_i)\sigma(r)\exp\left(\frac{-\Delta^2_{r_i,r}}{2L_{corr}}\right)\sum_j S^{-1}_{ij}(d_{j0} - G_j m_0) \right] \tag{7}$$

with

$$S_{ij} = C_{d0_{ij}} + \frac{1}{L_i L_j} \int_i \int_j \sigma(r_i)\sigma(r_j)\exp(\frac{-\Delta^2_{r_i,r_j}}{2L^2_{corr}}) ds_i ds_j \tag{8}$$

The solution of Eq. 7 for *m(r)* presents several difficulties. The first is the computation of the integral

$$A(r) = \frac{1}{L_i} \int ds_i \sigma(r_i)\sigma(r)\exp(\frac{-\Delta^2_{r_i,r}}{2L^2_{corr}}) \tag{9}$$

To determine *A(r)* at each geographical point r, it is necessary to compute the correlation between each geographical point r_i of path i and each model at point r. When the distance $\Delta_{r_i,r} >> L_{corr}$, the exponential term is negligible. Therefore, comparison between $\Delta_{r_i,r}$ and L_{corr} is performed and *A(r)* is computed only when the exponential term is significant. However, even with this assumption $\Delta_{r_i,r}$ must me evaluated for all pairs of points (r_i,r) and for each path i, and this can be very time consuming in case of large number of paths. [11] restrict all calculations to only the pairs of points that contribute significantly to *A(r)* by defining an influence zone about the ith path containing all points r which satisfy:

$$\Delta_{r_i,r} < 2.64 L_{corr} \tag{10}$$

The influence zone corresponds to the region where the exponential term in *A(r)* falls to about 3% of its maximum [12].

The second difficulty in solving for *m(r)* is the computation of S_{ij} (8) which requires the evaluation of the double integral

$$B = \frac{1}{L_i L_j} \int_i \int_j \sigma(r_i)\sigma(r_j)\exp(\frac{-\Delta^2_{r_i,r_j}}{2L^2_{corr}}) ds_i ds_j \tag{11}$$

The estimation of B requires the computation for each geographical point r_i of path i the correlation between r_i and each geographical point r_j belonging to path j. However, only the points of path j located in the influence zone of path i contribute significantly in B. To overcome this problem, the scheme of [11] considers only points in both influence zones of each pair of paths. $\Delta_{r_i,r}$ is not evaluated when r is outside the influence zone of r_i.

3.2. Development of the 3-D shear Velocity and Q_b^{-1} Models

The 36 1-D velocity models obtained were combined over a 1×1 degree grid into a single 3-D model for the lateral variations in shear velocity, using the continuous regionalization algorithm described above. In the continuous regionalization algorithm, the lateral smoothness of the inverted model is obtained by assuming a priori a Gaussian correlation between neighboring points with a specified scale length L_{corr} and a scale factor σ(r). The a priori standard deviation σ(r) controls the amplitude of the perturbation allowed in the process and the horizontal correlation length is a spatial filter that constrains the lateral smoothness of the model. The a priori standard deviation σ(r) was taken equal to 0.05 km/s, according to average phase velocity error estimate at periods 20-60 sec. The choice of an appropriate correlation length depends on the path coverage and on the period range of the inversion. After several trials L_{corr} was chosen equal to 150 km, thus favoring a smooth model considering our ray density and shortest wavelengths used. By trying larger values of L_{corr} the final model was smoother but the features of the model were essentially the same. Given the distribution of paths (Figure 3) a 150 km correlation length is a compromise that should also minimize contamination by diffraction effects.

The 17 path-average 1-D Q_b^{-1} models with depth obtained by the inversion of attenuation coefficient curves were empolyed to the continuous regionalization scheme in order to obtain local Q_b^{-1}. The path-average Q_b^{-1} models with depth were combined over a 1×1 degree grid into a 3-D model for the lateral variations of Q_b^{-1}. L_{corr} was chosen equal to 150 km. Conclusions about the distribution of Q should therefore be based only on the gross features of models and fine-scale features should be ignored. σ(r) was taken equal to 5×10^{-3}. The depth range of the computed tomograms was restricted between 30 and 120 km.

4. Resolution Assessment

The estimation of resolution is one of the main problems in seismic tomography. As the solution is not unique, the input data do not contain the information on velocities of seismic waves in every point of the medium. The size of the heterogeneity that can be reliably detected using a given data sampling can be estimated through resolution analysis. It also helps to distinguish artifacts detected in the solution that are caused by the applied method. Although the under study area is of small dimensions, the algorithm used is inverting for local velocity over a 1×1 degree global grid and consequently the a posteriori covariance matrix cannot be computed. This fact limits our ability to apply formal statistical theory to analyze the reliability and precision of the resulting tomograms. Two different approaches were applied, in order to produce a qualitative estimation of models constraint as a function of geographical position.

4.1. Resolution Assessment of the Shear Velocity Model

The model space was parameterized by means of constant velocity pixels with 1° dimensions and under the great circle approximation, ray paths and the corresponding tomographic matrix G, limited to the study area, were calculated. Singular Value Decomposition (SVD) on matrix G was performed to extract the significant feature components by decomposing the matrix into a set of basic patterns (singular vectors) with associated scaling factors (singular values). Through SVD any matrix G can be factored into a product of three other matrices [18]:

$$G = U \Lambda V^t \tag{12}$$

where for the n data and p parameters $U_{(nxp)}$ and $V_{(pxp)}$ are respectively the data space and parameter space singular vectors, and Λ is a $p \times p$ diagonal matrix containing at most $r \leq p$ singular values. When a singular value is zero or very small compared with the largest one, the associated singular vector in data space cannot be mapped in model space and vice versa. These vectors span the null space and cannot be resolved while their contribution is usually regularized in order to have a stable inversion. The limit to accept a singular value and the associated vectors as a significant part of the solution has been discussed by many authors. To define a singular value threshold we follow the route of [7] and include all singular values comprised in so many orders of magnitudes, with respect to their maximum, as many are available in the accuracy of our experiment.

SVD has been applied in the past as a robust mathematical tool to analyze the matrices involved in seismic tomography. [9, 51, 56 and 41] present singular values and their corresponding singular vectors both in model and data space and discuss their components. By definition the null space does not depend on the shear velocities and so we can not eliminate it by minimizing the uncertainties [6], since the mapping from data to model space is controlled by the acquisition geometry and model parameterization.

Following, several parameters were calculated to investigate for possible artifacts in the solution and the resolution limits of our path configuration. The familiar checkerboard test was avoided, since it seemed as a rather crude experiment for such an acquisition layout.

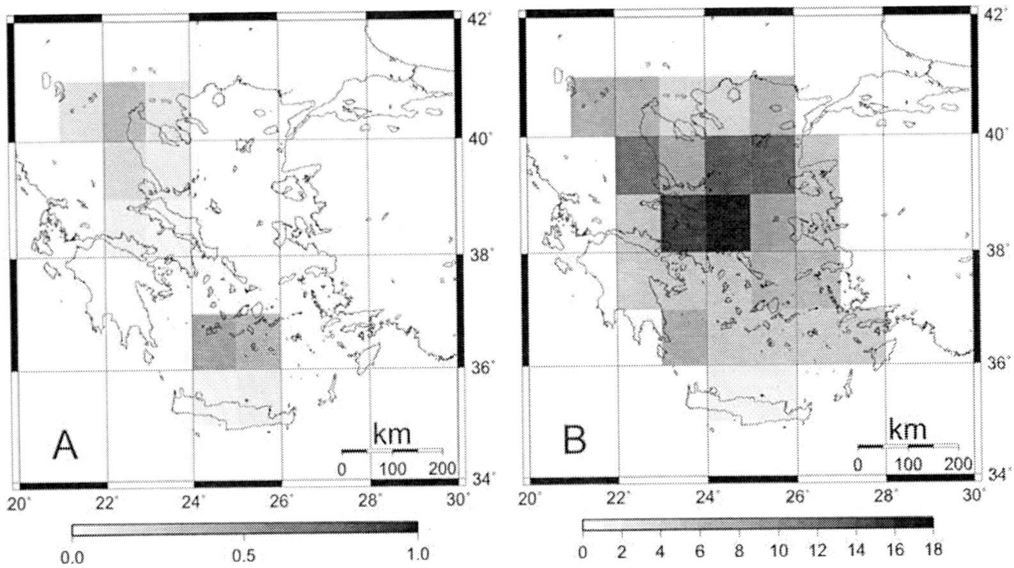

Figure 6. Null space energy (NSE) distributions (A) and the corresponding ray density map (B).

The first tool employed in the error analysis was the Null Space Energy indicator (NSE) that provides an estimate of the local reliability in model space displaying areas where the data cannot reconstruct features with the desired resolution [61, 62]. This statistical tool is a function of the ray distribution and the grid size. Each component n_j of the NSE is associated with the j_{th} pixel:

$$n_j = \sum_i V_{ij}^2 \tag{13}$$

where V_{ij} are the elements of the matrix V. The summation is performed for the indexes i of the singular values that are below the chosen threshold. We consider that two decimal points for the elements of matrix G reflect the desired actual accuracy and for this reason two orders of magnitude are the limit for the singular values that contribute positive to the solution. High values of NSE indicate solution vagueness at the corresponding pixels and vice versa. Figure 6 presents the NSE distributions along with the corresponding ray hits per cell.

From the above figure, it is obvious that the largest uncertainties occur at the lower boundary of the model as a direct effect of limited ray coverage. In general the NSE is quite low, while in some areas it is almost zero (North East part of Figure 6A).

Figure 7 allows the evaluation of the singular value spectra. The abrupt decrease in this graph occurs near the last 3 to 4 singular values which indicate that the data can adequately be resolved and that only few components of model space are difficult to be retrieved.

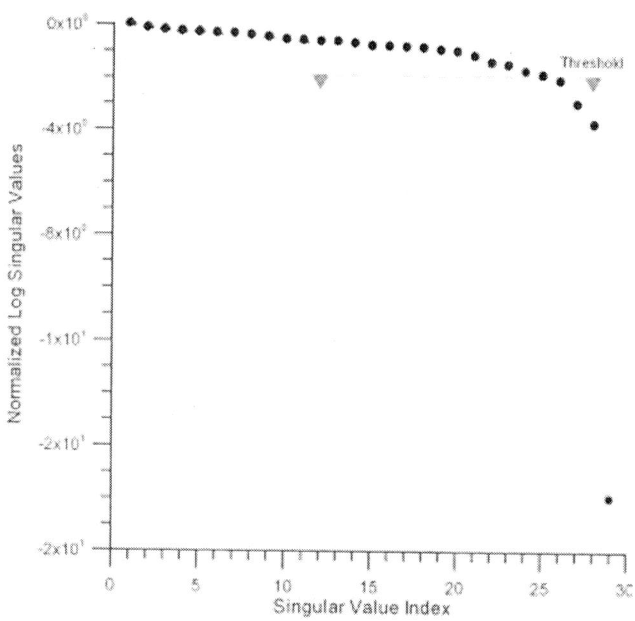

Figure 7. Singular value spectra.

Following [41] the singular vectors in model space (corresponding to singular values below the threshold) are represented as images. This way of mapping makes easier to understand the relations among the different components of the vector. Figure 8 presents models corresponding to the last singular vectors. The largest singular values correspond to vectors in model space (not shown here for efficiency) that contain almost horizontal layers (E-W orientation) or isolated features especially at the center of the model. As the singular values begin to drop the vectors in model space tend to contain more high frequency oscillations and isolated N-S features at the borders of the model. By inspecting Figure 8, distinctive patterns at the areas with limited ray coverage (southern part) and almost at the boundaries of the model are observed. If synthetic tests with models that were built only from null space model vectors were performed, the recovered images would be perfectly uniform pattern. These "forbidden" distributions would be entirely missed during the inversion, whereas the current data kernel would provide no information. Consequently, if the final velocity tomograms using real data reveal these kinds of structures they should be thoroughly investigated for artifacts due to the inversion scheme and the regularization chosen.

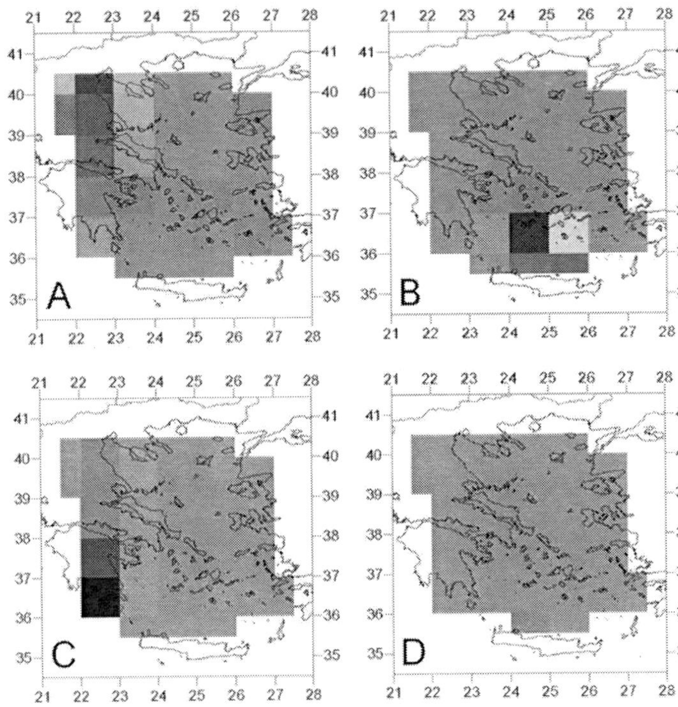

Figure 8. Singular vectors in model space that span the null space for the model. The gray scale goes from white (negative) to black (positive).

The last tools used as reconstruction estimates are the resolution matrix and the spread function associated with each parameter. The full resolution matrix describes the influence that each pixel of the velocity model has on every other one. In many cases the visualization of the resolution matrix is a problem, given the large size of the matrix. A 3D plot of the resolution matrix (Figure 9) is constructed to take into account not only the resolvability of each parameter but also the amount of smearing prevalent in the model. The elementary pixels have been numbered sequentially first from left to right then from top to bottom (the upper left pixel of the model relates to the lower right diagonal element of the resolution matrix).

In figure 9 it can be seen that the resolution of the central elements of the matrix is quite high, while in general the positive and negative side-lobes (off-diagonal elements) are of small amplitude. This implies that the resolution of the model parameters is fair enough to accept most of the anomalies in the solution tomogram.

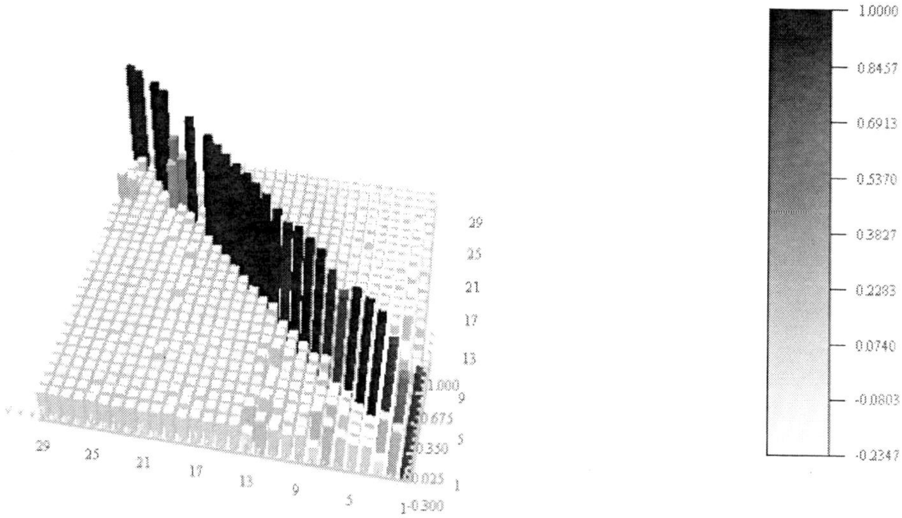

Figure 9. Perspective view of the resolution matrix.

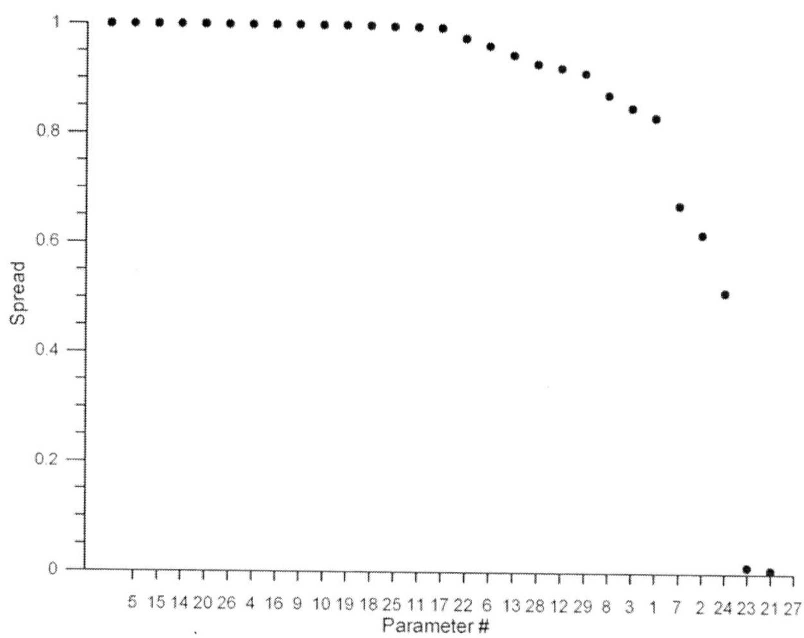

Figure 10. Spread values for the model.

A comprehensive view of each row of the resolution matrix is given by the spread function (e.g. [40]). The spread of each cell is defined as:

$$Spread_i = 1 - \sum_{j=1}^{p} (R_{ij} - I_{ij})^2 \tag{14}$$

where R is the resolution matrix and a value of 1 for a pixel implies that the unknown parameter can be estimated with good accuracy and that this estimate is not influenced by other pixels. Spread functions for the two cases are shown in figures 10 and 11, where for most of the well and reasonably resolved pixels (1<Spread<0.8) the diagonal element of the resolution matrix is the largest value of this row. Badly resolved parameters correspond to pixels at the borders of the models or cells with few ray hits or almost parallel rays. As a result, at these situations the pixel's values tend to remain close to the initial velocity model.

Consideration of all of the above measures of resolution implies that most of the model can be regarded as a well-resolved one. Velocity perturbations are reasonably reliable, except of two or three isolated regions of the model where the reconstruction can be inadequate.

4.2. Using Voronoi Diagrams to Assess Anelastic Model Constraint

In order to estimate the variation of the Q_b^{-1} model constraint provided by the ray path coverage we used Voronoi polyhedra, by applying a technique proposed by [11]. This approach makes use of computational geometry techniques, which have seen several applications to geophysical problems in recent times. In two dimensions a Voronoi diagram divides the plane into a set of polygons, one about each node, such that all points in a particular cell are closer to its defining node than any other node. To produce a qualitative measure of model constraint we aim to build an optimized parameterization of the model in which each geographical point belongs to the smallest cell for which an appropriate quality criterion is satisfied. The overall pattern of the optimized parameterization will reflect variations in model constraint.

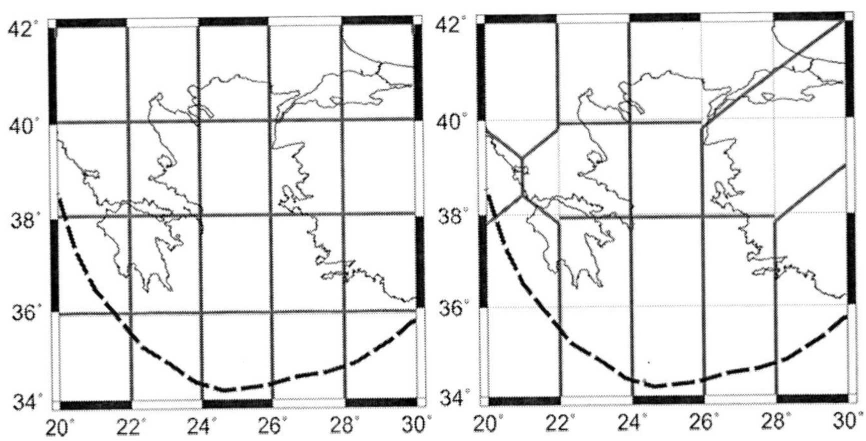

Figure 11. Voronoi diagrams to assess the anelastic model constraint as a function of geographical position. Left: Initial Voronoi diagram over a 2° × 2° grid. Right: Optimized Voronoi diagram.

As a starting point we built an initial Voronoi diagram falling on a 2° × 2° grid. From this starting Voronoi diagram we proceeded iteratively by building new diagrams in such a way that a particular quality criterion depending on the ray distribution is satisfied for each cell. The process of generating the new optimized Voronoi diagram is a matter of deleting nodes which do not match the criterion and then recalculating the Voronoi cells about the remaining nodes. When nodes are deleted, the neighboring Voronoi cells inherit the area and ray paths previously contained in the deleted Voronoi cell. This makes the new cells larger on average and more likely to satisfy the quality criterion. In the final optimized diagram (Figure 11) each cell denotes the size of the structures our inversion can model. As it can be seen in Figure 11, best resolution is achieved in the central region. In northern and southern region, resolution is not efficient, but it is better in the E-W direction.

5. Elastic Tomograms

The shear wave velocity model was obtained from an inversion with L_{corr} = 150 km, σ(r) =0.05 km/s. The a priori error in S wave velocity σ(r) defines to what extent the final model can vary from the reference one. Considering that the a priori initial model is a reference model for the region, this value is relatively large and should allow all realistic shear velocity variations. Variation of L_{corr} values does not affect the results, with respect to the conclusions we can draw. Velocity maps are shown in Figure 12 for 7 depths varying from 30 to 120 km. Velocity variations are relative to the average shear velocity at each depth.

At the depth of 30 km, shear velocities reflect variations of crustal thickness in the area. Our model shows a thin crust in South Aegean, while in North Aegean and continental area the crust appears thicker. These results are compatible with previous studies (e.g. [27]).

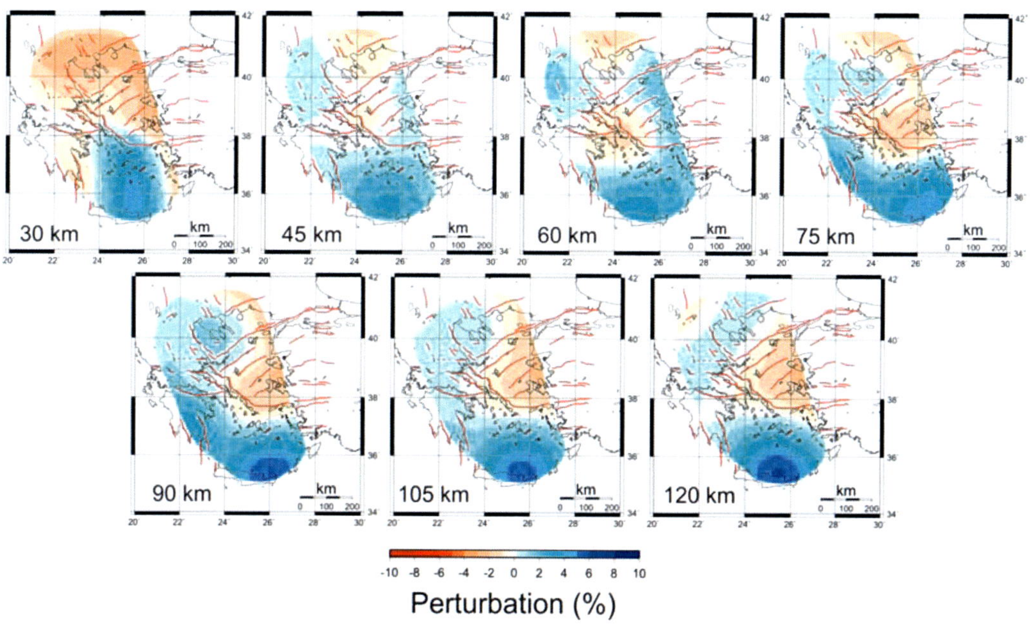

Figure 12. Tomographic slices of the shear velocity model at different depths. Tomograms start at 30 km depth. The velocity perturbations are indicated with a color scale in percent relative to a common reference model, derived from the average shear velocity at each depth (30 km, 3.97 km s^{-1}; 45 km, 4.05 km s^{-1}; 60 km, 4.23 km s^{-1}; 75 km, 4.46 km s^{-1}; 90 km, 4.43 km s^{-1}; 105 km, 4.44 km s^{-1}; 120 km, 4.40 km s^{-1}. The North Anatolian Trough and other major tectonic features are superimposed onto the maps. Lateral resolution of the model is assessed in section 4.1.

The amplitude of S-wave velocity variations (approximately -3% to +8%) is compatible with phase velocity variations derived in the study of [28]. This fact implies a good compromise between experimental measurements and the inverted model. The amplitude of the velocity variations is in agreement with previous studies in the Aegean based on local and teleseismic tomography [47, 59] and surface wave tomography [8]. The amplitude ranging is still compatible with studies performed elsewhere in regional (e.g. [34]) and global scale (e.g. [32]).

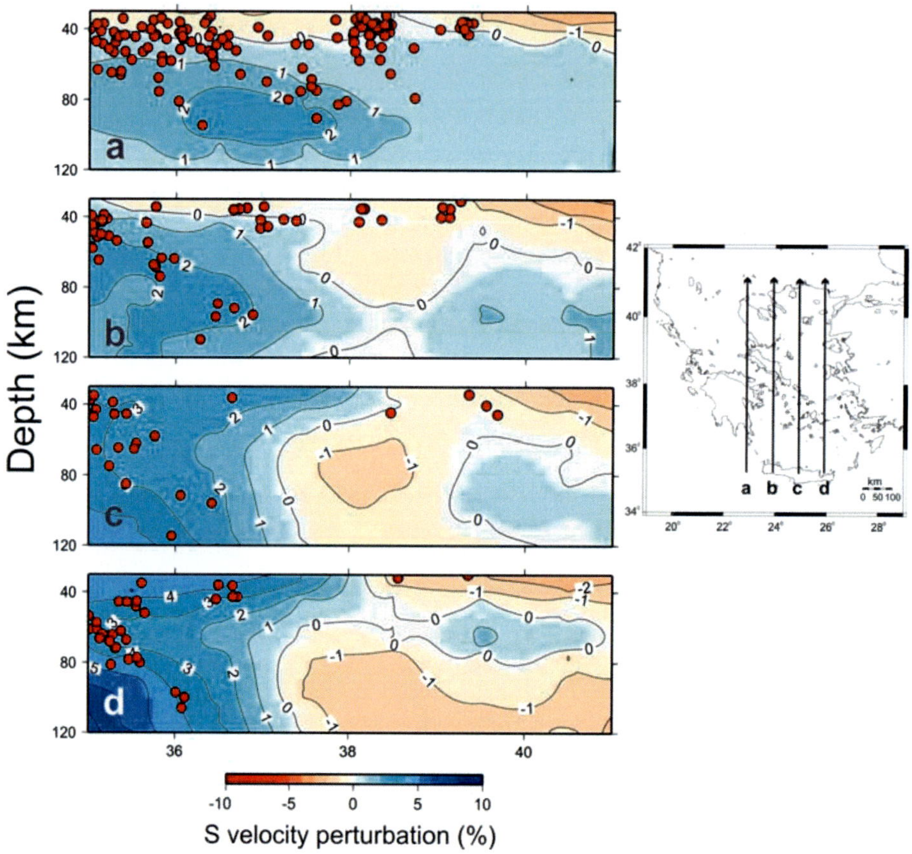

Figure 13. Vertical cross-sections across the 3-D shear wave velocity model. Velocity variations are in percentage with respect to the average shear velocity at each depth. Solid red circles represent earthquakes foci [13]. Right corner: Map showing the location of the cross sections (thick black lines) labeled from a to d.

Perhaps the most dominant feature of the model is the velocity contrast in Central and North Aegean. This tendency is evident in almost the whole depth range of the model. [47], observed a low velocity zone beneath North Aegean at about 100 km depth. However, our model can not be directly compared with velocities derived from body waves tomography because of anelastic effects. Low shear velocities were observed by [8] in this area at depths 50-120 km. The amplitude of the observed velocity contrast is between -1% and -3%. Interpretations of seismic velocity heterogeneities are generally based on thermal variations [17, 52]. Vs variations of 1% to 3% over distances of a few hundred kilometers, as observed in our model, can be obtained with temperature variations of 70^0 to 200^0 C [45, 17], which may result to melts and/or fluids circulation. Moreover, Pleistocene to Recent trachyandesitic volcanism and high heat flow, as expressed by hot springs in Central Greece (SE. Thessaly and N. Evoicos Gulf), can be the result of mantle-originated magmas differentiated during their ascent through the relatively thin crust of the area [26, 49, 35]. The southwestwards NAF propagation during the last 5 Ma, which has reached the eastern continental coast of Central Greece, may also be of significant importance to both volcanism and low velocity zone in this particular location.

High velocities are observed in continental Greece at depths 45-120 km. This is compatible with [47, 37, 8]. The observed high velocity zone extends from north-western Greece to the Turkish coast, going through western Greece and southern Aegean and subparaller to the Hellenic trench. [37] interpret this structure as the subducted lithospheric African plate, which is bounded to the east of the study area, in the region of southwestern coast of Turkey.

One of the main features of our model is a high-velocity anomaly in south Aegean, associated with the slab (Figure 13). However, a limited resolution and better path coverage is required to obtain constraints on the slab geometry. The absolute shear velocities in the slab at a depth of 100 km are up to 4.70 km s^{-1} corresponding to a P velocity of 8.14 km s^{-1} (assuming a Poisson ratio of 0.25) which is in good agreement with the P velocity of 8.1 km s^{-1} obtained in the same location by [47].

North of NAT, a high velocity anomaly is observed (Figure 13), possibly associated with the southern margin of the Eurasian continental lithosphere. Although null space energy is high at this part of our model (poor resolution) the pattern is compatible with a high velocity anomaly found at the same area by [47].

6. Anelastic Tomograms

The attenuation model was obtained from an inversion with L_{corr} =150 km, $\sigma(r)$ =0.5×10^{-3} according to a conservative experimental error estimate. The qualitative resolution of the model as a function of geographical position is discussed in section 4.2.

Attenuation maps are shown in Figure 14 for 7 depths varying from 30 to 120 km. Given the low sampling of the area for periods <20 sec and >60 sec we restrict our observations within the above depth range. Attenuation variations are relative to the average Q_b^{-1} at each depth.

As expected, the attenuation pattern follows that of shear velocities. Within the examined depth range (30-120 km) the amplification of perturbations in the anelastic tomograms is between 0-20 % with respect to the background model, about 2 times higher than shear velocity perturbations, magnitude compatible with attenuation observations elsewhere [42].

Except in the case of 30 km depth, where elastic tomography is highly connected with crustal thickness, high attenuation is associated with low shear velocities, apparently in the North Aegean region. As in elastic tomography this region is about 200 km wide and extends in depth possibly deeper than our model (Figure 15). The resolution assessment shows that in this particular area the structures which can be retrieved are of similar dimensions, thus the possibility of an artifact intrusion is quite low.

The presence of this zone could be attributed to several reasons, including melting in the lithospheric mantle or asthenosphere and distributed deformation. Indeed, the NAF in the area and the associated intense north-south extension caused rapid tectonic instability in the area expressed in both mantle and crustal levels. NAT and Orfanos Gulf basins are their crustal surface results. NE-SW mantle olivine anisotropy [19] and extension in about the same direction suggest upper mantle deformation or flow. Recent magmatic mantle processes are manifested in the small volcanic islets of Psathura and Kalogeri of Northern and Central Aegean, respectively (Figure 1). They consist of sodic basalts derived from oceanic island basalt (OIB) asthenospheric mantle source similar in geochemical character to Pliocene basalts of Western and Central Anatolia [15, 49, 25]. High heat flow in the area of N. Aegean, signified by many hot springs in its islands, as well as the above mentioned petrogenetic and tectonic processes, indicate a relatively thin crust (<28 km) or a lithospheric delamination [36, 1, 27]. Moreover, the high attenuation mantle zone depicted clearly in the northern part of the cross sections of Figure 15 pressumably suggests magma either underplates the crust or most possibly remains further below at lithospheric-asthenospheric mantle levels, and the lithosphere-asthenosphere boundary is within our model depths, that is less than 120 km deep beneath northern Aegean.

At 30 km depth, the high attenuation anomaly could be associated with thermal activity from hot fluids and/or magma residing in deep crustal to lithospheric mantle levels below the South Aegean active volcanic centers, and originated in the asthenospheric mantle wedge above the dehydrating African subducted slab. Reactions of the released fluids with peridotitic lithospheric mantle can also cause serpentinization, which futher may decrease the seismic velocity and increase attenuation. Low attenuation is now observed throughout the Hellenic mainland likely attributed to subducted features.

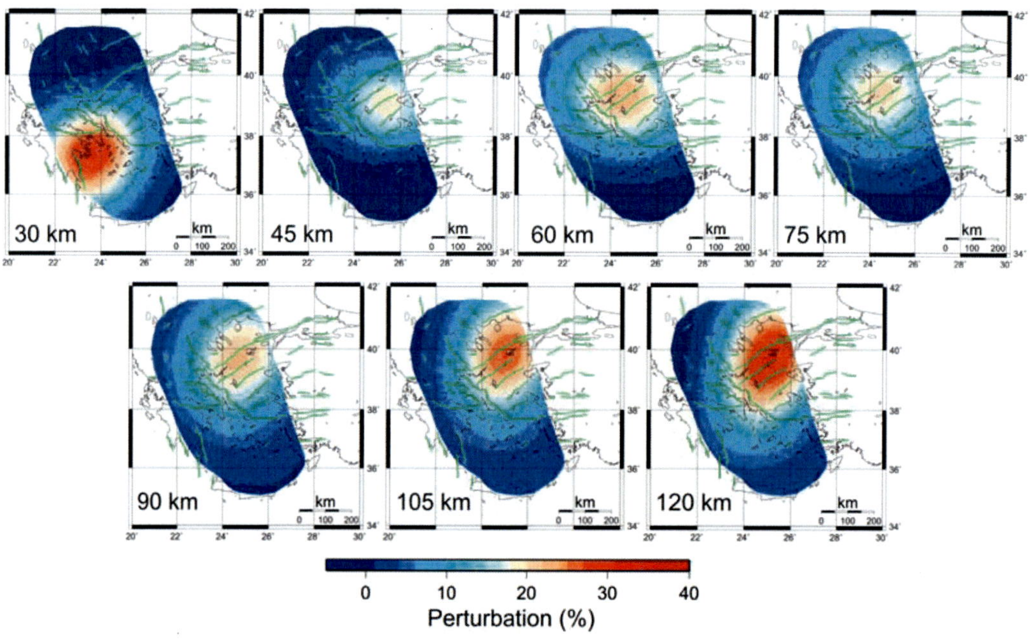

Figure 14. Tomographic slices of the Q_b^{-1} model at different depths. Tomograms start at 30 km depth. The Q_b^{-1} perturbations are indicated with a color scale in percent relative to a common reference model, derived from the average shear velocity at each depth (30 km, 4.0×10^{-2}; 45 km, 4.8×10^{-2}; 60 km, 5.4×10^{-2}; 75 km, 6.2×10^{-2}; 90 km, 6.5×10^{-2}; 105 km, 6.5×10^{-2}; 120 km, 6.3×10^{-2}.

7. Conclusion

Although the elastic structures of the Aegean have been submitted to several studies using seismic surface waves and tomographic techniques, this is the first time that surface wave anelastic parameters are obtained for the region. In this chapter we present a 3-D shear velocity and a 3-D attenuation model for the Aegean lithosphere, derived from the analysis of about 1100 fundamental mode Rayleigh wave seismograms. The two models are extending from 30 to 120 km depth. Local shear velocities and Q_b^{-1} of the models were obtained by the regionalization of 36 1-D path average shear velocity and 17 attenuation coefficient models. The continuous regionalization approach of [43] modified by [11] was used. A qualitative estimation of the uncertainty in the inverse solution of the elastic model was performed by means of the SVD method. The statistical measures that were used for the assessment

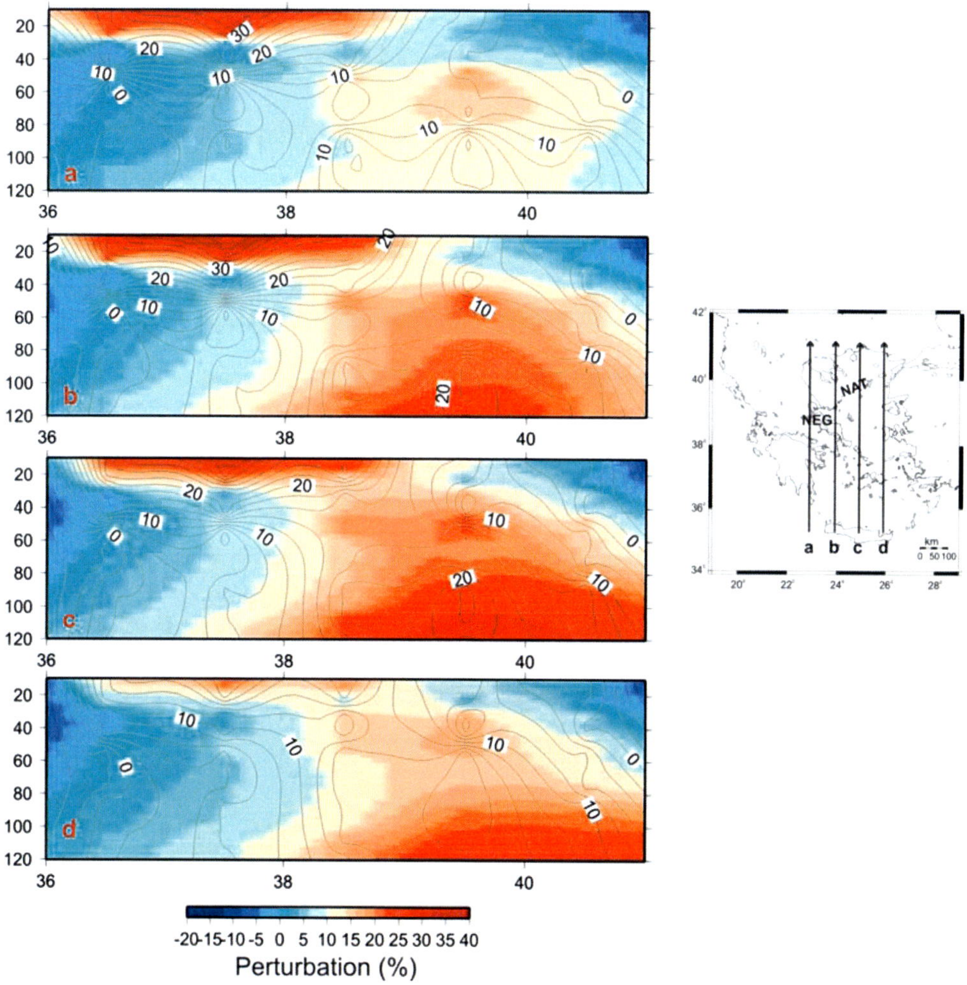

Figure 15. Vertical cross-sections across the 3-D Q_b^{-1} model. Attenuation variations are in percentage with respect to the average Q_b^{-1} at each depth. Cross-sections include shallow layers (10-30 km depth) but they are not discussed due to inhomogeneous data sampling. Right: Map showing the location of the cross sections (thick black lines) labeled from a to d.

confirmed that most of the model features were sufficiently resolved. Resolution of the anelastic model, assessed by optimized Voronoi polyhedra, is limited, as a consequence of the reduced ray coverage causing a considerable blurring at the final tomograms. However, anelastic tomograms are compatible with shear velocity variations. Hence, even though Q models derived from surface wave attenuation data may have large uncertainties, we are still be able to obtain information about the variability of Q and its relation to crustal and upper mantle structure and evolution.

One of the main features in the elastic model is the subducted slab in South Aegean and mainland Greece. In South Aegean variations exhibit a velocity contrast up to about 7% with respect to the background model. The amplitude of the slab heterogeneity is in good agreement with [47]. The precise shape and position of the slab cannot be resolved by the data used in the present study because of insufficient lateral resolution in the specific areas.

The most remarkable feature of the model is a low velocity anomaly (down to about 3% with respect to the background model) beneath Central and North Aegean. This feature extends laterally at more than 200 km and vertically possibly deeper than our model. This low velocity contrast can unlikely be associated with a narrow low velocity zone (70-100 km) which [47] determined beneath the volcanic arc and attributed to back-arc slab dehydration process.

An average joint elastic-anelastic model, representative for the Aegean lithosphere and upper mantle is derived by applying the concept of the uncoupled causal inversion. Average Q_b values at depths from 0 to 200 km range between 29±13 and 23±10 for the non-causal and uncoupled-causal inversion respectively. The observed low values contrast sharply with results from stable continental regions where Q_b may reach 200 or more. These low values likely indicate that fluids reside in lower crustal, as well as upper mantle depths.

One of the main features in the attenuation model is a low attenuation zone in South Aegean and East Peloponnesus likely associated with the subducted slab. High attenuation is observed beneath North Aegean. This high attenuation feature has a surface width of more than 200 km, whereas vertically perhaps continues deeper than our data suggests. It coincides with an area dominated by low shear velocities as the 3-D elastic model indicates. Zones of plate convergence and regional upper mantle heating are characterized by low Q extending from the surface to upper mantle depths [42]. As proposed by [42] Q of any wave propagating through the lithosphere in any region, is largely governed by elevated temperature from broad regions of the mantle which produces aqueous fluids. Upper mantle fluids, once released, propagate slowly upward and eventually either enter the crust and/or reduce the solidus of the upper mantle or lower crust rocks causing their partial melting. High attenuation in North Aegean may therefore be due to fluids reactions and involved partial melting at elevated temperatures produced by an upper mantle heat source. Decompression melting of an upwelling asthenosphere may also play a significant role.

On the other hand these elastic-anelastic anomalies are associated with an area with high extensional strain rates due to the westward propagation of the NAF system within the Aegean, slab roll back and back arc basin development [38]. Thus, such observations suggest a hot or perhaps partially molten uppermost asthenospheric mantle or distributed deformation of the upper mantle beneath this region, or a combination of both, that is an uppermost mantle with high-stress deformation zones associated with faulting and fluid or decompression induced magma generation [14].

Acknowledgments

We thank the GeoForschungsZentrum Potsdam, Germany for the GEOFON data from GVD, KRIS and SANT stations. We thank Eric Debayle for stimulating discussions and for providing the tomography algorithm. We thank Rolando Armijo for critical remarks. We thank all our colleagues who participated in the field experiment and data manipulation. Most of the figures presented in this chapter have been made using the GMT software (Wessel and Smith, 1995).

References

[1] Al-Lazki, A.I., Sandvol, E., Seber, D., Barazangi, M., Turkelli N. and Mohamad. R., 2004. Pn tomographic imaging of mantle lid velocity and anisotropy at the junction of the Arabian, Eurasian and African plates. *Geophys. J. Int.* **158**, 1024–1040.

[2] Alsina, D., and R. Snieder, 1996. Constraints on the velocity structure beneath the Tornquist-Teisseyre Zone from beam-forming analysis. *Geophys. J. Int.,* **126**, 205–218.

[3] Armijo, R., Meyer, B., King, G.C.P., Rigo, A., and Papanastassiou, D., 1999. Westward propagation of the North Anatolian fault into the northern Aegean: timing and kinematics, *Geology* **27**, 267–270.

[4] Armijo, R., Meyer, B., King, G.C.P., Rigo, A., and Papanastassiou, D., 1996. Quarternary evolution of the Corinth Rift and its implications for the later Cenozoic evolution of the Aegean. *Geophys. J. Int.,* **126**, 11- 53.

[5] Baumont, D., Paul, A., Beck, S., Zandt, G. and Pedersen, H., 2002. Lithospheric structure of the central Andes based on surface waves dispersion. *Geophys. J. Int.*, **107**, B12: 103-115.

[6] Bohm, G., and Vesnaver, A., 1999. *In quest of the grid: Geophysics*, **64**, 1116-1125.

[7] Bohm, G., Madrusani, G., Rossi, G., and Vesnaver, A., 2001. Ray footprint and redundancy in seismic tomography. *J of Seismic Explor*., **9**: 285-302.

[8] Bourova, E., Kassaras, I., Pedersen, H.A., Yanovskaya, T., Hatzfeld, D. & Kiratzi, A., 2005. Constraints on absolute S velocities beneath the Aegean Sea from surface wave analysis. *Geophys. J. Int*, **160**, 1006-1019.

[9] Bregman, N. D., Bailey, R. C., and Chapman, C. H., 1989. Ghosts in tomography: the effects of poor angular coverage in 2-D seismic travel time inversion. *Can. J. Expl. Geophys*., **25**, 7-27.

[10] Cotte, N., H. A. Pedersen, M. Campillo, V. Farra, and Y. Cansi, 2000. Off great circle propagation of intermediate period surface waves observed on a dense array in the French Alps. *Geophys. J. Int*., **142**, 825– 840.

[11] Debayle, E. and Sambridge, M., 2004. Inversion of massive surface wave data sets : model construction and resolution assessments, *J. Geophys. Res.,* **109**, B02316, doi:10.1029/2003JB002652.

[12] Dziewonski, A.M., Bloch, S. and Landisman, M., 1969. A technique for the analysis of transient seismic signals. *Bull. Seism. Soc. Am*., **59**, 427-444.

[13] Engdahl, E.R., Van der Hilst, R.D. and Buland, R.P., 1998. Global teleseismic earthquake relocation with improved traveltimes and procedures for depth determination. *Bull. Seism. Soc. Am*., Vol. 98, 722-743.

[14] Falus, G., Drury, M., Roermund, H., and Szabó, C., 2004. Magmatism-related localized deformation in the mantle: a case study. *Contrib. Miner. Petr.*, **146**, 493-505.

[15] Fytikas, M., Innocenti, F., Manetti, P., Mazzuoli, P., Reccerillo, A., and Villari, L., 1984. Tertiary to Quaternary evolution of volcanism in the Aegean region. In: Dixon, J.E., & Robertson A.H.F. (eds): The geological evolution of the eastern Mediterranean., *Geol. Soc., London, Spec. Publ.*, **17**, 687-699.

[16] Godey, S., 2002. Structure of the uppermost mantle beneath North America: Regional surface wave tomography and thermo-chemical interpretation, PhD thesis, Utrecht University.

[17] Goes, S., R. Gover, and P. Vacher, 2000. Shallow mantle temperatures under Europe from P and S wave tomography, *J. Geophys. Res.*, **105**, 11,153– 11,169.

[18] Golub, G. H., and Van Loan, C. F., 1989. Matrix computations. Johns Hopkins Univ. Press.

[19] Hatzfeld D., Karagianni, E., Kassaras, I., Kiratzi, A., Louvari, E., Lyon-Caen, H., Makropoulos, K., Papadimitriou, P., Bock, G. and Priestley, K., 2001. Shear wave anisotropy in the upper mantle beneath the Aegean related to internal deformation. *J. Geophys. Res.*, **106**, No. 12, 30737-30753.

[20] Hatzidimitriou, P., 1993. Attenuation of coda waves in Northern Greece, Pure *Appl. Geophys.*, **140**, 63–78.

[21] Herrin, E., and T. Goforth, 1977. Phase-matched filters: Application to the study of Rayleigh waves. *Bull. Seismol. Soc. Am.*, **67**, 1259– 1275.

[22] Herrmann, R.B., (ed. R.B. HERRMANN), 1991. Surface wave inversion program, Saint Louis University, Saint Louis, MO.

[23] Herrmann, R.B., 1973. Some aspects of band-pass filtering of surface waves, *Bull. Seism. Soc. Am.*, **62**, 129-139.

[24] Herrmann, R.B., 1994. Computer programs in Seismology, User's manual. St. Louis University, Missouri.

[25] Innocenti F., Agostini, S., Di Vincenzo, G., Doglioni, C., Manetti, P., Savascin, M.Y., Tonarini S., 2005. Neogene and Quaternary volcanism in Western Anatolia: Magma sources and geodynamic evolution. *Marine Geology*, **221**, 397– 421.

[26] Innocenti, F., Manetti, P., Reccerillo, A., and Poli, G., 1979. Inner arc volcanism in NW Aegean: geochemical and geochronological data. N. Jb.. Miner., Mn., 1979, 145-158.

[27] Karagianni, E.E., Papazachos, C.B, Panagiotopoulos, D.G., Suhadolc, P., Vuan, A. and Panza, G.F., 2005. Shear velocity structure in the Aegean area obtained by inversion of Rayleigh waves. *Geophys. J. Int.*, **160**, 127-143.

[28] Kassaras, I., Makropoulos, K., Bourova, E., Pedersen, H. & Hatzfeld, D., 2005. Upper mantle structure of the Aegean derived from two-station phase velocities of fundamental mode Rayleigh waves. The South Aegean Active Volcanic Arc, Developments in Volcanology, Volume 7, Hardbound, ISBN 0-444-52046-5, Imprint ELSEVIER: 19-45.

[29] Kennett, B., 1995. Approximations for surface-wave propagation in laterally varying media. *Geophys. J. Int.,* **122**, 470– 478.

[30] Kovachev, S.A., Kuzin, I.P., Shoda, O.Yu., 1991. Attenuation of S waves in the lithosphere of the Sea of Crete according to OBS observation, *Phys. Earth Planet. Inter.*, **69**, 101– 111.

[31] Lander, A. V., and A. L. Levshin, 1989. Recording, identification, and measurement of surface wave parameters, in Seismic Surface Waves in Laterally Inhomogeneous Earth, edited by V. I. Keilis-Borok, pp. 131–182, Kluwer Acad., Norwell, Mass.

[32] Laske, G. and Masters, G., 1996. Constraints on global phase velocity maps from long-period polarization data. *J. Geophys. Res.*, **101**, 16059-16075.

[33] Levshin, A., Ritzwoller, M and Ratnikova, L., 1994. The nature and cause of polarization anomalies of surface waves crossing Northern and Central Eurasia. *Geophys. J. Int.,* **117**, 577-591.

[34] Levshin, A., Ritzwoller, M. and Resovsky, J., 1999. Source effects on surface waves group travel times and group velocity maps. *Phys. Earth Planet. Int.*, **15**, 293-312.

[35] Makris, J., Papoulia, J., Papanikolaou, D. and Stavrakakis, G., 2001. Thinned continental crust below northern Evoikos Gulf, central Greece, detected from deep seismic soundings, *Tectonophysics*, **341**, 225–236.

[36] Makris, J., Vees, R., 1977. Crustal structure of the Aegean Sea and the Islands of Evia and Crete, Greece, obtained by refraction seismic measurements. *J. Geophys.* **42**, 329– 341.

[37] Marone, F., van der Lee, S. & Giardini, D., 2004. Three-dimensional uppermantle S-velocity model for the Eurasia–Africa plate boundary region, *Geophys. J. Int.,* **158**, 109–130.

[38] McClusky, S., Balassanian, S., Barka, A., Demir, C., Ergintav, S., Georgiev., I., Gurkam, O., Hamburger, M., Hurst, K., Kahle, H., Kastens, K., Kekelidze, G., King, R., Kotzev, V., Lenk, O., Mahmoud, S., Mishin, A., Nadariya, M., Ouzounis, A., Paradissis, D., Peter, Y., Prilepin, M., Reilinger, R., Sanli, I., Seeger, H., Tealeb, A., Toksöz, M.N. and Veis, G., 2000. Global Positioning System constraints on plate motions and dynamics in the eastern Mediterranean and Caucasus. *J. Geophys. Res.*, **105**, No B3, 5695-5719.

[39] McMechan, G. A., and Yedlin, M. J., 1981. Analysis of dispersive waves by wave field transformation. *Geophysics*, **46**, 869-874.

[40] Menke, W., 1984. Geophysical Data Analysis: Discrete Inverse Theory. Academic Press, Orlando, FL, 260p.

[41] Michelena, R. J., 1993. Singular value decomposition for Cross-well tomography: *Geophysics*, **58**, 1655-1661.

[42] Mitchell, B.J., 1995. Anelastic Structure and Evolution of the Continental Crust and Upper Mantle From Seismic Surface Wave Attenuation, *Rev. Geoph.*, vol. 33, no. 4, 441–462.

[43] Montagner, J., 1986. Regional three-dimensional structures using long period surface waves, *Ann. Geophys., B*, **4**, 283-294.

[44] Montagner, J.P. and Jobert N., 1988. Vectorial Tomography II. Application to the Indian Ocean. *Geophys. J.,* **94**, 309-344.

[45] Nataf, H.-C. and Ricard, Y., 1996. 3SMAC: An a priori tomographic model of the upper mantle based on geophysical modeling, *Phys. Earth Planet. Inter.*, **95**, 101– 122.

[46] Papazachos, B.C., Comninakis, P.E., 1971. Geophysical and tectonic features of the Aegean arc, *J. Geophys. Res.*, **76** (35), 8517–8533.

[47] Papazachos, C. and Nolet, G., 1997. P and S-wave velocity structure of the Hellenic area obtained by robust nonlinear inversion of travel times. *J. Geoph. Res.*, **102**, B4, 8349-8367.

[48] Papazachos, C.B., Hatzidimitriou, P.M., Panagiotopoulos, D.G. & Tsokas, G.N., 1995. Tomography of the crust and upper mantle in southeast Europe. *J. Geophys. Res.*, **100**, 405-422.

[49] Pe-Piper, G. and Piper, D.J.W., 2002. The igneous rocks of Greece, Gebruder Borntraegen, Berlin-Stuttgard, p. 573.

[50] Polatidis, A, Kiratzi, A, Hatzidimitriou, P and Margaris, B., 2003. Attenuation of shear-waves in the back-arc region of the Hellenic arc for frequencies from 0.6 to 16 Hz, *Tectonophysics*, **367**, 29– 40.

[51] Pratt, R. G., and Chapman, C. H., 1992. Traveltime tomography in anisotropic media: Application. Geophys. *J. Intl.*, **109**, 20-37.

[52] Rohm, A. H., R. K. Snieder, S. Goes, and J. Trampert, 2000. Thermal structure of the continental upper mantle inferred from S wave velocity and surface heat flow, *Earth Planet. Sci. Lett.*, **181**, 395– 408.

[53] Romanowicz, B., 1994. Anelastic Tomography: A New Perspective on Upper-Mantle Thermal Structure, *Earth. Planet. Sci. Lett.* **128**, 113–121.

[54] Sato, Y., 1955. Analysis of dispersed surface waves by means of Fourier transform: Part I, *Bull. Earthquake Res., Tokyo Univ.*, **33**, 33-47.

[55] Spakman, W., Van der Lee, S. and Van der Hilst, R.D., 1993. Travel-time tomography of the European-Mediterranean mantle down to 1400 Km. Phys. Earth Planet. Inter., 79, 3-74.

[56] Stork, C., 1992, Singular value decomposition of the velocity reflector depth tradeoff, Part 2: High resolution analysis of a generic model: *Geophysics*, **57**, 933-343.

[57] Stutzmann, E. and Montagner, J.P., 1994. Tomography of the transition zone from the inversion of higher-mode surface waves. *Phys. Earth Planet. Int.* **86**, 99–116.

[58] Tarantola, A. and Valette, B., 1982. Generalized nonlinear inverse problems solved using the least square criterion, *Rev. Geophys*., **20**, 219-232.

[59] Tiberi, C., Lyon-Caen, H., Hatzfeld, D., Achaeur, U., Karagianni, E., Kiratzi, A., Louvari, E., Panagiotopoulos, D., Kassaras, I., Kaviris, G., Makropoulos, K. and Papadimitriou, P., 2000. Crustal and upper mantle structure beneath the Corinth rift from a teleseismic tomography study. *J. Geophys. Res*., **105**, 28159-28171.

[60] Tselentis, G.-A., 1993. Depth-dependent seismic attenuation in western *Greece*, *Tectonophysics*, **225**: 523– 528.

[61] Vesnaver, A., 1994. Towards the uniqueness of tomographic inversion solutions. *J. of Seismic Explor*., **3**, 323-334.

[62] Vesnaver, A., 1996. The contribution of reflected, refracted and transmitted waves to seismic tomography: A tutorial: *First Break*, **14**, 159-168.

[63] Wessel, P. and Smith, W.H.F., 1995. New version of the Generic Mapping Tools released, EOS, *Trans. Am. Geophys. Un*., **76**, 329.

[64] Woodhouse, J.H., 1974. Surface waves in the laterally varying structure, *Geophys. J. R. Astr. Soc.* **90**, 713–728.

[65] Wortel, M.J.R., and Spakman, W., 2000. Subduction and slab detachment in the Mediterranean-Carpathian region. *Science*, **290**, 1910-1917.

[66] Yoshizawa, K. and Kennett, B.L.N., 2002. Determination of the influence zone for surface wave paths. *Geophys. J. Int*., **149**, 440– 453.

In: The Lithosphere: Geochemistry, Geology and Geophysics ISBN: 978-1-60456-903-2
Editors: J.E. Anderson et al, pp. 295-316

Chapter 8

SPACE AND TIME VARIATIONS OF ELASTIC AND ANELASTIC PROPERTIES IN THE SHALLOW LITHOSPHERE

C. Chiarabba and P. De Gori
Istituto Nazionale di Geofisica e Vulcanologia, Rome

Abstract

Rock fracturing, fluid migration and pore pressure build-up is the sequence of a phenomenon which is believed to occur within the brittle outer shell of the earth during the preparatory processes of magmatic intrusion and earthquake generation. This phenomenon produces transient variations of the elastic parameters within wide rock volumes. Since the lithosphere is largely pervaded by fluids, we expect that high frequency seismic waves interact with these liquids and store them as "earth breath" in locations yet to be discovered.

Time resolved seismic tomography (4D) progressively demonstrates that transient Vp or Vs anomaly variations can be identifiable by seismic data. On a local scale, oil industries pilot research projects for the 4D seismic tomography in order to retrieve the depletion of fossil fuels in reservoirs during extraction. The proposed approach allows us to recognize eventual rock fracturing or pore pressure variations, which are useful for magmatic intrusion or earthquake forecasting.

We present a series of new observations which testify that both the elastic and anelastic properties of the crust vary in time during the rapidly changing tectonic process. We address clear 4D variations of velocity and, for the first time, the attenuation of body waves during the development of normal faulting seismic sequence, seismic swarms and magmatic supply at Quaternary volcanoes. Our results support the theory that physical properties of the lithosphere are not (dramatically) static but follow predictable patterns generated by the tectonic process.

Introduction

In this paper, we discuss new ideas on the physical properties of the shallower part of the lithosphere (the crust), inferred from the analysis of seismological data. The lithosphere, the thin outer shell of the earth, does not significantly deform with stress applied on time scales of up to 10^9 years. Although it is remarkably thin compared to the Earth's ray, we base our knowledge on the indirect measurement of its physical properties, as well as the information provided by petrological and geochemical data.

For more than a century, the lithosphere is studied by the propagation of seismic waves. In all studies so far, the elastic and anelastic properties of the lithosphere are estimated by using the transmitted, reflected or converted energy of body or surface waves to create 1D to 3D velocity and attenuation models (from global to local, see Grand et al., 1997 Bijward et al., 1998; Thurber et al., 2004, among many others). We consider these images as static, since they are created by averaging the properties over a more or less wide earth volume, not taking into account that the physical properties may vary within the period spanned by the data. In many cases, such an approximation is valid, since we do not expect that the whole lithosphere sustains a physical process which varies the elastic parameters in a time scale of tens of years. However, in active faults or volcanoes, and more generally in the uppermost crust, these rapid variations are present and are invoked by physical models of earthquakes (Aggarwal et al., 1973; Scholz et al., 1973). The nucleation of a large earthquake is associated with a series of phenomena that starts from a progressive generation of microcracks in the fault damage zone and continues with the crack coalescence, fluid migration and pore pressure build-up (Nur and Booker, 1972; Gupta, 1973; Sibson, 1992; Cox, 1995; Segall and Rice, 1995; Caine et al., 1996; Miller et al., 2003). All of these events increase in a paroxistic way and culminate in an earthquake. Our goal is to verify if this process generates resolvable variations of elastic parameters and to extrapolate results for a better understanding of the lithosphere.

During the 70's, the possibility to detect elastic properties variations in the crust, related to rock dilatancy before impending earthquakes (see Scholz et al., 1973), created the first strong impulse in research and premature optimism in crowning the Vp/Vs ratio as a seismic precursor. This stimulated theoretical and laboratory data to provide background knowledge on how physical parameters affect the propagation of seismic waves (O'Connel and Budiansky, 1974; Ito et al., 1979; Christensen, 1985; Mavko et al., 1998). Since then, the dependence of body wave velocity to porosity, in situ pressure and temperature, pore pressure is progressively clarified by numerous studies, both theoretically and with laboratory experiments (Nur, 1972; Wang and Nur, 1989; Dvorkin et al., 1999).

Almost 30 years later, the improved efficiency, ensured by the continuous recording of digital three components seismic stations and the deployment of dense local networks directly targeting faults and volcanoes, make the recognition of time change in elastic properties more feasible. Following dramatic technology improvements, we now dispose of a fast growing amount of high quality data. The dense station distribution, close to the main targets and the use of abundant data, yield high resolution images of the uppermost part of the lithosphere. Seismic rays efficiently sample the volume over which magma intrusion or pore pressure variation on faults are presumed to have occured. P- and S-wave waveforms somehow store these variations, but their efficient extraction is difficult. Recent studies herald the capability to timely-resolve 4D variations (in space and time). Gunakasera et al. (2003) showed the first example of velocity variations in a geothermal field related to the depletion of fluids in the reservoir. Patané et al. (2006) showed an unambiguous transient variation of the Vp/Vs ratio associated with the emplacement of a gas enriched dyke underneath Mt. Etna.

In this study, we show examples of velocity and attenuation time-variations on seismic faults and Quaternary volcanoes in Italy. These variations are clearly related either to local seismicity increases or to the proximity of large earthquakes. Thus, we are confident that the variations can be resolved and directly related to the migration of pore pressure in the upper crust. We do not speculate that such observations will lead to earthquake or eruption previsions, but they can highlight that the transient phenomena within the shallow lithosphere can be recognized by seismic data.

Pioneeristic Studies of 4D Variations in Quaternary Volcanoes, Geothermal Fields and Active Faults

Although the P- and S-wave arrival times of local earthquakes are very common data in seismology, the repeated seismic tomography (4D) is still not widely applied. The main scepticism arises from the reliability of the time-variation of anomalies revealed by seismic data. Since these studies use natural sources of elastic waves (earthquakes), the ray sampling within the target volume differs through time and an exact repetition cannot be expected. Conversely, if artificial sources are used, results obtained by similarly repeated experiments can be compared, but loose information in between the experiments, neglecting the eventual high frequency variation of elastic parameters.

Considering these drawbacks, our effort is to try and understand how the temporal variations of elastic parameters can be efficiently retrieved and if transient signals affect the 3D velocity images frequently discussed in the literature. The 4D tomography's main requirement is a time-consistent resolution, i.e. the parameter resolution of different tomograms needs to be the same. At present, this requirement can only be addressed with a complete resolution analysis of velocity models and synthetic tests.

The first repeated seismic tomography study, revealed a variation of a low Vp/Vs anomaly located under the The Geyser geothermal field, featuring a change of about 50 % in a 7 years long period (Gunakasera et al., 2003). The Authors attributed the decrease of the Vp/Vs ratio to the depletion of pore liquid water and the replacement with steam in the reservoir. Similarly, Foulger et al. (2003) documented relevant changes in both Vp and Vs, consistent with the migration of CO_2 in the uppermost crust beneath the Mammoth Mountain volcano (Long Valley Caldera).

Recently, Patané et al. (2006) observed a clear low Vp/Vs anomaly in the shallow crust beneath Mt. Etna during the emplacement of a volcanic dyke. This anomaly did not exist before the intrusion, and is heralded by a seismic swarm which vanished after the eruption. The Authors hypothesized that this low Vp/Vs anomaly is a transient variation of elastic parameters associated with rock fracturing and gas released during the emplacement of the dyke. The close relation between the low Vp/Vs volume and the location and geometry of the dyke, as modelled by geodetic data, along with the gas-enriched eruption which occurred a few days later than the onset of the transient feature, are arguments in favour of the reliability of 4D seismic tomography.

More recently, Chiarabba, C., De Gori P., Boschi E., presented evidence for strong Vp/Vs variations during the Umbria-Marche 1997 seismic sequence in central Italy. They find a clear transient high Vp/Vs anomaly located on the fault segment which ruptured by two MW>5.6 earthquakes a few days later. Here, we will summarize some of these results and show further evidence for transient signal related to the anelastic properties of the medium.

Data and Inversion Methodologies

The data used in this study comes from the elaboration of local earthquakes waveforms from dense temporary, or permanent, seismic networks operating on active faults and volcanoes in Italy.

We use two different types of data: the body wave arrival times and the body wave high frequency spectral decay (t*). The extraction of P and S-wave arrival times from local earthquake waveforms is the most classic seismological analysis. While P-wave onset is almost always very clear, (depending on the signal to noise ratio) the accuracy of the direct S-waves onset determination is less objective and more difficult. Usually, data are read by experts and scientists, with a standard weighting scheme based on an estimate of the phase onset reading error. We emphasize that the use of data elaborated by scientists yields the high quality necessary to resolve the fine details of the structure, otherwise blurred by the use of routinely picked data, such as those usually found in seismic bulletins. This process is very time-consuming and limits the creation of a wide dataset. Since nowadays we are able to fully acquire background seismicity down to very small magnitudes with continuously recording digital dense networks, we believe that only an automatic procedure can feasibly scan the digital data to recognize and furnish consistently weighted P- and S-wave arrival times (see Di Stefano *et al.*, 2006). The advantages of using an automatic procedure are the reduced time for the analysis, the highly accurate first-arrival *P*-wave picks and the consistency in timing error assessment. Such procedure can make the extraction of abundant data feasible in the close future. Further implementations can derive from the use of cross-correlation of earthquake waveforms to extract P- and S-wave data more precisely (see Rowe et al., 2002, Waldhauser and Ellsworth, 2000). In this study we use body wave arrival times read on digital waveforms by scientists.

P-and S-wave arrival times are used to investigate the spatio-temporal variations of Vp and Vp/Vs anomalies in the target volume. We present results for different areas which have been monitored with dense seismic local networks.

First, we compute Vp/Vs values for each earthquake, by the Wadati diagram and subsequently we average the Vp/Vs values over moving windows adapted to each case.

Independently, we use the arrival times to compute high resolution Vp and Vp/Vs models with state of the art local earthquake tomography codes (Eberhart-Phillips and Reyners, 1997). We derive 3D velocity images including data from the entire period, in most cases results previously published, and by using sub-set of data for the repeated seismic tomography (4D). In this latter case, we subdivide the entire period into epochs, based on the temporal variation highlighted during the first analysis. The main requisite for this subdivision is to achieve a timely consistent model resolution among the different epochs (see Patané et al., 2006). We prefer to have the same resolution in each epoch, rather than achieving the best resolution, because the ''4D'' images need to be comparable.

To study the anelastic properties of the medium, i.e. how the seismic energy is dissipated during the propagation, two different approaches can be followed, in the time or frequency domains respectively. The measurements estimated on the earthquake waveforms are the pulse width or the spectral decay. We use these data to define the spatio-temporal variations of seismic attenuation. A quantitative measure is given by the Quality factor (Qp and Qs), which is inversely proportional to the attenuation, the higher the Q the smaller the attenuation.

In the frequency domain, we need to measure the high frequency amplitude decay of spectra (the t* operator, see Scherbaum, 1990). Assuming a source type (for example a spectral falloff equal to -2, see Brune, 1970), the spectral parameters (corner frequency, low frequency plateau, and t*) are computed by simultaneously inverting the velocity spectra, selecting time windows that enclose the P- or S-wave arrivals. The length of the time window should be large enough for a good resolution in the frequency domain, avoiding the contribution of scattered waves or later secondary arrivals. For a thorough review of the methodology, not as widely applied as the body wave arrival times, we refer to papers by Rietbrock (2001), Eberhart-Phillips and Chadwick (2002), De Gori et al. (2005), Hauksson and Shearer (2006).

The tomographic inversion for both velocity and attenuation models is a state-of-the-art technique largely applied to several relevant geologic environments over the past 30 years. The more diffuse inversion schemes are based on a linearized inversion consisting of several repeated iterations during which a reference model is perturbed until the convergence to a final solution is reached. This linearization holds efficiently, and avoids local minimal, when the choice of the a priori model is close to the reality (Aster et al., 2005). We used the inversion code developed by Thurber (1983) and modified by Eberhart-Phillips and Reyners (1997) and Rietbrock (2001). The model is parameterized by assigning starting values at a 3D grid of nodes. After the parameter decomposition, both hypocentral and velocity parameters are updated at each damped iteration until the convergence to the final solution, hopefully the global minimum in the model space. Seismic rays are traced in the 3D model using an approximate ray tracer and a pseudo-bending algorithm (Um and Thurber, 1987). In our case, we invert for Vp and Vp/Vs, instead of Vs, being the choice recommended to avoid bias on the Vp/Vs ratio due to the uneven resolution of Vp and Vs parameters. Once velocity models have been derived, they are used in the inversion for Qp and Qs (Rietbrock, 2001).

Model resolution is verified, analysing the whole resolution matrix (Menke, 1989). The most efficient way to visualize the well (or similarly) resolved volumes is with the spread function, which compactly describe how much each row of the resolution matrix is picked on the diagonal element. Well resolved parameters have small values of the spread function, which coincide with delta-like averaging vectors.

New results for Quaternary Volcanoes and Normal Fault Systems in Italy

A. The Umbria Marche Normal Faulting Region

The normal fault belt in central Italy is characterized by intense background seismicity, frequent clustered sequences and moderate magnitude events developing on adjacent NW-trending fault segments, whose dimensions barely exceed 10 km (maximum magnitude around 6.0, see). One of the more active seismic areas is the Umbria-Marche, struck in 1997 by several Mw>5.6 normal faulting earthquakes (see Figure 1). We investigate two different scenarios: the 1997 seismic sequence and the Alto Tiberina low angle normal fault system (figure 1). Both cases have been read with very dense local seismic networks.

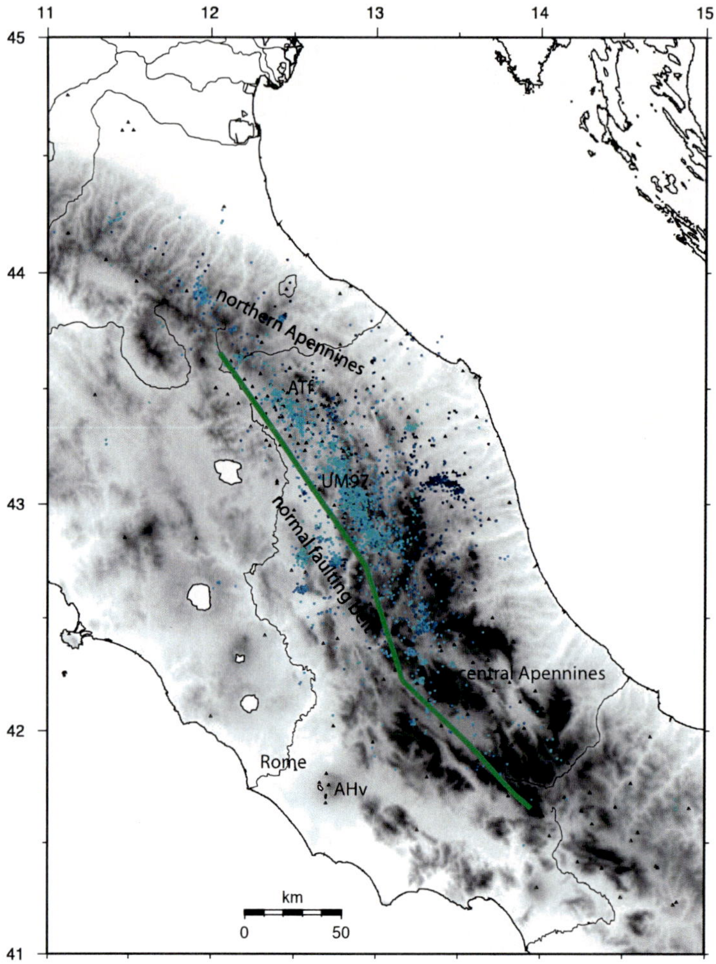

Figure 1. Seismicity of the northern-central Apennines in Italy. Hypocentral depths are differentiated by colours. The examples presented in the paper are relative to the normal fault seismic belt: the1997 Umbria-Marche seismic sequence (UM97) and the Alto Tiberina low angle normal fault system(ATf); the Alban Hills volcano (AHv), and Mt. Etna volcano, not shown in the figure.

The first example is the 1997 Umbria Marche seismic sequence, well described by a series of papers (see Chiaraluce et al., 2004 and references therein). The main peculiarities of the sequence are the seismicity migration over the fault system and the occurrence of five MW>5.2 main events during a one month period. Different models have been proposed to explain these features; the two end-members are the stress transfer between adjacent faults (Nostro et al., 2004) and the fluid and pore overpressure migration (Miller et al., 2003; Antonioli et al., 2005). While the former mechanism hardly explains the activation of large part of the aftershocks, the non-linear diffusion model by Miller et al (2003) indicates that a pressure pulse with amplitude up 10-20 MPa is very effective to drive and sustain the seismic sequence. The excellent data available from these dense local networks makes this one of the best locations to test whether the fluid overpressure modifies the elastic parameters in a way that is recoverable by the body waves.

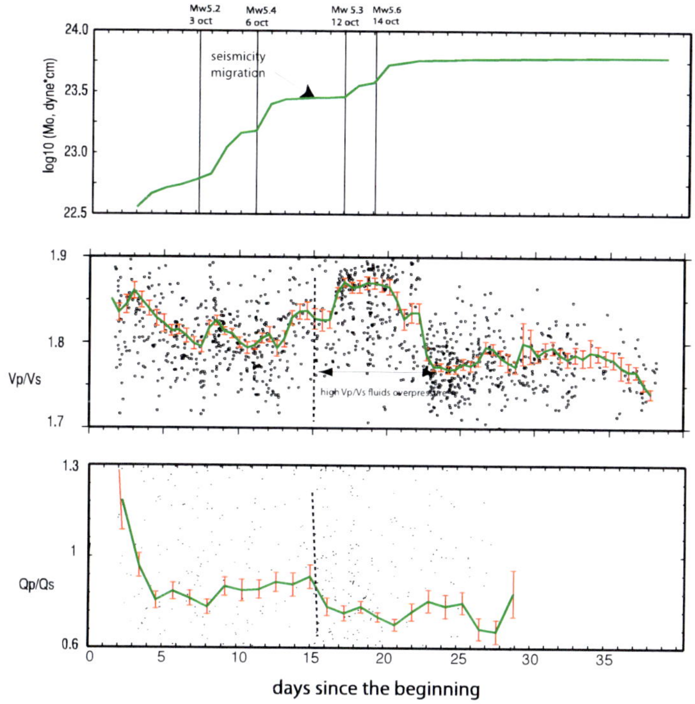

Figure 2. Variations of Qp/Qs (bottom panel), Vp/Vs (central panel), and the cumulative seismic moment (top panel) during the 1997 seismic sequence. Green lines are the moving windows averages with the standard deviations bars. The occurrence of large aftershocks (Mw>5.2) is marked by the solid lines on the top panel, while the dashed lines in the other panels mark the sharp Vp/Vs increase and Qp/Qs decrease that preceded the seismicity migration. Note a first Vp/Vs increase before the October 3 event, and the sharp increase preceding the October 12 and 14 shocks.

We extracted a total of 19484 P- and 18000 S-wave arrivals and computed a moving average of the Vp/Vs ratio for the one-month-long period monitored by the dense local array. The length of each window is one day, with an overlap of 12 hours between contiguous windows. We observe clear and strong variations of the Vp/Vs ratio during the sequence (figure 2). After the two first mainshocks, we note a Vp/Vs decrease and a first increase in early October (day 8), coinciding with the October 6, Mw 5.6 shock. After that, a second decrease of the Vp/Vs ratio anticipated the southward seismicity migration. Starting from the day 12, we observe a sharp and strong Vp/Vs increase concurrent with the activation of the southern fault segment and the occurrence of the two large earthquakes of October 12 and 14. In the final part of the sequence, the Vp/Vs ratio returns to values smaller than those observed at the beginning of the sequence. Both large aftershocks and earthquake migration ceased/ or came to an end. Chiarabba, C., De Gori P., Boschi E., documented that the sharp Vp/Vs increase is a transient feature occurring on the fault segment located at the south-eastern end of the fault system, over which the large magnitude earthquakes nucleated.

We verify whether the anelastic properties vary similarly to the elastic ones. Inverting velocity spectra of P- and S-waves, we obtained 6192 tp* and 10325 ts* measurements. Computing the P- and S-wave travel times with the 3D velocity models, we estimated the path averaged Qp and Qs from tp* and ts* (Rietbrock, 2001). Similarly to the Vp/Vs, we computed moving averages for the Qp/Qs ratio. We observe a well resolved Qp/Qs decrease, mostly related to a decrease of the S-waves attenuation, concomitant with the high Vp/Vs anomaly (figure 2). Seismicity migration and large aftershocks are related to the transient positive Vp/Vs and negative Qp/Qs anomaly along the fault system.

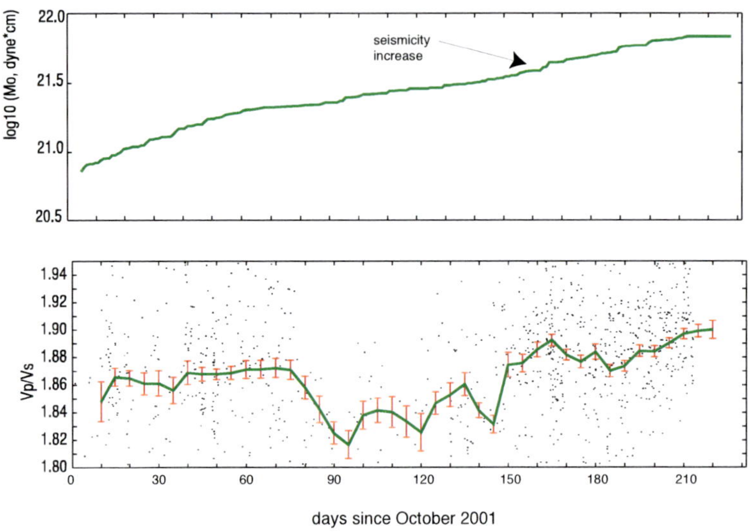

Figure 3. Variations of Vp/Vs (bottom panel), and the cumulative seismic moment (top panel) during the 8-months-long ATf experiment. Green lines are the moving window averages with the standard deviations bars. Note that periods of higher seismicity correlate with periods during which the Vp/Vs ratio is relatively higher. In particular, the increase of seismicity (around day 165) follows a well defined increase of the Vp/Vs ratio.

The second example is relative to a normal fault system, located a few kilometres to the northwest of the 1997 sequence (figure 1). During 2000-2001, a dense local network was installed to record the background seismicity of the Alto Tiberina fault (ATf), a low angle normal fault that is believed to control the extension in this portion of the Apennines (Chiaraluce et al., 2007). Throughout an 8 month long experiment, a total of 1159 M_L earthquakes were recorded providing a high quality dataset of 13077 and 12540 P- and S-wave arrivals. We have computed Vp/Vs averages on moving windows with a length of 10 days, with an overlapping of 5 days. The analysis, similar to that described above, indicates clear time variations of the Vp/Vs ratio (figure 3). This variation is not as large as those observed during the 1997 sequence and follows local variations in the level of background seismicity. The Vp/Vs ratio increases correlate with periods of higher seismicity.

Following the promising variations outlined by the data, we run the repeated seismic tomography, subdividing the 8 months long period into three sub-epochs. The subdivision takes account of both the different patterns in the Vp/Vs (figure 4) and the time-consistent resolution requirements (see section above). The thorough discussion of procedures and inversion parameters is beyond the scope of this paper. The tomographic inversion is carried out using model parameterization and inversion parameters similar to those published by Moretti M, De Gori P. and Chiarabba C. The resolution of velocity models in the volumes confined by the violet line (spread function = 2) is similar between the three epochs. We refer to the paper by Patanè et al. (2006) for details on the resolution analysis.

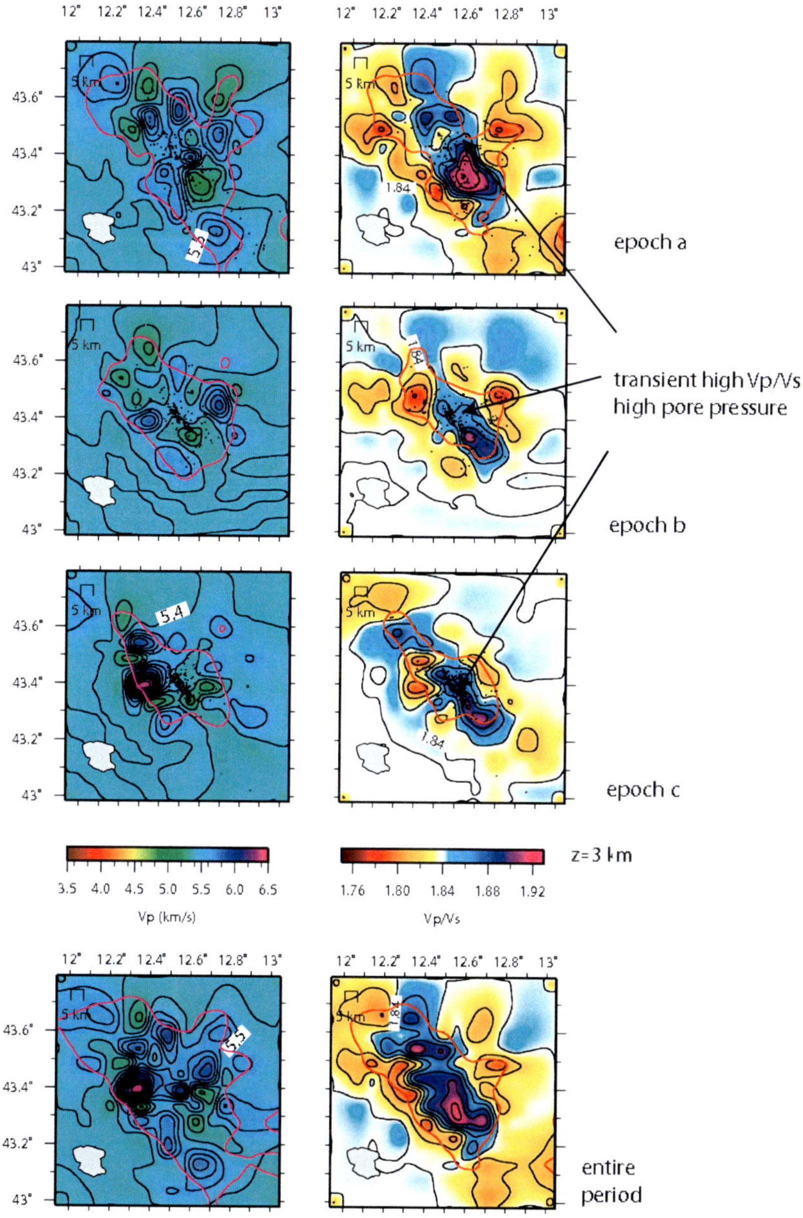

Figure 4. 4D seismic tomography of the ATf. Vp (left) and Vp/Vs (right) models in the same layer located at 3 km depth during the three epochs. Small dots are the earthquake hypocenters. The pink solid line encompasses the similarly resolved model volumes. The difference in the Vp and Vp/Vs patterns within the well resolved volume is reliable. Along the bottom, we show the Vp and Vp/Vs models obtained by using the data from the entire period. Note that the velocity patterns vary significantly during the epochs and the model derived with all data is an average of the features present in each epoch.

Here we only show the most significant features: two main high and low Vp/Vs anomalies located on the hanging wall of the ATf and mostly coinciding with the fault, respectively.

The shape and the absolute values of these anomalies change among the epochs.

Although we recognize that the volume sampling is not the same between the three epochs, the location of these anomalies in the central well resolved part of the model led us to be confident with the reliability of the 4D anomalies.

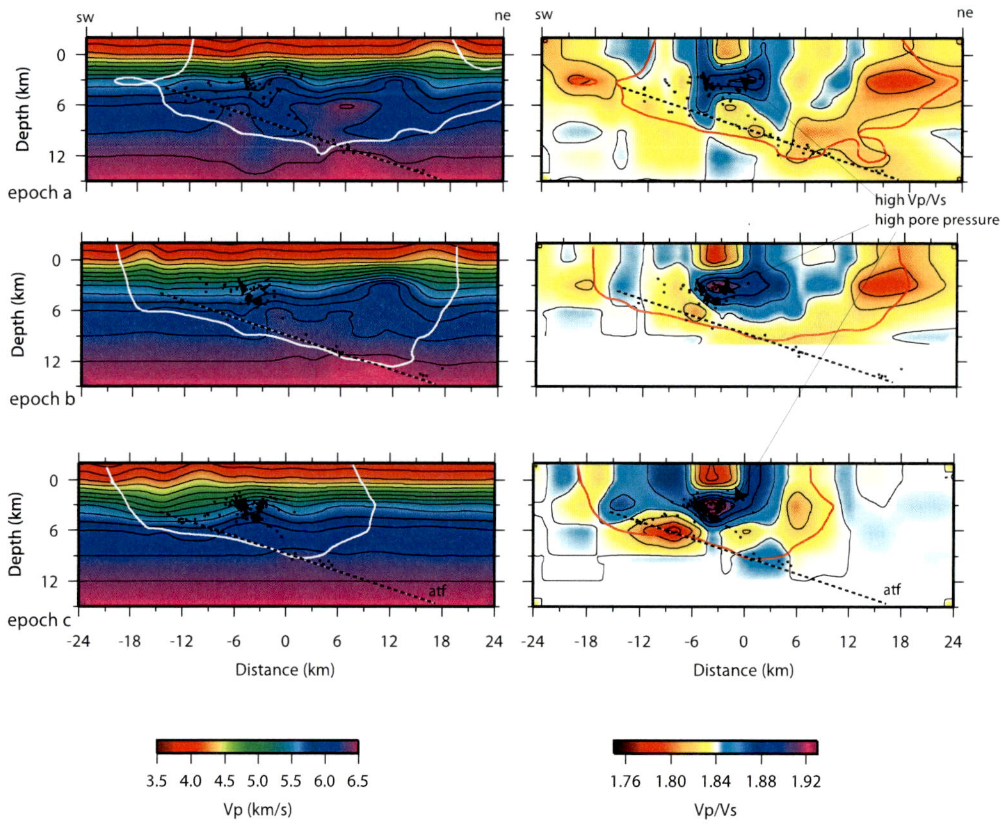

Figure 5. Vertical sections of the Vp (left) and Vp/Vs (right) models across the ATf in the three epochs. The white and red lines indicate the similarly resolved model volumes for Vp and Vp/Vs, respectively. Note the sharp high Vp/Vs increase during epoch c in the volume within which the seismicity on the two antithetic normal faults increased.

A vertical section across the normal faulting system shows a good example of two very shallow antithetic faults describing a small graben (figure 5). During the three epochs, the Vp/Vs ratio varied of about 4-5 % within the rock volume confining the graben structure, presenting a sharp increase in period c, where the seismic activity on the faults increased. The second low Vp/Vs anomaly is located along the ATf, and features a time variation of the values of about 3-4 % with a maximum reduction during epoch 3. Although the seismicity on the ATf is present in all the three epochs (see also Chiaraluce et al. 2007), it is more clustered around the low Vp/Vs anomaly during epoch c.

On the bottom of figure 4, the Vp and Vp/Vs results obtained by using data of the entire period (those published in Moretti M, De Gori P. and Chiarabba C.) are shown. It is evident that this latter inversion incorporates features present in the three epochs. The overall model is a volumetric average of all the features present separately at different times during the experiment.

The Alban Hills Quaternary Volcano

The Alban Hills is a Quaternary volcano located close to Rome and its last eruption dates back to 27 Ka. Retained for long time as extinct, studies in past years have revealed that unrest episodes occur frequently at the volcano (Chiarabba et al., 1997). The last swarm, which occurred in 1989-90, lasted for almost a year and experienced five earthquakes with M_L larger than 3.9 (Amato et al., 1994). This sequence has been monitored with a local seismic network, allowing for the detection of thousands of earthquakes and the computation of earthquake parameters for at least a 1000 events. Centred beneath the youngest volcanic centres, the seismic swarm ruptured a 2-3 km thick limestone layer. The seismicity is associated with pronounced ground uplift (Amato and Chiarabba 1995; Chiarabba et al., 1997). Using arrival times from the best subset of earthquakes, Feuillet et al. (2004) found a high Vp/Vs anomaly in the volume where the source of inflation is located, suggesting that fluid overpressure at 3-4 km depth is the driving mechanism of volcanic unrest (see figure 6).

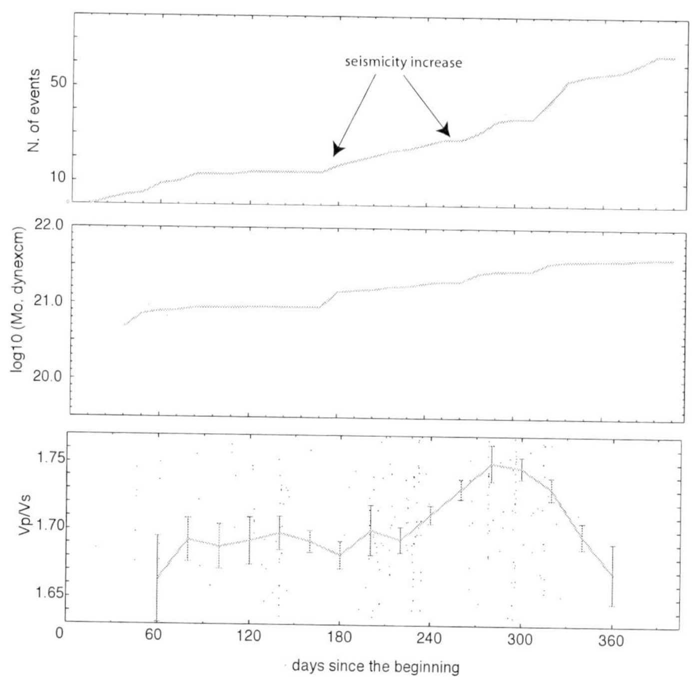

Figure 6. Vp/Vs variations (bottom), cumulative seismic moment (middle) and number of events (top) during the 1989-90 Alban Hills seismic swarm. Note the sharp Vp/Vs increase in the second part of the swarm coinciding with the increase of seismic activity.

While the sequence lasted for about one year, the seismicity floated and experienced some renewal, the most significant occurring in mid-April 1990 (day 220).

We used 2978 P- and 2749 S-wave arrival times from 372 events to derive the Vp/Vs versus time plot. The length of the moving window is 40 days, with an overlap of 20 days. We observe a sharp increase of the Vp/Vs ratio (from 1.69 to 1.75) starting from day 220. This feature coincides with the increase in the number of events and seismic moment release (figure 6).

Although the limited number of data prevents the possibility to compute resolved 4D Vp and Vp/Vs images, we hypothesize that the sharp temporal increase of Vp/Vs, occurred within the small central high Vp/Vs volume and is indirectly testifying pore pressure pulses in the inflation source volume, driving the seismic swarm.

The Mt. Etna Volcano

The frequent volcanic phenomena makes the Mt. Etna a natural laboratory to investigate time variations of physical parameters and to verify whether seismic data is able to outline the on-going processes. The abundance of earthquakes and data allowed the computation of velocity and attenuation models at different scales and for several recent eruptions (Chiarabba et al., 2000; Laigle et al., 2000; Patanè et al, 2002; De Gori et al., 2005, among many others). Recently, Patané et al. (2006) documented that the 2002-2003 explosive eruption has been accompanied by a transient low Vp/Vs anomaly located around the intruding dike. This anomaly, evident only during the eruption, has been attributed to gas-enriched cracks in the volume around the intruding dyke.

The presence of transient anomalies reconciles the difference in the Vp/Vs patterns revealed by previously published 3D models of the volcano plumbing system (Laigle et al., 2000; Patanè et al., 2002). Both models, computed by averaging data scanning different periods of volcanic activity, identify different patterns and locations of the positive and negative Vp/Vs anomalies. In a volcano which experiences frequent magma injection in the shallow crust, we are not surprised by the recognition of significant modification in the structure.

During the last 8 years, the monitoring of the seismic network has improved dramatically, promoting Mt. Etna as one of the best volcanoes where 4D studies can test their reliability. Here, we try to see if the attenuation of P-waves stores such variations too. At a large scale, the 3D Qp structure consists of a deep high Qp solidified body and a shallow low Qp volume, connected at depth by a low Qp conduit (De Gori et al., 2005). Shallow and local details of the attenuation have been revealed by studies that used subset of data (Martinez-Arevalo et al., 2005).

We analysed the same period during which the transient Vp/Vs anomaly was observed (the 2002-2003 eruption) and computed 1548 tp* values for a set of 300 earthquakes. The Qp values are estimated as described above and averaged on a moving window of 1 day with 12 hours of overlap. We observe a continuous decrease of the Qp after the dyke intrusion, coincident with the onset of the seismic swarm, and before the eruption (figure 7). The inversion for the Qp structure shows a low Qp anomaly centred in the same volume of the transient low Vp/Vs anomaly. The top panel of figure 7 shows the moving average of the Vp/Vs variations, averaged every 20 days with an overlap of 10 days. The low Qp anomaly occurs in the same period of the transient low Vp/Vs anomaly observed by Patanè et al. (2006).

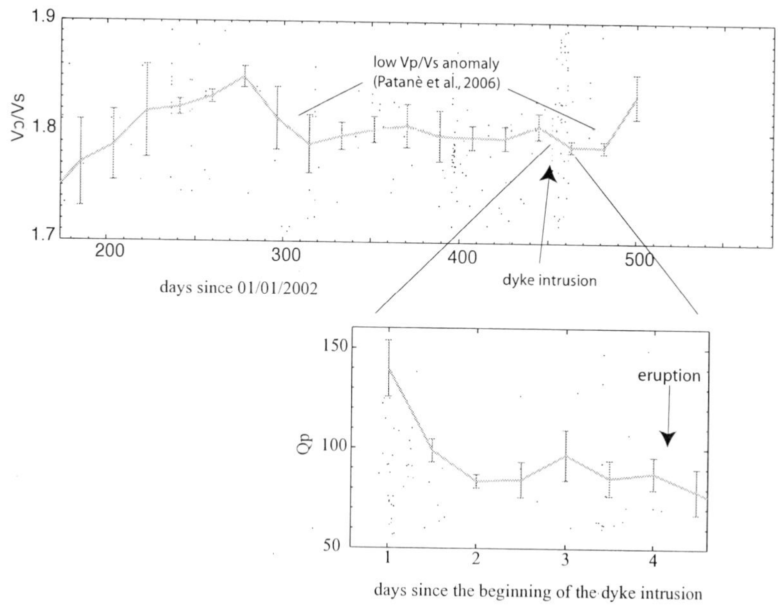

Figure 7. Vp/Vs (top) and Qp variations (bottom) during the 2002-2003 Mt. Etna eruption and its preparatory process. The decrease of low Vp/Vs is coincident with the transient low Vp/Vs anomaly observed by Patané et al. (2006). The Qp variation, note the progressive decrease, is relative to the small interval of days after the dyke intrusion.

Interpretation of Velocity and Attenuation Anomalies

Seismic velocities are an indirect measure of rock deformability, conditioned by in situ pressure, temperature, porosity, fluid saturation and pressure (Nur, 1972; Ito et al., 1979; Christensen, 1985; Wang and Nur, 1989; Watanabe, 1993; Mavko et al., 1998). Although laboratory experiments identify how the different factors influence the velocity and attenuation of rocks, the intrinsic difference between the laboratory and the in situ conditions still results in a not univocal transfer of results. In particular, fault zones are complex. The fault core and the damage zone (see Caine et al., 1996) may have a different architecture and degree of maturity, and possess different petrophysical properties, porosity and permeability (Agosta and Kirshner, 2003; Agosta et al., 2007). Fault core and damage zones, with typical dimensions around tens to hundreds of meters, act as seals or conduits for fluid flow. The velocity models computed by local earthquake tomography, with a spatial resolution of 1 to 3 km, are estimated over large volumes around such complex faults. Typically, we acquire smooth averages of rock physical properties around active faults.

The variation of bulk and shear modulus severely depends on the content and physical state of fluids, promoting the Vp/Vs ratio as a valid candidate to reveal saturated rock volumes (Dvorkin et al., 1999). The shear wave is mostly affected by bulk density change, the shear modulus being less affected by pore fluid substitution. Assuming the crack theory by O'Connel and Budianski (1974), we can estimate elastic moduli and fluid saturation from Vp and Vs. P- and S-wave velocity varies significantly as a function of the rock saturation. In general, wet rocks exhibit P-wave velocities higher than dry rocks and smaller effects on the Vs, resulting in high Vp/Vs ratio, accompany the increase of Vp with the saturation. This is due to the different compressibility of rocks when brines or gas occupy the pores. Moreover, fluid pressure acts inversely to the lithostatic pressure, preventing the increase of effective stress and reducing both P- and S-wave velocities. For flat pores, we expect that the Vp/Vs ratio decreases or increases depending on whether the fluid type is gas or brine, respectively (Dvorkin et al., 1999).

For the examples shown here, our target volume consists of the Meso-Cenozoic limestone rocks from the Adria continental margin. Since carbonates are one of the main oil reservoir rocks, several studies focused on the definition of their petro-physical and elastic properties (Anselmetti and Eberli, 1993; Wang, 1997, Assefa et al., 2003). The seismic properties of carbonates are influenced by many parameters, including pore shape and type and pore fluid and saturation (see Wang, 1997). The effect of fluid saturation on carbonate rocks velocity depends on whether the pore types are round or flat, being larger in the latter case due to the higher compressibility of flat pores. The importance of pore geometry has been clearly addressed by Takei (2002). Due to the increase of pressure with depth, the flat pores rapidly close, while round pores close slowly. Therefore, we should expect different values of the Vp/Vs ratio depending on the pore shape and aspect ratio (Kato et al., 2005). For rocks with very low aspect ratio pores, such as cracks, the velocities measured in laboratory is a better proxy of the velocity at seismic frequencies (Wang, 2001).

For the Triassic carbonate rocks of central Italy, Trippetta et al. (2007) found a porosity less than 2-3 % and a P–wave velocity of about 6.6 or 7.0 km/s, in dry or wet conditions, at confining pressure compatible with the burial depth. These Triassic dolostones and anhydrites form a more or less thick layer at the base of the younger carbonate succession, which has P-wave velocities smaller than 6.4 km/s (Agosta et al., 2007). The measurement of Vp/Vs for carbonate rocks in a range of effective pressure and temperature replicating the in situ conditions is programmed in the near future at our high P/T laboratory and will give tighter constraints to the physical meaning of our results.

The attenuation of body waves are generated by a viscous dissipation of the energy. In regions with low geothermal gradient, like the central Apennines and the Alban Hills volcano, the attenuation of body waves is mostly affected by lithology, rock permeability and the presence and state of the fluids. The role of fluids in controlling seismic wave attenuation has been explored in several studies. Pore pressure variations, in partially saturated rocks, generate an oscillatory liquid flow during which a viscous loss of energy occurs (Dvorkin and Mavko, 2006). Winkler and Nur (1979) reported strong variations of the P-wave attenuation close to the saturation point. The attenuation increases with the saturation of rocks until it drops at the full saturation point. Klimentos (1995) shows that the presence of gas in sandstones increases the P-wave attenuation but has negligible effect on the S waves attenuation. Conversely, in brine saturated conditions, the S-wave attenuation is higher than or equal to the P-wave attenuation. Like the Vp/Vs ratio, it is the differential attenuation of P- and S- waves which better describes the fluid saturation and physical state of fluids in the crust (Toksoz et al., 1979). Laboratory data show that low Q_P/Q_S ratio values ($R < 1$) are related to a partially saturated condition of the rock volume. Conversely, $Qp/Qs > 1$ might indicate a complete fluid saturation, or dry rock, both alternative features similarly explain the large reduction of shear rather than bulk attenuation. Considering that lithology cannot account for temporal changes of Qp or Qs since the sampled volume is the same, the observed Qs variations indicate that rock permeability and/or fluid saturation changed along the fault.

In igneous rocks resembling the Mt. Etna situation, a decrease of the Vp/Vs ratio is observed as the temperature increases (Sanders et al. 1995). This is at the base of the interpretation by Patanè et al. (2006) of the transient low Vp/Vs anomaly. The reduction of pore pressure and the increase in compressibility associated with the depletion of brines and substitution with gas, led to a decrease of Vp and an increase of Vs, resulting in low Vp/Vs anomalies (see Gunakasera et al., 2003, Vanorio et al., 2005). In high geothermal regions, the main variation of the body wave attenuation is certainly related to the increase of temperature, with the most dramatic effect close to the rock melting point (Mavko et al., 1980, Mizutani and Kanamori, 1988). Therefore, in the Mt. Etna example the strong Qp reduction (high attenuation) observed during the 2002-2003 eruptions indicates local increases of the volcanoes thermal state due to the injection of new magma in the shallow crust.

To summarise, high Vp, high Vp/Vs and low Qp/Qs anomalies suggest a strong but not complete fluid (liquid) saturation of the rock volume, while the opposite features are observable if the pores are occupied by gas (or in dry rocks). In the wet condition, and normal geothermal gradients, typical of the Apennines upper crust, a Vp reduction and a Vp/Vs increase indicate fluid overpressure even in case of incomplete saturation (low Qp/Qs). Conversely, in volcanic rocks a Vp/Vs decrease may indicate the presence of gas overpressure, as also documented at the Phlegrean Fields volcano by Vanorio et al. (2005).

What We Learn about the Lithosphere

We have shown that sudden variations of elastic and anelastic parameters occur in the crust, as a response to rock fracturing and pore pressure variations around dyke intrusion or along active faults. Our discovery indicates that the shallow part of the continental lithosphere is breathing with an irregular pattern, and a priori is not known, but resolvable by seismological data collected with dense seismic networks. In the reported examples, the variations occur at the scale of a few days or months, the smaller the periods the more pronounced the physical phenomena, i.e. large earthquakes or magma intrusion. Such periods are entirely similar to those that encompass the seismic data used to derive 3D velocity and attenuation models of the lithosphere. Thus, we cannot rule out that these models are biased by transient variations of the material properties. We can at first exclude that such variations are significant for the entire lithosphere, but not for the crust and in regions of active faulting or volcanism.

Due to the intrinsic nature of tomography, the spatio-temporal average of physical properties produces an image that is a melange of both transient and static features. As a result, the interpretation of velocity anomalies in terms of stationary features of faults or crust volumes may not be correct. The ATf case study can clarify this point. The 3D velocity model shows a clear high Vp/Vs anomaly on the fault (figures 4 and 5). The 4D tomography indicates that the high Vp/Vs anomaly appears with different shapes and values during the three epochs. The velocity image obtained using data of the entire period presents a mix of all the features attributable at each single epoch. Considering the temporal changes, we should interpret the anomaly as outstanding evidence for high pore pressure pulses on the fault, rather than a volume of high pore pressure stationary present on the fault (see also Moretti M,

De Gori P. and Chiarabba C.). This hypothesis may be of crucial importance while recognizing those fault patches which have a characteristic behavior during faulting (like in the asperity or barrier models, see Das and Aki, 1977; Das and Kostrov, 1983). The earthquake occurrence may be closely related to in situ material properties and heterogeneities, rather than stress conditions. Such properties may vary even rapidly; we documented variations occurring over a few days (like in the Umbria-Marche or Mt. Etna sequences) or a few months (like over the ATf).

The close relationship between the variation of velocity anomalies and seismicity strengthens the role of fluids in driving the earthquakes occurrence during episodic fluid overpressure pulses.

The high Vp/Vs anomaly recognized along the normal fault segments, suggests that the fluid pressure plays a dominant role in moulding the seismicity. In literature, the examples of high Vp/Vs volumes in the source region of earthquakes are widely reported (Lees, 1990; Eberhart-Phillips and Michael, 1993; Gupta et al., 1996; Zhao et al., 1996; Zhao and Negishi, 1996; Mishra and Zhao, 2003). Crack density, saturation rate and porosity, derived from Vp and Vs, all indicate the presence of localized micro-cracks and fluid at the earthquake hypocenter. High fluid pressure on faults reduces the frictional strength and its time variations may control the occurrence of earthquakes. In this study, we outline that it is possible to reveal the high pore pressure pulses invoked in fluid driven seismicity models (Johnson and McEvilly, 1995) with the 4D seismic tomography.

We interpret the transient Vp/Vs anomaly variations as rock dilatancy (generation of microcracks that produce a Vp/Vs decrease) followed by a sudden pore pressure pulse (Vp/Vs increase) on the fault. The contemporaneous reduction of the Qp/Qs observed in the Umbria-Marche seismic sequence suggests that the fluid pressure increases along with the fluid saturation of the rock volume. These results offer new evidence for the validity of the earthquake occurrence model proposed in the early '70 (Scholz et al., 1973) and are fully consistent with the model proposed by Miller et al. (2003) to explain the seismic sequence as driven by CO_2 overpressure.

Part of this lesson can be generalized to the lithosphere. The crust is indeed the part of the earth's interior for which we can finely explain the structure. Based on our knowledge from the surface down to greater depths, there is an almost continuous reduction in the resolution of geophysical models and a progressive defocusing of the structure imaging. This is in part related to the decrease of the elastic wave frequency used to sound different depths of the earth, but also due to the progressive complication of seismic waves travelling heterogeneous media. Nevertheless, we believe that the variation of elastic and anelastic parameters that we documented for the crust may interest the entire lithosphere at a large-scale even if at longer periods. All the processes that include the mineral dehydration, like the expulsion of fluids from evaporitic rocks or the serpetinization of the subducted lithosphere (Dobson et al., 2002; Hyndman and Peacock, 2003; Abers et al., 2006; Brudzinski et al., 2007), and the progressive melting of rocks and magma ascent may generate transient variation of physical parameters. The pressure fluctuations of slab-released fluids have been proposed as a possible mechanism to modulate the generation of episodic tremor and slip (ETS, Rogers and Dragert, 2003). Deep crustal volumes with high Vp/Vs are recognized at the hypocenter of ETS. These volumes may also be characterized by transient pulse of pore pressure. The extension to the entire lithosphere of 4D studies is obviously a long term scenario, being represented in present day, for data and methodologies, to the development of 3D models of subduction zones (Salah and Zhao, 2003; Schurr et al., 2003, Shelly et al., 2006; Chiarabba et al., 2008, among many others). As long as our capability to collect and analyze huge amount of data increases, we will be able to look at the lithosphere as a breathing portion of the earth.

Conclusion

We show clear examples of time changes of velocity and attenuation in active faults and volcanoes in central Italy. At present, a model that explains the cause for these transient anomalies still needs ad hoc laboratory tests able to replicate at the best the natural environment (lithology, effective pressure, temperature and fluid saturation). The most promising effect seems to be the local variation of fluid saturation and pressure. This transient variation may drive the occurrence of large earthquakes or sudden increases of background seismicity. Since seismological centres are accumulating more and more high quality data, the development of 4D studies will became feasible in the near future, allowing us to recognize or, possibly, forecast impending phenomena.

References

Abers, G. A., van Keken, P. E., Kneller, E. A., Ferris A., Stachnik J. C. (2006). The termal structure of subduction zones constrained by seismic imaging: Implications for slab dehydration and wedge flow, *Earth Planet. Sci. Lett.*, **241**, 387-397.

Aggarwal, Y. P., Sykes, L. R., Armbruster, J., Sbar, M. L. (1973). Premonitory changes in seismic velocities and prediction of earthquakes, *Nature*, **241**, 101.

Agosta, F., Kirshner, D. L. (2003). Fluid conduits in carbonate-hosted seismogenic normal faults of central Italy, *J. Geophys. Res.,* **108**, doi:10.1029/2002JB002013.

Agosta, F., Prasad, M., Aydin, A. (2007). Physical properties of carbonate fault rocks, fucino basin (Central Italy): implications for fault seal in platform carbonates *Geofluids* **7** (1) , 19–32, doi:10.1111/j.1468-8123.2006.00158.x.

Amato, A., Chiarabba, C., Cocco, M., Di Bona, M., Selvaggi, G. (1994). The 1989-1990 seismic swarm in the Alban Hills volcanic area, Central Italy, *Journal of Volcanology and Geothermal Res.*, **61**, 225-237.

Amato, A., Chiarabba, C. (1995). Recent uplift of the Alban Hills Volcano (Italy): evidence for a magmatic inflation? *Geophys. Res. Lett.,* **22**, 1985-1988.

Anselmetti, F. S., Eberli, G. P. (1993). Controls on sonic velocity in carbonates, *Pure and Applied Geophysics*, **141**, 287-323.

Antonioli, A., Piccinini, D., Chiaraluce, L., Cocco, M. (2005). Fluid flow and seismicity pattern: evidence from the 1997 Umbria-Marche (central Italy) seismic sequence. *Geophys. Res. Lett.,* **32**, doi:10.1029/2004GL022256.

Assefa, S., McCann, S., Sothcott, J. (2003).Velocities of compressional and shear wave in limestones, *Geophysical Prospecting*, **51**, 1-13.

Aster, R. C., Borchers, B., Thurber, C. H. (2005). Parameter estimation and inverse problems, 296 pp., Elsevier Academic Press, London, UK.

Bijward H., Spakman, W., Engdahl, E. R. (1998). Closing the gap between regional and global travel time tomography. *J. Geophys. Res.*, **103**, 30055-30078.

Brudzinsky, M.R., Thurber C. H., Hacker B. R., Engdahl E. R. (2007). Global prevalence of double Benioff zones. *Science*, **316**, 1472-1474.

Brune, J. N. (1970). Tectonic stress and the spectra of seismic shear waves from earthquakes. *Bull. Seism. Soc. Am.*, **60**, 4997-5009.

Caine, J. S., Evans, J. P., Forster C. B. (1996). Fault zone architecture and permeability structure. *Geology*, **24**, 11, 1025-1028.

Chiarabba, C., Amato, A., Delaney, P. T. (1997). Crustal structure, evolution, and volcanic unrest of the Alban Hills, central Italy, *Bull. of Volcanol.*, **59**, 161-170.

Chiarabba, C., Amato, A., Boschi, E., Barberi, F. (2000). Recent seismicity and tomographic modeling of the Mount Etna plumbing system. *J. Geophys. Res.*, **105**, 10,923-10,938.

Chiarabba, C., De Gori P., Boschi E., Pore pressure migration along a normal fault system resolved by time repeated seismic tomography, *Geology*, v. 37:, no. 1, 67-70, doi:10.1130/G25220A.1

Chiaraluce, L., Amato A., Cocco, M., Chiarabba, C., Selvaggi, G., Di Bona, M., Piccinini, D., Deschamps, A., Margheriti, L., Corboulex, F., Ripepe, M. (2004). Complex normal faulting in the Apennines thrust-and-fold belt: the 1997 seismic sequence in central Italy, *Bull. Seismol. Soc. Am.*, **94**, 99-116.

Chiaraluce, L., Chiarabba, C., Collettini, C., Piccinini, D., Cocco, M. (2007). Architecture and mechanics of an active low-angle normal fault: Alto Tiberina Fault, northern Apennines, Italy. *J. Geophys. Res.*, 112, B10310, doi:10.1029/2007JB005015.

Christensen, N. I. (1985). Measurements of dynamic properties of rocks at elevated temperatures and pressures. In Pincus, H.J., and Hoskins, E.R. (Eds.), Measurements of Rock Properties at Elevated Pressures and Temperatures. *Spec. Tech. Publ.—ASTM,* **869**:93–107.

Cox, S.F. (1995). Faulting processes at high fluid pressures: An example of fault valve behavior from the Wattle Gully Fault, Victoria, Australia, *J. Geophys. Res.*, **100**, 12841-12859.

Das, S., Aki, K. (1977). Fault plane with barriers: A versatile earthquake model, *J. Geophys. Res.*, **82**, 5658-5670.

Das, S., and Kostrov, B.V. (1983). Breaking of a single asperity: rupture process and seismic radiation, *J. Geophys. Res.*, **88**, 4277-4288.

De Gori, P., Chiarabba, C., Patanè, D. (2005). Qp structure of Mount Etna: constraints for the physics of the plumbing system. *J. Geophys. Res.,* **110**, B05303, 10.1029/2003JB002875.

Di Stefano, R., Kissling, E., Alderson, F., Baccheschi, P., Chiarabba, C. (2006). Automatic seismic phase picking and consistent observation error assessment: application to Italian Seismicity, *Geophys. J. Int*, **165**(1), 121-134.

Dobson, D.P, Meredith, P.G., Boon, S.A. (2002). Simulation of subduction zone seismicity by dehydration of Serpentine, *Science*, **298**, 1407-1409.

Dvorkin, J., Mavko, G., and Nur, A. (1999). Overpressure detection from compressional and shear wave data, *Geophys. Res. Lett.*, **26**, 3417-3420.

Dvorkin J.P., and Mavko, G. (2006). Modeling attenuation in reservoir and noreservoir rock, *The Leading Edge*, **25**, 194-197.

Eberhart-Phillips, D., Michael, A.J. (1993) Three-dimensional velocity structure, seismicity and fault structure in the Parkfield region, central California, *J. Geophys. Res.* **98**, 15737-15758.

Eberhart-Phillips, D., Reyners, M. (1999). Plate interface properties in the northeast Hikurangi subduction zone, New Zealand, from converted seismic waves. *Geophys. Res. Lett.*, **26**(16), 2565-2568, 10.1029/1999GL900567.

Eberhart-Phillips, D., Chadwick, M., (2002). Three-dimensional attenuation model of the shallow Hikurangi subduction zone in the Raukumara Peninsula, New Zealand. *J. Geophys. Res.*, **107**(B2), 2033, doi:10.1029/2000JB000046.

Feuillet, N., Nostro, C., Chiarabba, C., and Cocco, M. (2004). Coupling between earthquake swarms and volcanic unrest at the Alban Hills Volcano (central Italy) modeled through elastic stress Transfer, *J. Geophys. Res.*, **109**, doi:10.1029/2003JB002419.

Foulger, G. R., Julian, B. R., Pitt, A. M., Hill, D. P., Malin, P. E., Shalev, E. (2003). Three-dimensional crustal structure of Long Valley caldera, California, and evidence for the migration of CO2 under Mammoth Mountain. *J. Geophys. Res.*, **108**, 10.1029/2000JB000041.

Grand, S.P., van der Hilst, R.D., and Widiyantoro, S. (1997). Global seismic tomography: A snapshot of convection in the Earth, *GSA Today*, **7**, 1-7.

Gunasekera, R.C., Foulger, G.R., and Julian, B.R. (2003). Reservoir depletion at The Geysers geothermal area, California, shown by four-dimensional seismic tomography, *J. Geophys. Res.*, **108**, NO. B3, 2134, doi:10.1029/2001JB000638.

Gupta, I.N. (1973). Dilatancy and premonitory variations of P, S travel times, *Bull. Seismol. Soc. Am.*, **63**, 1157-1161.

Gupta, H.K., Sarma, S.V., Harinarayana, T., Virupakshi, G. (1996). Fluids below the hypocentral region of Latur earthquake, India: Geophysical indicators, , *Geophys. Res. Lett.*, **23**, 1569-1572.

Hauksson, E., Shearer, P.M. (2006). Attenuation models (Qp and Qs) in three dimensions of the southern California crust: Inferred fluid saturation at seismogenic depths. *J. Geophys.* Res., **111**, B05302, doi:10.1029/2005JB003947.

Hyndman, R. D., Peacock, S.M. (2003). Serpentinization of the forearc mantle, *Earth Planet. Sci. Lett.*, **212**, 417-432.

Ito, H., De Vibriss, J., Nur, A. (1979). Compressional and shear waves in saturated rock during water-steam transition. *J. Geophys. Res.*, **84**, 4731-4735.

Johnson, P.A., McEvilly, T.V. (1995). Parkfield seismicity: Fluid driven?, *J. Geophys. Res.*, 100, 12937-12950.

Kayal, J.R:, D. Zhao, O.P. Mishra, R. De, O.P., Singh, The 2001 Bhuj earthquake: Tomographic evidence for fluids at the hypocenter and its implications for rupture nucleation, *Geophys. Res. Lett.*, **29**, 10.1029/2002GL015177, 2002.

Kato, A., Kurashimo, E., Hirata, N., Sakai, S., Iwasaki, T., and Kanazawa, T. (2005). Imaging the source region of the 2004 mid-Niigata prefecture earthquake and the evolution of a seismogenic thrust-related fold, *Geophys. Res. Lett.*, **32**, doi:10.1029/2005GL022366.

Klimentos, T., (1995). Attenuation of P- and S-waves as a method of distinguishing gas and condensate from oil and water. *Geophysics*, **60**, 447–458.

Laigle, M., Hirn, H., Sapin, M., and Lepine, J.C. (2000). Mount Etna dense array local earthquake P and S tomography and implications for volcanic plumbing, *J. Geophys. Res.*, **105**, 21,633-21,646.

Lees, J.M. (1990). Tomographic P-wave velocity images of the Loma Prieta earthquake asperity, *Geophys.Res. Lett.*, **17**, 1433-1436.

Lees, J.M., Wu, H. (2000). Poisson's ratio and porosity at Coso geothermal area, California, *J. Volcanol. Geotherm. Res.*, **95**, 157-173.

Martinez-Arevalo, C., Patané, D., Rietbrock, A., and Ibanez, J.M.(2005). The intrusive process leading to the Mt. Etna 2001 flank eruption: Costraints from 3D attenuation tomography, *Geophys. Res. Lett.*, **32**, doi:10.1029/2005GL023736.

Mavko, G.M. (1980). Velocity and attenuation in partially molten rocks, *J. Geophys. Res.*, **85**, 5173-5189.

Mavko, G., Mukerji, T., Dvorkin, J., (1998). *The Rock Physics Handbook*, **329** pp., Cambridge University Press, Cambridge.

Menke, W. (1989). *Geophysical Data Analysis: Discrete Inverse Theory*. Academic Press, 289 pp.

Miller, S.A. (1996). Fluid-mediated influence of adjacent thrusting on the seismic cycle at Parkfield, *Nature*, **382**, 799-802.

Mishra, O.P., and Zhao, D. (2003). Crack density, saturation rate and porosity at the 2001 Bhuj, India, earthquake hypocenter: a fluid driven earthquake?, *Eart Planet. Sci. Lett.*, **212**, 393-405.

Mizutani, H., Kanamori, H., (1988). Variation of elasic wave velocity and attenuation property near the melting temperature. *J. Phys. Earth* **12**, 43-49.

Moretti M, De Gori P. and Chiarabba C. "Earthquake relocation and three-dimensional Vp and Vp/Vs models along the low angle Alto Tiberina fault (Central Italy): Evidence for fluid overpressure", Geophys. J. Int., doi: 10.1111/j.1365-246X.2008.03984.x

Nostro C., Chiaraluce, L., Cocco, M., Baumont, D., Scotti, O. (2004). Coulomb stress changes caused by repeated normal faulting earthquakes during the 1997 Umbria-Marche (central Italy) seismic sequence, *J. Geophys. Res.*, **110**, doi:10.1029/2004JB003386.

Nur, A. (1972). Dilatancy, pore fluids, and premonitory motions it ts/tp travel times, *Bull. Seismol. Soc. Amer.*, **62**, 1217-1222.

Nur, A. & Booker, J.R. (1972). Aftershocks caused by pore fluid? *Science*, **175**, 885-887.

O'Connell, R. J., Budiansky, B. (1974). Seismic velocity in dry and saturated cracked solids. *J. Geophys. Res.*, **79**, 5412-5426.

Patané, D., Chiarabba, C., Cocina, O., De Gori, P., Moretti, M., and Boschi, E. (2002). Tomographic images and 3D earthquake locations of the seismic swarm preceding the 2001 Mt. Etna eruption: Evidence for a dyke intrusion, *Geophys. Res. Lett.*, **29**, 10, 10.1029/2001GL014391.

Patanè, D., Barberi, G., Cocina, O., De Gori, P., Chiarabba, C. (2006). Time resolved seismic tomography detects magma intrusions at Mt. Etna, *Science*, **313**, 11, 821-823.

Prasad, M., Manghanani, M.H. (1997). Effects of pore and differential pressure on compressional wave velocity and quality factor in Bera and Michigan sandstones, *Geophysics*, **62**, 1163-1176.

Rietbrock, A. (2001). P wave attenuation structure in the fault area of the 1995 Kobe earthquake. *J. Geophys. Res.*, **106**, 4141-4154.

Rowe, C.A, Aster, R.C., Phillips, W.S., Jones, R.H., Borchers, B., and Fehler, M.C. (2002). Relocation of induced microearthquakes at the Soultz geothermal reservoir using automated, high precision picking, *Pure Appl. Geophys.*, **159**, 563-596.

Salah, M.K., Zhao, D. (2003). 3-D seismic structure of Kii Peninsula in southwest Japan: evidence for slab dehydration in the forearc, *Tectonophysics*, **364**,191-213.

Sanders, C.O., Ponko, S.C., Nixon, L.D., Schwartz, E.A. (1995). Seismological evidence for magmatic and hydrothermal structure in Long Valley caldera from local earthquake attenuation and velocity tomography, *J. Geophys. Res*, **100**, 8311-8326.

Scherbaum, F. (1990). Combined inversion for the three-dimensional Q structure and source parameters using microearthquake spectra. *J. Geophys. Res.*, **95**, 12423-12438.

Scholz C.H., Sykes, L.R., Aggarwal, Y.P. (1972). *Earthquake prediction: A physical basis, Science*, **181**, 803-809.

Schurr, B., Asch, G., Rietbrock, A., Trumbull, R., Haberland, C. (2003). Complex patterns of luid and melt transport in the central Andean subduction zone revealed by attenuation tomography, *Earth Planet. Sci. Lett.* **215**, 105-119.

Segall, P & Rice, J.R. (1995). Dilatancy, compaction and slip instability of a fluid infiltrated fault, , *J. Geophys. Res.*, **100**, 22155-22171.

Shelly, D.R., Beroza, G.C. Zhang, H., Thurber, C.H., and Ide, S. (2006). High resolution subduction zone seismicity and velocity structure beneath Ibaraki Prefecture, Japan, *J. Geophys. Res.*, **111**, doi:10.1029/2005JB004081.

Sibson, R.H. (1992). Fault valve behavior and the hydrostatic-lithostatic fluid pressure interface, *Earth Sci. Rev.*, **32**, 141-144.

Takei, Y. (2002). Effect of pore geometry on Vp/Vs: From equilibrium geometry to crack, *J. Geophys. Res.*, **107**, no. B2, 2043, doi: 10.1029/2001JB000522.

Thurber, C.H. (1983). Earthquake locations and three-dimensional crustal structure in the Coyote Lake area, Central California. *J. Geophys. Res.*, **88**, B10, 8226-8236.

Thurber, C. H., Roecker, S., Zhang, H., Baher, S., and Ellsworth, W.L. (2004). Fine-scale structure of the San Andreas zone and location of the SAFOD target earthquakes, *Geophys. Res. Lett.*, **31**, doi:10.1029/2003GL01398.

Toksoz, M.N., Johnston, D.H., and Timur, A. (1979). Attenuation of seismic waves in dry and saturated rocks:I, Laboratory measurements, *Geophysics*, **44**, 681-690.

Trippetta, F., Vinciguerra, S., Collettini, C., Meredith, P.G. (2008). Physical properties of the seismogenic Triassic evaporates in the northern Apennines (central Italy), *Geophysical Research Abstract*, EGU2008-A-09990.

Um, J., and Thurber, C.H. (1987). A fast algorithm for two-point seismic ray tracing, Bull. *Seismol. Soc. Am.*, **77**, 972-986.

Vanorio, T., Virieux, J., Capuano, P., Russo, G. (2005). Three-dimensional seismic tomography from P wave and S wave microearthquake travel times and rock physics characterization of the Campi Flegrei caldera, *J. Geophys. Res.*, **110**, doi:10.1029/2004JB003102.

Waldhauser, F., and Ellsworth, W.L. (2000). A double-difference earthquake location algorithm: Method and application to the northern Hayward fault, CA, Bull. *Seismol. Soc. Am.*, **90**, 1353-1368.

Wang, Z. (1997). Seismic properties of carbonate rocks, *in* Palaz, I., and Marfurt, K.J., eds., *Geophysical Developments Series*, no. **6** - Carbonate Seismology: Society of Exploration Geophysicists, p. 29-52.

Wang, Z. (2001). Fundamentals of seismic rock physics, *Geophysics*, **66**, 398-412.

Wang, Z., Nur, A.M. (1989). Seismic and Acoustic Velocities in Reservoir Rocks, Volume 2, *Theoretical and Model Studies: SEG Geophysics reprint series*, No. **10**, p. 457.

Watanabe, T. (1993). Effects of water and melt on seismic velocities and their application to characterization of seismic reflectors, *Geophys. Res. Lett.*, **20**, 2933-2936.

Winkler, K., Nur, A. (1979). Pore fluids and seismic attenuation in rocks, *Geophys. Res. Lett.*, **6**,1–4.

Zhao, D., Negishi, H. (1998). The 1995 Kobe earthquake: seismic image of the source region and its implications for the rupture nucleation, *J. Geophys. Res.*, **103**, 9967-9986.

Zhao, D., Kanamori, H., Negishi, H. (1996).Tomography of the source area of the 1995 Kobe earthquake: evidence for fluids at the hypocenter?, *Science*, **274**, 1891-1894.

In: The Lithosphere: Geochemistry, Geology and Geophysics ISBN 978-1-60456-903-2
Editors: J.E. Anderson et al, pp. 317-344

Chapter 9

The Magnetic Lithosphere: A Novel View

Mioara Mandea and Vincent Lesur
Helmholtz-Zentrum Potsdam, Deutsches GeoForschungsZentrum
Telegrafenberg, D–14473 Potsdam, Germany

Abstract

At the Earth's surface the lithospheric magnetic field is some two-three orders of magnitude smaller than that of the core's contribution. It is produced by magnetic material located within a thin layer (10 km–70 km thick) of the crust and upper mantle. This source of magnetic field is thus carried by the magnetic lithosphere, whose definition is slightly different than that used in the fields of geochemistry, seismology or geology, although in the end these definitions all converge. The Earth's lithosphere is generated by both induced and remanent magnetizations that depend on the chemical composition and crystal conformation of subsurface rocks as well as on the Earth's core field history. Addressing the question of which type of magnetization takes over at each depth would, in particular, partially unveil the nature of the upper mantle. Such an investigation would require consistent magnetic information provided by ground stations, aeromagnetic and marine data, and satellite measurements. A comprehensive understanding of the magnetic lithosphere depends on the ability to combine all these kinds of information. Here, we discuss the available magnetic data for characterizing the lithospheric field, the methods used in describing the lithospheric field at global and regional scales, and the remarkable achievements made in obtaining the first World Digital Magnetic Anomaly Map (WDMAM). We address also the difference between magnetic anomalies over the continents and the oceans, as well as comparing induced versus remanent magnetizations. Finally, we illustrate a few of the major and recent achievements that have been made using both the near-surface and satellite platforms, and products like WDMAM.

1. Introduction

The precise definition used here for the magnetic lithosphere is generally slightly different from that used in other disciplines like geochemistry, seismology or geology, even if in the end they all converge. Generally, the lithosphere includes the crust and the uppermost mantle, joined to the crust across the Mohorovičić boundary. The lithosphere is usualy considered as a rigid part situated on the top of the asthenosphere – the weaker, hotter, and deeper part of the upper mantle; the boundary between these two structures is defined by a difference in response to stress. The term magnetic lithosphere includes, in a thermal sense, the rocks of both the crust and upper mantle with temperatures below the Curie point of the dominant magnetic phase, typically about 600°C. However, it is often the case that the term *crust* is used instead of *lithosphere*, mainly when magnetizations are thought to be restricted to the crust.

A consistent description of the lithospheric field can help in evaluating different hypotheses of geologic origin in various regions. In addition, numerous scientific problems can be addressed with regard to the variable structure, temperature, and rheology of the lithosphere. Finally, a good knowledge of the lithospheric magnetic field can provide insights leading to a better separation of the internal field contributions (i.e. differentiating between the individual contributions of field originating from the core and the lithosphere).

A rock is considered magnetic if at least one of its constituent minerals is magnetic. The strength of a rock magnetization is related not only to the amount of magnetic material it contains, but also to the physical properties of those minerals (e.g. grain size). In order to understand the lithospheric signature imbedded within the Earth's magnetic field signal, it is useful to consider first a brief introduction of the geomagnetic field itself (for further details the reader is referred to Mandea and Purucker (2005), or for a more exhaustive description the recently published *Encyclopedia of Geomagnetism and Paleomagnetism* (Gubbins and Herrero-Bervera, 2007) and *Treatise of Geophysics* (Schubert, 2007)).

The remaining portion of this chapter deals with a description of the lithospheric magnetic field starting with a brief summary of the magnetic field sources, followed thereafter by a discussion of the available data, then its processing and subsequent use in modeling the lithospheric field. A specific section addresses the remarkable achievements obtained in producing global models and compilations, with an emphasis on the first World Digital Magnetic Anomaly Map (WDMAM). The issue of the continental and oceanic magnetic anomalies is then discussed, continuing with the very interesting topic of separation of the induced and remanent magnetization components. Finally, a few regional case studies are presented that demonstrate the utility of the magnetic lithospheric field.

2. The Lithospheric Signature in the Earth's Magnetic Field

Observing the Earth's magnetic field has had a long and diverse history, from its infancy linked to oceanic navigation during the age of exploration (Malin, 1987) to its more recent conquest of space via satellites (Olsen et al., 2006b). The long and careful collection of direct and indirect observations through history has allowed for the partial understanding of spatial and temporal variations of the Earth's magnetic field (Merrill et al., 1998). Both geomagnetism and paleomagnetism are required to understand the magnetic field of the Earth's lithosphere with geomagnetism focusing on the magnetic field from the recent past to the present and paleomagnetism reconstructing the past history of the magnetic field.

2.1. Contributions to the Geomagnetic Field

The Earth's magnetic field is the sum of several contributions including both internal (the core/main field and the crustal/lithospheric field) and external sources (the ionospheric and magnetospheric fields and the field generated by the field aligned currents). It is well know that the Earth's magnetic field is continuously subject to temporal variations (Mandea and Purucker, 2005). These variations originate from solar activity and variability from the outer core, as well as from the Earth's environment itself. Temporal variations in the geomagnetic field cover a large range of timescales, from seconds to hours (external in origin), from years to decades (overlapping between external and internal changes), and from millennia to reversals (internal variations) (Olsen et al., 2006a).

2.1.1. The External Fields

Ascending from the lithosphere out to the outer part of the Earth's atmosphere, there is a region lying between about 50 km and 600 km altitude, called the ionosphere. In this region circulate strong electric currents, mainly in the sunlit hemisphere, generating their own magnetic fields with intensities up to 80 nT at the Earth's surface (Richmond, 2002). This system of currents is dominated by two vortices, called the S_q system, separated at the magnetic equator by the equatorial electrojet. More energetic features such as magnetic storms, may occur, which are linked to the activity of the Sun and may create magnetic fields as high as a few thousands nT (Mc Pherron, 1991).

There exists a region surrounding the Earth at about 10-20 Earth radii called the magnetosphere which normally prevents solar wind particles from penetrating into the lower atmosphere. A complicated system of currents flows in this region (Vennerstrom et al., 2006) which geometry is partly controlled by the core geomagnetic field – approximately a tilted dipole situated near the center of the Earth and inclined by 11° with respect to the rotation axis. The equatorial circulation of the system constitutes an electric current, known as the ring current that is the dominating contribution at the Earth's surface. It can be mathematically described as a uniform, continuously varying field with two harmonic overtones of a period of 6 months and 1 year. This varying external field induces within the Earth electric currents, which in turn produce an induced magnetic field. This induced magnetic field depends on the electrical conductivity of the upper lithospheric layers, the level of the magnetic field activity and extends to a depth controlled by the skin effect (Backus et al., 1996).

2.1.2. The Internal Fields

The core field

The commonly held belief that the magnetic field can be pictured as a gigantic axial magnet is not accurate. As the Earth rotates and cools, the liquid outer core spins and convects generating the main part of the Earth's magnetic field. The core's magnetic field contribution and the flow responsable for it can be investigated at the core-mantle boundary (at a depth of about 2900 km). The core field, referred to as the main field, represents more than 90% of the full geomagnetic field measured at the Earth's surface. Its intensity ranges in magnitude from approximately 30,000 nT at the equator (even less in the region known as the South Atlantic Anomaly) up to more than 60,000 nT in the polar regions, as indicated by the GRIMM magnetic field model (Lesur et al., 2008) in Figure 1.

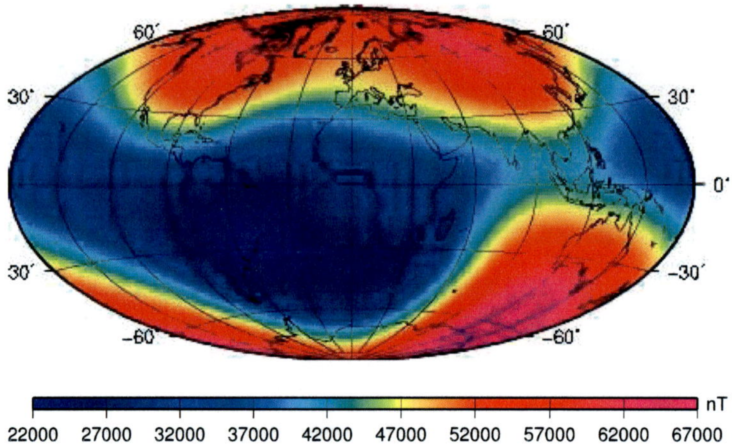

Figure 1. The magnetic core field strength for epoch 2003.5 calculated using the GRIMM model at the Earth's surface.

The core field is not static and undergoes both spatial and temporal variation. This is known as secular variation and is clearly observable in vector field time-series measured at magnetic observatories. For lithospheric studies, the most dramatic and important changes in the core field are the excursions (large swings in intensity) or the reversals (during which the direction of the magnetic field is flipped). In recent epochs the field has reversed on average four times in 1 million years. However, there is considerable time variability in these polarity changes during the geomagnetic field's history (Merrill et al., 1998). These core magnetic field changes are recorded in the lithosphere and thus can inform us about the Earth's dynamic past.

The lithospheric field

The Earth's lithosphere has an associated magnetic field, commonly referred to as the crustal or lithospheric field, which is much weaker than that of the core field (its mean value being some two-three orders of magnitude smaller). Some huge, well-known anomalies such as Kursk in Russia and Bangui in Central Africa extend over large areas with amplitudes reaching thousands of nT. Banded iron formations and other rock types peculiar to the Proterozoic era are common in the surface exposures of these two regions. In the oceans, characteristic anomalies are associated with sea-floor spreading; these patterns are characterized by stripes of nearly constant magnetic anomaly field strength along many hundred km. These stripes can be correlated with the dipole reversal sequence over millions of years (see Figure 2).

The magnetic field of the lithosphere arises partly from the remanent magnetism of magnetic minerals in the crust and upper mantle. The main magnetic mineral is magnetite (Fe_3O_4, with varying amounts of titanium content) – a common mineral found disseminated throughout most rocks in differing concentrations. The dominant source of remanent magnetism is then associated with igneous and metamorphic rocks, whereas sedimentary rocks are usually characterized as being less magnetic (yet still high enough to be measured). This field source is spatially limited to a relatively thin layer (10–70 km thick

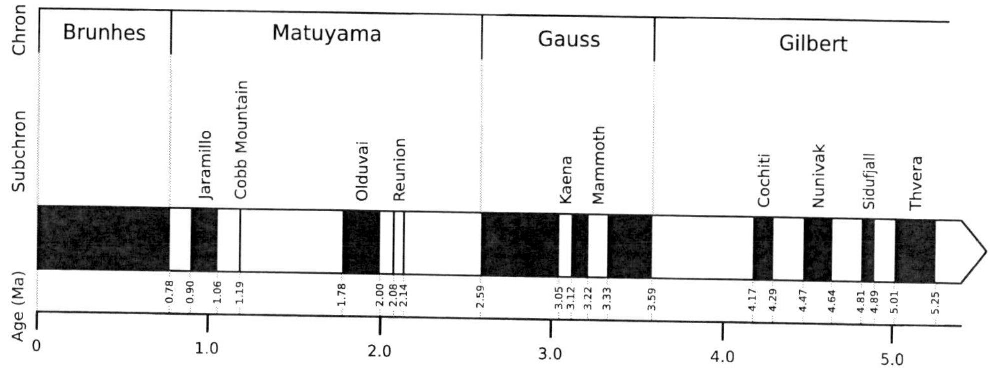

Figure 2. Recent geomagnetic reversals.

depending upon the local heat flow) since the internal temperature of the Earth rises with depth up to a level above the Curie temperature. This temperature rise limits the formation and maintenance of remanent magnetic fields (titano-magnetites have Curie temperatures of 400-600°C). The thermodynamic conditions do not allow rocks to sustain magnetization. This is the reason why no magnetic sources exist between say the bottom of the lithosphere and the core-mantle boundary.

Depending upon its composition, a rock placed in a magnetic field, i.e. the Earth's magnetic field, may become magnetized. This magnetization is a result of the re-orientation of atoms and the spin of uncoupled electrons. The degree to which a given body acquires induced magnetization depends on its magnetic susceptibility, a fundamental parameter in magnetic exploration. The magnetic susceptibility expresses the proportionality relating the Earth's ambient magnetic field to the induced magnetization.

The total field intensity, called also field magnitude, is the total magnetic field strength without vector field directions. This quantity is typically measured for lithospheric magnetic field studies. The total field anomaly is then defined as:

$$\Delta B = |B_{obs}| - |B_{mod}| \tag{1}$$

where $|B_{obs}|$ is the measured magnetic field intensity and $|B_{mod}|$ is the estimated large-scale magnitude of the non-crustal field, determined, for instance, from a global or regional model. The total anomaly field geometry, dominated by induced magnetization, is dependent upon the geometry of the induced field. It is expected that the average strength of the anomalies produced by an induced magnetization at high latitudes is larger than the average strength at low latitudes, due to the core field geometry.

2.2. Geomagnetic Field Representation

Global modeling

The most commonly employed method for representing the global geomagnetic field is using Spherical Harmonics (SH). In the abscence of magnetic sources the geomagnetic induction field $\vec{B}$ can be expressed as the gradient of a potential:

$$\vec{B} = -\vec{\nabla} V \tag{2}$$

where V is a scalar potential satisfying the Laplace equation $\Delta V = 0$. At the Earth's surface this potential is the sum of an internal and external scalar potential, V_i and V_e, respectively :

$$V = V_i + V_e. \tag{3}$$

In spherical coordinates (r, θ, ϕ), the Laplace equation becomes:

$$\frac{\partial}{\partial r}\left(r^2 \frac{\partial V}{\partial r}\right) + \frac{1}{r^2 sin\theta}\frac{\partial}{\partial \theta}\left(sin\theta \frac{\partial V}{\partial \theta}\right) + \frac{1}{r^2 sin^2\theta}\frac{\partial^2 V}{\partial \phi^2} = 0. \tag{4}$$

This equation can be solved by separation of variables and the potential represented in terms of spherical harmonics $Y_n^m(\theta, \phi)$, characterised by a degree n and an order m. In geomagnetic applications, the angle θ denotes the geocentric co-latitude, and angle ϕ denotes the longitude (usually positive East). The Schmidt (semi)-normalized spherical harmonics are given by the expression:

$$Y_n^m(\theta, \phi) = P_n^m(\cos\theta) \left\{ \begin{array}{l} \cos(m\phi) \\ \sin(m\phi) \end{array} \right. \tag{5}$$

where $P_n^m(\cos\theta)$ are the Schmidt-normalized associated Legendre functions of degree n and order m. Spherical harmonics are widely used in mathematics and physics to represent complex, usually natural, functions in spherical geometry. These harmonics are periodic functions with a period characterised by the degree n. The more harmonics are used, the closer to reality the representation is.

Following Gauss (1838), the internal potential V_i, used to represent the lithospheric field, can be decomposed into a "spherical harmonic expansions":

$$V_i = a \sum_{n=1}^{N_i^{max}} \left(\frac{a}{r}\right)^{n+1} \sum_{m=0}^{n} \left(g_n^m \cos(m\phi) + h_n^m \sin(m\phi)\right) P_n^m(\cos(\theta)) \tag{6}$$

where a is the radius of the Earth, $r \geq a$ denotes the radial distance of the observation from the center of the Earth. Measurements of the magnetic field, taken either on the ground or in space, are used to estimate the so-called Gauss spherical harmonic coefficients (g_n^m, h_n^m). Equation 6 is truncated to N_i^{max} yielding $N_i^{max}(N_i^{max} + 2)$ real internal coefficients.

Using the internal coefficients from a spherical harmonic analysis is it possible to define the power spectrum of the internal field (Lowes-Mauersberger spectra), following Lowes (1966):

$$W(n) = (n + 1) \sum_{n=1}^{N_i^{max}} \left((g_n^m)^2 + (h_n^m)^2\right). \tag{7}$$

For example, by applying this procedure to MAGSAT vector data Langel and Estes (1985) were able to show clearly for the first time that there is a major break in the power spectrum near spherical harmonic degree $n = 13$. This break is interpreted to represent the change from dynamic core processes to quasi-static lithospheric ones. In Figure 3 the power spectrum obtained from the very recent GRIMM field model (Lesur et al., 2008) is shown. The

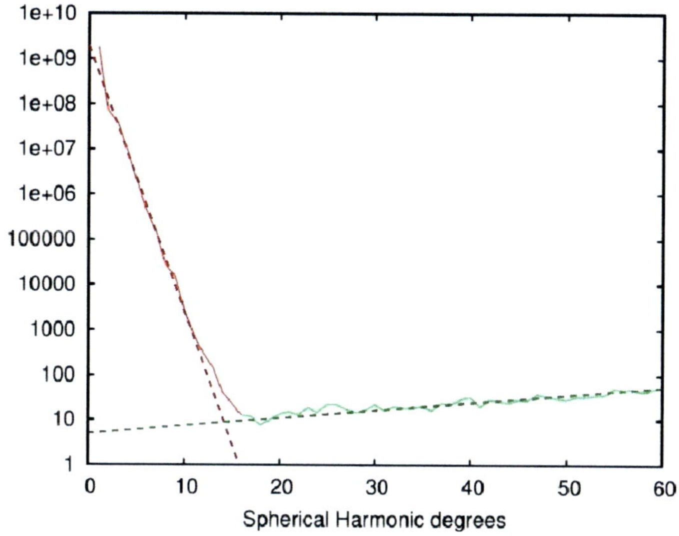

Figure 3. The power spectrum (nT^2) of the core (red) and lithospheric (green) contributions computed using the GRIMM model for epoch 2003.5 at the Earth's surface.

clearly steep part of the spectrum ($n \leq 13$) represents the signature from the long wavelengths of the core field, the transition degrees ($n = 13 - 15$) can be attributed to signals from both the core and the crustal sources, and the higher degrees ($n \geq 16$) are dominantly lithospheric in origin.

Regional modeling

Traditionally, over small areas, the data processing is done using Fourier analysis (Blakely, 1995). For larger areas, methods such as Spherical Cap Harmonic Analysis (SCHA) (Haines, 1985) and the related Translated Origin Spherical Cap Harmonic Analysis (De Santis, 1991) have been developed to model the magnetic field. Korte and Holme (2003) have pointed out that the basis functions used in these methods are not orthogonal, so it is not possible to simultaneously represent the potential for the vertical and horizontal field components exactly. Thébault et al. (2005) have reformulated SCHA as a boundary value problem within a cone extending above the reference surface, thereby allowing satellite data to be downward continued to the Earth's surface. This representation has been successfully applied to regions as small as the greater French Metropolitan area Thébault et al. (2006).

Basis functions such as harmonic splines (Shure et al., 1982), or with localized support, like wavelet functions (Holschneider et al., 2003), are better suited to data sets with variable resolution over the globe. This has been achieved, for example, by (Lesur and Maus, 2006): in the obtained model of the lithosphere the resolution at mid- and low latitudes reaches SH degree 90, whereas at high latitudes stays below SH degree 60.

3. Magnetic Surveys and Data Processing

3.1. Aeromagnetic Surveys

Systematic surveys are necessary in order to describe magnetic lithospheric variations. Indeed, the detailed features of the lithospheric magnetic field are so complicated that observatory or repeat station measurements are insufficient. Moreover, dense ground measurements, when existing, are generally inappropriate because of irregular surface topography and surface anthropogenic activity that can create artificial magnetic sources. Therefore, aeromagnetic surveys play an important role in providing information about the distribution of magnetic material buried under thin layers of sedimentary rock, needed to understand geological and tectonic structures.

In order to describe lithospheric magnetic features, measurements at fixed aircraft altitude and along regular profiles are required. Since it is extremely difficult to maintain a precisely fixed direction and plane attitude while in flight, aeromagnetic surveys generally record only scalar measurements (i.e. the strength of the magnetic field). These survey profiles usually extend over distances from a few kilometers to hundreds of kilometers. They are carried out for a variety of purposes. For instance, fields of research utilizing such data include investigation of crystalline basement and mineral exploration, for fault and fracture zones or for imaging volcanic structure. Large aeromagnetic surveys with wider profile spacing are devoted to both regional and detailed geological investigations over landmass and continental shelves. One example is the United States (U.S.) Navy's Magnet programme from 1951 through 1994, during which continuous vector aeromagnetic survey data were collected to support U.S. world magnetic modeling and charting needs[1].

Aeromagnetic surveys are relatively fast, which reduces the secular variation problem for a given aeromagnetic survey dataset. The external field variations are reduced using the nearest observatory or a fixed base station set up especially for a given survey. Once aeromagnetic data have been collected, a correction must be applied in order to remove estimates of the Earth's magnetic core field as well as any field changes. As such, aeromagnetic data are devoted exclusively to the detection of magnetization patterns in the lithosphere at various scales, and in essence take an image of the lithospheric field that is assumed to be static.

Figure 4 shows that the global coverage of aeromagnetic data is not only especially sparse over the oceans, but also over Africa and South America (where data exist, but are not freely accessible). The available compilations vary between the Northern and the Southern hemisphere according to regional characteristics. It is difficult to estimate the data quality over each region as very few compilations have a complete set of metadata. However, when available, metadata information indicate that different compilations can have different coordinate systems and projections. Generally, large-scale compilations result from the stitching together of small-scale surveys, which have been carried out at various altitudes and epochs with individual data grids upward continued or not to a common altitude.

[1]http://www.ngdc.noaa.gov/seg/geomag/proj_mag.shtml

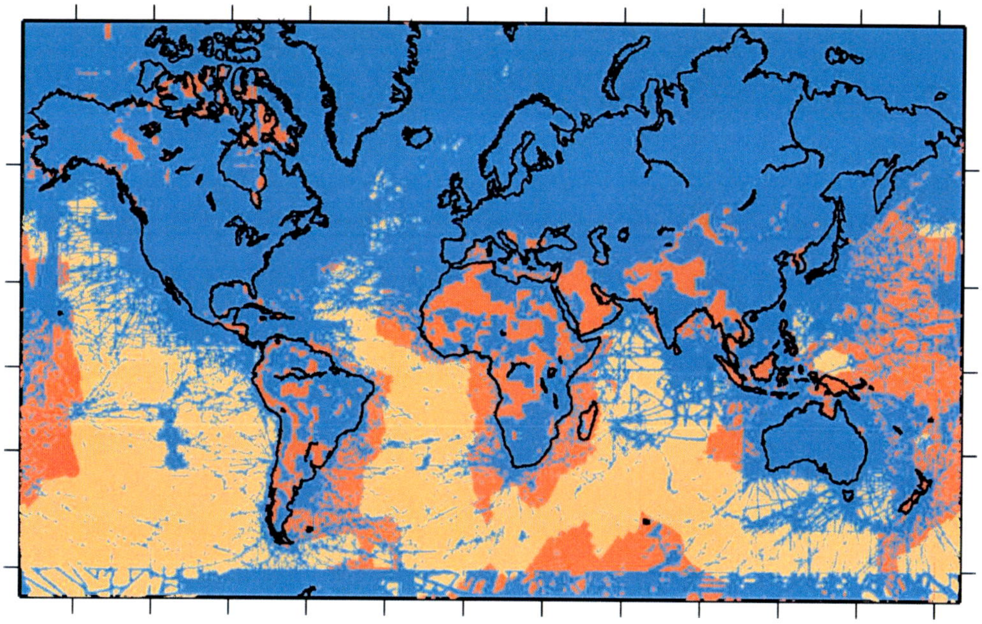

Figure 4. The global coverage of survey data available for scientific research in 2008 (blue). For regions without coverage from survey magnetic measurements, data synthetized from bottom ocean age maps (yellow) or from satellite-based models (red) are used in building the World Magnetic Anomaly Map.

3.2. Marine Magnetic Data

Marine magnetic measurements are crucial for studying the oceanic lithosphere, which covers over two-thirds of the Earth. Marine magnetic data are gathered during geophysical oceanographic campaigns. Generally, the trackline geophysics databases contain bathymetry, magnetics, gravity and seismic data in addition to navigation information. Marine magnetic data can be acquired on board seismic vessels, vessels dedicated to gravity and magnetic surveys, or on hydrographic and multi-beam bathymetry vessels. The most common way of measuring magnetic fields at sea is to install a scalar magnetometer at the end of a non-magnetic line behind a ship (i.e. ship-towed). The acquisition method can be automated to take measurements every few minutes. The intensity of the magnetic field is measured for each location; however, these measurements have to be corrected mainly for the external field variations, but also for the magnetic noise of the ship, which must be determined beforehand.

Currently, a large worldwide database is available that contains magnetic marine measurements collected during marine cruises from 1953 to the present[2]. This database, including data from both U.S. and foreign oceanographic institutions and government agencies, is distributed via the NOAA National Geophysical Data Center's GEODAS search and retrieval system.

[2]http://www.ngdc.noaa.gov/mgg/geodas/trackline.html

Figure 5. CHAMP (CHAllenging Minisatellite Payload) is a German near-Earth satellite mission for geoscientific and atmospheric research and applications.

3.3. Satellite Magnetic Data

Very important results obtained in recent decades are due to a new observational strategy involving the measurement of the geomagnetic field from satellite altitude. The first satellite to carry a magnetometer in orbit around the Earth was the Russian Sputnik 3 (1958) spacecraft. However, global mapping of the intensity of the Earth's magnetic field only began with the U.S. POGO satellites (1967-1971). In 1979, the U.S. MAGSAT satellite ushered in the era of highly accurate magnetic field component mapping. MAGSAT data have been used to map the geomagnetic field from its largest scale (dipolar) down to smaller scales (less than 1000 km). Unfortunately, no other MAGSAT-class mission flew during the next 20 years. In February 1999, the Danish Ørsted satellite was launched[3], soon followed by the German CHAMP (July 2000). Both of these missions map the vector field to an accuracy of a few nT. Other missions in this epoch include the Argentinian-U.S. SAC-C (November 2000) satellite (magnitude only). The most successful mission for lithospheric field study is the CHAMP satellite (see Figure 5), still fully operational as of 2009[4].

CHAMP's low-Earth, quasi-circular orbit is particularly suitable for crustal field mapping. The initial average satellite altitude was 454 km but the satellite is slowly descending and is in 2009 at about 330 km altitude. The lithospheric maps obtained from this satellite experiment are the most robust and have an unprecedented resolution. Recent lithospheric models mainly rely on data from this satellite mission.

[3]http://www.spacecenter.dk/research/solarphysics/orsted

[4]http://www.gfz-potsdam.de/portal/-?$part = sec23$

3.4. Processing Magnetic Data

The first important step in processing an aeromagnetic survey dataset, so that it is suitable for further processing, modeling and interpretation, is to remove, as much as possible, the other magnetic contributions from the profile measurements. These contributions come from both external fields and the main field. Another important step for building large-scale compilations is the merging of small-scale compilations. These aspects are summarized in the following.

The external field contributions can be minimized by taking measurements during magnetically quiet times. However, this constraint is difficult to apply for two resons: 1) it is difficult to predict the magnetic field evolution over a period comparable with that for collecting aeromagnetic data over a region; 2) for high magnetic latitudes, the external field variations are large, even during quiet times. The generally applied technique is to continuously measure the magnetic field variations at a fixed station, situated somewhere near the center of the surveyed area, and if possible, far from magnetic anomalies (crustal or anthropogenetic). The magnetic field variations recorded by these stations represent a good approximation of the external field variations over the area, and can be used to correct the profile data.

A commonly used method for removing the core field contributions is to estimate them from main field models, like the IGRF[5]. Using these sets of coefficients (available on a 5-year interval basis), the core field can be computed for each location and epoch of the survey measurements. However, if measurements over two adjacent areas are done at different epochs, merging them for large compilations can create mismatches, as the IGRF/DGRF representation is not perfect. A possibility for dealing with the problems arising from the use of the IGRF/DGRF models is to consider more complex models, such as the comprehensive models. The CM4 model (Sabaka et al., 2004), for example, has a large number of parameters expressing the core, crustal, external, and induced fields, which are all co-estimated. This model, covering the time-span 1960-2002, has been used extensively and so better estimates for the lithospheric field can be obtained from the aeromagnetic data (Nabighian et al., 2005). For measurements after 2002, there are other models based on satellite and observatory data available, such as CHAOS (Olsen et al., 2006b), xCHAOS (Olsen and Mandea, 2008) and GRIMM (Lesur et al., 2008).

Until recently, grids inside each individual compilation were derived for different epochs and reduced with either local polynomials or the IGRF/DGRF models. In most cases the corrections applied have been lost and therefore, it is difficult to find out which model was used to reduce the data. Furthermore, the resolution is usually not homogeneous within the compilations themselves and some regions artificially appear devoid of small magnetic anomalies. As a result, the final grids could contain an anomalous signal that may easily be confused with real magnetic anomalies. Checking the consistency between two overlapping grids is therefore challenging in some areas where the actual resolution is not known (Figure 6).

In Figure 6, the European grid shows anomalies that are smoother than the neighboring grid. Two reasons may be involved. Firstly, the European grid is upward continued to 3 km altitude, while the Eurasian grid is produced by combining many disparate aeromagnetic

[5]http://www.iugg.org/IAGA/iaga_pages/pubs_prods/igrf.htm

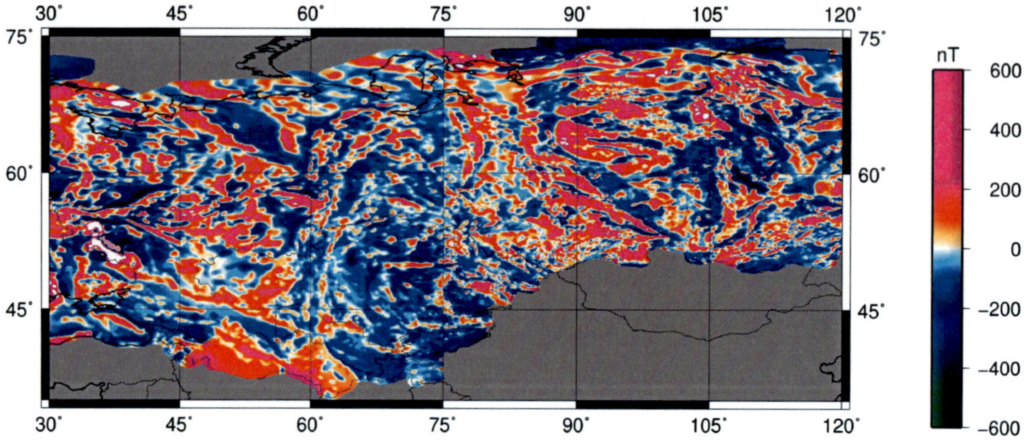

Figure 6. Direct merging of the European and Eurasian grids illustrates problems that can occur at the junction of neighboring regions. This is very clearly seen around the 75° longitude.

surveys of smaller dimension without upward continuing the final grid. Secondly, a spectral analysis carried out on the European grid showed sudden power decay for wavelengths lower than 30 km and it seems that the European grid was filtered in order to produce a map of homogeneous resolution. The anomaly features across the boundary of the European and Eurasian grids also show a partial disagreement. Comparing the total anomaly field strength over profiles crossing the boundaries between the two grids indicates discrepancies exist up to some 200 nT (see the area around 75° longitude in Figure 6).

4. Global Models and Compilations

4.1. Lithospheric Models

On a global scale, the lithospheric magnetic field is commonly represented by a spherical harmonic analysis of data gathered by satellites. The onboard satellite measurements are firstly selected considering various criteria. Generally, only measurements taken during night-time and during magnetically quiet times are used to represent the internal field in order to minimize, as much as possible, the external contributions. Recent models of this type include the CM4 model (Sabaka et al., 2004), the MF series (Maus et al., 2007), the CHAOS model (Olsen et al., 2006b), and the GRIMM model (Lesur et al., 2008). These models reflect somewhat different design philosophies and hence have different strengths and weaknesses. A short description of these models is given below.

The Comprehensive Models (CMs)[6] are constructed from ground-based and satellite geomagnetic observations, facilitating the separation of the various field sources and their spatial and temporal sampling extents. The last published model, denoted CM4 (Sabaka et al., 2004), solves simultaneously for the core and crustal field, for ionospheric and magnetospheric contributions (plus their induced counterparts), and for the toroidal magnetic

[6]http://core2.gsfc.nasa.gov/CM/CM4_data.html

field produced by electrical currents at satellite altitude. The internal (core and lithospheric) fields are represented by a degree and order 65 internal spherical harmonic expansion, with the secular variation represented by cubic B-splines up to degree and order 13 having a knot spacing of 2.5 yr extending through mid-2002.

The MF series represents the lithospheric field only. The MF5 model (Maus et al., 2007) was produced using a new scalar and vector data product from the latest three years of measurements from the CHAMP magnetometers. The modeling methodology remained largely unchanged from the previous versions. The improvements come mainly from the increase of the lithosphere signal strength in the collected magnetic data as the altitude of the satellite decreases. The lastest model MF6[7] extends from degrees 16 to 120 and follows essentially the same methodology.

The CHAOS model[8] is a spherical harmonic field model of the near-Earth region (Olsen et al., 2006b). Its dataset originates solely from satellite measurements (Ørsted, SAC-C and CHAMP) and covers the years 1999–2005. This model provides a reliable representation of the internal fields up to spherical harmonic degree 40, and of the first time derivative of the core field up to degree 15. A new version of this model, xCHAOS, has been developed recently (Olsen and Mandea, 2008).

The GRIMM model[9] has been derived from nearly six years of CHAMP satellite data and five years of observatory hourly means (Lesur et al., 2008). At high latitudes, full vector satellite data are used at all local times, which allows a separation between the fields generated by ionosphere and field aligned currents, and the fields generated in the Earth's core and lithosphere. The static core field and the lithospheric field are expressed up to degree 60.

Because of the different design philosophies of these models, the estimated lithospheric contributions have differing amounts of power. This may be explained by data selection and processing or by the differences in the direct damping applied (or not) to crustal coefficients. Figure 7 shows a comparison of the Lowes-Mauersberger spectra at the surface of the Earth, for the crustal fields represented by these models.

Both the GRIMM and xCHAOS models present similar spectra and high correlation up to spherical harmonic degree 45. The correlation is also good when comparing with the CM4 model. The figure shows that the MF5 model lacks power and presents spurious features at low spherical harmonic degrees.

Figure 8 shows the lithospheric contribution at the Earth's surface obtained from the GRIMM model. The magnetic lithospheric signal derived from this model is on the order of some ± 200 nT. At first glance it becomes clear that the large magnetic anomalies are situated over continents, where the extrema values are enhanced. Indeed, the maps reproduce some of the large well-known magnetic anomalies like Bangui (Central Africa) and Kursk (Russia). The oceans are mainly characterized by low magnetic anomalies of the order ± 50 nT.

When the differences between MF5 and GRIMM are mapped, they do not exceed about ± 50 nT and generally stay below ± 10 nT at mid- and low- latitudes. These differences are relatively large and they have the characteristic shape of spurious anomalies generated by

[7] http://geomag.org/models/MF6.html

[8] http://www.spacecenter.dk/research/solarphysics/magnetic-models

[9] http://www.gfz-potsdam.de/portal/-?$part = sec23$

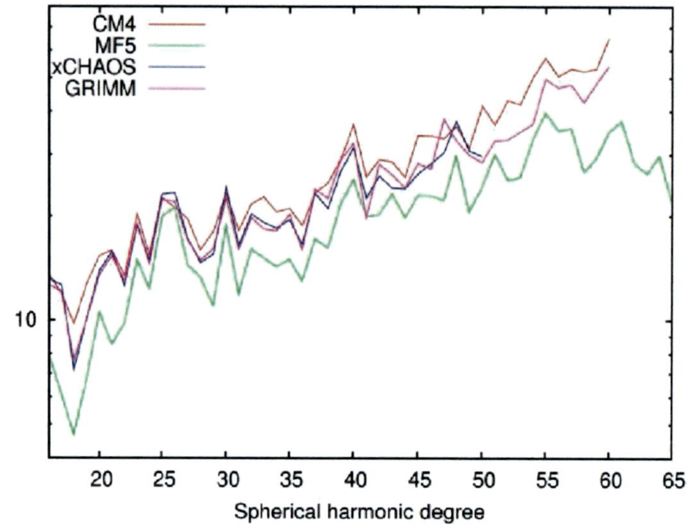

Figure 7. Power spectra (nT^2) for spherical harmonic degrees 16 to 65 for the CM4, MF5, xCHAOS and GRIMM models.

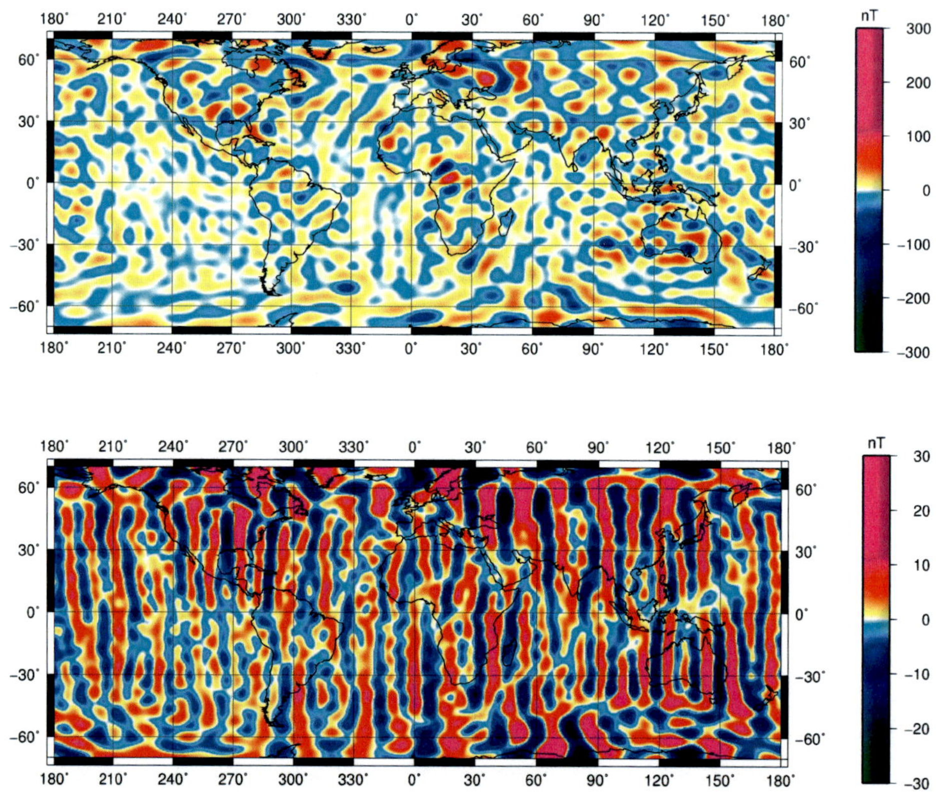

Figure 8. Vertical down component of the GRIMM lithosphere field for spherical harmonic degrees 16 to 45, at the Earth's reference radius (top). The difference between the vertical down component given by the MF5 and GRIMM models for spherical degrees 16 to 45, at Earth's reference radius. (bottom)

along-track filtering.

4.2. World Digital Magnetic Anomaly Map

Recently, a large effort has been made to compile a world magnetic anomaly map by combining ground-based, airborne and marine magnetic data, under the auspices of a working group of the International Association of Geomagnetism and Aeronomy (IAGA). Note that this effort has been ongoing over the last three decades. In 2006, using available ground, aero- and satellite magnetic data, five candidate models were produced by teams at the Geological Survey of Finland (GTK), GeoForschungsZentrum (GAMMA), GSFC-NASA, NOAA and the University of Leeds[10].

The WDMAM committee has worked to produce a compiled magnetic anomaly map consisting of all possible wavelengths that would be useful for geological and tectonic mapping of the Earth's crust. The authors' contribution to the WDMAM project is summarized in Hamoudi et al. (2007), and some of the difficulties associated with obtaining a self-consistent global map are outlined. Indeed, merging aeromagnetic data of different quality, which were reduced using different procedures and core field models, causes significant differences and leveling errors between separately collected and processed adjacent magnetic surveys. These aspects are discussed below, so for more detail the reader is referred to the next two sections and related references therein.

However, it is important to underline that a careful examination of data suggests that near-surface compilations contain long wavelengths ($>$ 400 km) that differ significantly, both in strength and shape, with adjoining compilation and global lithospheric field models. The difference in maps indicates that discrepancies are of the order of the strength of long wavelength anomalies of the merging grids. In contrast, the wavelengths smaller than 400 km are naturally attenuated in satellite data, therefore the satellite-based model is used to replace incorrect long wavelengths in the near-surface compilations. For filtering and smoothing the final grid a spherical harmonic analysis, approximating the total intensity anomly grid by the projection of the lithospheric field onto the reference core field (the CM4), was done. The grid has been interpolated on Gauss-Legendre knots to avoid aliasing and the parameters, up to spherical harmonic degree 500, are obtained by least-squares (maximum resolution of 80 km). Above degree 500, the coefficients are difficult to obtain as the signal/noise ratio is very low. This is also due to data gaps and heterogeneous resolutions at the global scale.

Finally, considering the recommendation of the WDMAM committee, a global grid at 5 km altitude above the World Geodetic System 1984 (WGS84) reference ellipsoid has been produced, on a $3' \times 3'$ latitude-longitude spacing (about 5 km). The map based on the GAMMA candidate model is plotted in Figure 9. A few characteristics of this map, which are also found in the official published map by CCGM/CGMW–UNESCO, are discussed below.

[10]http://projects.gtk.fi/WDMAM/

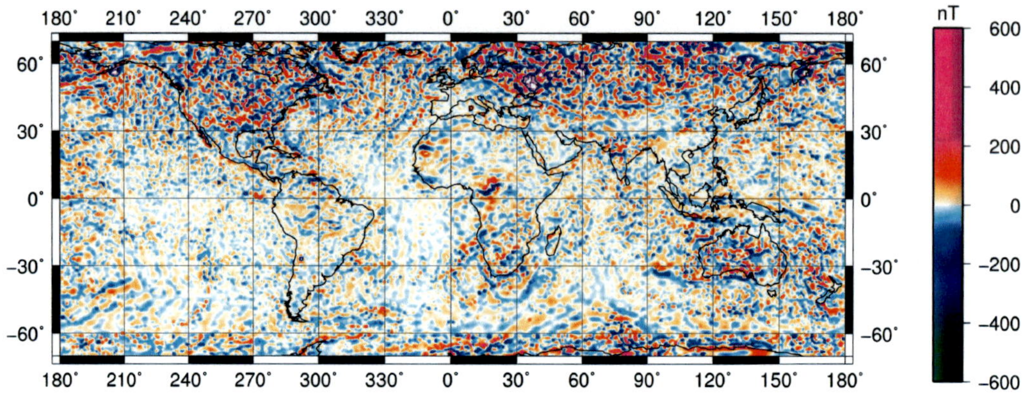

Figure 9. Magnetic anomaly map of the world. The anomaly field is shown at an altitude of 5 km above the WGS84 ellipsoid and is computed from the GAMMA candidate model.

5. Magnetic Anomalies: A Global View

5.1. From Continents to Oceans

At first glance, Figure 9 illustrates that the magnetic anomalies over the world clearly show different behavior over the continental and oceanic regions. The anomaly distribution is heterogeneous with strong anomalies lying mostly over the continents (for instance, the strongest being situated over Kursk (Russia) and Bangui (Central Africa)).

The lithospheric magnetic field is stronger over the main continental regions: Africa, Antarctica, Asia, Australia, Europe, Greenland, India, North and South America. A comparison with geological maps indicates that large magnetic anomaly features correlate well with Precambrian and Paleozoic provinces, although a significant portion of old crust is probably hidden by Phanerozoic cover. If the strongest and largest positive magnetic anomalies are associated with shields and cratons, large subduction zones are also distinct (Indonesia) showing that the magnetic anomalies also correlate with seismotectonic maps of the world. Moreover, each large structure includes diverse geological features of smaller extension-like mobile belts or collision zones with comparatively lower magnetization. In some places, continental magnetic anomalies seem to extend in oceanic regions along continental margins and island arcs, as in Southern Argentina.

Oceans cover some 70% of the Earth and half of them are deeper than 3000 m. Ocean floors are thus mostly unknown and remote measurements play a key role in understanding their dynamics. Oceanic areas offer a large contrast and seem to carry weaker magnetism overall. One of the outstanding features shown in Figure 10 is that of seafloor spreading, which is well defined in the North Atlantic, for example.

Indeed, these oceanic linear magnetic anomalies are composed of linear segments of inverse polarity, first recognized in the 1950's, but a satisfactory explanation was missing. In the beginning of the 1960s, Vine and Matthews (1963) noticed the pattern similarity between continental vertical magnetostratigraphy and horizontal oceanic lineaments. They proposed that seafloor linear anomalies result from a combination of reversals in the Earth's magnetic field and the seafloor spreading symmetrically parallel to ocean ridge crests. As

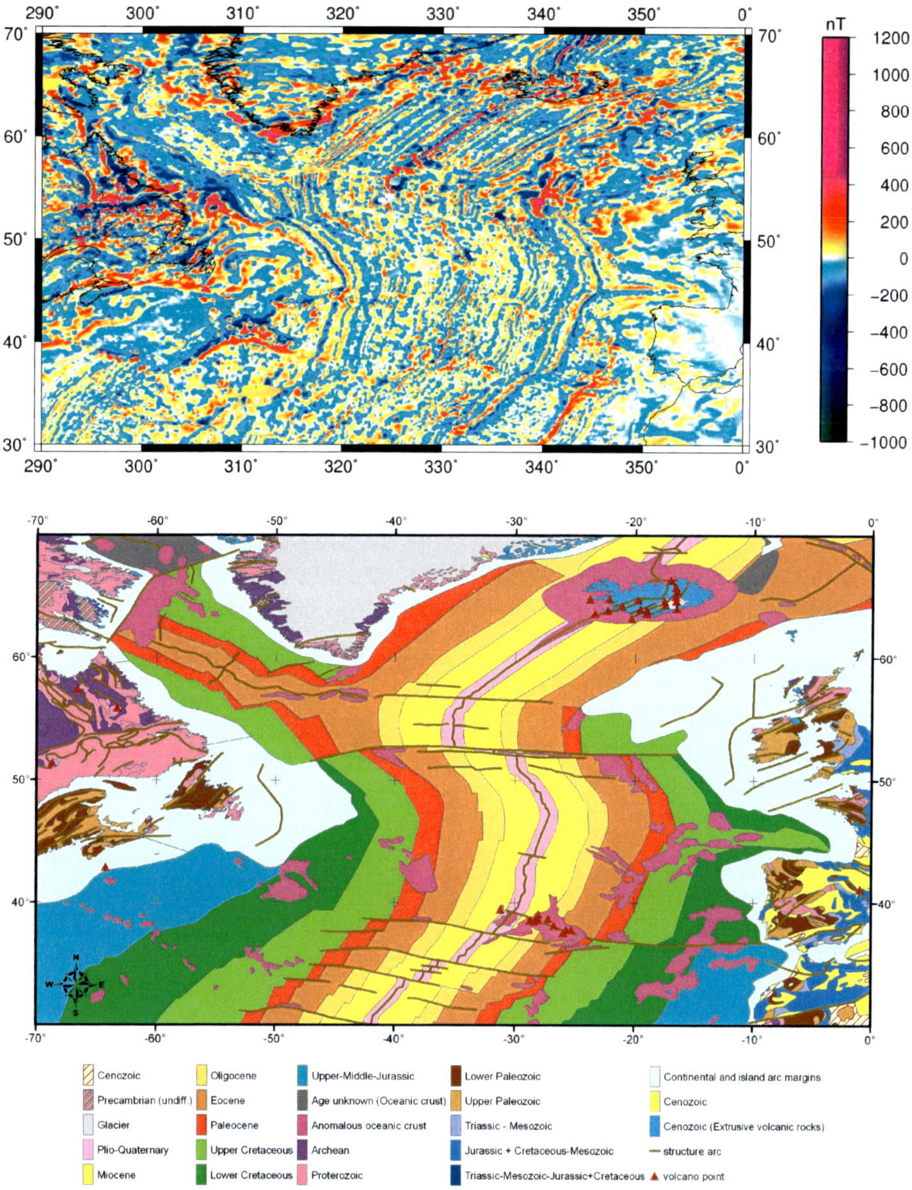

Figure 10. Magnetic field anomalies described by the GAMMA model over the North Atlantic region. (top) The geology of the same region as given in a geological map of the world. (bottom)

magma ascends along the ridges it cools and solidifies. As the rocks form, they acquire a thermo-remanent magnetization aligned along the ambient magnetic field. As seafloor spreading continues, the magnetic field is symmetrically recorded in the rocks along the mid-ocean ridge, creating these remarkable continuous lines that are occasionally displaced by perpendicular fracture systems that reveal variations in the seafloor spreading and tectonic movement, as shown in Figure 10.

These lines, the magnetic isochrons or chrons, are observed all around the world, but the coverage is rather sparse on the global scale. It is possible to interpolate along the direction of spreading in order to estimate the age of the oceanic floors. Based on the magnetostratigraphic scale, the oceanic extrapolation method gives us an age estimate of unsurveyed areas giving for the first time a global view of ocean floors from sparse near-surface measurements. For instance, an interesting example is the Carlsberg ridge separating the African and Indian plates, which has a poorly known spreading history. The tiny wiggles observed due to magnetic anomalies suggest complex local tectonics as changes in isochron direction are noticeable. The physical mechanisms inducing a ridge reorientation and spreading rate modification can be thermal or mechanical. Although it is difficult to connect precisely the Carlsberg ridge with the collision history of India and Eurasia, Dyment and Arkani-Hamed (1998) suggest the two following scenarios. The first scenario is related to the "hard collision" between India and Eurasia (43 Ma ago) that could have induced regional kinematic modifications. Some precursors of the collision, explaining the tiny wiggles offsets prior to this event, may have occurred earlier, 47-53 Ma ago. The second proposed hypothesis is that the nearby Deccan-Reunion hotspot induces this spreading asymmetry.

More intense positive anomalies are noticeable over oceanic plateaus while basins or abyssal plains are associated with negative anomalies. Other anomalies of larger extension and weaker magnitude are difficult to interpret as a result of our poor understanding of oceanic crust. The magnetic mapping of ocean floor is an essential tool for reconstructing the history of the Earth. Regional ship-borne magnetic measurements show more detail than these global first order models and small intensity variations inside a single isochron give evidence of more complex kinematics.

5.2. Induced versus Remanent Magnetizations

As discussed above the strength of the lithospheric field depends on the underlying geology and the magnetization of the underlying rocks. Two types of magnetizations are possible: induced and remanent magnetizations. Induced magnetization is generated by an applied field (for the Earth a good approximation is the core field) and is proportional to the magnetic susceptibility of the rocks. Although a detailed study of the rock magnetization often shows that the susceptibility is anisotropic, when interpreting magnetic surveys the susceptibility is generally assumed isotropic. For this latter case, the induced magnetization is aligned with the inducing field and slowly varies with it. In contrast, the remanent magnetization is "frozen" in rocks and has generally been acquired during their formation. Typically a lava flow acquires the magnetization in the direction of the ambient field while cooling down below its Curie temperature. Remanent magnetization can also be acquired by other processes like lightning or chemical reactions. The remanent magnetization can also be viscously modified, but only on geological time-scales.

When samples of rock are acquired, the ratio of remanent to induced magnetization can be measured. This ratio is called the Koenigsberger ratio and varies between rock types, depending on their locations and history. Typically, the ratio is around 1 for slowly crystallized igneous and thermally metamorphosed rocks in continental areas, rises to 10 for volcanic rocks and reaches up to 30-50 for many rapidly quenched basaltic rocks. The ratio is generally smaller than 1 for sedimentary rocks. Table 1 gives typical values for this ratio.

When a magnetic survey is done and an estimate of the lithospheric field made, it is not known which part of this field is due to the induced magnetization and which part is due to the remanent magnetization of rocks. This has been a longstanding question in geomagnetism. For oceanic rocks, the well-known stripes of the magnetic anomaly field associated with the reversal of the core field is a clear evidence that remanent magnetization dominates. Over continental areas, the techniques developed by geophysicists for interpretation of the magnetic data generally assume an induced magnetization. The known Koenigsberger ratio of the underlying rocks does not provide any information since on the large scale the contribution of rock samples, each with its own remanent magnetization direction and strength, may average to zero. It is therefore often assumed that for the large-scale crustal field the contribution of the induced magnetization dominates whereas for small scales, it is the remanent magnetization contribution that dominates.

The problem of separating the remanent and induced magnetization can be studied by looking at the dependence on time of the lithospheric field. One can consider long time-series of observatory data, subtract a core field model, and try to find in the residuals a signal coherent with an induced field (Lesur and Gubbins, 2000). The main difficulty in this approach is the accuracy of the core field model. Another possible approach is to look at the time evolution of maps of the lithospheric field. However, accurate aeromagnetic surveys have been made only recently and in the rare places where two surveys were flown over the same area at different epoch, no obvious sign of variation in the lithospheric strength or orientation has been observed. Alternatively, one can process survey data and try to find the minimum amount of magnetization necessary in the direction orthogonal to the core field to fit the data. This approach has been already suggested by Whaler and Langel (1996) and gives an estimate of a "minimum amount of remanent magnetization".

6. Magnetic Anomalies: Regional Case Studies

In the following, three case studies are used to illustrate the utility of the magnetic lithospheric field. The first one is the Chicxulub meteorite impact structure (Mexico), showing how this geological complex can be seen in magnetic data. Thereafter, one of the most elongated magnetic continental anomalies is presented, the Beattie anomaly (South Africa), with some proposed further investigations. Finally, the structural and tectonic interpretation of the magnetic lithospheric field obtained over the Java trench region is discussed, with some caveats that must be considered in order to properly understand geological hazards.

6.1. Chicxulub Impact Structure

With a diameter of about 200 km, Chicxulub, located below and straddling the coastline of the northwest section of the Yucatan (Mexico) peninsula, is one of the largest meteorite impact craters on Earth. The Chicxulub crater is also one of the best preserved impact structures, although it lacks a significant topographic signature (it is buried under about 1 km of sediment and is mostly undersea (Figure 11)). Indirect investigation methods, such as airborne magnetic surveys, show the circular anomaly shape of this hidden structure.

Table 1. Koenigsberger ratios for selected rocks (Hunt et al., 1995, after).

Rocks	Koenigsberger Ratio
Sedimentary rocks	
marine sediments	5
red sediments	1.6 - 6
siltstone	0.02 - 2
silty shale	5
avg sedimentary rocks	0.02 - 10
Igneous rocks	
granite	0.1 - 28
granodiorite	0.1 - 0.2
dolerite	3 - 3.5
diabase	0.2 - 4
gabbro	1 - 9.5
oceanic gabbro	0.1 - 58.4
intrusions	0.1 - 20
volcanics	30 - 50
subaerial basalt	1 - 116
seamounts	8 - 57
avg. igneous rocks	1 - 40
Metamorphic rocks	
granulites	0.003 - 50
Others	
magnetite ore	1 - 94
manganese ore	1 - 5
lunar rocks	0.001 - 1

This impact is considered to be responsible for the environmental changes at the end of the Cretaceous Period, some 65 Ma ago (Hildebrand et al., 1991). The magnetic signature consists of three clear concentric zones with radii of 20, 45 and 80 km, which extend to 210 km (Pilkington and Hildebrand, 2000). The innermost zone has a positive magnetic anomaly, suggesting the presence of a single source, while the second-ring zone anomaly, is less structured, and suggests evidence of numerous smaller dipolar anomalies.

The available magnetic data have to be modeled for a single or multi-ring structure. This is due to the fact that large structures, such as Chicxulub, tend to exhibit a central high amplitude anomaly. Recently, some new imaging techniques have been used to emphasize the edges of magnetic bodies via derivatives, or via artificial illumination in one or more directions.

Some other similar magnetic features have been found using remote sensing techniques (satellite imagery, aerial photography and airborne geophysics), but only ground analy-

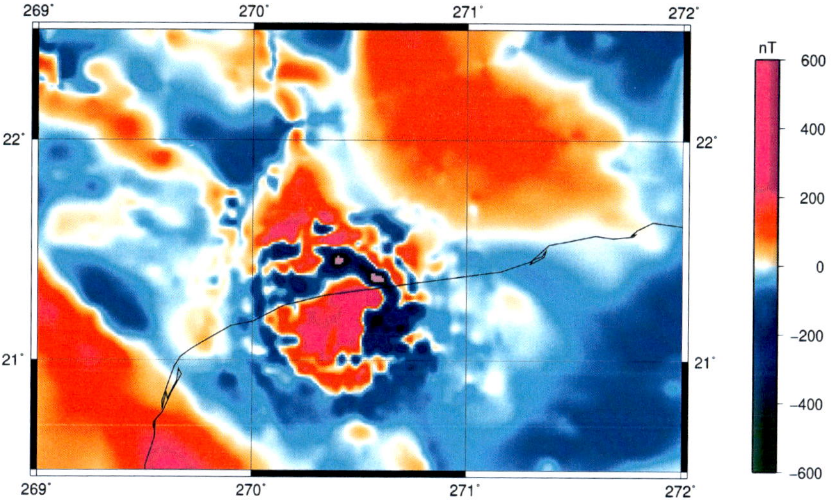

Figure 11. The Chicxulub impact crater is partly covered by sediments and water, but is still easily detectable in aeromagnetic survey data.

sis can be conclusive for recognizing shock geological structures. For instance, the large anomaly in Central Africa, the Bangui Anomaly, has been subjected to different interpretations from very different models. Near the magnetic equator, a purely dipolar induced magnetic source could produce the Bangui anomaly; i.e, a magnetic low with two positive side lobes to the north and south (Figure 9). The source can be explained considering a pure geological origin involving emplacement of basalts in the lower crust as a result of Pan-African orogeny creating the anomaly. The second hypothesis involves iron meteoritic materials that impacted more than 1 Ga ago during the Precambrian period. The Bangui anomaly was investigated using gravity, topography and forward modeling of an idealized circular source that suggested an 800 km diameter impact structure. If this is confirmed, it will be the largest impact feature yet found on Earth. Field work has identified shock related deformation features and iron alloys associated with iron meteorites, which supports the impact. Nevertheless, the evidence is tenuous as collateral surface and subsurface data are too scarce to completely assess the validity of this model. Moreover, it is difficult to understand how the apparent topographic ring has survived the processes of erosion and crustal deformations since the Precambrian time, unless the dating is erroneous. Note that the estimated age of the Bangui crater cannot be correlated with fossils or a mass extinction as the fossil record does not become significant until the Cambrian age, an age of major life diversification, almost 500 Ma later.

6.2. Beattie Magnetic Anomaly

One of the most prominent magnetic anomalies corresponds to the very elongated Beattie magnetic anomaly in South Africa. The origin of this anomaly is unclear, and an important issue revolves around the width, depth and magnetization of its source.

The Beattie magnetic anomaly extends across some 1000 km in the east-west direction, with a width ranging from about 10 to 100 km (Figure 12). In its western part, this anomaly

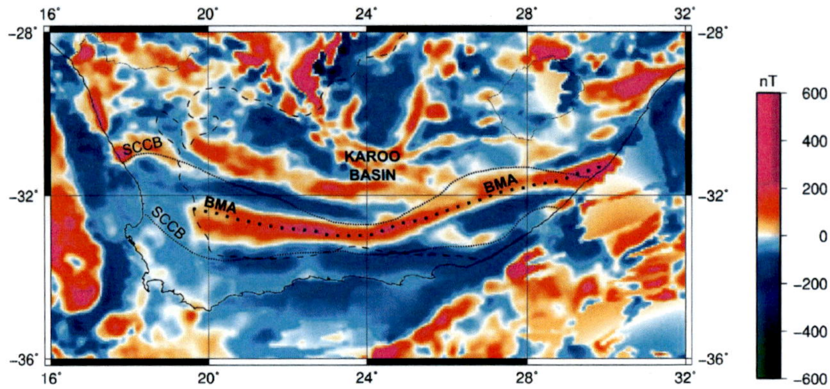

Figure 12. Magnetic field anomalies described by the regional scale aeromagnetic survey over the Southern African continent.

seems to be separated in two branches, both of them ending over the Paleozoic Cape Fold Belton. In the eastern part, the east-west direction changes northeastward before reaching the Indian Ocean. Some similarly extended anomalies are observed in South America and Antarctica, at locations that surrounded South Africa during Gondwana-times, suggesting the same type of source (Corner and Groenewald, 1991; Ghidella et al., 2002; Jokat et al., 2003).

This very large anomaly is not clearly visible on lithospheric models derived from satellite data (see Figure 8). To characterize this magnetic anomaly , data from a regional aeromagnetic survey have been used.

Recently, (Quesnel et al., 2009) have studied this anomaly and found that the associated magnetization seems to be mostly induced, although a weak remanent component is required to exactly fit the shape of the Beattie magnetic anomaly over the investigated area. Their results have been compared with the very new interpretation of seismic and magnetotelluric measurements along the same profile, by Stankiewicz et al. (2007); Weckmann et al. (2007). A coherence has been found for a source mostly located in the middle crust, possibly extending to the lower crust and cut by a shear zone or a fault. One key remaining question concerns the equivalence between the sources of the Beattie magnetic anomaly and the Southern Cape Conductive Belt. A joint interpretation of recent geological and geophysical data is in progress to enhance our understanding of the crustal structure in this region, the origin of this intriguing anomaly, and the geodynamics of the Gondwana-breakup.

6.3. The Java Trench

Magnetic methods are a relatively inexpensive way to learn about geological hazards, which can be examined at different scales using ground and satellite measurements. Magnetic anomaly maps can also give information about geodynamics at geological time-scales. This information can thus be used to understand geological hazards, but cannot be employed for hazards predictions.

An interesting case is the Java trench, which, even with its large dimensions, shows the limits of magnetic investigations. In theory, aeromagnetic images of the region before and

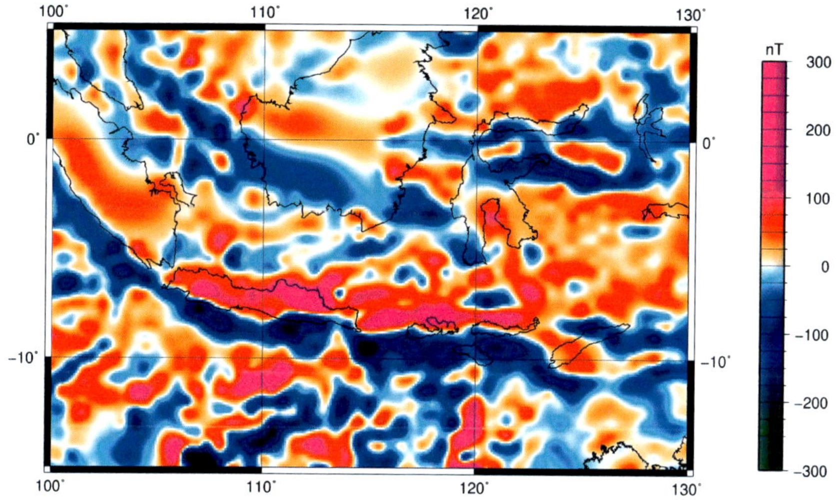

Figure 13. The magnetic field anomalies described by aeromagnetic survey over the Java Trench.

after the Great Sumatran earthquake of 26 December 2004 could provide a historical and current view of the active subduction in the region. Although not well known, the history of this plate margin region is probably buried beneath arc sediments and volcanics located at the north of the current trench.

The subduction can be seen on the gravity anomaly map by an increase of the gravity field along the subduction zone. In the same direction, a lithospheric magnetic field satellite image confirms the compression boundary between the Eurasian and Indo-Australian plates (Figure 13). Even if this conception is oversimplified, a subduction zone near the magnetic equator like the Java trench has a typical positive magnetic signature above the location where the "cold" magnetic ocean crust penetrates into the relatively non-magnetic mantle. However, the magnetic anomaly map realized at sea-level appears noisy and the large-scale structures are hidden or distorted. One reason is that data distribution is not homogeneous and that no data are available inland. The more physical reason is that a magnetic map made using ship-borne and airborne data is mostly sensitive to shallow and small dimension magnetic materials. In contrast, the far-field imaging from satellite data has a better sensitivity to the deep and large magnetic anomalies. Both views are complementary and the use of each data depends on the spatial scale at which the study is realized.

7. Conclusion

In this chapter the static part of the magnetic lithospheric field is described, using the latest available methods and data. The huge interest to this component of the magnetic field has arisen mainly because of two important factors: the World Digital Magnetic Anomaly Map (WDMAM) project and the new low-altitude CHAMP magnetic satellite data. The WDMAM project also benefits from improved global magnetic field models. Analyzing the compatibility of ground survey and satellite data is helpful for correcting data and filling-in the spectral gap. This could be done using regional modeling to jointly invert aeromagnetic and satellite data. In the near future, attention has to be paid to the magnetic lithospheric field variations in time, again at global but especially regional scales. Indeed, if the lithospheric field is mainly induced, it also varies with the secular changes of the inducing field. To which proportion remains a debated question, and this requires long time monitoring of the magnetic field.

Moreover, all magnetic contributions overlap and even assuming an exact correction for all non lithospheric field sources, features larger than 2500 km originating from the crust are inevitably masked by the core field. In addition, far field and near field measurements correlated only partially because measurements themselves provide us with partial views. Currently, some essential information for a comprehensive understanding are missing, unless the remote sensing devices and the high-precision and high-resolution of data are improved, together with the mathematical description of magnetic data.

The next important step in describing the lithospheric field will come from the high-resolution magnetic field monitoring from space, which will continue after the end of current missions. The European Spatial Agency (ESA) selected Swarm as the fifth Earth Observation Opportunity mission[11]. This constellation mission consists of three CHAMP-like spacecraft, flying in a dedicated formation and providing simultaneously measurements from different local times and altitudes (Figure 14). Their launch is planned for 2011 with an expected duration of 4 years. High-precision and high-resolution measurements complemented by precise navigation, accelerometer and electric field measurements, will provide the necessary observations that are required to separate and model various sources of the geomagnetic field. With the new highly accurate satellite measurements at low altitudes a great improvement in the representation of the intermediate wavelength lithospheric field is expected, with possible regularly updates of the WDMAM without excessive effort.

Finally, let us note that the discipline of comparative planetology is a fruitful domain teaching us about the history and the possible future of our own planet. All planets represent a natural laboratory with different physical conditions that lead to various planet evolutions and compositions. To that respect, the Earth is unique, but it is necessary to understand how fragile its stability is. Common to all telluric planets (Earth and its Moon, Mars, Mercury and Venus) is the existence of several magnetic field contributions of internal and external origin. The magnetic field of a planet can be measured remotely and can be used to probe the planet's interior and composition. The most intriguing comparison is between the terrestrial and martian magnetic fields lithospheric in origin: the Mars' one is a factor of 10 larger than the Earth's one, when directly measured at comparable altitudes above the planet (some 400 km). The most intense sources are observed in the Cimmeria region, where fields in excess of 1500 nT were observed near Mars periapsis at 100 km altitude (Langlais et al., 2004). The terrestrial lithospheric field contains a significant component of induced magnetization from the large terrestrial core field. This component is largely absent at Mars, where remanent magnetizations dominate. If the comparison is restricted to regions of Earth where remanent magnetization is dominant (the oceanic basins), then the Earth-Mars contrast increases by a factor of 5, to 50.

We finish by stating that the magnetic lithospheric field stands here to help in understanding our planet life and to be itself better understood!

[11]http://www.esa.int/esaLP/ESA3QZJE43D_LPswarm_0.html

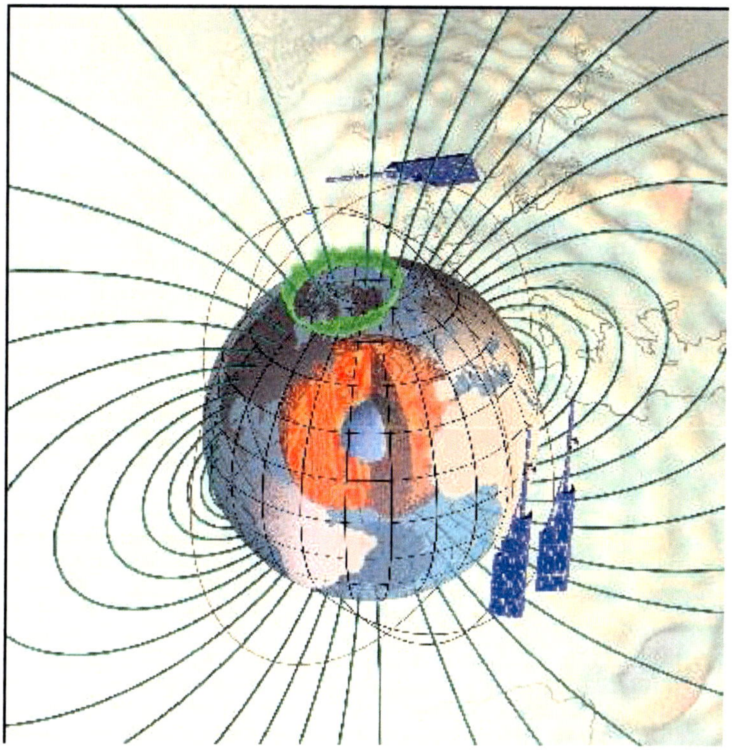

Figure 14. The Swarm magnetic constellation.

Acknowledgements

This chapter was improved by conversations with a large number of people who helped us. Particular thanks to Yoann Quesnel and Mohamed Hamoudi. We would like to thank Alexander Jordan for help in graphics, and David Mozonni for his comments in improving the writing of this manuscript.

References

Backus, G., Parker, R., and Constable, C. (1996). *Foundations of Geomagnetism*. Cambridge Univ. Press, New York.

Blakely, R. J. (1995). *Potential Theory in Gravity and Magnetic Applications*. Cambridge Press, Cambridge.

Corner, B. and Groenewald, P. (1991). Gondwana reunited. *South Afr. Trans. Nav. Antarkt.*, **21**(2):172.

De Santis, A. (1991). Translated origin spherical cap harmonic analysis. *Geophys. J. Int.*, **106**:253–263.

Dyment, J. and Arkani-Hamed, J. (1998). Contribution of lithospheric remanent magnetization to satellite magnetic anomalies over the world's oceans. *J. Geophys. Res.*, **103**:15423–15442.

Gauss, C. F. (1838). *Allgemeine Theorie des Erdmagnetismus*. Resultate aus den Beobachtungen des magnetischen Vereins im Jahre 1838. Göttingen & Leipzig.

Ghidella, M., Yanez, G., and LaBrecque, J. (2002). Revised tectonic implications for the magnetic anomalies of the western weddell sea. *Tectonophysics* **347**:65–86.

Gubbins, D. and Herrero-Bervera, E., editors (2007). *Encyclopedia of Geomagnetism and Paleomagnetism* ISBN: 978-1-4020-3992-8, Springer, Heidelberg.

Haines, G. (1985). Spherical cap harmonic analysis of geomagnetic secular variation over Canada 1960-1983. *J. of Geophys. Res.*, **90**(B15):12563–12574.

Hamoudi, M., Thb̀ault, E., Lesur, V., and Mandea, M. (2007). Geoforschungszentrum anomaly magnetic map (gamma): A candidate model for the world digital magnetic anomaly map. *Geochemistry Geophysics Geosystems (G3)*, **8**:Q06023.

Hildebrand, A. R., Penfield, G. T., Kring, D. A., Pilkington, M., Camargo-Zanoguera, A.and Jacobson, S., and Boynton, W. V. (1991). Chicxulub crater: A possible cretaceous/tertiary boundary impact crater on the yucatan peninsula, Mexico. *Geology*, **19**:867–871.

Holschneider, M., Chambodut, A., and Mandea, M. (2003). From global to regional analysis of the magnetic field on the sphere using wavelet frames. *Physics of the Earth and Planetary Interiors*, **135**(2):107–124.

Hunt, C., Moskowitz, B. M., and Banerjee, S. (1995). Magnetic properties of rocks and minerals. *Rock Physics and Phase Relations, A Hanbook of Physical Constants*, AGU Reference Shelf 3.

Jokat, W., Boebel, T., König, M., and Meyer, U. (2003). Timing and geometry of early gondwana breakup. *J. Geophys. Res.*, **108**(B9):2428, doi:10.1029/2002JB001802.

Korte, M. and Holme, R. (2003). Regularization of spherical cap harmonics. *Geophys. J. Int.*, **153**(1):253–262.

Langel, R. A. and Estes, R. H. (1985). The near-Earth magnetic field at 1980 determined from MAGSAT data. *J. Geophys. Res.*, **90**:2495–2509.

Langlais, B., Purucker, M., and Mandea, M. (2004). *Crustal magnetic field of Mars. J. Geophys. Res.*, **109**. doi:10.1029/2033JE002048.

Lesur, V. and Gubbins, D. (2000). Using geomagnetic secular variation to separate remanent and induced sources of the crustal magnetic field. *Geophys. J. Int.*, **142**:889–897.

Lesur, V. and Maus, S. (2006). A global lithospheric magnetic field model with reduced noise level in the polar regions. *Geophys. Res. Lett.*, **33**. doi:10.1029/2006GL025826.

Lesur, V., Wardinski, I., Rother, M., and Mandea, M. (2008). GRIMM- the GFZ reference internal magnetic model based on vector satellite and observatory data. *Geophys. J. Int.*, **173**. doi:10.1111/j.1365-246X.2008.03724.x.

Lowes, F. J. (1966). Mean-square values on sphere of spherical harmonic vector fields. **71**:2179.

Malin, S. (1987). Historical introduction to geomagnetism. In Jacobs, J. A., editor, *Geomagnetism*, volume 1, pages 1–49. Academic Press, London.

Mandea, M. and Purucker, M. (2005). Measurements of the earth's magnetic field from space. *Surveys in Geophysics*, **26**:415–459.

Maus, S., Lühr, H., Rother, M., Hemant, K., Balasis, G., Ritter, P., and Stolle, C. (2007). Fifth generation lithospheric magnetic field model from CHAMP satellite measurements. *Geochem. Geophys. Geosyst.*,.**26**. doi:10.1029/2006GC001521.

Mc Pherron, R. L. (1991). Physical processes producing magnetospheric substorms and magnetic storms. In Jacobs, J. A., editor, *Geomagnetism*, pages 593–739. Academic Press, London.

Merrill, R., McFadden, P., and McElhinny, M. (1998). *The Magnetic Field of the Earth: Paleomagnetism, the Core, and the Deep Mantle*. Academic Press.

Nabighian, M., Grauch, V., Hansen, R., LaFehr, T., Li, Y., Peirce, J., Phillips, J., and Ruder, M. (2005). The historical development of the magnetic method in exploration. *Geophysics*, **70**:33–61.

Olsen, N., Haagmans, R., Sabaka, T. J., Kuvshinov, A. V., Maus, S., Purucker, M. E., Lesur, V., Rother, M., and Mandea, M. (2006a). The *Swarm* End-To-End mission simulator study: Separation of the various contributions to Earth's magnetic field using synthetic data. *Earth, Planets and Space*, **58**:359–370.

Olsen, N., Lühr, H., Sabaka, T. J., Mandea, M., Rother, M., Tøffner-Clausen, L., and Choi, S. (2006b). CHAOS - a model of Earth's magnetic field derived from CHAMP, Ørsted, and SAC-C magnetic satellite data. *Geophys. J. Int.*, **166**:67–75.

Olsen, N. and Mandea, M. (2008). Rapidly changing flows in the earth's core. *Nature Geosciences*, **1**:390–394, doi:10.1038/ngeo203.

Pilkington, M. and Hildebrand, A. (2000). Three-dimensional magnetic imaging of the chicxulub crater. *Lunar Planetary Science*, **31**:1190–1191.

Quesnel, Y., Weckmann, U., Ritter, O., Stankiewicz, J., Lesur, V., Mandea, M., Langlais, B., Sotin, C., and Galdéano, A. (2009). Simple Models for the Beattie Magnetic Anomaly in South Africa. *Tectonophysics*, doi:10.1016/j.tecto.2008.11.027, in press.

Richmond, A. D. (2002). Modeling the geomagnetic perturbations produced by ionospheric currents, above and below the ionosphere. *J. Geodynamics*, **33**:143–156.

Sabaka, T. J., Olsen, N., and Purucker, M. (2004). Extending comprehensive models of the Earth's magnetic field with Ørsted and CHAMP data. *Geophys. J. Int.*, **159**:521–547, doi: 10.1111/j.1365–246X.2004.02421.x.

Schubert, G., editor. (2007). *Treatise on Geophysics*, volume 5.

Shure, L., Parker, R. L., and Backus, G. E. (1982). Harmonic splines for geomagnetic modelling. *Physics of the Earth and Planetary Interiors*, **28**:215–229.

Stankiewicz, J., Ryberg, T., Schulze, A., Lindeque, A., Weber, M., and de Wit, M. (2007). Initial results from wide-angle seismic refraction lines in the southern cape. **110**:407–418.

Thébault, E., Mandea, M., and J.-J., S. (2006). Modeling the lithospheric magnetic field over France by means of revised spherical cap harmonic analysis (R-SCHA). *J. Geophys. Res.*, **111**:B05102.

Thébault, E., Schott, J. J., and Mandea, M. (2005). Revised Spherical Cap Harmonic Analysis (R-SCHA): Validation and Properties. *J. Geophys. Res.*, **111**, doi:10.1029/2005JB003836.

Vennerstrom, S., Moretto, T., Raststätter, L., and Raeder, J. (2006). Modeling and analysis of solar wind generated contributions to the near-Earth magnetic field. *Earth, Planets and Space*, **58**:451–461.

Vine, J.-V. and D. H. Matthews, (1963). Magnetic anomalies over oceanic ridges. *Nature*, **199**:947–949.

Weckmann, U., Ritter, O., Jung, A., Branch, T., and de Wit, M. (2007). Magnetotelluric measurements across the beattie magnetic anomaly and the southern cape conductive belt, south africa. **112**(B11). doi:10.1029/2005JB0039755.

Whaler, K. and Langel, R. (1996). Minimal crustal magnetization from satellite data. *Phys. Earth Planet. Int.*, **48**:303–319.

In: The Lithosphere: Geochemistry, Geology and Geophysics ISBN: 978-1-60456-903-2
Editors: J.E. Anderson et al, pp. 345-368

Chapter 10

PALEOSHORELINES AND THE EVOLUTION OF THE LITHOSPHERE OF MARS

***Javier Ruiz*[1], *Rosa Tejero*[1], *David Gómez-Ortiz*[2] *and Valle López*[3]**
[1]Departamento de Geodinámica, Facultad de Ciencias Geológicas, Universidad Complutense de Madrid, 28040 Madrid, Spain
[2]Escuela Superior de Ciencias Experimentales y Técnicas-Área de Geología, Universidad Rey Juan Carlos, 28933 Móstoles, Madrid, Spain
[3]Servicio de Sistemas de Información Geográfica, Instituto Geológico y Minero de España, Ríos Rosas 23, 28003 Madrid, Spain

Abstract

The existence of features indicative of shorelines of ancient oceans on Mars has been proposed for several authors. In this chapter we revise the topography of possible Martian paleoshorelines, and their consequences for the amount of water infilling ocean basins when water load is considered. We show that a re-evaluation of paleoshorelines is need. For example, the putative Meridiani shoreline could be the same feature as some portions of the Arabia shoreline. Indeed, elevations in the Meridiani shoreline are roughly similar to that of the Arabia shoreline in northeast Arabia, Utopia (not taken into account the Isidis basin), Elysium, and Amazonis regions. This is still far of an equipotential surface, but a paleoshoreline through these regions and the Meridiani shoreline would be better candidate to represent a paleoequipotential surface than the Arabia shoreline *sensu strito*. Moreover, the elevation of the Arabia shoreline in northern Arabia Terra after is intriguingly close to the mean elevation of the Deuteronilus shoreline, and it cannot be discarded a "mixed" Arabia/Deuteronilus shoreline, which would include the Arabia shoreline in northern Arabia Terra, and the Deuteronilus shoreline elsewhere.

If reality of global shorelines is accepted, as increasing evidence suggests, then present-day topographic variations in these features postdate shorelines formation. So, their topographic range should provide information on large-scale vertical movement of the lithosphere, which in turn provides information on the thermal evolution of Mars. We describe the application of thermal isostasy concept to constraint the ancient thermal state of the lithosphere from present-day paleoshoreline topography. For the ~1.1 km total elevation range of the Deuteronilus shoreline, the relative amplitude of heat flow variations (the ratio between maximum and minimum heat flow) is ≤1.6. This value is clearly lower than that presently observed on continental areas on the Earth. If heat flow variations on Mars are currently greatly disappeared, then the obtained heat flow variations upper limits must be mostly related to the paleoshoreline formation time: the present-day elevation range along Deuteronilus shoreline suggests that differences in the thermal state of the lithosphere in regions along this putative paleoshoreline have been relatively small since the feature was formed, and therefore the absence of lithospheric tectonothermal events by the latest ~3 Gyr, at least. If the Deuteronilus shoreline is a combination of portions of several paleoshorelines, then the total elevation range, and the implied heat flow variations, would be lower, and the lithosphere stability higher.

Introduction

A plenty of studies about the characteristics, thermal state, and evolution of the lithosphere of Mars has been previously performed from several lines of work.

On a hand, numerous efforts have focused on the mechanical properties of the lithosphere, which serve to calculate its effective elastic thickness and thermal structure (e.g., Comer et al., 1985; Solomon and Head, 1990; Anderson and Grimm, 1998; Zuber et al., 2000; Nimmo, 2002; McGovern et al., 2002, 2004; McKenzie et al., 2002; Kieffer, 2004; Ruiz et al., 2006), and the topography and geometry or greath faults, and hence the brittle-ductile transition depth (e.g., Schultz and Lin, 2001; Schultz and Watters, 2001; Vidal et al., 2005). These studies inform about the thermal structure and heat flow in the time when the structures were formed.

On the other hand, thermal history models make predictions about the evolution of surface heat flow and lithospheric thickness, or even crustal growth (e.g., Stevenson et al., 1983; Schubert and Spohn, 1990; Schubert et al., 1992; Grasset and Parmentier, 1998; Nimmo and Stevenson, 2000; Choblet and Sotin, 2001; Spohn et al., 2001; Hauck and Phillips, 2002; Breuer and Spohn, 2003). Obviously, results from thermal history models should be consistent with constraints imposed by the studies of the mechanical properties of the lithosphere.

A third line of information about the evolution of the Martian lithosphere is the analysis of present-day topography of features interpreted as paleoshorelines. Indeed, the presence of features indicative of shorelines of ancient oceans on Mars has been proposed (Parker et al., 1989, 1993; Edgett and Parker, 1997; Clifford and Parker, 2001). If reality of global paleoshorelines is accepted, then present-day topographic variations in these features postdate shorelines formation. So, their topographic range should provide information on large-scale vertical movement of the lithosphere, which in turn would provide information on the thermal evolution of Mars (Ruiz, 2003): total elevation differences along paleoshorelines impose constraints to the differences in the evolution of the thermal structure of the lithosphere in shoreline-crossed regions. In fact, it is possible to make an approximate calculation of the amplitude of the ancient heat flow variations necessary to compensate, through thermal isostasy, elevation differences, and transform the paleoshorelines into equipotential surfaces. Like other geological processes could have produced vertical movements, the results so obtained suppose an upper limit.

Diverse efforts have been carried out to test reality of the proposed paleoshorelines. Specifically targeted MOC images have been interpreted as not supporting the shoreline hypothesis (Malin and Edgett, 1999, 2001), although these results have been disputed (Parker et al., 2001; Clifford and Parker, 2001; Fairén et al., 2003). Moreover, recent works using high resolution images have found clear evidences of erosion in places located in putative paleoshorelines (Webb and McGill, 2003; Webb, 2004), which support its formation in relation to coastal processes. The observed present-day Martian topography (Smith et al., 1999, 2001) has also been used to analyze elevations along of the main proposed paleoshorelines, in order to test their chance to represent true paleoshorelines (Head et al., 1998, 1999; Carr and Head, 2003, Webb, 2004). These analyses obtain that the Late Hesperian Deuteronilus shoreline is a viable paleoshoreline, since this feature slightly deviates from an equipotential surface. Otherwise, the putative older and higher-standing Arabia shoreline deviates substantially from an equipotential surface, indicating that it may not be representative of a true shoreline.

The evaluation of possible paleoshorelines through assessment of present-day topography must be made cautiously, because it is not necessarily true that a paleoequipotential surface must fit well a present-day equipotential surface. Lithosphere rebound due to water unloading associated with the disappearance of an ocean with irregularly shaped margins could result in deviations of equipotentiality of up to several hundreds of meters (Leverington et al., 2003; Leverington and Ghent, 2004). Different thermal histories among regions may have appreciably contributed to the deformation of the original large wavelength topography of putative paleoshorelines: variations in Martian heat flow, similar in relative amplitude to those observed in terrestrial continental tectonothermally stable areas, could result in large wavelength elevation differences of kilometric scale through differential thermal isostasy (Ruiz, 2003, Ruiz et al., 2003, 2004), an important amount of deformation for any possible paleoshoreline. Tectonic processes could also produce vertical movements, and erosional or sedimentary activity could affect the original paleoshoreline signatures (Clifford and Parker, 2001). Moreover, lateral continuity of paleoshorelines is not well established, and diverse division and mixing of the originally proposed features are likely required (Ruiz et al, 2003; Webb, 2004; Ruiz, 2005).

In this chapter we revise and re-evaluate the topography of possible Martian paleoshorelines, and the consequences of the water load for the amount of water infilling ocean basins. We also describe the application of thermal isostasy concept to constraint the ancient thermal state of the lithosphere from present-day paleoshoreline topography, and we present the results obtained and their implications for the evolution of the lithosphere of Mars.

Paleoshorelines

In this section we first revise elevation ranges in the paleoshorelines originally proposed by Parker et al. (1989, 1993) and Edgett and Parker (1997), which were revised, redrawn and renamed (as Deuteronilus, Arabia and Meridiani) by Clifford and Parker (2001) (Figure 1 shows a map of Mars with some of the geographical features mentioned in the text, and Figure 2 shows the paleoshorelines after Clifford and Parker (2001) represented on the Martian topography). Later, we use MOLA topography (Smith et al., 1999, 2001) to refine calculations of the water volume infilling the ancient oceanic basins, by taking into account both present-day topography and the maximum possible effect caused by the water weight on the oceanic floors topography. Finally, we re-evaluate these features through of comparing their respective elevations (geomorphologic revisions or re-evaluations are beyond the scope of this chapter), using the mapping of Clifford and Parker (2001) and MOLA topography.

In this point it is necessary recall that Martian elevations are given with respect to an arbitrary zero elevation level, defined as the equipotential surface (gravitational + rotational) whose average value at the equator is equal to the mean radius (see Smith et al., 2001).

Table 1. Martian chronology after Hartmann and Neukum (2001) and Hartmann (2005).

Period	Age (Gyr)
Late Amazonian	
	0.2-0.6
Middle Amazonian	
	1.4-2.1
Early Amazonian	
	2.9-3.2
Late Hesperian	
	3.2-3.6
Early Hesperian	
	3.5-3.7
Late Noachian	
	3.6-3.9
Middle Noachian	
	3.8-4.1
Early Noachian	

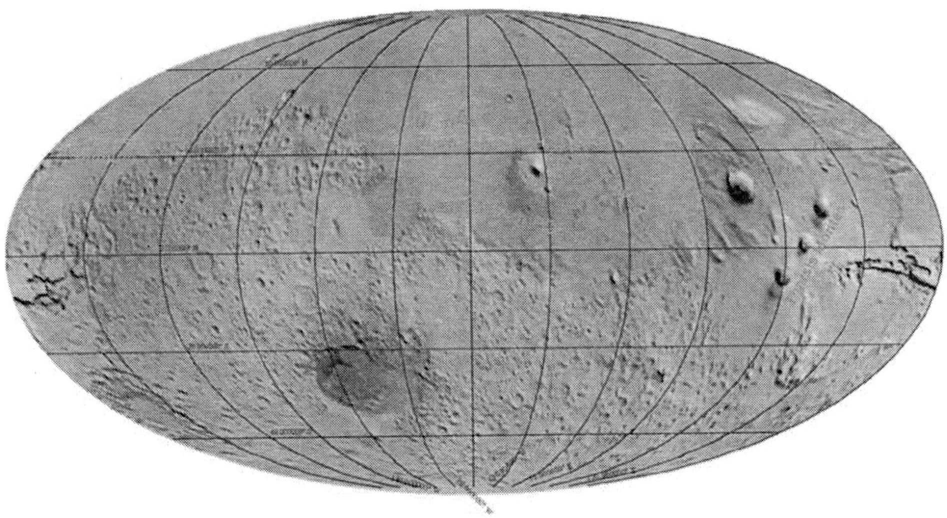

Figure 1. Map of Mars with some of the geographical features mentioned in the text, and other representative features indicated on the Martian topography.

Elevations Range along Deuteronilus, Arabia, and Meridiani Shorelines

The better evidence for a Martian paleoshoreline (dating from the Late Hesperian; temporal equivalences, as obtained from crater counts statistical, are given in Table 1) is the presence in the northern lowlands of a mapping contact (originally named Contact 2) marking the outer boundary of the northern plains, which was interpreted to be the shoreline of an ancient Martian ocean (Parker et al., 1989, 1993). This "contact" was later redrawn and renamed Deuteronilus shoreline by Clifford and Parker (2001). Head et al. (1998, 1999) and Carr and Head (2003), using MOLA data, have shown that this putative shoreline represents a relatively good approximation to an equipotential surface: its mean altitude is –3.792 ± 0.236 km, and its whole topographic range is ~1.1 km, from –3.2 to –4.3 km (Carr and Head, 2003). The elevation of the base levels of the Chryse outflow channels, –3.742 ± 0.153 km (Ivanov and Head, 2001), is close to the mean level of Deuteronilus shoreline, could also indicate that they debouched into a large standing body of water (Head et al., 1999; Ivanov and Head, 2001); this seem also be the case for other channels termini (Salamuniccar, 2004). Additionally, there is clear evidence for erosion along the Deuteronilus shoreline (Webb, 2004).

Alternatively (or complementarily), Carr and Head (2003) considered that the Late Hesperian Vastitas Borealis Formation, which extends for a great part of the northern lowlands, represents better support for the past existence of a large standing body of water on Mars. The Vastitas Borealis Formation has been interpreted as a sedimentary veneer at least 100 m thick on the East Hesperian ridged plains (Head et al., 2002), which could have originated as a sublimation residue from a large (probably frozen) water body (Kreslavsky and Head, 2002). The outer contact of the Vastitas Borealis Formation is coincident with the trace of the Deuteronilus shoreline in the Deuteronilus, Nilosyrtis, Isidis, Tempe, and Chryse regions, but not in Elysium or the Olympus Mons aureole. If the outer contact of the Vastitas Borealis Formation in the Utopia basin is ignored (where it is covered by younger Amazonian Elysium materials, and therefore, the original contact trace is not visible), the outer contact of the Vastitas Borealis Formation has a mean altitude of –3.658 ± 0.282 km, with a whole elevation range of ~1.0 km, from –3.3 to –4.3 km (Carr and Head, 2003; see their Figure 12).

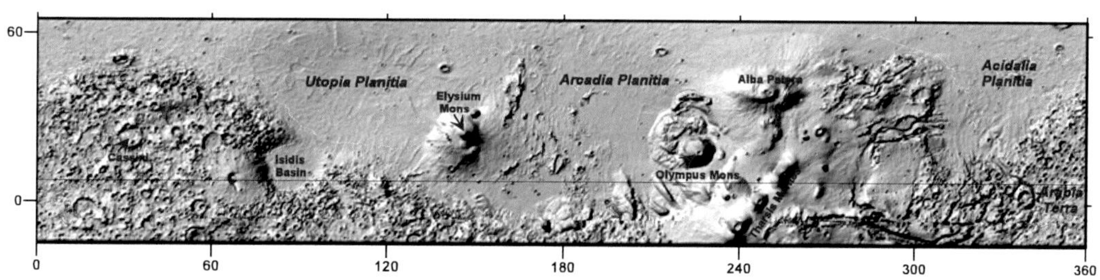

Figure 2. The Deuteronilus (yellow), Arabia (green) and Meridiani (red) shorelines after Clifford and Parker (2001), represented on the Martian topography.

Parker et al. (1989, 1993) also proposed an older, higher-standing Contact 1, later on renamed Arabia shoreline (Clifford and Parker, 2001). This shoreline, which would be of Noachian age (see Clifford and Parker, 2001), is roughly coincident with the Martian dichotomy separating the lowlands from the highlands, and the elevation along its outline highly deviates from an equipotential surface (Head et al., 1998, 1999), and thus it is not a good candidate to paleoshoreline. The topography along the Arabia shoreline is characterized (Carr and Head, 2003) by a mean altitude of –2.090 ± 1.400 km, and a total elevation range of ~5.6 km (from 1.6 to –4.0 km).

Finally, the existence of a pre-Arabia shoreline has been proposed in northern Sinus Meridiani and western Arabia Terra (Edgett and Parker, 1997; Clifford and Parker, 2001); precisely, the MER Opportunity has recently found evidences for an aqueous, maybe sea-related, environment at Meridiani Planum, close to this possible paleoshoreline (Squyres et al., 2004). The mean elevation of this Meridiani shoreline (as named by Clifford and Parker, 2001), would be about –1.5 km (Parker et al., 2000), although its topography has not been examined in previous works. Figure 3 shows constant elevation contours (0.5 km spaced) crossed by the Meridiani shoreline. It can be seen that elevations along the mapped paleoshoreline mostly range between 0 and –2 km. If Hesperian chaos materials (see Tanaka et al., 1992), impact craters, and an isolated peak are not taken into account (see also Figure 6), then total elevation range is ~1 km, from –0.5 to –1.5 km, a reasonable amount for a very old paleoshoreline.

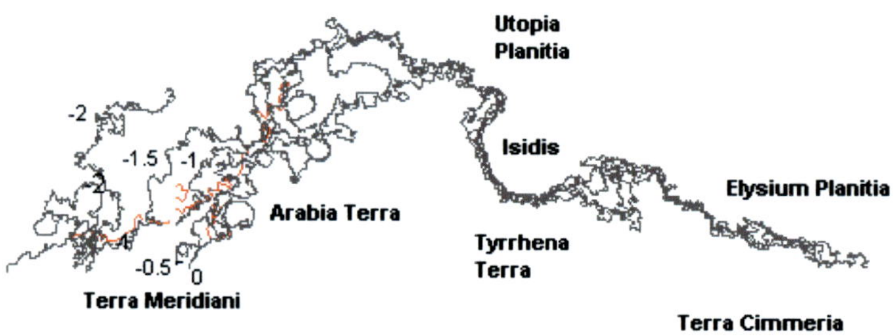

Figure 3. Constant elevation contours in km (black) crossed by the Meridiani shoreline (red). Elevation contours are spaced 0.5 km.

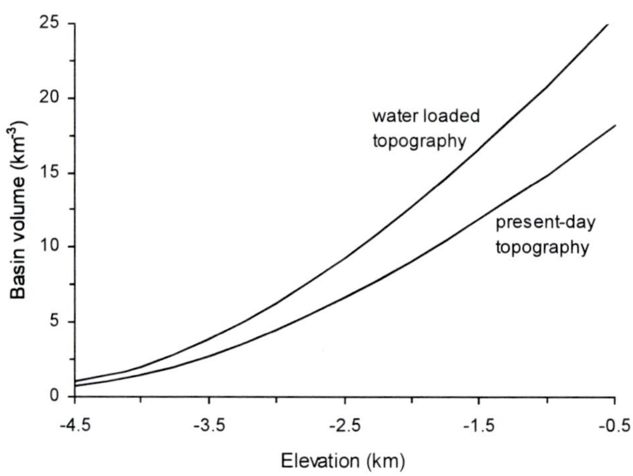

Figure 4. Basin volume as a function of the elevation level, for both, present-day topography and water loaded, Airy compensated, topography (below the -4.35 km level volume includes the contribution from both North Polar and Utopia basins).

Water Volumes in the Ancient Oceans

MOLA topography and mean elevations of proposed paleoshorelines has been previously used to calculate the water volumes contained in the basin related to these putative coastal limits (Head et al., 1998, 1999; Carr and Head, 2003; Öner et al., 2004). Similar calculations are presented in Table 2, and Figure 4 shows basin volume as a function of the elevation level. For the Meridiani shoreline Results in Table 2 are very lower than that in Carr and Head (2003) due to the fact that these authors used a value of 0 for the mean paleoshoreline elevation, whereas we use a mean elevation of −1.5 km following Parker et al. (2000).

Table 2. Basin characteristics at the mean elevations of the proposed Deuteronilus, Arabia and Meridiani shorelines. North polar cap contribution is included in every case.

	Deuteronilus shoreline	Arabia shoreline	Meridiani shoreline
Mean elevation (km) [a]	−3.792	−2.090	−1.5
Basin area enclosed	2.47	4.66	5.34
Present-day basin volume [b] (10^7 km^3)	1.93	8.66	11.88
Total water volume [c,d] (10^7 km^3)	2.00-2.80	8.77-12.28	11.99-16.79
GEL (km) [c,d]	0.14-0.19	0.61-0.85	0.83-1.16
Mean depth (km) [c,d]	0.81-1.13	1.88-2.63	2.25-3.14
Maximum depth (km) [c]	1.46-2.04	3.16-4.42	3.75-5.25

[a] Mean elevations for Deuteronilus and Arabia shoreline after Carr and Head (2003), and for Meridiani Shoreline after Parker et al. (2000),

[b] North polar cap contribution not included.

[c] Lower limit: present-day topography; Upper limit: water loaded, Airy compensated, topography.

[d] North polar cap contribution included.

These calculations assume that the topography of the northern plains has changed little since the putative shorelines were formed. Irrespective of the accuracy of this assumption, estimations based on observed present-day topography can only provide lower limits to the basins volume. Indeed, the existence of oceans in the northern lowlands would imply that the water column provided an additional load over the lithosphere in the regions covered by water (Leverington et al., 2003; Leverington and Ghent, 2004). The effects of variations of the water load on topography are well known for Earth (for a review see Watts, 2001). Therefore, the weight of the water column would have produced the subsidence of the sea floor during the possible periods in which an ocean occupied the lowlands, increasing the basins volume; the subsequent desiccation of the ocean would results in the opposite effect (Hiesinger and Head, 2000; Thomson and Head, 2001; Kreslavsky and Head, 2002; Leverington et al., 2003; Leverington and Ghent, 2004).

We take into account the effect of water load by assuming Airy compensation to calculate upper limits to the basins volume for Meridiani, Arabia and Deuteronilus shoreline. To assume Airy compensation achieved in the mantle implies that elevation variation in the sea floor due to changes in the height of the overlying water column is

$$\Delta y_{\text{floor}} = \frac{y_w \rho_w}{\rho_m},$$

where y_w is the height of the water column (i.e., the ocean depth), and ρ_w and ρ_m are the ocean and mantle densities, respectively. If y_w is taken as the depth of an ancient ocean, and y is the altitude difference between the ocean basin floor for the empty basin and the sea level when the basin was filled, obviously $\Delta y_{\text{floor}} = y_w - y$, and the isostasy principle requires

$$\frac{y_w}{y} = \frac{\rho_m}{(\rho_m - \rho_w)}.$$

So, if y is taken as the present-day depth of the basin under the sea level, then this implies that estimations for the water volume enclosed in the Martian oceans should be increase by a factor y_w/y (with higher ocean density and lower mantle density increasing this factor). If we assume $\rho_w = 1000$ kg m^{-3} and $\rho_m = 3500$ kg m^{-3}, it is obtained $y_w/y = 1.4$. Figure 4 shows basins volume as a function of the elevation level for water loaded, Airy-compensated, topography.

It is important to note that, although water load of putative ancient oceans would result in a substantial increase in the basin volume with respect to calculations based on present-day topography, the assumption of Airy isostasy would imply that the lithosphere has not rigidity, and therefore the increasing factor y_w/y is an upper limit. So, values presented in Table 2 for water volume, GEL (the Global Equivalent Layer if the water is homogeneously distributed on the surface of Mars), mean depth and maximum depth are given as intervals between estimated results for present-day topography and results for water load compensated by Airy isostasy.

In the calculations of total water volumes the north polar cap volume under the appropriate mean shoreline elevation is taken into account (the total volume of the north polar cap is 1.14×10^6 km^3; Smith et al., 2001), since this contributes to the present-day topography. Lithospheric flexure due to north polar cap loading is not considered here, since that the assumption of not flexure represents the lower limit for north polar contribution to the basins volumes, and upper limits are here calculate assuming Airy compensation.

Re-evaluation of Paleoshorelines

The original global mapping of the putative paleoshorelines was limited by resolution of Viking images. Besides this, it is fairly evident that diverse degradational processes could have affected the original morphology (and topography) of any putative paleoshoreline. Thus, reevaluations of paleoshorelines mapping are probably guaranteed. These revaluations would be of great interest to improve the knowledge the hydrogeological history, but also for the tectonothermal history of Mars, because they could affect the elevation range attributed to a given paleoshoreline, and hence the information derived from elevation ranges.

The possibility that the putative Meridiani shoreline could be the same feature as some portions of the Arabia shoreline was first suggested by Ruiz et al. (2003) related to preliminary work about thermal isostasy applied to Mars. Indeed, the mean elevation in the Meridiani shoreline (−1.5 km following Parker et al. (2000)) is roughly similar to that of the Arabia shoreline in northeastern Arabia, Utopia (not taken into account the Isidis impact basin), Elysium, and Amazonis regions. Thus, a possible paleoshoreline might follow the outline of the Arabia shoreline in these regions, but including the outline of Meridiani Shoreline in western Arabia Terra and Sinus Meridiani. The elevation range of this "mixed" Meridiani/Arabia shoreline, although not examined, would be mostly about 2 km, from −1 to −3 km after Ruiz et al. (2004) on the basis of the topography analysis of Arabia shoreline in Carr and Head (2003). This is still far of an equipotential surface, but this Meridiani/Arabia shoreline would be better candidate to represent a paleoequipotential surface than the Arabia shoreline *sensu strito*: for that reason, it was incorporated to the hypothesis for the Martian hydrogeological history suggested by Fairén et al. (2003), in order to represent the boundary of a putative Noachian ocean.

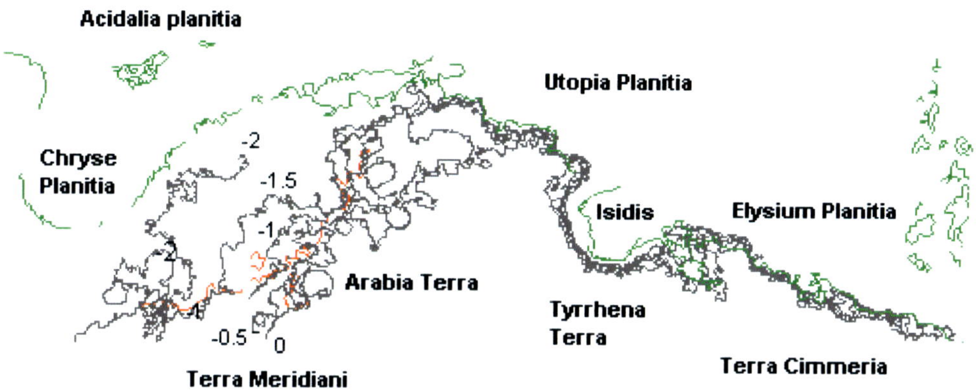

Figure 5. Meridiani (red) and Arabia (green) shorelines, and constant elevation contours in km (black) crossed by the Meridiani shoreline, which are spaced each 0.5 km.

Figure 5 shows Meridiani and Arabia shorelines, as mapped by Clifford and Parker (2001), and constant elevation contours (0.5 km spaced) crossed by the Meridiani shoreline. It can be seen that elevations of the Arabia shoreline at northeastern Arabia, Utopia, and Elysium regions are similar to these along the Meridiani Shoreline, which support the "mixed" Meridiani/Arabia shoreline as a true paleoshoreline. Also is evident that elevations in the Arabia shoreline are much lower in northwestern Arabia Terra, as well as further to the west. Figure 6a shows Meridiani and Arabia shorelines and the contour of the −1.5 km elevation level (the mean level of the Meridiani shoreline after Parker et al. (2000)) superimposed on MOLA topography. The Arabia shoreline at northeastern Arabia, Utopia, Elysium, and Amazonis regions is well close to the −1.5 km elevation level at the majority of places, further supporting the "mixed" Meridiani/Arabia shoreline as a true paleoshoreline. Figure 6b is similar, but showing the contour of the −2.09 kn elevation, corresponding to the mean elevation of the Arabia shoreline *sensu strito*: the Arabia shoreline at Arabia, Utopia, Elysium, and Amazonis regions is at, or generally above, the −2.09 kn elevation level (with the exception of Isidis impact basin, which probably postdates paleoshoreline formation). Thus, the whole topographic range in the Meridiani/Arabia shoreline would be ~1.6 km, between –0.5 and –2.1 km. Moreover, if the Arabia shoreline *sensu strito* is not a true paleoshoreline, then areas, volumes, mean depths and GELs obtained for this "feature" from MOLA topography are not representative of any Martian oceanic stage, but estimations for the –1.5 km elevation level would be roughly appropriate for the Meridiani/Arabia shoreline.

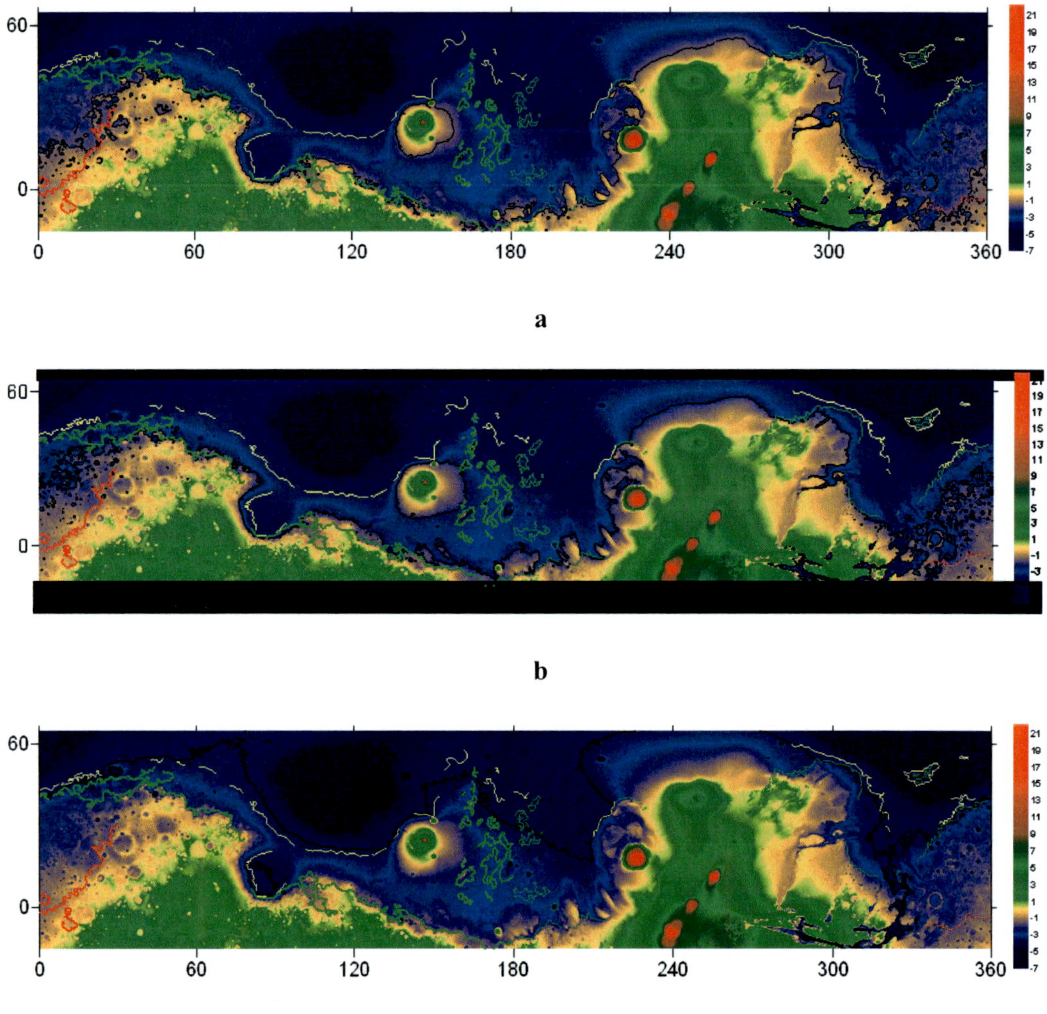

Figure 6. The Deuteronilus (yellow), Arabia (green) and Meridiani (red) shorelines after Clifford and Parker (2001), represented on the Martian topography (scale in km). Also represented (black) are the contour of the (a) -1.5 km, (b) -2.09 km, and (c) -3.792 km elevation levels.

The elevations along putative shorelines on northern Arabia Terra has been recently analyzed in higher resolution (Webb, 2004), finding a elevation of 3707 ± 21 m, for the Arabia shoreline, and two different elevations, 4000 ± 14 m and 4200 ± 12 m, for two separate portions of the Deuteronilus shoreline, which could therefore represent two distinct shorelines. Elevation difference between the two separate portions of the Deuteronilus shoreline is difficultly due to post-formation processes, because elevation in the two portions is nearly constant along distances of about 500 km, and to the clear bimodality in the elevation values. On the other hand, the elevation of the Arabia shoreline in northern Arabia Terra after (Webb, 2004) is intriguingly close to the mean elevation of the Deuteronilus shoreline, and it the possibility of a "mixed" Arabia/Deuteronilus shoreline, which would include the Arabia shoreline in northern Arabia Terra and the Deuteronilus shoreline elsewhere, has been mentioned (Ruiz, 2005).

Figure 6c shows the Arabia and Deuteronilus shorelines, and the contour of the –3.792 km elevation level, the mean elevation of the Deuteronilus shoreline, superimposed on MOLA topography. It is obvious that the Arabia shoreline is very close to –3.792 km elevation level at northwestern Arabia Terra and northeastern Tempe Terra. In turn, Deuteronilus shoreline elevation at Tempe Terra is close to –4 km (see also Carr and Head, 2003), a similar elevation to that found for this feature by Webb (2004) for northern Arabia Terra. Thus, locally at northern Arabia Terra and northeastern Tempe Terra putative shorelines fit well equipotential surfaces, but they are suggesting a complex scenario for the possible evolution of Martian oceans.

Elevation Ranges along Paleoshorelines and Vertical Movements of the Lithosphere

Implications of paleoshoreline reevaluations are evident: the lower the true elevation range of a paleoshoreline the lower the magnitude of vertical movements postdating its formation. For example, elevation range of 1.1 km along the Deuteronilus shoreline is an upper limit, because this range probably includes portions of several different paleoshorelines. Vertical movements deduced from the differences of this feature with respect to an equipotential surface should also be an upper limit.

The fit, although rough, of paleoshorelines to equipotential surfaces would imply a relatively calmed history for the Martian lithosphere, at least since these features were formed. This is clearly the case for the Deuteronilus shoreline *sensu strito*, if this feature is a true paleoshoreline, and moreover for the "revised" Deuteronilus shorelines, which are dating from the Late Hesperian, ~3 Gyr ago (see Table 1). The evidences for a reasonably equipotential Noachian paleoshoreline are attractive, but they must be taken more carefully. If the Meridiani/Arabia shoreline represents a true paleoshoreline, then its whole elevation range of 1.6 km would imply a quite stable lithosphere since 3.5 Gyr ago at least.

By contrast, Gratton et al. (2003) find that the Meridiani, Arabia and Deuteronilus shorelines could fit equipotential surfaces if the reference ellipsoid for the planet has changed with the time, maybe due to the dichotomy formation or Tharsis volcanism. Although this possibility is attractive, it is based on the interpretation of the Meridiani, Arabia, and Deuteronilus *sensu strito* shorelines as true paleoshorelines, which is unlikely (see above).

Thermal isostasy and Thermal Evolution of the Lithosphere

A significant relation exists between Earth's surface elevation and the thermal state of the lithosphere: the warmer the lithosphere, the lower its mean density, and the higher its buoyancy with respect to the underlying fluid materials. This principle (known as thermal isostasy) has been broadly applied to the thermal subsidence of the cooling oceanic lithosphere (e.g., Turcotte and Schubert, 2002), but it can be also applied in a general way to the continental lithosphere (Lachenbruch and Morgan, 1990), including a tectonothermally stable one. This allows the use of topography and surface heat flow data to constrain Earth's continental lithospheric thermal structure (e.g., Tejero and Ruiz, 2002; Lewis et al., 2003). This relation between surface elevation and thermal state of the lithosphere has been also applied to Mars (Ruiz, 2003; Ruiz et al., 2003, 2004).

In this section we present a complete description of the application to Mars of the thermal isostasy concept. We also revise and extend the results obtained about the evolution of the Martian lithosphere.

Thermal Isostasy

Because of thermal expansion and contraction, the elevation of the surface, referenced to the free height of the asthenosphere, depends on the thermal state of the lithosphere and has contributions from the lithospheric mantle and crust,

$$H = H_m + H_c, \tag{1}$$

where H_m and H_c are the lithospheric mantle and crust contribution to the elevation of the surface respectively. (The term lithosphere is used here to define a thermally conductive layer, in which base isostatic compensation can be achieved.) The contribution due to the lithospheric mantle is given by (Lachenbruch and Morgan, 1990)

$$H_m = \alpha(\overline{T}_m - T_a)b_m, \tag{2}$$

where α is the volumetric thermal expansion coefficient, T_a is the temperature of the asthenosphere, is $\overline{T}_m$ the mean temperature of the lithospheric mantle, and b_m is the lithospheric mantle thickness; density differences between asthenospheric and lithospheric mantle are taken as solely due to temperature differences, which is a very reasonable approximation. A similar equation can be written to describe the crustal contribution, but taking into account a correction factor for the lesser crustal density,

$$H_c = \frac{\alpha\rho_c}{\rho_a}(\overline{T}_c - T)b_c, \tag{3}$$

where ρ_c and ρ_a are respectively a reference crustal density and the density of the asthenosphere, and b_c is the crustal thickness. In turn, the mean temperatures of each lithospheric layer is given by

$$\overline{T} = \frac{1}{z_2 - z_1}\int_{z_1}^{z_2} T(z)\mathrm{d}z. \tag{4}$$

where z, z_1 and z_2 are the depth, the depth in the layer top, and the depth at the layer base, respectively. The component of the topography due to thermal isostasy, expressed in terms of heat flow, is (Ruiz, 2003)

$$h = H_m(F_h) + H_c(F_h) - H_m(F_o) - H_c(F_o), \tag{5}$$

where h is the local elevation with respect to a reference elevation, F_h is the local surface heat flow, and F_o is the heat flow for the reference elevation.

Temperature Profiles

Temperature profiles in this chapter are calculated by assuming radioactive heat sources homogeneously distributed in the crust, and linear thermal gradients for the lithospheric mantle mantle. The temperature at a depth z within the crust is given by (Roy et al., 1968)

$$T_z = T_s + \frac{Fz}{k_c} - \frac{Az^2}{2k_c}, \tag{6}$$

where T_s is the surface temperature, F is the surface heat flow, k_c is the thermal conductivity of the crust, and A is the volumetric heat production rate. Ruiz (2003) calculate b_m, $\overline{T}_m$, and $\overline{T}_c$ in terms of the surface heat flow, and the proportion f of the heat flow originated from crustal heat sources, assuming that crustal heat sources are homogeneously distributed. The factor f can be formally defined as

$$f = \frac{Ab_c}{F}; \tag{7}$$

so, within the crust, the temperature at a depth z is

$$T_z = T_s + \frac{Fz}{k_c}\left(1 - \frac{fz}{2b_c}\right). \tag{8}$$

The existence of heat sources in the lithospheric mantle is not take into account, because on Earth radiogenic sources are sparse beneath the near-surface radioactive element-rich layer, and within the lithospheric mantle, the heat flow can be assumed constant (e.g., Turcotte and Schubert, 2002); so, within the mantle lithosphere, the temperature at a depth z is

$$T_z = T_c + \frac{F(1-f)(z-b_c)}{k_m}, \tag{9}$$

where T_c is the temperature at the crust base, calculated taking $z = b_c$ in equation (8); in turn, b_m is calculated from

$$b_m = b_c + \frac{k_m (T_a - T_c)}{F(1-f)} \quad (10)$$

Mean lithospheric mantle and crust temperatures are respectively given by

$$\overline{T}_m = \frac{T_a + T_c}{2}, \quad (11)$$

and

$$\overline{T}_c = T_s + \frac{F b_c (1 - f/3)}{2 k_c} \quad (12)$$

Parameter Values

The calculations have been performed using $\alpha = 3 \times 10^{-5}$ °C^{-1}, $k_c = 2.5$ W m^{-1} °C^{-1}, $k_m = 3.5$ W m^{-1} °C^{-1}, $\rho_c = 2900$ kg m^{-3}, and $\rho_a = 3500$ kg m^{-3} for material properties. The surface temperature is taken as 0°C, maybe more appropriate for a time in which an ocean is assumed than the current mean surface temperature of about –50°C (although the results are relatively insensitive to the election of this parameter). The asthenosphere temperature is taken as 1300°C, which is a value typically used for the Earth's asthenosphere (e.g., Ranalli, 1997). Crustal thickness is assumed to be 40 km, in accordance with the typical mean crustal thickness below the northern lowlands derived from topography and gravity data (Zuber et al., 2000). Two possibilities have been taken for the value of f in equation (8), although we note that this value could locally vary: $f = 0$ (corresponding to a linear thermal gradient through the crust) and $f = 0.5$. The latter value is in accordance with the proposal (made from geochemical arguments drawn from the materials on Mars' surface) that perhaps over 50% (or even 75%) of radioactive heat sources in this planet are placed in its crust (McLennan, 2001, 2003); similarly, in the Earth, the 40-60% of the heat flow lost in continental areas originates from crustal heat sources (Pollack and Chapman, 1977; Turcotte and Schubert, 2002).

Deformation of Paleoshorelines

Ruiz et al. (2004) have used the thermal isostasy concept to show that it is not necessarily true that a paleoequipotential surface implies a good fit to a present-day equipotential surface. A similar conclusion has been obtained taking into account the lithosphere rebound due to water uploading associated to the disappearance of an ocean of irregularly shaped margins (Leverington et al., 2003; Leverington and Ghent, 2004). Here we present the first-order calculations of the magnitude of the relation between possible variations in the thermal state of the lithosphere and elevation differences, which could cause concrete deviations in equipotentiallity along the proposed paleoshorelines. Note that for the purposes of this chapter, only large wavelength topographic differences are relevant, since that the rigidity of the Martian lithosphere could prevent small-scale isostatic adjustment.

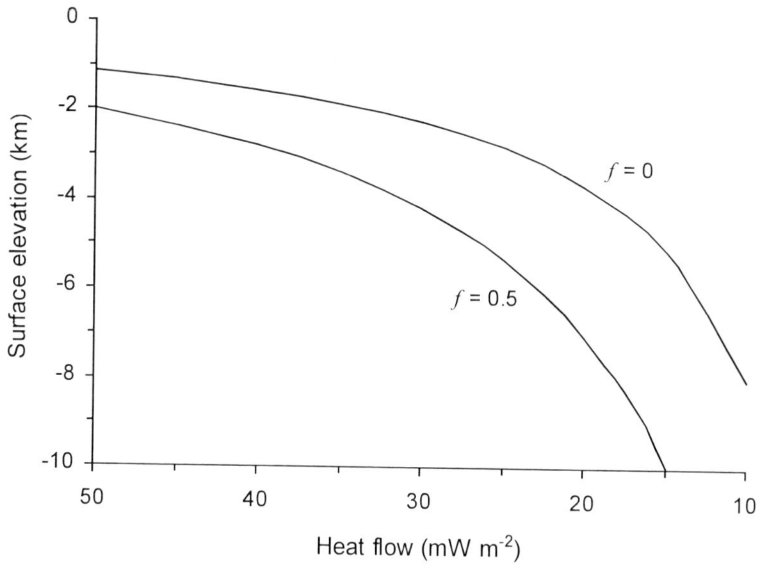

Figure 7. Elevation over the free height of the asthenosphere in terms of surface heat flow for $f = 0$ and $f = 0.5$. Surface heat flow is represented in reverse order (describing the topographic evolution of a cooling region isostatically compensated). Adapted from Ruiz et al. (2004).

Figure 7 shows H in terms of surface heat flow for $f = 0$ and $f = 0.5$. Calculations have been made for a range of F values between 10 and 50 mW m^{-2}, which roughly correspond to the whole range of surface heat flows proposed for diverse regions and times using estimates of the elastic thickness of the lithosphere (McGovern et al., 2002, 2004). It is important to note that the f value can change along a possible paleoshoreline (for example, due to local variations in crustal heat sources, mantle heat flow, or both), or with time (due to waning of radiogenic dissipation intensity or to changes in the efficiency of convective heat transfer). In any case, these possibilities are not important for the purpose of this first-order calculation, which is show the feasibility of differential thermal isostasy histories to affect the large wavelength topography of possible Martian paleoshorelines. For the purposes of this analysis the interesting point is the relative differences of H, and not the absolute values obtained for this parameter (planetary topographies are referred to arbitrary datum). Figure 7 indicate that variations in thermal state of the lithosphere can result in differential thermal isostasy, which in turn can result in important elevation differences, even of kilometric scale as occur on the Earth (Lewis et al., 2003).

Assuming a Martian paleoshoreline, the posterior attenuation, disappearance (as is expected with the waning of internal heat sources), or formation (if reheating of the lithosphere postdating the shoreline formation occurred) of heat flow variations must result in the deformation of the paleoshoreline topography, deviating it from an equipotential surface.

Figure 8 shows the relative amplitude of surface heat flow variations that can produce elevation ranges of 1 and 0.5 km centered on the *H* value corresponding to a reference heat flow in the range from 10 to 50 mW m^{-2}. The relative amplitude of heat flow variations is obtained as the quotient between the maximum and minimum heat flow that can produce positive and negative elevations, respectively, of 0.5 and 0.25 km with respect to the reference *H* value. In Figure 8 it can be seen that ancient surface heat flow variations less than a factor ~2 may account for differences of elevation of 1 km. An elevation range of 0.5 km could be produced by surface heat flow variations less than a factor of ~1.5. These values are further lowered if a substantial amount of the Martian heat sources are located within the crust.

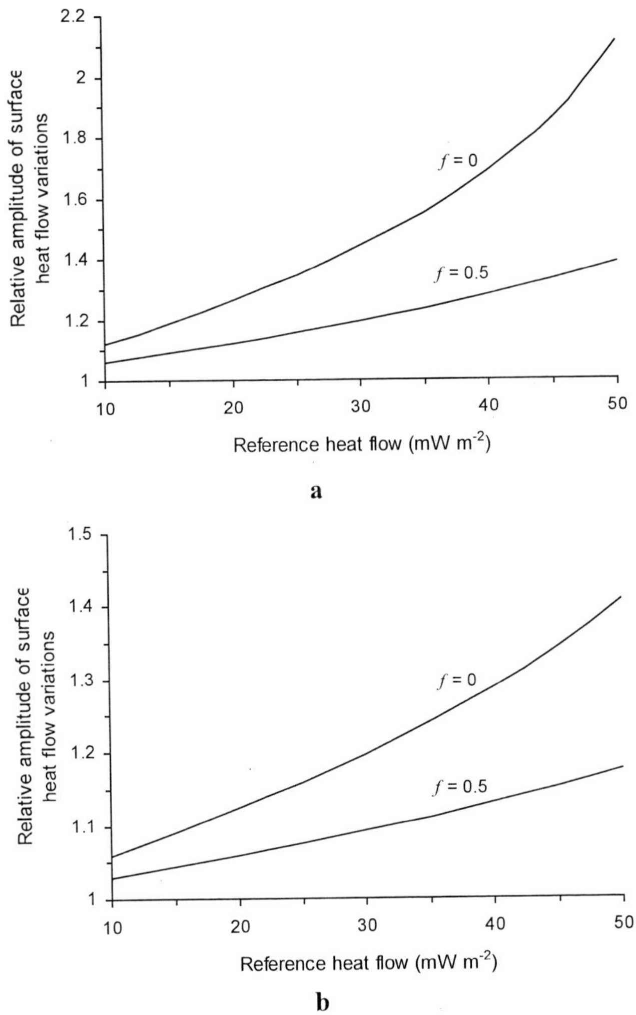

Figure 8. Relative amplitude of surface heat flow variations that can produce elevation ranges of 1 (a) and 0.5 (b) km centered on the *H* value corresponding to a reference heat flow in the range from 10 to 50 mW m^{-2}. As in Figure 1 surface heat flow is represented in reverse order. Adapted from Ruiz et al. (2004).

On the Earth, present-day surface heat flow variations expand by more than one order of magnitude (e.g. Pollack et al., 1993). The major part of these variations is due to plate tectonics, but for Mars an early phase of plate tectonics is controversial. Variations in surface heat flow in Earth's continental regions, from contoured maps (Cermak, 1993; Pollack et al., 1993), can be higher than a factor of 2 or 3, sufficient for important deformation of paleoshorelines (see Figure 7). Those areas include terrains of different ages, and it is known for continental areas that an inverse relation exists between surface heat flow and age of the last tectonothermal stabilization (e.g., Hamza, 1979; Vitorello and Pollack, 1980; Cermak, 1993). Moreover, continental heat flow depends also on a wide array of factors, as for example radioactive heat production in the crust, local mantle heat flow, or tectonic or erosive redistribution of crustal heat producing elements (for reviews see Beardsmore and Cull, 2001; Sandiford and McLaren, 2002).

In any case, heat flow variations on old and tectonothermally stable terrestrial continental areas can be as high as a factor ~1.5-2 (e.g., Cermak, 1993; Roy and Rao, 2000; Rolandone et al., 2002). If local variations of surface heat flow of at least similar amplitude existed in Mars during any moment of its history, then these results indicate that differential thermal isostasy should result in important deformation, and deviation of equipotentiallity, along putative shorelines. Moreover, it is significant that if the half of the surface heat flow was originated from crustal heat sources (as it is the case on the Earth) when the paleoshorelines were formed, then heat flow variations lower than a modest factor of ~1.2-1.4 may account for present-day elevation ranges of 0.5-1 km (if, as it seem reasonable, these heat flow variations are currently greatly attenuated). These elevation ranges are respectively similar to the ±1 standard deviation and whole elevation ranges in the Deuteronilus shoreline, but they represent an important amount of deformation along any possible paleoshoreline.

It is important remind that thermal isostasy is only a contributor to the topography. Though it is not our intentions to discuss the many other factors that may have contributed to the modification of an equipotential surface, we highlight that other possibilities may be degradation by wind, water, tectonic, and volcanic modification (e.g., Clifford and Parker, 2001; Fairén et al., 2003), rebound of the lithosphere due to dissipation of a water body (Leverington et al., 2003; Leverington and Ghent, 2004; for a review of isostatic and flexural effects related to changes in sea level see Watts, 2001), flexure (non-thermal) isostasy due to surface loading, erosion, or subsurface magmatic intrusions. In fact, endogenic-driven geologic activity (probably implying vertical movements) and exogenic activity clearly postdates the possible paleoshorelines in Arabia Terra and Tharsis and Elysium (e.g., Head et al., 1999; Anderson et al., 2001). Together, all those possibilities make more pressing the main argument in this section: any paleoequipotential surface dating of the ancient Mars must be importantly deformed at present, even in a range of elevations of a kilometric scale.

Ancient Heat Flow Variations

If the existence of paleoshorelines is accepted, then elevation differences along these paleoshorelines impose constraints to the differences in the evolution of the thermal structure of the lithosphere in shoreline-crossed regions (Ruiz, 2003). Indeed, it is possible to calculate the amplitude of heat flow variations necessary to compensate present-day topography and transform a paleoshoreline into an equipotential surface. If surface heat flow variations on

Mars are currently almost disappeared (as it would expectable for an efficiently cooled planet), then the heat flow variations deduced from shoreline topography must be mostly related to the time when this feature was formed. Like other than thermal isostasy processes could have produced vertical movements in the shoreline-crossed regions, the results obtained in this way suppose an upper limit to the amplitude of heat flow variations. Diverse processes, including geomorphologic evolution, can affect small-scale topographic variations, and the rigidity of the Martian lithosphere could also prevent small-scale isostatic adjustment; therefore, large wavelength topographic variations, in which isostatic adjustment can work, are more relevant again for the purposes of this argument.

The calculations have been performed for the Deuteronilus shoreline, but the argument and results are also valid for the outer contact of the Vastitas Borealis Formation (if this is considered to be an ancient oceanic limit) since the ranges of elevations are similar in both features. So, h is taken as ±0.55 km, for a total elevation range of 1.1 km for the Deuteronilus shoreline. This corresponds to a reference elevation (for which $h = 0$ and $F_h = F_o$) of –3.75 km, close to the mean elevation of the Deuteronilus shoreline and to the mean elevation of the termini of the Chryse outflow channels. Like elevation differences along the shoreline must be compensated by heat flow variations, h should have sign minus in equation (5) for mathematical consistency. The F_o value is not known, and for that reason, the calculations have been performed for a range of F_o values between 15 and 35 mW m^{-2}, a range based on estimates (uncertainty included) of the late Hesperian/early Amazonian elastic lithosphere thickness (Zuber et al., 2000; McGovern et al., 2002, 2004).

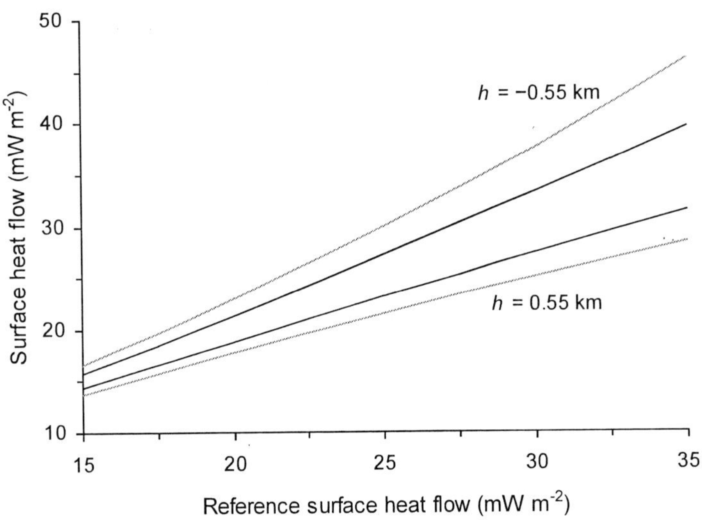

Figure 9. F_h values for h = 550 m (lower curves) and h = –550 m (upper curves) in terms of F_o. Gray and black lines indicate f = 0 and f = 0.5, respectively. In each case, the difference between the upper and lower curves gives the maximum surface heat flow variations permitted assuming the Deuteronilus shoreline as a paleo-equipotential surface. Adapted from Ruiz (2003).

Figure 9 shows F_h values for h = ±0.55 km in terms of F_o. In each case, the difference between the upper and lower curves gives the maximum heat flow variations allowed, taking into account the Deuteronilus shoreline topography. Figure 10 shows upper limits to the relative amplitude of surface heat flow variations on Deuteronilus shoreline locations; these relative amplitude upper limits are the ratio between the maximum and minimum values shown in Figure 9.

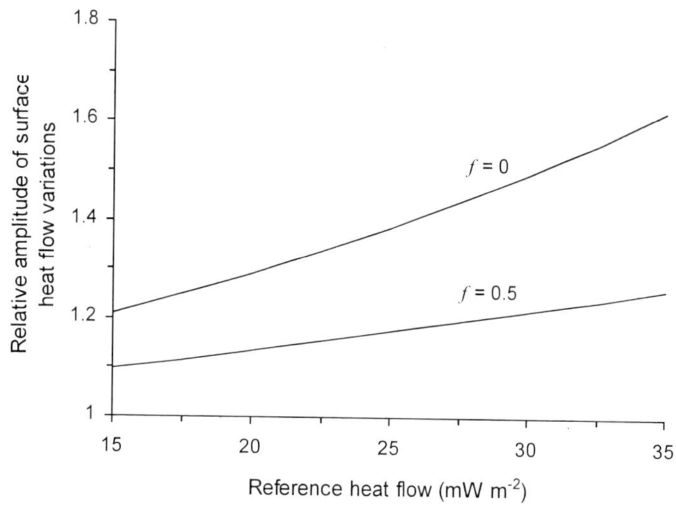

Figure 10. Upper limits to the relative amplitude of surface heat flow variations on the Deuteronilus shoreline locations. These relative amplitude upper limits are given by the ratio between upper and lower values. Adapted from Ruiz (2003).

It can be seen in Figures 9 and 10 that variations in heat flow in regions through the Deuteronilus shoreline were small in the Late Hesperian. In fact, the obtained upper limits for the relative amplitude of these variations are, at most, a factor of 1.6. If crustal heat sources are taken into account, the magnitude of the relative amplitude variations decreases for each given value of F_0. For the F_0 range used here, the depth to the 1300°C isotherm is ~100-300 km for $f = 0$ and ~200-600 km for $f = 0.5$. Since the small radius of Mars the calculations for the case $f = 0.5$ (at low heat flows values) should take in account the spherical shape of Mars. In addition, the range of F_0 values used here is based on works assuming linear thermal gradients. Calculation of surface heat flows from elastic thicknesses would result in higher values if heat sources are present in the crust (Solomon and Head, 1990; Ruiz et al., 2005). This, in turn, decreases the depth to the 1300°C isotherm and also increases the relative amplitude of variations in F_h for the case $f = 0.5$. As the relative amplitude of F_h variations is clearly lower in the $f = 0.5$ case than in the $f = 0$ one, the main conclusions so obtained are not altered.

The upper limits for the relative amplitude of heat flow variations obtained here are clearly lower than those presently observed on Earth. On our planet, the higher heat flows are associated with sea floor spreading centers, but there is not clear evidence for a phase of plate tectonics in the Mars's history (and, in any case, not for the late Hesperian or later on), and so, those very high heat flows are not relevant for the purposed of is chapter. As above mentioned, heat flow show variations in continental areas can be higher than a factor 2-3, and as high as a factor ~1.5-2 in tectonothermally stable terrestrial continental areas. Those areas include terrains of different ages, and it is known for continental areas that an inverse relation exists between surface heat flow and age of the last tectonothermal stabilization (e.g., Hamza, 1979; Vitorello and Pollack, 1980; Cermak, 1993).

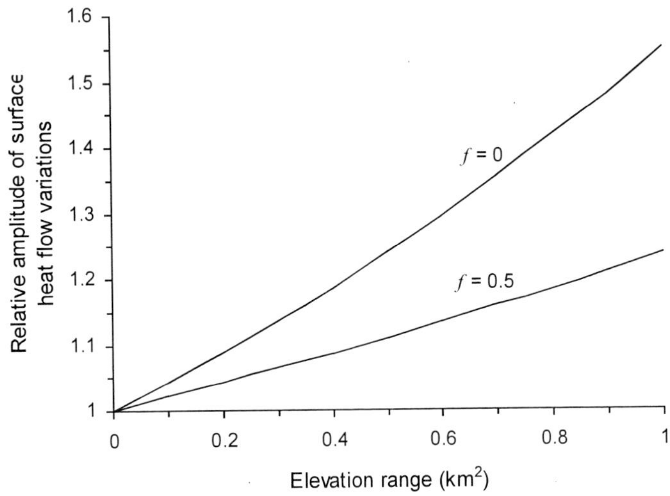

Figure 11. Upper limits to the relative amplitude of surface heat flow variations in terms of the total elevation range of a paleoshoreline, calculated for F_0 = 35 mW m^{-2} (the upper limits in the range of heat flows assumed for analyze the Deuteronilus shoreline case) at the *H* value corresponding to the central value in the elevation range.

If surface heat flow variations on Mars are currently almost disappeared, then the upper limits to the heat flows variations deduced of Deuteronilus shoreline topography are related to the time when this feature was formed (i.e., the Late Hesperian, ~3 Gyr ago; Hartmann and Neukum, 2001). In this case, the present-day elevation range along Deuteronilus shoreline suggests that differences in the thermal state of the lithosphere in the "Deuteronilus shoreline regions" have been relatively small since the feature was formed, and therefore, that very large areas of the Martian lithosphere has been tectonothermally stable since the Late Hesperian. This is consistent with near complete building of the Tharsis rise by the end of the Noachian (Phillips et al., 2001), with a significant decrease in volcanic resurfacing rates following the Hesperian's end (e.g., Hartmann and Neukum, 2001) and with the localization of the waning Amazonian magmatic and tectonic activity at areas in Tharsis and Elysium (e.g., Anderson et al., 2001; Dohm et al., 2001; Head et al., 2001).

Alternatively, reheating of the lithosphere postdating the Deuteronilus shoreline could have caused, or contributed to, the distortion of the topography. In this case, the reheating should have been maintained (at least partially) until the present time, since the dissipation of the thermal anomalies should lead to the disappearance of their effect on the topography. But, as above mentioned, Amazonian geological activity represents the waning and localized magmatic and tectonic activity on Mars, and for that reason, the obtained upper limits to the heat flow variations more probably refer to the thermal state of the Martian lithosphere when the Deuteronilus shoreline was formed.

In summary, if the Deuteronilus shoreline (or equivalently, the outer contact of the Vastitas Borealis Formation) represents a Late Hesperian paleo-equipotential surface, then three conclusions can be deduced from these calculations. First, the relative variations in Late Hesperian surface heat flow in shoreline regions were lower than relative variations in present-day surface heat flow in continental areas on Earth. Second, if substantial amounts of radiogenic heat sources are located in the Martian crust, those relative variations are likely lower. Finally, very large areas of the Martian lithosphere have been tectonothermally stable since (at least) the Late Hesperian.

If the Deuteronilus shoreline is a feature combining portions of several paleoshorelines, then the total elevation range would be lower, and the arguments above more pressing, as clearly illustrated by Figure 11. This figure shows upper limits to the relative amplitude of heat variations, for $f = 0$ and $f = 0.5$, as functions of the total elevation range of a paleoshoreline. As only upper limits for a given elevation range are represented, the calculations have been performed for a F_0 value of 35 mW m^{-2} (the upper limits in the range of heat flows assumed for analyze the Deuteronilus shoreline case) at the H value corresponding to the central value in the elevation range, and therefore $h = \pm\Delta H/2$.

Conclusion

The range of elevations of the Deuteronilus shoreline is strongly indicating the great tectonothermal stability of the Martian lithosphere since the Late Hesperian, at least. If an important proportion of radioactive heat sources are located in the crust, as geochemical evidences suggest, the lithosphere should have been greatly thermally homogeneus in "Deuteronilus shoreline" regions when coastal processes creating this paleoshoreline were working.

There are evidences suggesting that the lateral continuity of the originally proposed paleoshorelines is not well established: diverse division and mixing of the originally proposed paleoshorelines seem be required. These revaluations are of great interest for the understanding of the evolution of the Martian lithosphere, because they would modify the elevation range attributed to a given paleoshoreline. Thus, it is necessary a careful geomorphologic reassessment of the diverse features interpreted as paleoshorelines and of the relation among them. The Deuteronilus shoreline (*sensu strito*) seems integrate portions of several separate paleoshorelines, which would imply a still higher thermal stability and homogeneity of the lithosphere by the latest ~3 Gyr. The confirmation of the Meridiani/Arabia shoreline would be indicative of a longer lithospheric stability.

Acknowledgments

The authors thank Frank Columbus for your invitation to prepare this chapter. JR was supported by a grant of the Spanish Secretaría de Estado de Educación y Universidades.

References

[1] Anderson, S.; Grimm, R. E. *J. Geophys. Res.* 1998, 103, 11,113-11,124.
[2] Anderson, R. C.; Dohm, J. M.; Golombek, M. P.; Haldemann, A. F. C.; Franklin, B. J.; Tanaka, K. L.; Lias, J.; Peer, B. *J. Geophys. Res.* 2001, 106, 20,563-20,585.
[3] Beardsmore, G. R.; Cull, J. P. *Crustal Heat Flow. A Guide to Measurement and Modelling*; University Press: Cambridge, UK, 2001; pp 1-324.
[4] Breuer, D.; Spohn, T. *J. Geophys. Res.* 2003, 108 (E7), 8.1-8.13.
[5] Carr, M. H.; Head, J.W. *J. Geophys. Res.* 2003, 108 (E5), 8.1-8.28.
[6] Cermak, V. *Phys. Earth Planet. Inter.*1993, 79, 179-193.
[7] Choblet, G.; Sotin, C. *Geophys. Res. Lett.* 2001, 28, 3035-3038.
[8] Clifford, S. M.; Parker, T. J. *Icarus* 2001, 154, 40–79.
[9] Comer, R. P.; Solomon, S. C.; Head, J. W. *Rev. Geophys.* 1985, 23, 61-92.
[10] Dohm, J. M.; Ferris, J. C.; Baker, V. R.; Anderson, R. C.; Hare, T. M.; Strom R. G.; Barlow N. G.; Tanaka, K. L.; Klemaszewski, J. E.; Scott D. H. *J. Geophys. Res.* 2001, 106, 32,943-32,958.
[11] Edgett, K. S.; Parker, T. J. *Geophys. Res. Lett.* 1997, 24, 2897-2900.
[12] Fairén, A. G.; Dohm, J. M., Baker V. R.; de Pablo M. A.; Ruiz J.; Ferris J. C.; Anderson, R. C. *Icarus* 2003, 165, 53-67, 2003.
[13] Grasset, O.; Parmentier, E. M. *J. Geophys. Res.* 1998, 103, 18,171-18.181.
[14] Gratton, R.; Mareschal, J.; Sotin, C. *Eos Trans. AGU Fall Meet. Suppl.* 2003, 84(46), abstract G41B-0033.
[15] Hamza, V. M. *Pure Appl. Geophys.* 1979, 117, 65-74.
[16] Hartmann, W. K. *Icarus* 2005, 174, 294-320.
[17] Hartmann, W. K.; Neukum, G. *Space Sci. Rev.* 2001, 96, 165-194.
[18] Hauck, S. A.; Phillips, R. J. *J. Geophys. Res.* 2002, 107 (E7), 6.1-6.19.
[19] Head, J. W.; Kreslavsky, M.; Hiesinger, H.; Ivanov, M. A.; Pratt, S.; Seibert, N.; Smith, D. E.; Zuber, M. T. *Geophys. Res. Lett.* 1998, 25, 4401-4404.
[20] Head, J. W.; Hiesinger, H.; Ivanov, M. A.; Kreslavsky, M. A.; Pratt, S.; Thomson, B. J. *Science* 1999, 286, 2134-2137.
[21] Head, J. W.; Greeley R.; Golombek, M. P.; Hartmann, W. K.; Hauer, E.; Jaumann, R.; Masson, P.; Neukum, G.; Nyquist, L. E.; Carr, M. H. *Space Sci. Rev.* 2001, 96, 263-292.
[22] Head, J. W.; Kreslavsky, M. A.; Pratt, S. *J. Geophys. Res.* 2002, 106 (E1), 3.1-3.29.
[23] Hiesinger, H.; Head, J. W. *J. Geophys. Res.* 2000, 105, 11,999-12,022.
[24] Ivanov, M. A.; Head, J. W. *J. Geophys. Res., 106*, 3275-3295, 2001.
[25] Kiefer, W. S. *Earth Planet. Sci. Lett.* 2004, 222, 349-361.
[26] Kreslavsky, M. A.; Head, J. W. *J. Geophys. Res.* 2002, 107 (E12), 4.1-4.25.
[27] Lachenbruch, A. H.; Morgan, P. *Tectonophysics* 1990, 174, 39–62.
[28] Leverington, D. W.; Ghent R. R. *J. Geophys. Res.* 2004, 109 (E01005), 1-10.
[29] Leverington, D. W.; Ghent, R. R.; Irwin R. P.; Craddock R. A.; Maxwell T. A. *Lunar Planet. Sci.* 2003, 34, abstract 1282.
[30] Lewis, T. J.; Hyndman, R. D.; Flück P. *J. Geophys. Res.* 2003, 108 (B6), 16.1-16.18.
[31] Malin, M. C.; Edgett, K.C. *Geophys. Res. Lett.* 1999, 26, 3049-3052.
[32] Malin, M. C.; Edgett, K.C. *J. Geophys. Res.* 2001, 106, 23,429-23,570.

[33] McGovern, P. J.; Solomon, S. C.; Smith, D. E.; Zuber, M. T.; Simons, M.; Wieczorek, M. A.; Phillips, R. J.; Neumann, G. A.; Aharonson, O.; Head, J. W. *J. Geophys. Res.* 2002, 107 (E12), 19.1-19.25.
[34] McGovern, P. J.; Solomon, S. C.; Smith, D. E.; Zuber, M. T.; Simons, M.; Wieczorek, M. A.; Phillips, R. J.; Neumann, G. A.; Aharonson, O.; Head, J. W. *J. Geophys. Res.* 2004, 109 (E07007), 1-5.
[35] McKenzie, D.; Barnett, D. N.; Yuan, D. L. *Earth Planet. Sci. Lett.* 2002, 195, 1-16.
[36] McLennan, S. M. *Geophys. Res. Lett.* 2001, 28, 4019-4022.
[37] McLennan, S. M. *Sixth International Conference on Mars* 2003, abstract 3099.
[38] Nimmo, F.; *J. Geophys. Res.* 2002, 107 (E11), 27.1-27.16.
[39] Nimmo, F.; Stevenson, D. J. *J. Geophys. Res.* 2000, 105, 11,969-11.979.
[40] Öner, A. T.; Ruiz, J.; Fairén, A. G.; Tejero, R.; Dohm J. M. *Lunar Planet. Sci.* 2004, 35, abstract 1319.
[41] Parker, T. J.; Saunder, R. S.; Schneeberger, D. M. *Icarus* 1989, 82, 111-145.
[42] Parker, T. J.; Gorsline, D. S.: Saunders, R. S.; Pieri, D. C.; Schneeberger, D. M. *J. Geophys. Res.* 1993, 98, 11,061-11,078.
[43] Parker, T. J.; Clifford, S. M.; Banerdt, W. B. *Lunar Planet. Sci.* 2000, 31, abstract 2033.
[44] Parker, T. J.; Grant, J. A.; Franklin B. J.; Rice J. W. *Lunar Planet. Sci.* 2001, 32, abstract 2051.
[45] Phillips R. J. et al. *Science* 2001, 291, 2587-2591.
[46] Pollack, H. N.; Chapman, D. S. *Earth Planet. Sci. Lett.* 1977, 34, 174-184.
[47] Pollack, H. N.; Hunter, S. J.; Johnson, J. R. *Rev. Geophys.* 1993, 31, 267-280.
[48] Ranalli, G., *Geol. Soc. Spec. Publ.* 1997, 121, 19-37.
[49] Rolandone, F.; Jaupart, C.; Mareschal, J. C.; Gariépy, C.; Bienfait, G.; Carbonne, C., Lapointe, R. *J. Geophys. Res.* 2002, 107 (B12), 7.1-7.19.
[50] Roy, R. F.; Blackwell, D. D.; Birch, F. *Earth Planet. Sci. Lett.* 1968, 5, 1-12.
[51] Roy, S.; Rao, R. U. M. *J. Geophys. Res.* 2000, 105, 25,587-25,604.
[52] Ruiz, J. *J. Geophys. Res.* 2003, 108 (E11), 8.1-8.5.
[53] Ruiz, J. *Lunar Planet. Sci.* 2005, 36, abstract 1135.
[54] Ruiz, J.; Fairén, A. G.; de Pablo, M. A. *Lunar Planet. Sci.* 2003, 34, abstract 1090.
[55] Ruiz, J.; Fairén, A. G.; Dohm, J. M.; Tejero, R. *Planet. Space Sci.* 2004, 52, 1297-1301.
[56] Ruiz, J.; McGovern, P. J.; Tejero, R. *Earth Planet. Sci. Lett.* 2006, 241, 2-10.
[57] Salamuniccar, G. *Lunar Planet. Sci.* 2004, 35, abstract 1992.
[58] Sandiford, M.; McLaren, S. *Earth Planet. Sci. Lett.* 2002, 204, 133-150.
[59] Schultz, R. A.; Lin, J. *J. Geophys. Res.* 2001, 106, 16,549-16,566.
[60] Schultz, R. A.; Watters, T. R. *Geophys. Res. Lett.* 2001, 28, 4659-4662.
[61] Solomon, S. C.; Head, J. W. J. Geophys. Res. 1990, 95, 11,073-11,083.
[62] Schubert, G.; Spohn, T. *J. Geophys. Res.* 1990, 95, 14,095-14,104.
[63] Schubert, G.; Solomon, S. C., Turcotte, D. L.; Drake, M. J., Sleep, N. H. In *Mars*; Kieffer H.; Jakosky, B. M.; Snyder, C. W.; Matthews, M. S.; Eds.; University of Arizona Press: Tucson, AZ, 1992; pp 249-297.
[64] Smith, D. E. et al. *Science* 1999, 284, 1495-1503.
[65] Smith, D. E. et al. *J. Geophys. Res.* 2001, 106, 23,689-23,722.
[66] Spohn, T.; Acuña, M. H.; Breuer, D.; Golombek, M.; Greeley, R., Halliday, A.; Hauber, E.; Jaumann, R.; Sohl, F. *Space Sci. Rev.* 2001, 96, 231-262.
[67] Squyres, S. W. et al. *Science* 2004, 306, 1709-1714.

[68] Stevenson, D. J.; Spohn, T.; Schubert, G. *Icarus* 1983, 54, 466-489.

[69] Tanaka, K. N.; Scott, D.H.; Greeley, R. 1992. In *Mars*; Kieffer H.; Jakosky, B. M.; Snyder, C. W.; Matthews, M. S.; Eds.; University of Arizona Press: Tucson, AZ, 1992; pp 345-383

[70] Tejero, R.; Ruiz, J. *Tectonophysics* 2002, 350, 49-62.

[71] Thomson, B. J.; Head, J. W. *J. Geophys. Res.* 2001, 106, 23,209-23,230.

[72] Turcotte, D. L.; Schubert, G. *Geodynamics. Second edition*; Cambridge University Press: Cambridge, UK, 2002; pp 1-456.

[73] Vidal, A.; Mueller, K. M.; Golombek, M. P. *Lunar Planet. Sci.* 2005, 36, abstract 2333.

[74] Vitorello, I.; Pollack, H. N. *J. Geophys. Res.* 1980, 85, 983-995.

[75] Watts, A. B. *Isostasy and flexure of the lithosphere*; Cambridge University Press: Cambridge, UK, 2001; pp 1-458.

[76] Webb, V. E. *J. Geophys. Res.* 2004, 109 (E09010), 1-12.

[77] Webb, V. E.; McGill, G. E. *Lunar Planet. Sci.* 2003, 34, abstract 1132.

[78] Zuber, M. T. et al. *Science* 2000, 287, 1788-1793.

INDEX

A

B

C

D

E

F

G

J

K

L

M

N

O

Q

R

S

T

U

V

W

X

Y

Z